D1477045

Small Antennas

About the Authors

John L. Volakis is Chope Chair professor at The Ohio State University and the director of the ElectroScience Laboratory. He obtained his Ph.D. from The Ohio State University in 1982, is a fellow of the IEEE, and was the president of the IEEE Antennas and Propagation Society in 2004. His career includes 2 years at Boeing Phantom Works, 19 years as a faculty member at the University of Michigan, and 7.5 years on the Ohio State University faculty. His publications include five books (*Approximate Boundary Conditions in Electromagnetics*, 1995; *Finite Element Methods for Electromagnetics*, 1998; and the classic 4th ed. *Antenna Engineering Handbook*, 2007), over 260 journal papers and over 400 conference papers. He has graduated/mentored nearly 60 doctoral students/postdocs with 12 of them having coauthored papers that won awards at international conferences. Over the years, he served as the principal or coprincipal investigator on numerous major US government and industry projects relating to antennas, RF materials, metamaterials, computational electromagnetics, scattering and diffraction, electromagnetic compatibility and interference, and RFID studies.

Chi-Chih Chen received his B.S. degree in electrical engineering from National Taiwan University in 1988, and M.S. and Ph.D. degrees from The Ohio State University, Electrical and Computer Engineering Department in 1993 and 1997, respectively. He has been with The Ohio State University ElectroScience Laboratory (ESL) since 1993, where he is currently a research scientist. Dr. Chen's research activities have been focused on novel radar and antenna technology for applications in ground penetrating radars, vehicle radars, RFID, radar target detection and discrimination, UWB antennas, dielectric antennas, low-profile antennas, miniature antennas, UWB antenna feeds for reflectors, tapered chambers, and near-field ranges.

The McGraw-Hill Companies

Cataloging-in-Publication Data is on file with the Library of Congress

McGraw-Hill books are available at special quantity discounts to use as premiums and sales promotions, or for use in corporate training programs. To contact a representative, please e-mail us at bulksales@mcgraw-hill.com.

Small Antennas: Miniaturization Techniques & Applications

1234567890 WFR WFR 109876543210

ISBN 978-0-07-162553-1
MHID 0-07-162553-4

Sponsoring Editor
Wendy Rinaldi

Editorial Supervisor
Janet Walden

Project Manager
Vipra Fauzdar, Glyph International

Acquisitions Coordinator
Joya Anthony

Copy Editor
Nupur Mehra

Proofreader
Megha R C

Indexer
Jack Lewis

Production Supervisor
Jean Bodeaux

Composition
Glyph International

Illustration
Glyph International

Art Director, Cover
Jeff Weeks

Cover Designer
spark-to-ignite.com

Small Antennas: Miniaturization Techniques & Applications

John L. Volakis
Chi-Chih Chen
Kyohei Fujimoto

New York Chicago San Francisco
Lisbon London Madrid Mexico City
Milan New Delhi San Juan
Seoul Singapore Sydney Toronto

Kyohei Fujimoto received his Ph.D. degree in electrical engineering from Tokyo Institute of Technology in 1967. His research and development experience includes mobile communication systems and antennas, especially on integrated and small antennas. He has previously authored/coauthored/edited books on small antennas and mobile antenna systems, and gave lectures in Germany, China, Bulgaria, and USA on these topics. Currently, he is a consultant to several companies, a professor emeritus at University of Tsukuba, Japan, a consultant professor at the Northwestern Polytechnical University, China, and a research scientist at the Foundation of Advancement for International Science, Japan. Dr. Fujimoto is a fellow of IEICE and a life fellow of IEEE.

Contents

Acknowledgments

The realization of this book is based on the work of many of our students whose contributions are integrated into the text. Several of our current and former students (Nil Apaydin, Kennie Browne, Jeff Chalas, Faruk Erkmen, Erdinc Irci, Justin Kasemodel, Bradley Kramer, Gil Young Lee, Ming Lee, Matilda Livadaru, Haksu Moon, William Moulder, Gokhan Mumcu, Ugur Olgun, Ioannis Tzanidis, Jing Zhao, Yijun Zhou) also became coauthors to the chapters that formed this book. Prof. G. Mumcu's and Prof. K. Sertel's cowriting of the metamaterial chapters is a work of love. We are also indebted to Prof. Koulouridis for his chapter on impedance matching and negative (non-foster) circuits to improve low-frequency matching. Prof. Yunqi Fu made a heroic effort to generate the extensive survey of small antenna examples. His coauthored work in Chap. 3 should serve the antenna community for years to come (as a single source of small antenna designs).

Jeff Chalas deserves special thanks for taking on the huge responsibility of integrating the small antenna theory into a singe large chapter (Chap. 1). Jeff also served the critical role of ensuring copyediting and figure quality of the entire book.

We would be amiss in not thanking McGraw Hill's Editor Wendy Rinaldi. Her enthusiasm about the project was contagious, and was a catalyst to making the process enjoyable.

Introduction

Over the next decade, wireless devices and connectivity are likely to have even more transformational impact on everyday life. Key to the wireless revolution is the implementation of multifunctionality and broadband reception at high data rates. This was a neglected area for several years as the industry was focusing on microelectronics, compact low noise circuits, and low bit error modulation techniques. Not surprisingly, the need for high performance small antennas and RF front ends has emerged as a key driver in marketing and realizing next generation devices.

The challenge in miniaturizing the RF front end was already highlighted by Harold Wheeler, a pioneer of small-size antennas. He noted that "... [Electrical Engineers] embraced the new field of wireless and radio, which became a fertile field for electronics and later the computer age. But antennas and propagation will always retain their identity, being immune to miniaturization or digitization." However, novel materials, either natural or synthetic (metamaterials), and a variety of synthesized anisotropic media are changing the status quo.

Also, materials such as modified polymers (friendly to copper) for silicon chip integration, high conductivity carbon nanofibers and nanotubes, and electric textiles and conductive fibers, all coupled in 3D packaging, are providing a new integration paradigm attractive to the IC industry. Certainly, low loss magnetics, or synthetic structures emulating magnetic structures, when and if realized, will provide a transformational impact in the wireless industry.

This book aims to provide the reader with a single stop on small antennas: theory and applications, performance limits subject to bandwidth, gain and size; narrowband and wideband; conformal and integrated; passive and negative (non-foster) impedance matching; materials (polymers, ceramics, and magnetics) and shape optimization; high impedance, artificially magnetic and electromagnetic bandgap (EBG) ground planes; techniques for miniaturizing narrowband and wideband antennas; metamaterial and photonic crystal antennas; RFIDs and power harvesting rectennas; and a multitude of real-world applications that represent an extensive literature survey over the past several years.

The book contains nine chapters as described next.

Chapter 1: Survey of Small Antenna Theory This extensive chapter starts with a presentation of key antenna parameters that defines their performance in terms of size, bandwidth, quality factor, and gain. As the title implies, the chapter goes on to provide step-by-step development of small antenna theory starting with the classic developments of Chu, Harrington, and Wheeler. The contributions of several other authors to small antennas performance limits is also included in a systematic and didactic manner. Among the works presented are those of Collin and Rothschild, Fante, Hansen, McLean, Foltz, Thiele, Geyi, Best and Yaghjian, Kwon and Pozar, Gustafsson and Thal. A summary of these works is provided at the end of the chapter along with their differences and implied assumptions. We hope that this chapter will provide the reader with an invaluable coverage and fundamental understanding of small antenna theory.

Chapter 2: Fundamental Limits and Design Guidelines for Miniaturizing Ultra-Wideband Antennas This chapter builds on the small antenna theory in Chap. 1 and extends it to wideband antennas. Specifically, this chapter develops the theory that leads to the maximum gain at the lowest operational frequency of a given size aperture. Key to this development is the representation of the antenna by a set of cascaded equivalent circuit blocks. Subsequently, the Fano-Bode theory is applied to demonstrate how the antenna's electrical size impacts impedance matching for narrowband (bandpass response) and ultra-wideband antennas (high-pass response). In doing so, a set of performance limits is derived for UWB antennas.

Chapter 3: Overview of Small Antenna Designs This lengthy chapter attempts to fulfill a tall order in presenting a literature survey of small antennas (in practical settings). This extensive survey focuses on subwavelength antennas and covers (a) a survey of miniaturization techniques based on shape design such as bending, folding, meandering, lumped circuit loading, etc., (b) example small antennas based on material loading, (c) antennas based on formal design optimization, and (d) low-profile and conformal UWB antenna examples.

Chapter 4: Antenna Miniaturization via Slow Waves This chapter presents a number of practical antenna miniaturization techniques. Particular emphasis is on wave slowdown techniques based on inductive and capacitive loading. Techniques such as circuit loading, volumetric coiling as well as slot loading and meandering are covered with practical examples.

Chapter 5: Spiral Antenna Miniaturization This chapter employs the spiral as an example UWB antenna to demonstrate practical approaches in designing miniature conformal antennas. By building on miniaturization techniques presented in Chaps. 3 and 4, spirals with volumetric coiling, and dielectric and magnetic material loadings are developed. The chapter presents a step-by-step design process in realizing optimal performance for conformal wideband antennas by exploiting the third dimension of the structure and by incorporating treated ground planes using a variety of materials.

Chapter 6: Negative Refractive Index Metamaterial and Electromagnetic Band Gap Based Antennas Over the past decade, concepts that realize negative reflective index (NRI) propagation were exploited to design small narrowband antennas. This chapter begins by describing the classic periodic arrangements of materials (isotropic or anisotropic dielectrics, magnetic materials, conductors, etc.) and their equivalent circuits used to form metamaterial structures. The classic Bloch K-ω diagrams are described and the NRI transmission lines leading to slow and leaky waves are explained. An extensive survey of literature examples based on NRI concepts are collected in this chapter. In addition, an extensive discussion of electromagnetic bandgap (EBG) structures and frequency selective surfaces (FSS) is provided along with example designs to lower the antenna's profile.

Chapter 7: Antenna Miniaturization Using Magnetic Photonic and Degenerate Band Edge Crystals This chapter covers recent work on using periodic material structures to realize anisotropic periodic media that exhibit extraordinary wave slowdown to achieve antenna miniaturization. The concept is associated with the so-called *frozen modes*. After describing the frozen mode theory, a practical realization based on the concept of planar coupled microstrip transmission lines is presented.

This simple structure is shown to realize in-plane anisotropy, introducing several additional design parameters. These are used to develop a systematic way to design miniature low-cost printed antennas. Designs on substrates with magnetic inserts are presented for further functionality and greater bandwidth antennas that reach the optimal Chu-Harrington limits. The described coupled transmission line concept has recently been adapted to develop small wideband conformal antennas.

Chapter 8: Impedance Matching for Small Antennas Including Passive and Active Circuits This chapter begins with passive impedance matching, and provides explicit formulae for the circuit elements needed to bring sub-wavelength antennas to resonance. Broadband antenna matching is discussed next with the inclusion of specific practical examples. The latter half of the chapter covers the topic of matching using negative impedances with examples demonstrating their potential in reducing the antenna's operational frequency. This is an ongoing research topic with the potential to overcome the performance limits associated with small antenna theories in Chaps. 1 and 2.

Chapter 9: Antennas for RFID Systems With the explosive growth of RFIDs, bound to become ubiquitous, there is a concurrent need for small antennas as front ends to these wireless devices. RFIDs are already used from tracking products in supermarkets to monitoring information and critical components in ground and airborne vehicles. This chapter begins with a didactic description of RFID functionality and related components. Both passive and active RFIDs are described with particular emphasis on passive tags as these are more prevalent. Expanding on Chap. 3, several RFID tag antennas are described from the literature. The latter portion of the chapter is focused on rectennas (antennas with integrated Schottky diodes) for RF power harvesting. These rectennas are used as front ends to RFID tags and serve to power and turn-on the tag to transmit its stored data. The chapter describes critical components of the power harvesting subsystem, and provides design examples.

CHAPTER 1

Survey of Small Antenna Theory

Jeffrey Chalas, Kyohei Fujimoto, John L. Volakis, and Kubilay Sertel

1.1 Introduction

Antenna miniaturization has long been discussed as one of the most significant and interesting subjects in antenna and related fields. Since the beginning of radio communications, the desire for small and versatile antennas has been ever increasing. Today's needs for more multifunctional systems further drive requirements for small mobile terminals, including cell phones, handheld portable wireless equipment for internet connection, short- and long-range communication devices, RFIDs (radio frequency identification), etc. Similarly, small equipment and devices used for data transmission and navigation (GPS systems) require small antennas. These applications and continuing growth of wireless devices will continue to challenge the community to create smaller and more multifunctional antennas.

This chapter is intended to provide a chronological review of past theoretical work crucial to antenna miniaturization. Throughout, we shall refer to the small antennas as "electrically small antennas," or ESAs, implying that their size is much smaller than a wavelength at the operational frequency.

Wheeler [1] proposed the ESA definition as an antenna whose maximum dimension is less than $\lambda/2\pi$, referred to as a "radianlength." Another commonly used (and equivalent) definition of an ESA is an antenna that satisfies the condition

$$ka < 0.5 \tag{1.1}$$

where k is the wave number $2\pi/\lambda$, and a is the radius of the minimum size sphere that encloses the antenna (see Fig. 1.1). We shall refer to this spherical enclosure as the "*Chu sphere.*" Small antennas fitting the Wheeler definition radiate the first order spherical modes of a Hertzian dipole (see Fig. 1.2) and have radiation resistances, efficiencies, and bandwidths. As is well known, these parameters typically decrease with electrical size ka.

Another commonly accepted definition of a small antenna is $ka < 1$, [2]. This definition can be interpreted as an antenna enclosed inside a sphere of radius equal to one radianlength. Such a sphere is referred to as a "radiansphere" [33], and represents the boundary between the near- and far-field radiation for a Hertzian dipole. Hansen [2] notes that for antennas of this size, higher order spherical modes ($n > 1$) are evanescent.

In the sections to follow, the small antenna performance will be characterized by their size ka, quality factor Q, fractional bandwidth B, and gain G. It is therefore important to have an understanding of these parameters. Of particular interest is how antenna bandwidth (or Q) is related to the antenna size. As will see, there is an optimum Q

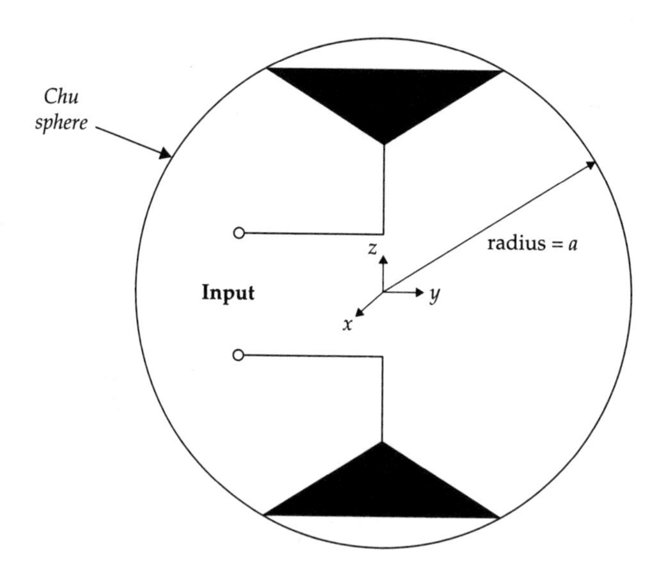

FIGURE 1.1 *Chu sphere* of radius "a" centered about the origin. The *Chu sphere* is the minimum circumscribing sphere enclosing the antenna of maximum dimension $2a$.

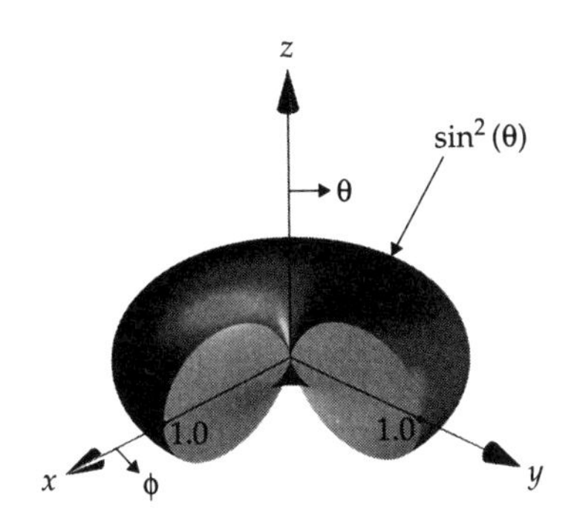

FIGURE 1.2 TM_{10} or TE_{10} mode power pattern with the region ($0° < \phi < 90°$, $0° < \theta < 90°$) omitted for clarity.

(smallest possible Q) for a given antenna size. Following a review of some basic antenna parameters, a short discussion on lumped resonant circuits and circuit Q is presented, which lays the foundation for small antenna analysis. A chronological review of the significant contributions to small antennas will then be presented, with a focus on the theoretical development of the field.

1.2 Small Antenna Parameters

To establish a foundation for discussing small antennas, an overview of their most important characteristics is presented below.

1.2.1 Directivity

It is often stated that small antennas have the familiar doughnut-shaped (see Fig. 1.2) omni-directional radiation pattern of a Hertzian dipole of directivity $D = 1.5$. This pattern may be also thought as the radiation of TE_{10} or TM_{10} spherical modes. However, this is not the only possible pattern for a small antenna, as seen in the work presented by Harrington [3], Kwon [4, 5], and Pozar [6]. By superposing various electric and magnetic Hertzian dipole arrangements, unidirectional and bidirectional patterns are theoretically possible, along with directivities ranging from $D = 1.5$ to 3 (see Sec. 1.3.11 in this chapter). We can state that antennas having significant spherical TE_{nm} and TM_{nm} mode radiation with $n > 1$ are generally not of the small type. Small antennas are also classified as superdirective antennas, since for decreasing size ka, their directivity D remains constant [2,7].

1.2.2 Radiation Efficiency

Radiation efficiency is a critical topic for small antennas but has not been studied rigorously. Antenna radiation efficiency factor η is simply the ratio of the power radiated by the antenna to the power delivered to the input terminals of the antenna. Often the efficiency factor is seen in the formula $G = \eta(1 - |\Gamma|^2)D$ where G is the realized gain that includes the mismatches between the source and matching network (see Fig. 1.3). We assume the matching network of Fig. 1.3 is lossless. The losses in the antenna apart from radiation are frequently modeled through a series loss resistor R_{loss}, in which case the radiation efficiency η can be represented as

$$\eta = \frac{R_{\text{rad}}}{R_{\text{rad}} + R_{\text{loss}}} = \frac{R_{\text{rad}}}{R_A} \tag{1.2}$$

where R_A is the total antenna input resistance $R_{\text{rad}} + R_{\text{loss}}$ (see Fig. 1.3).

It has been observed that as antenna size ka decreases, R_{rad} decreases and the loss resistance R_{loss} dominates the efficiency expression of Eq. (1.2). This decrease in efficiency is primarily due to frequency-dependent conduction and dielectric losses within the antenna. As mentioned later, Harrington [3] quantified the efficiency for an ideal spherical antenna, showing that losses are extremely prominent for smaller ka values.

A simple method to find the radiation efficiency η and separate R_{rad} from R_A is to use the Wheeler Cap [8] method (see Fig. 1.4). The Wheeler Cap (shown in Fig. 1.4) is a hollow perfectly electric conducting (PEC), enclosing sphere of the same size as the radiansphere. Wheeler noted that the size and shape of the Wheeler Cap is not critical. However, it must be electrically large enough so that the near-zone-antenna fields are not disturbed while still preventing radiation, and small enough so that cavity resonances are not excited. Indeed,

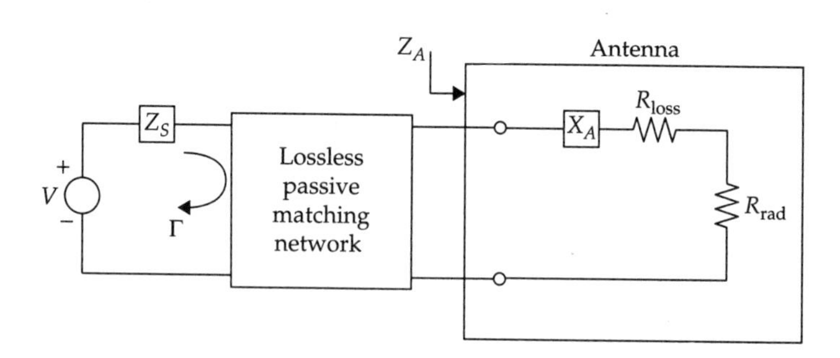

FIGURE 1.3 Lossless passive matching network with antenna load and input reflection coefficient Γ.

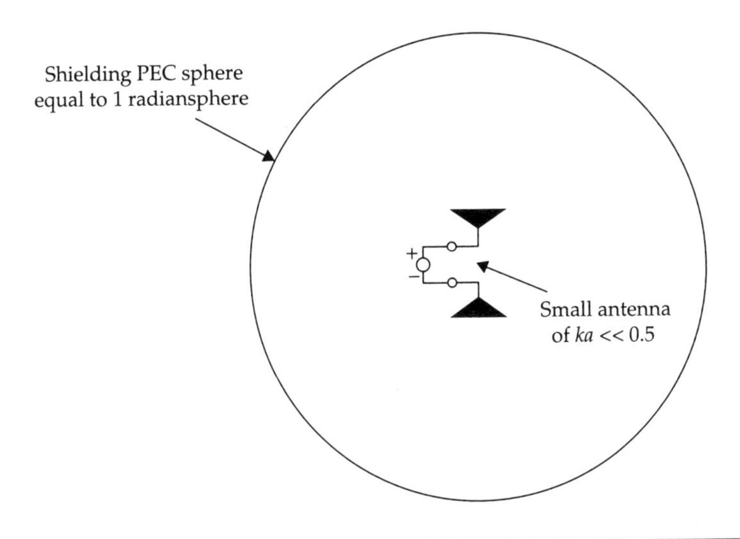

FIGURE 1.4 Wheeler Cap for small antenna efficiency measurement. *(After Wheeler* [8].)

Huang [9] proved rigorously that using the radiansphere-sized spherical Wheeler Cap does not significantly affect the near fields excited.

To measure the radiation efficiency η using the Wheeler Cap method, first a computation or measurement is done at the antenna resonant frequency in the absence of the Wheeler Cap to obtain R_A. If the antenna is not self-resonant, it must be tuned to resonance by a reactive element at the input. The tuned antenna is then enclosed inside the Wheeler Cap, and the measured input resistance then yields R_{loss}. Substitution of R_A and R_{loss} in Eq. (1.2) then gives $\eta = R_{\text{rad}}/R_A$.

1.2.3 Quality Factor

1.2.3.1 Antenna Quality Factor

An intrinsic quantity of interest for a small antenna is the Q factor, defined in [3] as

$$Q = \frac{2\omega_0 \max(W_E, W_M)}{P_A} \tag{1.3}$$

W_E and W_M are the time averaged stored electric and magnetic energies, and P_A is the antenna received power. The radiated power is related to the received power through $P_{\text{rad}} = \eta P_A$, where η is the antenna efficiency. It is assumed in Eq. (1.3) that the small antenna is tuned to resonance at the frequency ω_0, either through self-resonance or by using a lossless reactive tuning element.

Antenna Q is a quantity of interest and can be also evaluated using equivalent circuit representations of the antenna. Another important characteristic of Q is that it is inversely proportional to antenna bandwidth (approximately). A commonly used approximation between Q and the 3 dB fractional bandwidth B of the antenna is

$$Q \approx \frac{1}{B} \quad \text{for} \quad Q \gg 1 \tag{1.4}$$

Equation (1.4) is based on resonant circuit analysis and tends to become more accurate as Q increases. An explicit relationship between Q and bandwidth is given later (see Sec. 1.3.10 or Yaghjian and Best [10]). For the moment let us review the lumped resonant circuit analysis used for computing Q in this chapter.

The reader may wonder why Q is the quantity of interest rather than bandwidth itself. One practical reason is that bandwidth remains an ambiguous term. Though it is often implied that bandwidth refers to the 3 dB bandwidth, this is not always the case for antennas. It is desirable to find an independently derived quantity Q that is also related to bandwidth. This idea is given in Sec. 1.3.10 by Yaghjian and Best [10]. However, the most important reason that Q remains of interest for small antennas is that a fundamental lower limit on Q can be found using a number of techniques (and consequently the max bandwidth of a small antenna). This fundamental limitation on Q (or max bandwidth) drives the majority of the work examined later on small antennas.

1.2.3.2 Quality Factor for Lumped Circuits

Wheeler [1] recognized that a small antenna radiating the single spherical TE_{10} mode can be accurately represented as a RLC combination of Fig. 1.5*a*. We note the series capacitor represents the ideal tuning element in Eq. (1.3) which brings the antenna to resonance. Similarly, a small antenna radiating only a spherical TM_{10} mode can be accurately represented by the parallel RLC combination as in Fig. 1.5*b*, where the shunt inductor represents the ideal tuning element in Eq. (1.3) that brings the antenna to resonance. More complicated, high-Q circuits can be accurately represented as a series [for $X'_{\text{in}}(\omega_0) > 0$] or parallel [for $X'_{\text{in}}(\omega_0) < 0$] RLC circuits within the neighborhood of their resonant frequencies.

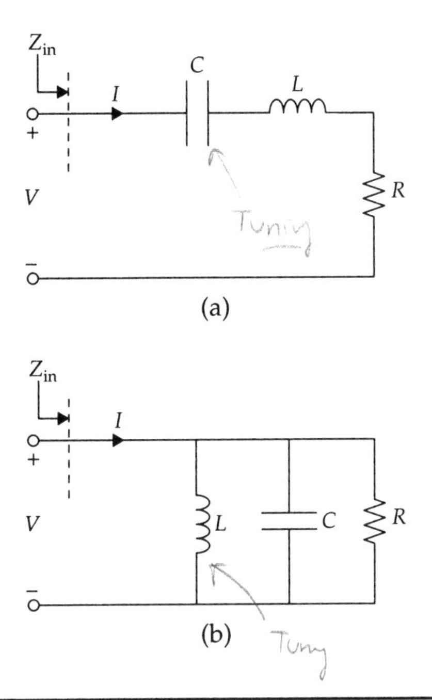

FIGURE 1.5 Series and parallel RLC circuits. (a) Series RLC, (b) Parallel RLC.

Series RLC Circuit For the series RLC circuit of Fig. 1.5*a*, the input impedance is

$$Z_{in} = R + j\omega L - \frac{j}{\omega C} = R + j\omega L\left(\frac{\omega^2 - \omega_0^2}{\omega^2}\right) \text{ with } \omega_0 = \frac{1}{\sqrt{LC}} \quad (1.5)$$

where ω_0 represents the resonant frequency at which the input impedance is purely real. This resonance occurs when the average stored electric energy is equal to the average stored magnetic energy in the circuit. Using the general definition of Q in Eq. (1.3) and recognizing the current is the same in all circuit components, we find that

$$Q = \frac{2\omega_0 W_H}{P_a} = \frac{2\omega_0\left(\frac{1}{4}LI^2\right)}{\frac{1}{2}I^2 R} = \frac{\omega_0 L}{R} = \frac{1}{\omega_0 RC} \quad (1.6)$$

where I is the current through the series RLC circuit in Fig. 1.5*a*. The bandwidth of the series RLC circuit can be estimated after introducing the approximation

$$F(\omega) = \omega^2 - \omega_0^2 \approx F(\omega_0) + (\omega - \omega_0)F'(\omega_0) = 2\omega\Delta\omega \quad (1.7)$$

valid for small $\Delta\omega = \omega - \omega_0$. With this Taylor series, Eq. (1.5) becomes

$$Z_{\text{in}} \approx R + j\omega L\left(\frac{2\omega(\omega - \omega_0)}{\omega^2}\right) = R + j2L\Delta\omega \tag{1.8}$$

From Eq. (1.8), it is then evident that the 3 dB points occur when

$$2L\Delta\omega_{3\,\text{dB}} = \pm R \tag{1.9}$$

where $\Delta\omega_{3\,\text{dB}}$ is the difference between the 3 dB frequency and resonant frequency. Using Eqs. (1.6) and (1.9) we can now write

$$2Q\frac{\Delta\omega_{3\,\text{dB}}}{\omega_0} = QB = 1 \tag{1.10}$$

since by definition, $2\Delta\omega_{3\,\text{dB}}/\omega_0 = B$ for an antenna having ω_0 as its operational frequency. From this result, we then have the relationship $B = 1/Q$ as mentioned in Eq. (1.4). Figure 1.6 depicts the impedance as a function of frequency for a typical series RLC circuit for various $Q \gg 1$ values.

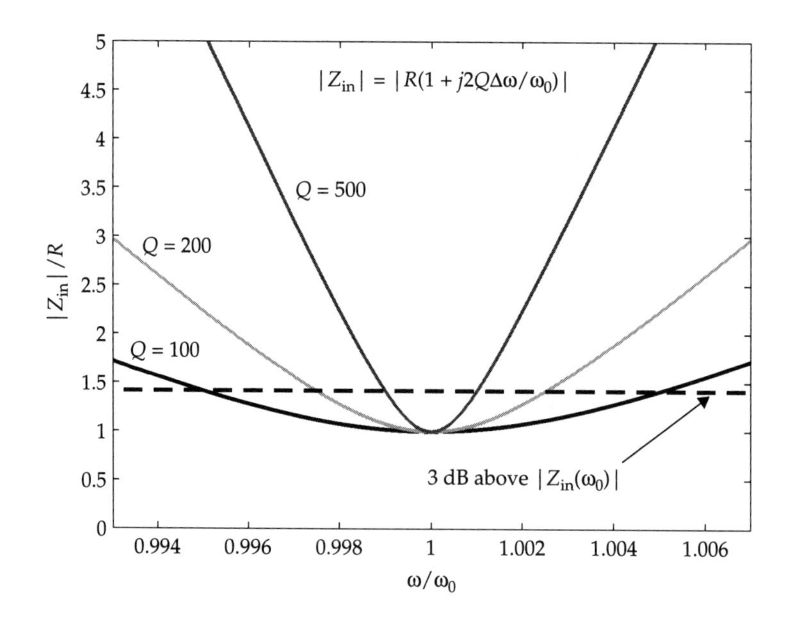

FIGURE 1.6 Normalized impedance magnitude for a series RLC circuit near resonance.

Parallel RLC Circuit For the parallel RLC circuit of Fig. 1.5*b*, the input admittance is

$$Y_{\text{in}} = G + j\omega C - \frac{j}{\omega L} = G + j\omega C\left(\frac{\omega^2 - \omega_0^2}{\omega^2}\right) \text{ with } \omega_0 = \frac{1}{\sqrt{LC}} \tag{1.11}$$

where ω_0 is again the resonant frequency for which the input admittance is purely real. Using the general definition of Q and recognizing for the parallel RLC circuit the voltage V across each component is the same, the Q for the parallel RLC circuit at resonance is found to be

$$Q = \frac{2\omega_0 W_E}{P_a} = \frac{2\omega_0\left(\frac{1}{4}CV^2\right)}{\frac{1}{2}V^2 G} = \frac{\omega_0 C}{G} = \frac{1}{\omega_0 G L} \tag{1.12}$$

From the dual nature of the series and parallel RLC circuits, the same bandwidth relations obtained in Eq. (1.10) hold for the parallel RLC circuit. Figure 1.7 depicts the impedance as a function of frequency for a typical parallel RLC circuit having $Q \gg 1$.

Arbitrary Lumped Networks In many cases, tuning the antenna impedance to resonance using a single lossless reactive element does not give a suitable value for the input resistance to match the transmission line. To minimize mismatches (reflections) and deliver maximum power to the antenna, two degrees of freedom are needed to provide

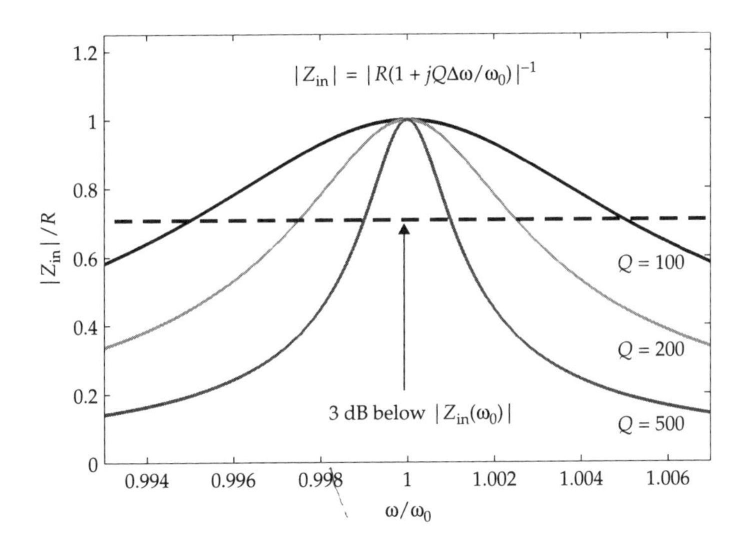

FIGURE 1.7 Normalized impedance magnitude for a parallel RLC circuit near resonance.

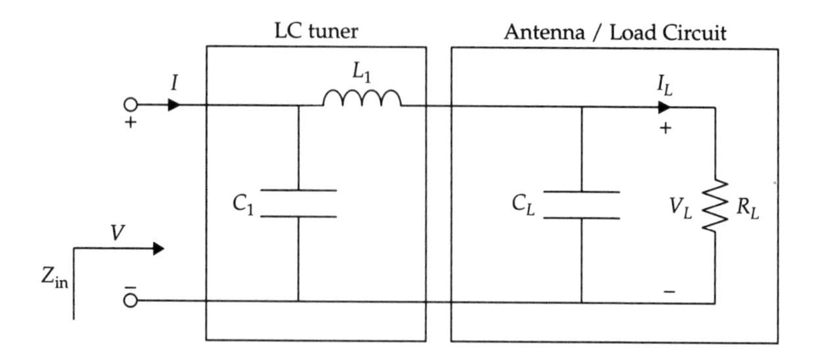

FIGURE 1.8 Antenna circuit with LC tuner.

an impedance match to a transmission line. Figure 1.8 shows an example of a lumped matching network with two degrees of freedom—one series and another shunt element. With these two degrees of freedom, an arbitrary load impedance can be transformed to a real impedance value, and matched to the transmission line.

To find the Q for the circuit configuration in Fig. 1.8, we note that at resonance

$$Z_{in} = \mathrm{Re}(Z_{in}) = R_{in}$$
$$P_{in} = P_L = \frac{1}{2}\,|I_L|^2\,R_L$$

where Z_{in} is the input impedance, P_{in} is the input power, and P_L is the power at the load. Using these conditions, we can find Q from Eq. (1.3). To employ Eq. (1.3), we note

$$P_{in} = \frac{1}{2}\frac{|V|^2}{R_{in}} = P_L = \frac{1}{2}\frac{|V_L|^2}{R_L} \tag{1.13}$$

$$W_{E1} = \frac{1}{4}\,|V|^2\,C_1 \tag{1.14}$$

$$W_{EL} = \frac{1}{4}\,|V_L|^2\,C_L = \frac{1}{4}C_L\frac{R_L}{R_{in}}\,|V|^2 \tag{1.15}$$

Substituting these quantities into Eq. (1.3), we get

$$Q = \frac{2\omega_0 W_E}{P_L} = \frac{2\omega_0\,(W_{E1} + W_{EL})}{P_L} = \omega_0 C_1 R_{in} + \omega_0 C_L R_L \tag{1.16}$$

Finding Q for arbitrary circuit topologies can become a cumbersome procedure. Formulating the topology using approximate RLC circuits can therefore be beneficial. One method used by Chu [11] is to equate the input resistance, reactance, and reactance frequency derivative of a more complicated passive circuit to those of a series RLC circuit at the resonant frequency.

1.2.3.3 External and Loaded Q

The Q described thus far is an intrinsic circuit quantity known as the unloaded Q, and is independent of the source. Figure 1.9 shows the more realistic situation of a source of impedance R_S driving a resonant circuit characterized by the unloaded quality factor Q. If the resonant circuit is a series RLC, the source impedance R_S combines in series with the input resistance R. Thus, to find the overall Q in Eq. (1.6), R is replaced with $R + R_S$. Similarly, for a parallel RLC circuit, the value of G in Eq. (1.12) need be replaced with $G + G_S$. It is nevertheless clear from Eqs. (1.6) and (1.12) that the source resistance has the effect of reducing the overall quality factor. To account for this Q reduction, the external quality factor Q_S is introduced and defined by

$$Q_S = \begin{cases} \dfrac{\omega_0 L}{R_S} & \text{Series resonator} \\[2ex] \dfrac{R_S}{\omega_0 L} & \text{Shunt resonator} \end{cases} \tag{1.17}$$

The total circuit Q, now referred to as the loaded quality factor Q_L, can then be defined by the known relation (adding Qs in parallel)

$$\frac{1}{Q_L} = \frac{1}{Q} + \frac{1}{Q_S} \tag{1.18}$$

We observe that when the source is matched to the resonant circuit (typically at resonance) then $Q_L = Q/2$, implying double the bandwidth.

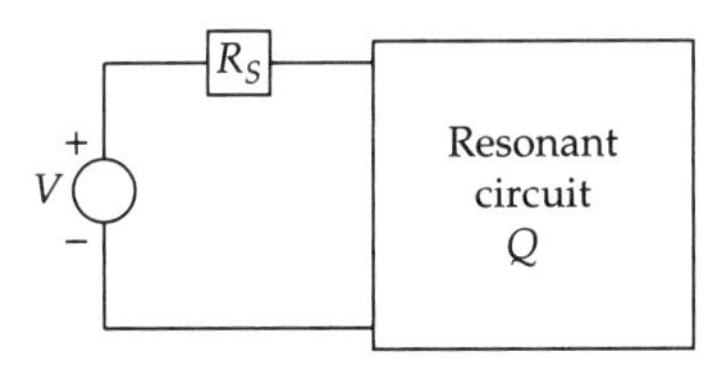

FIGURE 1.9 Resonant circuit with unloaded Q driven by a source of impedance R_S.

1.2.4 Input Impedance and Matching

In general, the input impedance of small antennas is typically characterized by low resistance and high reactance [12]. As the antenna size decreases, the radiation resistance R_{rad} decreases, causing the antenna reactance X_A to dominate. With respect to Fig. 1.10, a convenient proportionality relationship for small monopole antennas was given by [12] and [13] as

$$R_{\text{rad}} \propto \left(\frac{h}{\lambda}\right)^2 \tag{1.19}$$

Here, h is the monopole height and λ is the wavelength, implying that the input resistance decreases quadratically with electrical size. Matching circuits are therefore needed to improve the small antenna efficiency over a wide range of frequencies. But this imposes a significant practical design challenge as the matching network must also be physically small. It is therefore important to design self-resonant antennas with high radiation resistances that can be matched to standard transmission lines. In this context, certain self-resonant (folded) antenna structures have been shown to have input resistances that approach those of standard transmission lines [13].

Several texts [14–16] have sections dedicated to both lumped and distributed matching networks, and in Chap. 2 we consider matching as an essential component to achieving optimal size narrowband and wideband antennas. For the narrowband antennas, the fundamental bounds on lossless passive matching networks were derived by Fano [17] as

$$BQ \leq \frac{\pi}{\ln\left(\dfrac{1}{\Gamma_{\max}}\right)} \tag{1.20}$$

where B is the 3 dB fractional bandwidth, $\Gamma_{\max}$ is the maximum allowable reflection coefficient in the passband, and Q is the quality factor of the load to be matched.

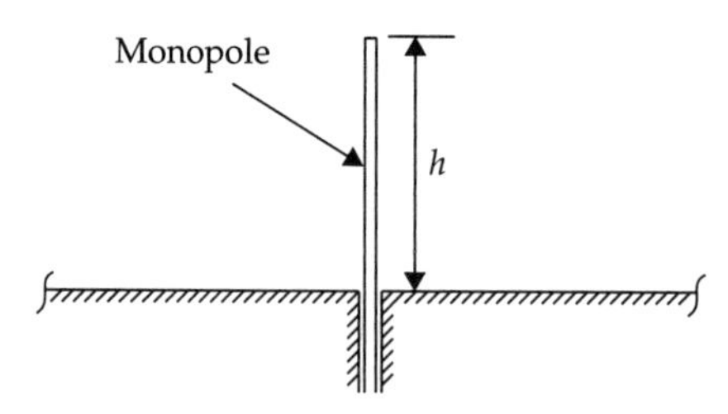

FIGURE 1.10 Monopole antennas.

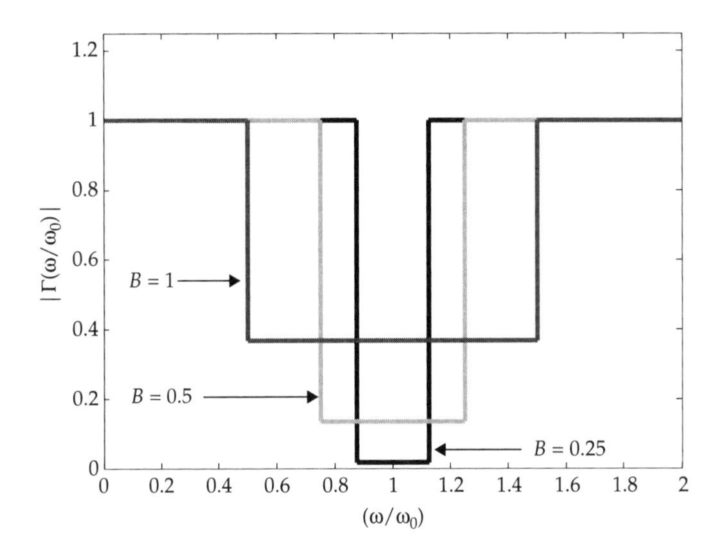

FIGURE 1.11 Bode limit on reflection coefficient Γ for various B values.

The interpretation of Eq. (1.20) is depicted in Fig. 1.11 (see Chap. 2 as well) for $B = 0.25, 0.5$, and 1, with $Q = \pi$. The fundamental limitation given by Eq. (1.20) indicates that greater bandwidth can only be achieved at the cost of increased maximum reflection coefficient (less realized gain).

1.3 Small Antenna Theory

Work on small antennas can be first traced back to the treatments done by Wheeler in 1947 [1]. Wheeler later discussed the fundamental limitations of small antennas using a simple model that approximates the small antenna with a lumped capacitance or inductance and a radiation resistance. He applied the concept of the radiation power factor (RPF)—a ratio of the radiated power to reactive power—to the antenna model [1] in an attempt to find a relationship between antenna size and radiation properties. He was the first to note that the reduction in the antenna size imposes a fundamental limitation on bandwidth. Wheeler concluded that the RPF was directly proportional to the physical antenna volume. To a degree, we can say that the RPF was the precursor to the commonly used quantity, Q. In Wheeler's circuit model, Q is the inverse of RPF (implying that RPF represents the small antenna bandwidth).

Wheeler's work was a rough approximation that was accurate only for extremely small antenna sizes. This is because it did not take into account the radiated spherical modes as the antenna size increased. Nevertheless, his treatment can be recognized as the first intense study of small antennas, and encouraged many followers to investigate fundamental properties and limitations proposed in his work. It also inspired advances in the theory of the small antennas and also the development of practical small antennas.

In 1948, Chu derived the minimum possible Q for an omnidirectional antenna enclosed in a *Chu sphere* (see Fig. 1.1 for definition) along with the maximum G/Q [11]. He accomplished this by using spherical mode wavefunction expansions outside the mathematical sphere enclosing the antenna (by expressing the radiated field as a sum of spherical modes). Each mode was then expressed in terms of an equivalent circuit, and through lumped circuit analysis the Q for each mode was found. Though Chu restricted his analysis to a specific type of omni-directional antenna, his contributions would lay the groundwork for many future authors that refined these limits. Chu's expression for calculating the minimum Q was later simplified by Hansen [2] (1981). Harrington [3] followed much of Chu's analysis (1960), and was the first to consider the antenna as radiating both TE and TM modes. As a result, his work led to lower minimum Q values.

As noted, the analysis done by Chu and Harrington involved equivalent circuit approximations for representing each generated mode. Collin and Rothschild (1964) pursued a field-based technique calculating the exact radiation Q [18] for an antenna radiating only TM or TE modes, and found the minimum Q possible for such an antenna enclosed in a *Chu sphere*. Their analysis was later generalized by Fante [19] to include both TE and TM modes. Fante [19] derived an exact Q expression for an arbitrary TM and TE mode configuration.

McLean (1996) used another method to calculate Q, and his results were in agreement with those presented by Collin and Rothschild [37]. McLean felt that Wheeler's method was too rough an approximation, and Chu's equivalent circuit approximations were not accurate enough to establish a fundamental limit on Q. Later, Foltz and McLean (1999) recognized that their lower Q bounds were not close to the verifiable values for many antennas, especially dipoles. As a result they repeated Chu's analysis using a prolate spheroidal wavefunction expansion outside a prolate spheroid. Their intention was to more accurately represent the geometry of many practical antennas such as dipoles [20]. They discussed the minimum possible Q associated with antennas enclosed in such a prolate spheroid, assuming only TM or only TE modes. Foltz and McLean concluded that Q increases as the spheroid becomes thinner, further reinforcing the concept that Q is inversely proportional to physical antenna volume.

Thiele (2003) observed that the exact lower limit for Q derived in earlier works was far from that of actual antennas and conjectured that the current distribution on the antenna had strong effect on the value of Q [7]. His method for determining Q utilized an extension of the superdirective ratio concept. Thus, his analysis was unique in determining Q from the far-field pattern and not a modal expansion. Thiele applied his approach to a dipole antenna having non-uniform current distribution, and obtained favorable results as compared to practical dipole antennas with a similar current distribution.

Geyi (2003–2005) thought, as previous authors, that Chu's treatment was not adequately accurate, and pointed out that Collin's analysis involved integrations which were not feasible for many applications. Geyi began by reinvestigation the antenna Q and directivity limits [21]. He also presented an approximate method for calculating Q using less taxing integrations than those derived by Collin and Rothschild.

More recently, Best (2003–2008), Yaghjian, and their colleagues [10,13,23–26] carried out extensive work both on the theory of small antennas and their design. Among the theoretical topics pursued by Best are the exact and approximate expressions for Q in terms of fields and impedance, and the relationship between Q and bandwidth [10]. He also considered specific self-resonant wire antennas that approach the Q limits. His work explored the effect of wire geometry, wire folding, and volume utilization on radiation resistance and Q. Best's [24] folded spherical helix antenna was shown to deliver nearly the lowest possible Q for a single radiating mode antenna. This design of the folded spherical helix is reminiscent of the spherical inductor antenna proposed decades earlier by Wheeler [27].

While previous theoretical work focused on finding the physical limitations on antenna Q, Kwon and Pozar (2005–2009) noted that many authors were not consistent in defining the TM and TE modes, the antenna gain, Q, and directionality. With this motivation, Kwon [4,5] performed extensive analytical work on the gain and Q associated with electric and magnetic dipoles in varying configurations. Pozar [6] summarized the results of Kwon and others, and pointed out inconsistencies among previous small antennas authors.

Thal (2006–2009) set out to determine a more restrictive Q than previously derived for a class of antennas represented by a surface current distribution over a sphere and radiating both TE and TM modes [28]. His work extended the equivalent mode circuit method of Chu by introducing an additional equivalent circuit to account for energy storage inside the sphere (previously assumed by Chu and others to be zero). The Q of Best's spherical helix antenna [24] was found to be nearly identical to the minimum Q limit found by Thal. Thal also considered the relationship between gain, Q, and the energy inside the *Chu sphere* for a small antenna, and concludes that these quantities are not independent of one another [29].

The more recent work of Gustafsson and associates (2007) [30] provided Q and gain expressions for small antennas having arbitrary shape. Specifically, Gustafsson took an approach radically different than previous authors using scattering theory and representing the antenna in terms of material dyadics. His work seems to indicate that accurate gain and Q limitations on antenna geometries such as prolate and oblate spheroids, disks, needles, and cylinders are now possible through numerical computations.

In the following sections we proceed to examine the aforementioned works in more detail.

1.3.1 Work of Wheeler (1947–1975)

The first widely published paper (in 1947) on small antennas was by H. A. Wheeler entitled "Fundamental Limitations of Small Antennas" [1]. Wheeler approximated the small antenna as a lumped capacitance or a lumped inductance, with a radiation resistance. He claimed that the physical limitations on small antenna design could then be found based on this simplified representation.

Wheeler explored the limitations of small antennas in [27] providing an analysis of the properties for a spherical coil antenna. Wheeler introduced the fundamental limitations of small antennas using the lumped element models which closely match the work of later authors based on rigorous field analysis. Wheeler's 1975 review paper [8] on small antennas stressed the importance of the small antenna's spherical "effective volume." He interpreted the latter as a "sphere of influence" for nonspherical antenna structures.

Much of Wheeler's work was reviewed and verified through computer simulation in Lopez's work [31], and later by many authors.

1.3.1.1 The Radiation Power Factor

In [1], Wheeler recognized that small antennas had far-field patterns and stored energy properties similar to those of electric and magnetic dipoles. Figure 1.12 presents a simplified lumped circuit for small L-type (magnetic) antenna and a small C-type (electric) antenna. Wheeler recognized that a small antenna used as an electric dipole would have far greater capacitively stored energy (as compared to inductive energy). Thus, the antenna could be represented

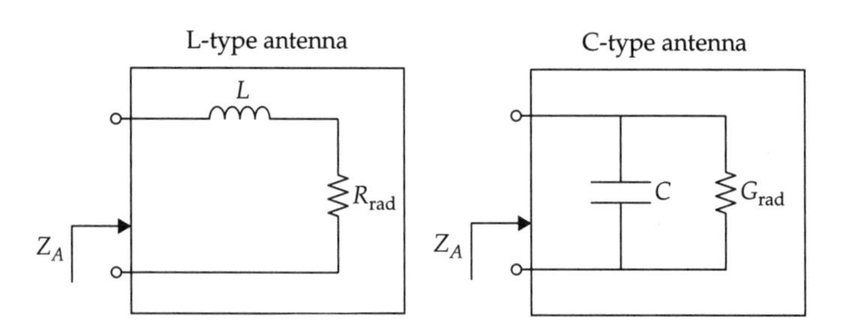

FIGURE 1.12 Small L-type antenna and C-type antenna models. (*See* [1].)

as a capacitance in parallel with a radiation conductance G_{rad}. Similarly, if the small antenna behaves as a magnetic dipole, the stored inductive energy is far greater. Thus it can be represented by a circuit having an inductance in series with a radiation resistance R_{rad}. Ohmic losses R_{loss} were neglected in his analysis as the goal was to gain insight on the behavior of the antenna Q. As such, in Wheeler's analysis the only real power dissipation was due to the radiation resistance, representing the radiated power.

With respect to the circuits in Fig. 1.12, Wheeler introduced the quantity he referred to as the *radiation power factor* (RPF). RPF is defined as the ratio of the radiated power P_{rad} to the reactive power P_{react} at the feed point of the antennas. He also introduced the definitions p_e = RPF for the capacitive (C-type) antenna and p_m = RPF for the inductive (L-type) antenna. His specific definitions are

$$p_m = \frac{R_{\text{rad}}}{\omega L} \tag{1.21}$$

$$p_e = \frac{G_{\text{rad}}}{\omega C} \tag{1.22}$$

Using the expressions (1.6) and (1.12) for the lumped RLC resonators, one can express the quality factors Q_m and Q_e of the magnetic and electric dipole circuits of Fig. 1.12 as

$$Q_m = \frac{2\omega W_M}{P_{\text{rad}}} = \frac{\omega L}{R_{\text{rad}}} = \frac{1}{p_m} \tag{1.23}$$

$$Q_e = \frac{2\omega W_E}{P_{\text{rad}}} = \frac{\omega C}{G_{\text{rad}}} = \frac{1}{p_e} \tag{1.24}$$

By inspection, one can see that Wheeler's RPF is equal to the inverse of the Q for the circuit models in Fig. 1.12. As Q has become the commonly used parameter today, it is appropriate to employ Q (rather than p_e and p_m) from here on.

In order to analyze the effect of antenna size to Q, Wheeler examined the circuit parameters for a cylindrically shaped capacitor and inductor representing the C-type or L-type dipole antennas (as depicted in Fig. 1.13), respectively. Wheeler also considered the cases where the C-type antenna may be filled with dielectric material of relative permittivity ε_r, and the L-type antenna may be filled with magnetic material of relative permeability μ_r.

Wheeler approximated the capacitance and inductance of the C-type and L-type antennas in Fig. 1.13 using the formulae

$$C = \varepsilon_0 \frac{k_a\, A}{b} \tag{1.25}$$

and

$$L = \mu_0\, n^2\, \frac{A}{k_b\, b} \tag{1.26}$$

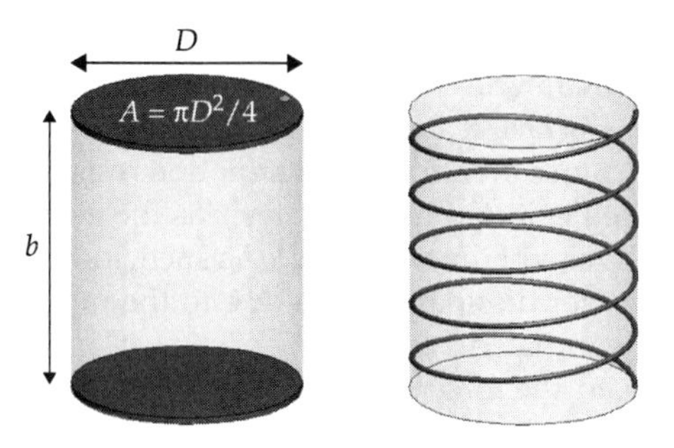

FIGURE 1.13 Small C-type and L-type cylindrical antennas occupying equal volume. (*See* [1].)

where n indicates the number of turns in the coil. We note that the Eqs. (1.25) and (1.26) differ from the standard formulas found in most texts through the additional factors k_a and k_b (which Wheeler defines as the shape factors). The capacitor shape factor k_a multiplies the physical area A to yield the effective area, taking into account the electric field outside the capacitor volume on the overall capacitance. The inductor shape factor k_b multiplies the inductor physical length to provide an effective length, taking into account the return path of the magnetic flux produced by the inductor. We note that these effective capacitance and inductances are larger and smaller, respectively, than the idealized ones ($k_a = k_b = 1$). Of course, if basic capacitor (no electric flux outside cylinder volume) and inductor models (no magnetic flux outside the inductor interior) are used, the shape factors will both become unity and the presented models will no longer be accurate. Therefore, it is critical to include the contributions of the fields outside the cylindrical structure to modify the L and C values to their more accurate representations.

Wheeler gave the radiation resistances derived from [32] of the C-type and L-type antennas as

C-type antenna | L-type antenna

$$R_{\text{rad}} = 20\left(\frac{b}{(\lambda/2\pi)}\right)^2 \qquad R_{\text{rad}} = 20\left(\frac{nA}{(\lambda/2\pi)^2}\right)^2 \tag{1.27}$$

$$G_{\text{rad}} = \frac{1}{6\pi\, Z_0}\left(\frac{k_a\, A}{(\lambda/2\pi)^2}\right)^2 \qquad G_{\text{rad}} = \frac{1}{6\pi\, Z_0 n^2}\left(\frac{k_b\, b}{(\lambda/2\pi)}\right)^2 \tag{1.28}$$

From the radiation resistance and reactance values in Eqs. (1.25) to (1.27), the radiation conductances G_{rad} can be found by representing the series RC (C-type antenna) or RL (L-type antenna) circuits as shunt GC (C-type) or GL (L-type) circuits. We note that there are no correction factors k_a and k_b in the radiation resistance formulas (1.27) and (1.28). Wheeler notes that this is because the electric current radiation for both antenna types is confined to the physical dimensions of the small antenna. This is unlike the electric and magnetic flux paths which extend beyond the antenna structure and modify the reactance values.

Substituting the parameters of Eqs. (1.25) to (1.28) into the definitions of Q given in Eqs. (1.23) and (1.24) yields

$$Q_e = \frac{\omega C}{G_e} = \frac{6\pi}{k_a\, Ab}\left(\frac{\lambda}{2\pi}\right)^3 = \frac{9}{2}\frac{V_{RS}}{V_{eff}} \tag{1.29}$$

$$Q_m = \frac{\omega L}{R_m} = \frac{6\pi}{k_b\, Ab}\left(\frac{\lambda}{2\pi}\right)^3 = \frac{9}{2}\frac{V_{RS}}{V_{eff}} \tag{1.30}$$

where V_{RS} is the volume of the radiansphere

$$V_{RS} = \frac{4\pi}{3}\left(\frac{\lambda}{2\pi}\right)^3 \tag{1.31}$$

and V_{eff} is the effective volume, defined as the physical volume of the antenna scaled by a physical and material dependent shape factor

$$V_{eff} = \sigma V_{physical} \quad \text{where} \quad \sigma = k_a, k_b \tag{1.32}$$

The final forms of Eqs. (1.29) and (1.30) provide two essential relations between Q and all small antennas:

(1) Q is inversely proportional to the effective antenna volume V_{eff} (and therefore physical antenna volume $V_{physical}$).

(2) Q is inversely proportional to the cubic power of frequency.

The effective volume is a concept frequently used in small antennas. In Eq. (1.32) it is seen that the effective volume is simply the physical volume multiplied by the material and structurally dependent shape factor σ. To explore the effect of shape factors have on Q (for the cylindrical antennas in Fig. 1.13), Wheeler introduced the shape factor for the C-type cylindrical antenna with radius a and height b as

$$k_a = \frac{k_{SC}^2}{k_{SC} + \varepsilon_r - 1} \qquad k_{SC} = 1 + \frac{8}{\pi}\frac{b}{D} \tag{1.33}$$

The corresponding shape factor for the L-type cylindrical antenna is

$$k_b = \frac{k_{SL}^2}{k_{SL} + \dfrac{1}{\mu_r} - 1} \qquad k_{SL} = 1 + 0.45\frac{D}{b} \tag{1.34}$$

Material filling effects can be seen in Eqs. (1.33) and (1.34), where filling the C-type antenna with a dielectric ε_r has the effect of decreasing the effective volume and concurrently increasing Q. However, filling the L-type antenna with a magnetic material μ_r results in an increase in effective volume, concurrently decreasing Q. Wheeler also notes in [1] that the C-type and L-type shape factor physical dependences are inversely related. As an example, a short and wide coil ($D/2 \gg b$, a spiral is such an example) has a higher effective volume as compared to a short and wide dipole. Similarly, a long and thin dipole ($b \gg D/2$) has a higher effective volume as compared to a long and thin coil.

Figure 1.14 gives another interpretation of the effective volume as described by in [8]. Wheeler states that the effective volume can be thought of as a "sphere of influence," providing a convenient reference to the radiansphere. From Eqs. (1.29) and (1.30), the effective volume is thus a sphere of radius

$$a' = \frac{\lambda}{2\pi}\left[\frac{9}{2Q}\right]^{1/3} \tag{1.35}$$

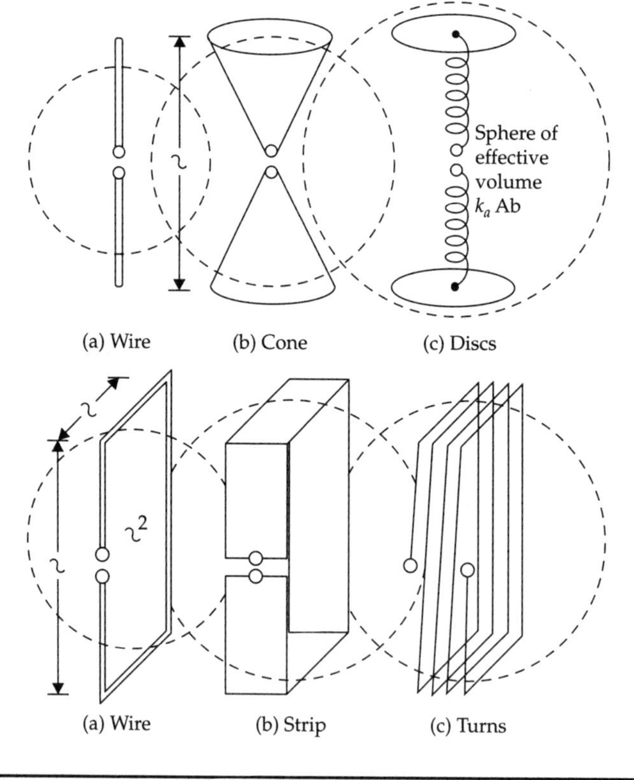

FIGURE 1.14 Effective volumes for C-type and L-type small antennas. (*After Wheeler ©IEEE, 1975* [8].)

1.3.1.2 The Spherical Coil

Constant Pitch Spherical Coil To determine which small antennas utilize their volumes most efficiently (or equivalently have the highest effective volume and lowest Q), Wheeler explored the constant pitch spherical coil antenna shown in Fig. 1.15, where the excitation voltage is across the poles of the coil.

The inductance of the spherical coil in Fig. 1.15 was derived starting with the familiar formula for an n-turn, air-filled cylindrical coil of radius a and length $2a$, surrounded by a perfect magnetic medium

$$L_{\text{cylinder}}(\mu_r = 1, \mu_{r,\text{external}} = \infty) = \frac{\pi}{2}\mu_0 a n^2 \tag{1.36}$$

Wheeler states that similarly to the cylindrical coil, a constant magnetic field is present inside the spherical volume of Fig. 1.15 when a ϕ-directed constant surface current is assumed over the sphere surface [27] (for a detailed analysis see Simpson [33,34]). Consequently, he then notes that since sphere volume $= 2/3$ cylinder volume, the stored magnetic energy is also two-thirds that of the cylinder. He then proceeds to correct the inductance formula by considering mediums other than air or perfectly magnetic. By integrating over the coil currents, he found the magnetic field along the axis perpendicular to the loops, and used this field to find the magnetic potential as a function of position along the z-axis. Wheeler concluded that two-thirds of the magnetic potential is inside the coil and one-third is external to it. This indicates that the magnetic reluctance outside the sphere is half the magnetic reluctance inside the sphere.

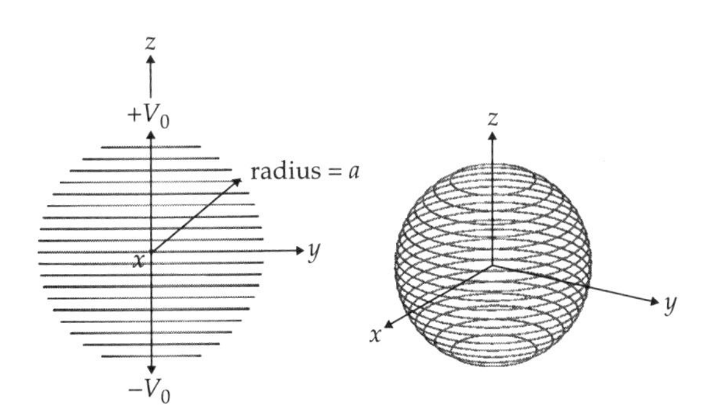

FIGURE 1.15 Wheeler's spherical coil antenna with constant pitch. (*See Wheeler* [27].)

From the above analysis, Wheeler proceeded to give the inductance formula for the spherical coil antenna in Fig. 1.15 with constant pitch as

$$L = \frac{2\pi}{9}\mu_0 a n^2 \frac{1}{\dfrac{2}{3\mu_r} + \dfrac{1}{3\mu_{r,\text{external}}}} = \frac{2\pi}{9}\mu_0 a n^2 \frac{3}{\dfrac{2}{\mu_r} + \dfrac{1}{\mu_{r,\text{external}}}} \tag{1.37}$$

As usual, μ_r is the relative permeability inside the spherical coil and $\mu_{r,\text{external}}$ is the relative permeability outside the spherical coil. Wheeler frequently includes shaping factors such as these to modify inductance and capacitance to account for shape and material variances. This was also noted in the cylindrical C-type and L-type antennas.

To determine the quality factor of the spherical coil antenna, initially Wheeler assumed an air surrounding medium and a magnetic medium (relative permeability μ_r) filling. The feed-point reactance of this antenna can be written as

$$\omega L = \frac{4\pi^2}{9} Z_0 \frac{an^2}{\lambda} \frac{3}{1 + 2/\mu_r} \tag{1.38}$$

For the radiation resistance, Wheeler again started with the corresponding formula for a cylinder coil of n turns, radius a, and length $2a$. He then used the same volume and material correction factors as in Eq. (1.37) to obtain

$$R = 20\left(\frac{2\pi\sqrt{A}}{\lambda}\right)^4 \tag{1.39}$$

where

$$A = \frac{2}{3}\pi a^2 n \frac{3}{1 + 2/\mu_r} \tag{1.40}$$

From Eq. (1.30) it then follows that

$$Q = \frac{\omega L}{R} = \frac{1}{(ka)^3}(1 + 2/\mu_r) = \frac{V_{\text{RS}}}{\sigma V_{\text{physical}}} \tag{1.41}$$

We observe that this expression for Q is simply the ratio of the volume of the radiansphere to that of the structure multiplied by the shape factor $\sigma = (1 + 2/\mu_r)^{-1}$. Clearly, the minimum Q for this spherical coil antenna is obtained when $\mu_r \to \infty$. This is also the same conclusion noted earlier for the L-type cylindrical coil antenna in Fig. 1.13. The condition of $\mu_r \to \infty$ corresponds to zero magnetic energy storage inside the volume, and since no capacitive component exists in Wheeler's model, the total energy storage inside the spherical volume is zero. Further, as the radiated fields are those of a TE_{10} mode, the spherical coil antenna with constant pitch and infinite permeability core can be thought of as giving the lower bound Q for single mode operation. From Eq. (1.41), this limit can be expressed as

$$Q_{\mathrm{min,Wheeler}} = \frac{1}{(ka)^3} \tag{1.42}$$

Self-Resonant Spherical Coil Wheeler [27] gave an example of a self-resonant spherical coil antenna depicted in Fig. 1.16, excited by a voltage across the coil poles. Wheeler gave a new wire arrangement on the sphere to realize constant electric and magnetic fields inside the structure for self-resonance. This is accomplished by using a tapered coil pitch arrangement (see Fig. 1.16).

Wheeler corrected the constant pitch inductance formula (1.37) by introducing a shape factor to account for the change in voltage along the wire due to tapering. He also determined the capacitance of the structure using the excitation potential difference between the two poles of the sphere. The equivalent inductance of the tapered spherical coil and coil length (necessary for self-resonance) are derived by Wheeler as

$$L = \frac{2}{9\pi^3}\mu_0\frac{l^2}{a}\frac{3}{1+2/\mu_r} \tag{1.43}$$

$$l = \frac{\lambda}{2}\sqrt{\frac{1+2/\mu_r}{1+\varepsilon_r/2}} = \frac{\lambda}{2}\alpha \tag{1.44}$$

As seen, the equivalent length of the tapered coil is equal to $l = \alpha\lambda/2$, where α is the scale factor dependent on the enclosed material. For the case when no energy is stored within the sphere

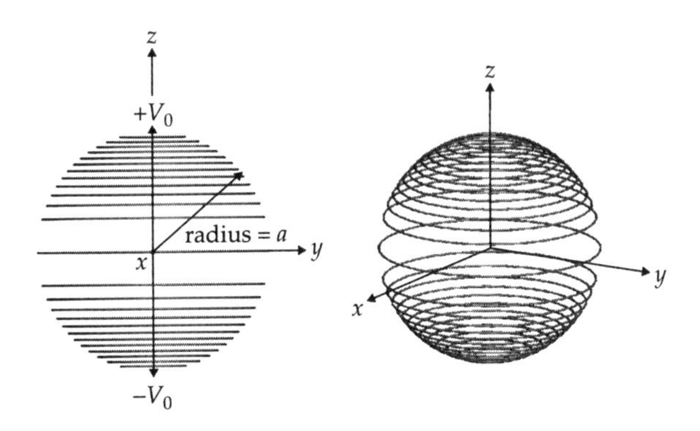

FIGURE 1.16 Wheeler's self-resonant tapered pitch spherical coil antenna. (*See* [2].)

($\varepsilon_r = 0$, $\mu_r = \infty$), the resonant length is exactly $\lambda/2$. This zero energy self-resonant spherical coil presents an important limiting case for a small antenna radiating both TM_{10} and TE_{10}. As in [27], it is stated that the Q of this self-resonant coil is half of the Q for constant pitch spherical coil (with zero internal stored energy). Wheeler states that this is a result of equal electric and magnetic stored energy outside the sphere, and the radiated power is now twice the single mode case. The Q for this self-resonant coil with a tapered pitch is therefore equal to

$$Q(\varepsilon = 0, \mu = \infty) = \hat{Q}_{\text{min,Wheeler}} = \frac{1}{2(ka)^3} \tag{1.45}$$

In terms of the definition of Q given in Eq. (1.3), this means that the $max(W_M, W_E)$ remains the same in the single mode case, but P_A is twice its original value.

Equivalent Spherical Coil Wheeler states that any coil formed by a loop winding in the same direction and with the loops located parallel to one another can be represented by an equivalent spherical coil as in Fig. 1.15. The equivalent spherical coil will have the same inductance and the radiation pattern as that of the original coil. He then gave the appropriate formulas for the equivalent spherical coil of radius a_{eq} and turns n_{eq} as

$$a_{\text{eq}} = \left(\frac{\mu_0 A^2}{2\pi L}\right)^{1/3} \tag{1.46}$$

$$n_{\text{eq}} = \left(\frac{6L^2}{\pi\mu_0^2 A}\right)^{1/3} \tag{1.47}$$

In this, $A = A_1 + A_2 + \cdots + A_n$ where A_i is the area of the ith turn in the original coil, and L is the inductance of the original coil. These results illustrate another key conclusion made by many authors and seen in much empirical evidence but never explicitly shown. That is, Wheeler's results imply that the diameter of the equivalent spherical coil is always less than the maximum dimension of the original coil. Concurrently, the equivalent spherical coil will retain the same Q and radiation resistance as that of the original coil. This demonstrates that the spherical coil makes more efficient use of the *Chu sphere* as compared to the class of coils described above.

Though this conclusion was only shown comparing different coils represented by an equivalent spherical coil, it holds for all small antennas. That is, *antennas which utilize more of their Chu sphere tend to have smaller values of Q*. Most of the empirical work in demonstrating this property is with surface current distributions over the *Chu sphere*.

1.3.1.3 Comments

Using the concept that a small antenna behaves as a lumped capacitor or inductor with a radiation resistance component, Wheeler expressed these parameters in terms of modified versions of well known capacitance, inductance, and radiation resistance formulae. This representation is, of course, an approximation as the single reactance lumped element model breaks down as antenna size increases. As the antenna size increases, other techniques for representing the properties of small antennas including multimode analysis are more appropriate.

The primary conclusion reached by Wheeler is that the Q of a small antenna is inversely proportional to its physical volume. A shape factor σ was also introduced to account for variances in the effective area or length of the antenna, giving

$$Q = \frac{V_{\text{RS}}}{\sigma V_{\text{physical}}} \tag{1.48}$$

Wheeler obtained two important limiting cases for the small antenna Q. For the coil in Fig. 1.15 with infinite permeability core, Wheeler gave the radiation characteristics as a series RL circuit. Recognizing that the mode radiated by this structure is a TE_{10} type and that an infinite permeability core results in zero stored energy inside the sphere, he reached the limiting value for Q given by

$$Q_{\text{min, Wheeler}} = Q_{\text{TE10}}\left(\mu_r = \infty\right) = \frac{1}{(ka)^3} \tag{1.49}$$

For the second limiting case, Wheeler considered the self-resonant coil of Fig. 1.16 supporting TE_{10} and TM_{10} modes, with zero stored internal energy. He found the limiting value for Q as

$$\hat{Q}_{\text{min, Wheeler}} = Q_{\text{TE10+TM10}}\left(\varepsilon_r = 0,\ \mu_r = \infty\right) = \frac{1}{2(ka)^3} \tag{1.50}$$

Wheeler's conclusions on small antenna limitations were later verified analytically by Thal [28] and experimentally by Best [13]. It is important to note that Wheeler's analysis is all based on equivalent circuit model of the small antenna. That is, he did not perform any full-wave analysis. Through his circuit analysis, Wheeler was the first to show that the Q of a small antenna is decreased with the increasing volume. The associated formula is in Eq. (1.48) and leads to the conclusion that *antennas which best utilize their minimum enclosing sphere tend to have small Q values as compared to other geometries within the same volume.* Wheeler also concluded that as the antenna size decreases, the ratio of reactance to radiation resistance increases even more rapidly. This reality exemplifies the issue of impedance matching for very small antennas.

1.3.2 Work of Chu (1948)

While Wheeler carried out his analysis using lumped circuit models for the small antenna itself, Chu [11] came to similar conclusions using spherical vector wavefunctions to evaluate the gain and bandwidth of omnidirectional antennas. Figure 1.1 shows the mathematical sphere of radius a enclosing the small antenna structure. This is the minimum sphere that encloses the antenna structure, and will be referred to hereafter as the *"Chu Sphere."* Chu [11] notes that the field configuration external to the *Chu sphere* is not uniquely defined by the interior source. Also he notes that an infinite number of source distributions are possible inside the *Chu sphere* for a single external field configuration.

1.3.2.1 Spherical Waves

Chu begins his analysis by assuming a vertically polarized, omnidirectional antenna of maximum dimension $2a$. Thus, the structure can be enclosed by the *Chu sphere* of radius a. With a spherical coordinate system (r, θ, ϕ), the transmitted/received fields are TM_{n0} waves of order n (the azimuthal index m is zero due to the omni-directional field nature). The nonzero electromagnetic field components associated with TM_{n0} modes are

$$H_\varphi = \sum_n A_n P_n^1(\cos\theta) h_n^{(2)}(kr) \tag{1.51a}$$

$$E_r = -jZ_0 \sum_n A_n n(n+1) P_n(\cos\theta) \frac{h_n^{(2)}(kr)}{kr} \tag{1.51b}$$

$$E_\theta = jZ_0 \sum_n A_n P_n^1(\cos\theta) \frac{1}{kr}\frac{d}{dr}\left(h_n^{(2)}(kr)\right) \tag{1.51c}$$

where A_n are constants determined by the source distribution of the antenna

$P_n^1(\cos\theta)$ is the associated Legendre polynomial of order $m = 1$ and degree n. This polynomial has a behavior similar to that of Fourier series with n terms. Several associated Legendre polynomials $P_n^1(x)$ are plotted in Fig. 1.17.

$P_n(\cos\theta)$ is the Legendre polynomial of degree n.

$h_n^{(2)}(kr)$ is the spherical Hankel function of the second kind, representing the outward traveling wave.

$Z_0 = \sqrt{\frac{\mu}{\varepsilon}}$ is the plane wave impedance of the homogeneous, isotropic medium represented by (μ, ε).

Chu then further restricted his analysis to an omni-directional antenna whose maximum gain lies in the equatorial plane $\theta = \pi/2$.

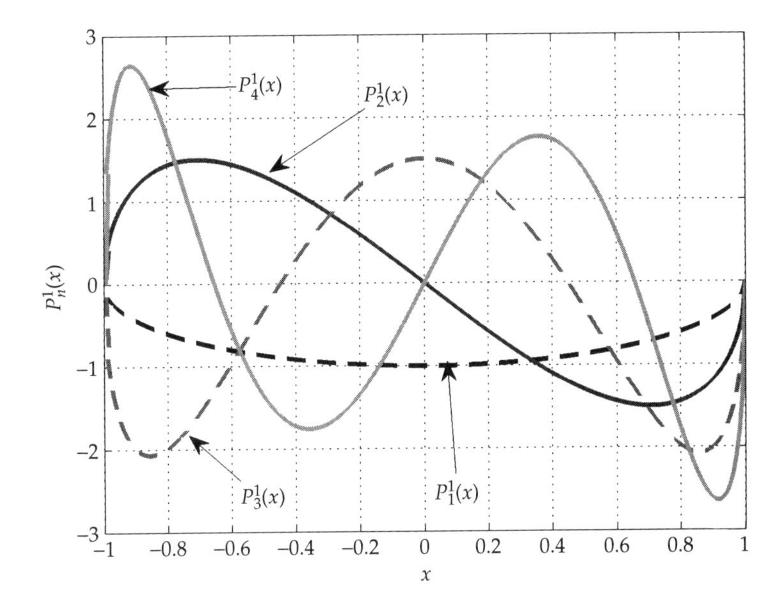

FIGURE 1.17 Associated Legendre polynomials $P_n^1(x)$ for $n = 1$ to 4.

We can see from Fig. 1.17 that $P_n^1(0)$ vanishes for n even and is finite for n odd. Therefore, for maximum gain in the equatorial plane it is necessary that A_n is zero for n even, and the A_n values for n odd combine in phase.

1.3.2.2 Chu Equivalent Circuit

To separate the total antenna energy into components associated with radiation, nonpropagating electric energy, and nonpropagating magnetic energy, Chu devised an equivalent circuit for the wave impedance of each propagating TM_{n0} wave. This circuit approach to determine the radiated and stored fields provides many benefits over a field approach using the Poynting theorem. Among them, the passive circuit provides greater insight into the nature of the spherical mode and their contribution to the overall antenna performance. Chu notes that the significant drawback in using a field approach to find the radiated power and stored energies is the nonlinear nature of the field components. As such, superposition cannot be directly applied to separate the electric and magnetic stored energy components in the near field.

To derive the equivalent circuits for spherical TM_{n0} mode, Chu begun by recognizing mode orthogonality among the modes. As such, the overall antenna power or energy can be viewed as a superposition of the corresponding powers and energies for each mode. Figure 1.18

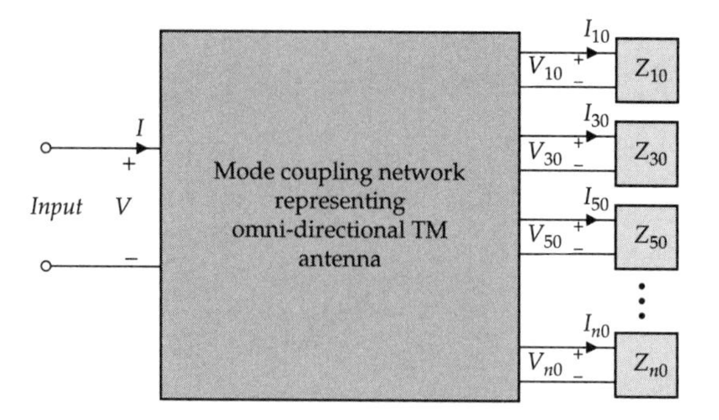

FIGURE 1.18 Chu's equivalent circuit network to represent the omni-directional TM_{n0} antenna. (*See Chu* [11].)

represents the equivalent circuit used to represent the modes generated by the omni-directional TM_{n0} antenna.

Chu proceeds to define the equivalent voltage and current for each TM_{n0} mode network on the following assumptions:

1. The input impedance of the TM_{n0} network is equal to the normalized wave impedance in the outward radial direction of the TM_{n0} mode at the surface of the *Chu sphere* of radius $= a$.
2. The complex power at the input terminals of the TM_{n0} mode circuit is equal to the complex power of the TM_{n0} mode exiting the *Chu sphere*.

The normalized wave impedance in the outward radial direction for the TM_{n0} mode can be written using the recurrence relations among Bessel functions. Specifically,

$$
\begin{aligned}
Z_{n0} &= \frac{j\left(kah_n^{(2)}(ka)\right)'}{kah_n^{(2)}(ka)} \\
&= \frac{n}{jka} + \cfrac{1}{\cfrac{2n-1}{jka} + \cfrac{1}{\cfrac{2n-3}{jka} \cdots \cfrac{1}{\cfrac{3}{jka} + \cfrac{1}{\cfrac{1}{jka} + 1}}}}
\end{aligned}
\tag{1.52}
$$

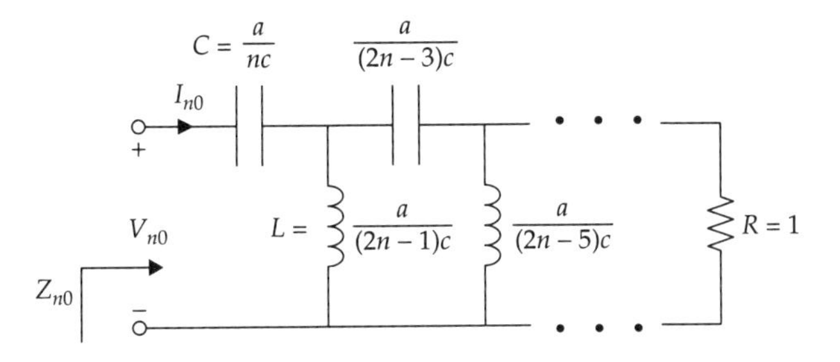

FIGURE 1.19 Equivalent impedance circuit for the TM_{n0} mode of the *Chu antenna* with a = sphere radius and c = velocity of light. (*See Chu* [11].)

This continued fraction expansion implies that Z_{n0} can be represented in terms of the circuit network shown in Fig. 1.19.

It is important to recognize that the equivalent circuit for any TM_{n0} wave (Fig. 1.19) behaves as a high-pass filter and stores capacitive energy. Thus, increasing the radius a of the *Chu sphere* has the same effect as increasing the frequency. By itself, the circuit models the impedance seen by a wave propagating outward from the *Chu sphere* surface. As the distance from the *Chu sphere* becomes larger, the wave impedance (for the TM_{n0} mode) approaches that of the intrinsic impedance Z_0 (as would be expected). Since there are an infinite set of source configurations that can generate any mode, *Chu chose the excitation leading to zero dissipation for the antenna and zero stored energy inside the Chu sphere*. This type of spherical, zero internal stored energy antenna will be referred to as a "*Chu antenna.*"

With respect to Fig. 1.19, the Q of the omni-directional TM_{n0} *Chu antenna* can be determined rather easily. Since the antenna is assumed to have no losses other than those due to radiation, the total power delivered (P_A) is the sum of the real powers delivered to each of the mode circuits. Further, since we assume zero energy inside the *Chu sphere*, W_E is the total stored electric energy in each of the mode circuits. With these considerations in mind, the total Q of the omni-directional TM_{n0} *Chu antenna* can be expressed as

$$Q = \frac{2\omega W_E}{P_a} = \frac{\displaystyle\sum_{n=odd} |A_n|^2 \frac{n(n+1)}{2n+1} Q_n}{\displaystyle\sum_{n=odd} |A_n|^2 \frac{n(n+1)}{2n+1}} \tag{1.53}$$

where Q_n is the quality factor of the TM_{n0} mode circuit (see Fig. 1.19). As Chu did not have access to efficient methods for computing the total electric energy in each of the mode circuits, he approximated each mode circuit in Fig. 1.19 as an RLC series circuit, a valid approximation near the operating frequency.

In summary, Chu restricts his analysis to an antenna with the following properties:

- Vertically polarized
- Omni-directional
- Even n modes are not excited ($A_n = 0$ for n even)
- For n odd, A_n has the same phase angle for each mode
- No conducting losses in the antenna structure
- No stored energy inside the *Chu sphere* (*Chu antenna*)

Chu examines three cases for this type of antenna: (1) maximum gain antenna, (2) minimum Q antenna, and (3) maximum G/Q antenna. These quantities are examined in the following sections.

1.3.2.3 Maximum Gain Omni-directional Chu Antenna

Using the standard antenna gain definition, the gain G (in the equatorial plane of the *Chu antenna*) is given by

$$G\left(\theta = \frac{\pi}{2}\right) = \left.\frac{4\,\pi\,|E_\theta|^2}{\displaystyle\int\!\!\int |E_\theta|^2 \sin\theta\, d\theta\, d\varphi}\right|_{\theta=\frac{\pi}{2}}$$

$$= \frac{\left|\displaystyle\sum_{n=1,odd}^{N} A_n (-1)^{\frac{n+1}{2}} P_n^1(0)\right|^2}{\displaystyle\sum_{n=1,odd}^{N} |A_n|^2 \frac{n(n+1)}{2n+1}} \qquad (1.54)$$

For maximum achievable gain, the A_n coefficients remain to choose appropriately. To do so we differentiate Eq. (1.54) with respect to A_n and set the result to zero. Solving the resulting $N \times N$ system yields the A_n values. Table 1.1 gives the maximum gain in the equatorial plane when $N = 1, 3,$ and 5.

From Table 1.1, we can conclude that in theory any desired gain can be realized independent of antenna size as long as the source

N	1	3	5 -------------------- N
G_{max} $(\theta = \pi/2)$	1.5	3.81	4.10 -------------------- $2N/\pi$

TABLE 1.1 G_{max} Versus N in the Equatorial Plane (After Chu [11])

distribution can be constructed. Equation (1.54) shows how the antenna gain increases with the inclusion of higher order modes (again with proper excitation coefficients). However, in order to excite these higher order modes, the source distribution complexity increases dramatically. As such, the needed excitation configurations may not be realizable in practice.

1.3.2.4 Minimum *Q* Omni-directional Chu Antenna

To determine the set of mode coefficients A_n that give minimum Q for the *Chu antenna,* we proceed to differentiate Eq. (1.53) with respect to A_n and set the result to zero, giving

$$Q_n \sum_{n=1,odd}^{N} |A_n|^2 \frac{n(n+1)}{2n+1} = \sum_{n=1,odd}^{N} \left\{ |A_n|^2 \frac{n(n+1)}{2n+1} Q_n \right\} \tag{1.55}$$

From Eq. (1.55) it is obvious that the only nontrivial solution to minimizing Q is to excite only a single mode (i.e., $A_n = 0$ except for $n = 1$). From Fig. 1.19, it is clear that the TM_{10} mode (see Fig. 1.20) gives the lowest Q for the omni-directional *Chu antenna,* and we observe that it also represents the mode generated by an infinitesimal (Hertzian) electric dipole.

1.3.2.5 Maximum *G/Q* Omni-directional Chu Antenna

The maximum G/Q ratio can be determined by using the derived formulas for G and Q in Eqs. (1.54) and (1.53), respectively. After dividing (1.54) by (1.53), the resultant G/Q ratio is again differentiated with respect to A_n and set to zero to create a system of equations. Solving these yields the A_n coefficients giving the max G/Q as plotted in Fig. 1.21.

Figure 1.21 shows the plot of the maximum G/Q ratio versus ka $(2\pi a/\lambda)$. The asymptotic nature of the curves in Fig. 1.21 demonstrates another fundamental limitation of omni-directional antennas. Specifically, for a given antenna supporting N modes, there is an upper bound

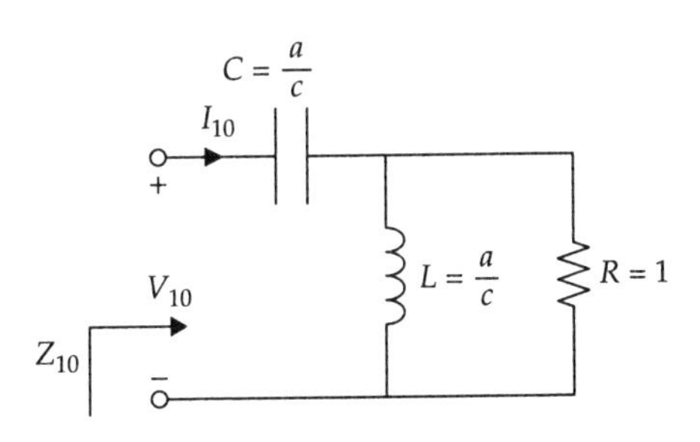

FIGURE 1.20 Equivalent impedance network for the TM_{10} mode with a = *Chu sphere* radius and c = velocity of light. (*See Chu* [11].)

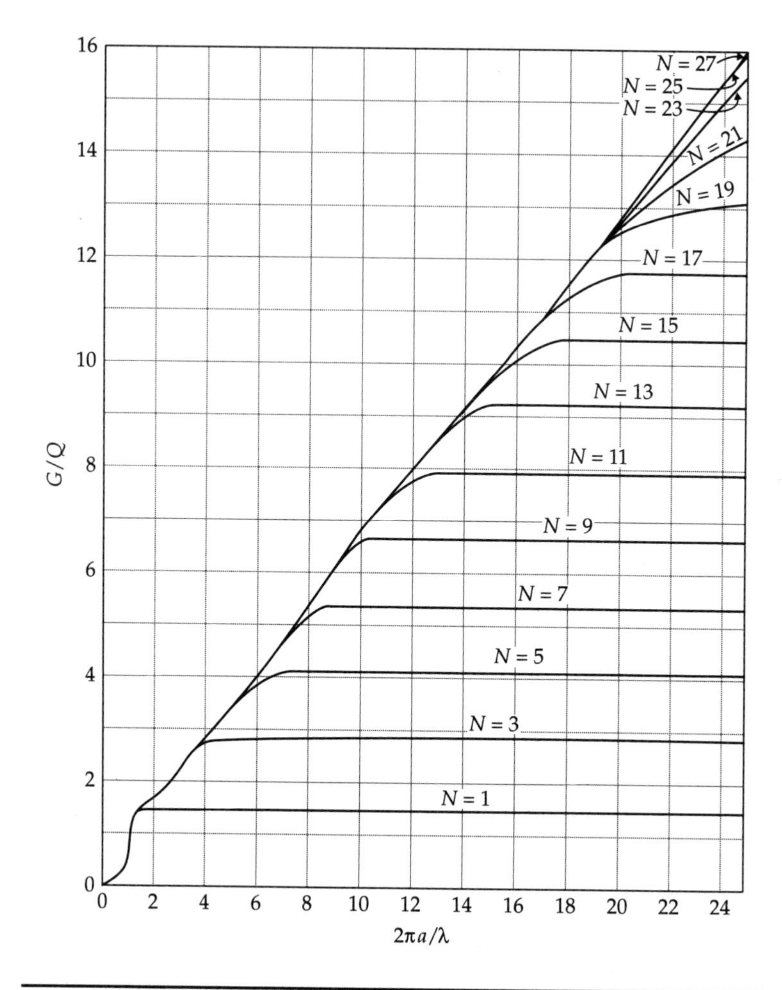

FIGURE 1.21 Maximum G/Q ratio for vertically polarized *Chu antenna* versus antenna size ka for various N. (*After Chu ©J. App. Phys., 1948* [11].)

of the G/Q ratio. Alternatively, we can state that as Q is reduced, the G/Q ratio reaches a maximum bound.

1.3.2.6 Horizontally Polarized Omni-directional Chu Antenna

The previous analysis was restricted to a *Chu antenna*, vertically polarized and having an omni-directional pattern with maximum gain in the equatorial plane. If this *Chu antenna* is now horizontally polarized as well as omni-directional and with maximum gain in the equatorial plane, the transmitted/received waves will be the superposition of the TE_{n0} waves. Once again, the m index is zero due to

TM	TE
H_ϕ	$-E_\phi$
E_R	H_R
E_θ	H_θ
Z_0	$1/Z_0$

TABLE 1.2 TM-TE Duality

the omni-directional nature of the fields. It is a straightforward task to identify the three nonzero field components associated with the TE_{n0} waves. They are listed in Table 1.2 along with their dual TM mode quantities [36].

The horizontally polarized TE_{n0} *Chu antenna* can also be represented by an equivalent circuit similar to Fig. 1.19. However, in this case each circuit outside the sphere must represent the TE_{n0} modes generated by the sources inside the *Chu sphere.* Chu noted that the normalized wave admittance of the TE_{n0} mode is equal to the normalized wave impedance of the TM_{n0} modes as given in Eq. (1.3.2.2). The TE_{n0} mode equivalent circuits can similarly be found by defining the unique mode voltages and currents such that the input admittance of each TE_{n0} equivalent circuit is equal to the normalized wave admittance of the corresponding TE_{n0} mode. This can be done by equating the complex power at the input terminals of each TE_{n0} circuit to the complex power of the TE_{n0} mode crossing the *Chu sphere.* The equivalent admittance circuit for the TE_{n0} mode is shown in Fig. 1.22.

It is important to note that the equivalent admittance circuit for any TE_{n0} wave behaves as a high-pass filter with inductive energy storage. Increasing the radius a of the *Chu sphere* has the same effect as increasing the frequency. It is also evident that as the distance from the *Chu sphere* increases, the intrinsic wave admittance approaches that of free space. Invoking the dual nature between the TE_{n0} and TM_{n0} fields, all

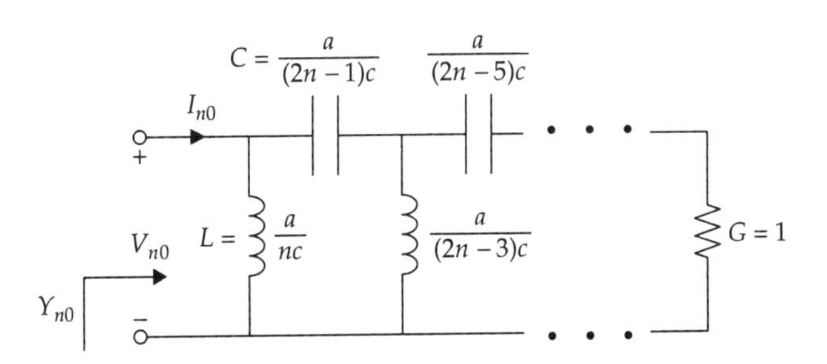

FIGURE 1.22 Equivalent admittance circuit for the TE_{n0} mode of the *Chu antenna* with a = sphere radius and c = velocity of light. (*See* Chu [11].)

previously derived results for a *Chu antenna* with vertical polarization also apply to the *Chu antenna* with horizontal polarization.

1.3.2.7 TE and TM Circularly Polarized Omni-directional Chu Antenna

Chu stated that a circularly polarized, omni-directional *Chu antenna* with maximum gain in the equatorial plane can be realized with 90° phasing between the TM_{n0} and TE_{n0} modes (provided the TE_{n0} and TM_{n0} pairs radiate equal power). Under these conditions, the stored electric and magnetic energies are also equal. The nth mode Q for such a *Chu antenna* can be found using the standard definition of Q given in Eq. (1.3). That is,

$$Q_n = \frac{2\omega W_n}{P_n} \tag{1.56}$$

with W_n being the total stored electric (see Fig. 1.19) *or* magnetic energy (see Fig. 1.22) in the networks, and P_n is the total real power delivered to both TE_{n0} and TM_{n0} circuits. For this type of circularly polarized antenna, we recognize that for small ka the TM_{10} equivalent circuit has predominantly electric stored energy. Likewise, the corresponding TE_{10} network has predominantly magnetic stored energy. However, since the total real powers delivered to both circuits are equal, Chu states that the resulting quality factor for the TM_{10}/TE_{10} circularly polarized antenna $Q_{CP,TM_{10}-TE_{10},Chu}$ is appropriately given by

$$Q_{CP,TM10-TE10,Chu} \approx \frac{1}{2} Q_{TM10,Chu} = \frac{1}{2} Q_{TE10,Chu} \tag{1.57}$$

Here, the *Chu* subscript indicates we refer to a *Chu antenna.*

1.3.2.8 Comments

Chu's work introduced many of the fundamental ideas used in modern analysis of small antennas. The concept of the *Chu antenna* (an antenna with zero energy storage inside its *Chu sphere*) is central to finding a lower bound on Q. Practical antennas exciting a given mode configuration must, of course, have nonzero energy storage inside their *Chu sphere*, to give a Q value larger than that of the *Chu antenna.* However, since Chu used a series RLC for each of his mode circuits there are inherent approximations, and much of the later work in small antennas is dedicated to finding more accurate and simpler closed form expressions for the Q limits.

The absolute lower bound on Q for a linearly polarized antenna (denoted as Q_{min}) radiating *only* TM or *only* TE modes was shown to be the Q of a *Chu antenna* radiating a pure TM_{10} or TE_{10} mode. The corresponding Q for a circularly polarized (CP) antenna radiating equal parts TM_{10} and TE_{10} modes was shown to have its lowest bound equal to half of the single mode Q

$$\hat{Q}_{min} = Q_{CP,TM10-TE10,Chu} \tag{1.58}$$

Chu also noted that in theory any gain can be realized with proper mode excitation coefficients. Additionally, Chu noted that there is a theoretical upper limit on G/Q, indicating that higher gain is possible at the cost of reduced bandwidth and increased antenna complexity.

The TM_{n0} and TE_{n0} mode high-pass networks proposed by Chu in Figs. 1.19 and 1.22, respectively, show the difficulty in radiating higher order modes using smaller antennas. For small antennas these modes are not practically realizable for two reasons: (1) the source complexity increases and (2) the large Q associated with higher order modes leads to large amounts of stored energy as compared to radiated power. This can be seen in the TM_{n0} and TE_{n0} mode networks as they operate well into their stopband for small antenna sizes. As is well known, in most cases, it is desirable to have minimal higher order mode excitation to maximize bandwidth. However, higher order modes can serve to tune out reactance (see Sec. 1.3.12).

1.3.3 Work of Harrington (1960)

Harrington [3] expanded Chu's [11] work, focusing on the gain and Q properties of a unidirectional *Chu antenna* radiating equally excited TE and TM modes along the $\theta = 0°$ direction. He also introduced the dissipation factor d_F in an attempt to quantify antenna losses as a function of size and number of modes excited.

1.3.3.1 Maximum Gain for a Directional Chu Antenna

As stated in Sec. 1.3.2 (Chu), the fields outside the *Chu sphere* can be written as the superposition of orthogonal TE_{nm} and TM_{nm} modes. Harrington [3] used the known TE_{nm} and TM_{nm} wavefunction representation

$$\Psi_{TE} = \sum_{m,n} A_{nm} h_n^{(2)}(kr) P_n^m(\cos\theta)\cos(m\varphi + \alpha_{TEnm}) \tag{1.59a}$$

$$\Psi_{TM} = \sum_{m,n} B_{nm} h_n^{(2)}(kr) P_n^m(\cos\theta)\cos(m\varphi + \alpha_{TMnm}) \tag{1.59b}$$

$$\boldsymbol{E} = -\nabla \times (\hat{\boldsymbol{r}}\Psi_{TE}) + \frac{1}{j\omega\varepsilon}\nabla \times \nabla \times (\hat{\boldsymbol{r}}\Psi_{TM}) \tag{1.60a}$$

$$\boldsymbol{H} = \nabla \times (\hat{\boldsymbol{r}}\Psi_{TM}) + \frac{1}{j\omega\mu}\nabla \times \nabla \times (\hat{\boldsymbol{r}}\Psi_{TE}) \tag{1.60b}$$

In these equations, $h_n^{(2)}(kr)$ is the spherical Hankel function of the second kind, and $P_n^m(\cos\theta)$ is the associated Legendre polynomial. Harrington used these expressions to derive the directive gain in the $\theta = 0°$ direction. His derivation finds that $G(\theta = 0)$ is independent of the phase constants $\alpha_{\mathrm{TE_{nm}}}$ and $\alpha_{\mathrm{TE_{nm}}}$, implying that $G(\theta = 0)$ is independent of polarization [3]. Harrington's gain expression has the A_{n1} and B_{n1} coefficients on the numerator, and all of the mode coefficients in the denominator. Thus, G increases when all mode coefficients with $m \neq 1$ vanish. The gain is further maximized when the A_{n1} and B_{n1} coefficients are chosen so that the TE_{n1} and TM_{n1} powers are equalized, implying

$$A_{n1} = Z_0 B_{n1} \tag{1.61}$$

Even higher gain is obtained when the phases of A_{n1} are chosen to maximize the numerator. Upon applying each of these conditions, Harrington proceeded to maximize G by differentiating it with respect to A_{n1} and solving for the A_{n1} values. His resulting maximum gain expression for a unidirectional *Chu antenna* for $n \leq N$ is

$$G_{\max} = N^2 + 2N \tag{1.62}$$

1.3.3.2 Minimum *Q* for a Directional Chu Antenna

Using the same equivalent circuits employed by Chu [11], Harrington recognized that the Q of the nth mode (denoted as Q_n) for a linearly polarized, unidirectional, equally excited TE and TM mode *Chu antenna* is the same as that of the circularly polarized, omni-directional, equally excited TE and TM mode antenna in [11]. As usual, Q_n can be computed from

$$Q_n = \frac{2\omega W_{\mathrm{nm}}^{\mathrm{Electric}}}{P_{\mathrm{nm}}} = \frac{2\omega W_{\mathrm{nm}}^{\mathrm{Magnetic}}}{P_{\mathrm{nm}}} \tag{1.63}$$

where P_{nm} denotes the power contained by the $\mathrm{TE_{nm}}$ or $\mathrm{TM_{nm}}$ modes and $W_{\mathrm{nm}}^{\mathrm{Electric,Magnetic}}$ refer to the corresponding electric or magnetic energy densities. He found that

$$Q_{\mathrm{LP,TE}_{n1}-\mathrm{TM}_{n1},\mathrm{Chu}} = Q_{\mathrm{CP,TE}_{n0}-\mathrm{TM}_{n0},\mathrm{Chu}} \approx \frac{1}{2} Q_{\mathrm{TE}_{n0},\mathrm{Chu}} = \frac{1}{2} Q_{\mathrm{TM}_{n0},\mathrm{Chu}} \tag{1.64}$$

where the notation $Q_{\mathrm{LP,TE1n-TM1n,Chu}}$ refers to the *Chu antenna Q* radiating LP polarized waves due to equally excited TE_{n1} and TM_{n1} modes. Similarly to Chu [11] the minimum Q, denoted by $\hat{Q}_{\min}$, is

$$\hat{Q}_{\min} = Q_{\mathrm{LP,TE11-TM11,Chu}} = Q_{\mathrm{CP,TE10-TM10,Chu}} \tag{1.65}$$

1.3.3.3 Antenna Losses and Efficiency

Harrington also investigated the losses in the antenna structure as a function of antenna size. To do so, he used a figure of merit referred to as the dissipation factor d_F. He considered a spherical antenna of radius a whose radiation can be represented by the equivalent surface magnetic currents

$$\boldsymbol{M} = \boldsymbol{E} \times \hat{\boldsymbol{r}}|_{r=a} \tag{1.66}$$

He also represented the $\mathrm{TE_{nm}}$ and $\mathrm{TM_{nm}}$ normalized wave impedances looking in the $-\hat{r}$ direction by

$$Z_{\mathrm{nm}}^{\mathrm{TE}-} \approx Z_{\mathrm{nm}}^{\mathrm{TM}-} \approx \frac{Z_c}{Z_0} = \frac{(1+j)}{Z_0}\sqrt{\frac{\omega\mu}{2\sigma}} \tag{1.67}$$

That is, he included the conductivity σ into his expression to account for losses.

Harrington proceeded to define an equivalent network with $\boldsymbol{M}$ representing a series voltage discontinuity at the sphere's surface. For each mode, the normalized wave impedance looking into the $+\hat{\boldsymbol{r}}$ direction is given by the Chu equivalent circuit. Additionally, the impedance in the $-\hat{\boldsymbol{r}}$ direction (interior to the sphere) is given by Eq. (1.67). This equivalent network problem is depicted in Fig. 1.23.

Denoting the powers dissipated inside the *Chu sphere* by $P_{\mathrm{diss}}^{\mathrm{TE_{nm}},\mathrm{TM_{nm}}}$ and the radiated mode powers as $P_{\mathrm{rad}}^{\mathrm{TE_{nm}},\mathrm{TM_{nm}}}$, the dissipation factor for each equally excited $\mathrm{TE_{nm}}/\mathrm{TM_{nm}}$ mode pair is

$$d_{\mathrm{F}_n} = \frac{P_{\mathrm{diss}}^{\mathrm{TE_{nm}}} + P_{\mathrm{diss}}^{\mathrm{TM_{nm}}}}{P_{\mathrm{rad}}^{\mathrm{TE_{nm}}} + P_{\mathrm{rad}}^{\mathrm{TM_{nm}}}} \tag{1.68}$$

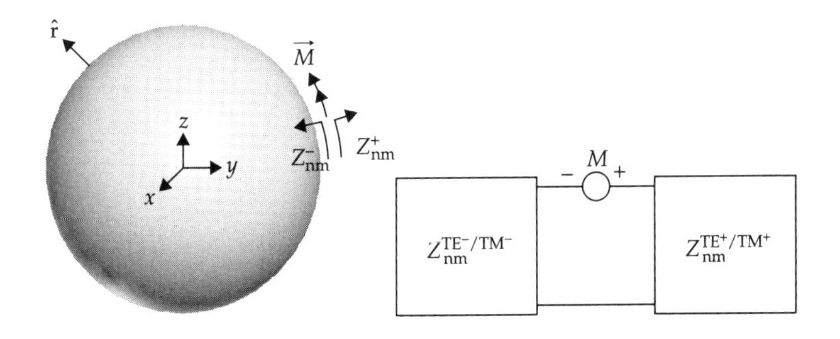

FIGURE 1.23 Circuit model for a lossy antenna structure used to compute the dissipation factor d_F.

The overall dissipation factor is then given by

$$d_F = \frac{\sum_{m,n} \left[\left(P_{\text{rad}}^{\text{TE}_{mn}} + P_{\text{rad}}^{\text{TM}_{mn}} \right) d_{F_n} \right]}{\sum_{m,n} \left(P_{\text{rad}}^{\text{TE}_{mn}} + P_{\text{rad}}^{\text{TM}_{mn}} \right)} \tag{1.69}$$

As expected, higher order modes are evanescent for small antennas. So, the dominant dissipation factor is d_{F1}. The efficiency is then given by

$$\mathit{Efficiency}(\%) = \frac{100}{1 + d_F} \tag{1.70}$$

and plotted in Fig. 1.24.

Though simplified models were used for the conductor losses, the dissipation factor provides an insight in the behavior of small antennas. We can state that *as antenna size decreases, its efficiency also decreases.* This is also the reason that low Q values for small antennas are primarily the result of material losses rather than radiation loss.

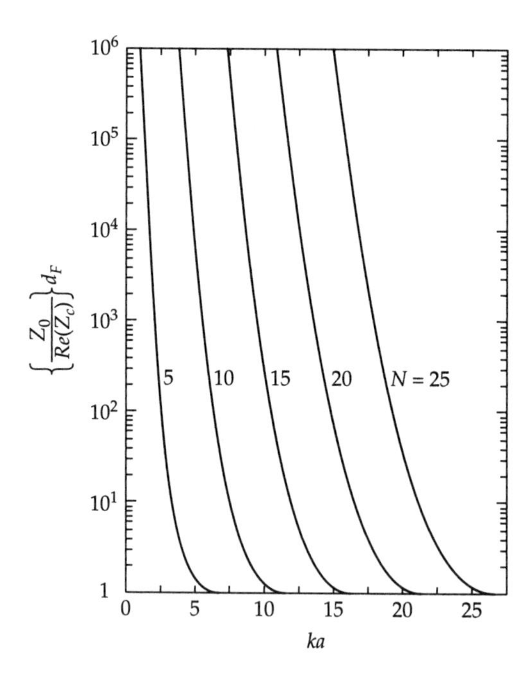

FIGURE 1.24 Dissipation factor d_F versus ka for various $N = \max(n)$ in Eq. (1.69). (*See Harrington* [3].)

1.3.3.4 Comments

Unlike Chu, Harrington focused on linearly polarized, unidirectional *Chu antennas* with equal excitation of TE_{nm} and TM_{nm} modes. In the small antenna limit, Eq. (1.1), the maximum gain or directivity of this antenna was derived to be 3. Also, the lower bound on Q was found to be approximately half that of the same *Chu antenna* supporting only TM or TE modes. This further reinforces the idea that reduction of Q from a linearly polarized antenna is a result of introducing TE and TM modes and not due to change in polarization. The reader may wonder why it was stressed in earlier sections that small antennas have the same pattern as that of a Hertzian dipole, yet Harrington's work focused on a unidirectional antenna. It will later be shown [4, 5] that these patterns can be realized in the small antenna limit using certain arrangements of electric and magnetic monopoles.

We close this discussion by nothing that Fig. 1.24 clearly shows the inverse relationship between antenna size and antenna efficiency. Though this analysis was performed for one specific case, the principle generally holds for all small antennas.

1.3.4 Work of Collin, Rothschild, and Fante (1964–1969)

Collin and Rothschild [18] sought to examine the antenna limits on Q by using a different method than Chu [11], based purely on the spherical TM_{nm} and TE_{nm} fields generated by a *Chu antenna*. Their work [18] was restricted to an antenna radiating either TM_{nm} modes or TE_{nm} modes. The authors also extended their analysis to cylindrical modes. However, the cylindrical mode analysis is not of relevance to small antennas since they focused on antennas of finite radius but of infinite extent.

1.3.4.1 Exact *Q* for a TM or TE Propagating Chu Antenna

Collin and Rothschild considered a *Chu antenna* of radius a (as defined in Sec. 1.3.2). The authors note that all antenna radiated fields have stored, nonpropagating reactive energy and a radiation component to carry real power out to infinity. Since the components cannot be easily separated, to evaluate Q Collin and Rothschild devised a method to separate the radiation component from the total stored electric and magnetic energy. The former can, of course, be found using Poynting's theorem.

Without loss of generality, Collin and Rothschild choose to analyze only the TM_{n0} mode of the *Chu antenna*. As discussed in Secs. 1.3.2 and 1.3.3, Q is independent of the azimuthal mode index m (the second index of the mode). So, it was assumed zero for simplicity. Also, the TE_{nm} modes, being the dual of the TM_{nm} modes, do not need to be considered. The field components of the TM_{n0} mode are given in Eqs. (1.51*a*) to (1.51*c*).

Central to determining the antenna Q, the authors note that the power flow from an antenna can be represented by an energy density multiplied by the energy velocity. That is,

$$P_{rad} = \textit{Radiated Antenna Power} = (\textit{Energy density})\ (\textit{Speed of light})$$
$$= \rho_{rad} \cdot c \tag{1.71}$$

From Poynting's theorem, the power exiting the *Chu antenna* surface S is given by (see Fig. 1.1 for coordinates)

$$P_{out} = \frac{1}{2}\oint_S (\boldsymbol{E} \times \boldsymbol{H}^*) \bullet d\boldsymbol{S} = \frac{1}{2}\int_0^{2\pi}\int_0^{\pi} E_\theta H_\phi^* a^2 \sin\theta\ d\theta\ d\phi$$
$$= P_{rad} + 2j\omega(W_M - W_E) \tag{1.72}$$

As usual, $W_{E,M}$ refer to the total stored electric and magnetic energy. Using the energy density ρ_{rad} in Eq. (1.71), we can subtract the total radiation energy from the total energy to yield the total stored energy

$$W_M + W_E$$
$$= \int_a^\infty \left\{ \left[\int_0^{2\pi}\int_0^{\pi} (w_e + w_m)\, r^2 \sin\theta\ d\theta\ d\phi \right] - \rho_{rad} \right\} dr$$
$$= \int_a^\infty \left\{ \left[\int_0^{2\pi}\int_0^{\pi} \left(\frac{1}{4}\varepsilon_0 |\boldsymbol{E}|^2 + \frac{1}{4}\mu_0 |\boldsymbol{H}|^2 \right) r^2 \sin\theta\ d\theta\ d\phi \right] - \frac{P_{rad}}{c} \right\} dr \tag{1.73}$$

where $\rho_{rad} = P_{rad}/c$ refers to radiated power density. Since the stored electric energy is greater than the magnetic one for TM_{nm} modes, W_M can be neglected. Thus, we evaluate on the W_E to get

$$Q_n = \frac{2\omega_0 W_{E,n}}{P_{rad,n}} \tag{1.74}$$

This is the Q for the TM_{n0} mode. Collin and Rothschild evaluate Eq. (1.74) to get the following expressions

$$Q_1 = \frac{1}{ka} + \frac{1}{(ka)^3} \tag{1.75a}$$

$$Q_2 = \frac{3}{ka} + \frac{6}{(ka)^3} + \frac{18}{(ka)^5} \tag{1.75b}$$

$$Q_3 = \frac{6}{ka} + \frac{21}{(ka)^3} + \frac{135}{(ka)^5} + \frac{675}{(ka)^7} \tag{1.75c}$$

We note that the first three of these are the same as those obtained by Chu using the equivalent circuits in Fig. 1.19.

1.3.4.2 Exact *Q* for a TM and TE Propagating Chu Antenna

Fante [19] expanded Collin and Rothschild's work by assuming a *Chu antenna* exciting TE and TM modes (similar to the case considered by Harrington in Sec. 1.3.3). Using Eq. (1.3) and the same procedure as Collin and Rothschild, Fante expressed the quality factor Q for an arbitrarily excited *Chu antenna* with TM and TE mode coefficients A_{nm} and B_{nm}, respectively, as

$$Q = \max \left\{ \frac{\sum_{n=1}^{\infty} \left[a_n^2 Q'_n + b_n^2 Q_n\right]}{\sum_{n=1}^{\infty} \left(a_n^2 + b_n^2\right)} \quad \text{or} \quad \frac{\sum_{n=1}^{\infty} \left[a_n^2 Q_n + b_n^2 Q'_n\right]}{\sum_{n=1}^{\infty} \left(a_n^2 + b_n^2\right)} \right\} \tag{1.76}$$

where

$$a_n^2 = \sum_{m=0}^{n} \lambda_{nm} \left|A_{nm}\right|^2 \tag{1.77}$$

$$b_n^2 = \sum_{m=0}^{n} \lambda_{nm} \left|B_{nm}\right|^2 \tag{1.78}$$

$$\lambda_{nm} = \frac{2\pi\hat{\varepsilon}_m}{2n+1} n(n+1) \frac{(n+m)!}{(n-m)!} \tag{1.79}$$

$$\hat{\varepsilon}_m = \begin{cases} 2 & \text{for} \quad m = 0 \\ 1 & \text{for} \quad m = 1 \end{cases} \tag{1.80}$$

In the equations, Q_n is the quality factor derived by Collin and Rothschild for the TM_{nm} or TE_{nm} modes, and Q'_n is an additional contribution. Closed form expressions for Q_n and Q'_n are given in [19] in terms of spherical Bessel functions with arguments of ka, and are plotted in Fig. 1.25 for several n. For the equally excited case, $a_n^2 = b_n^2$

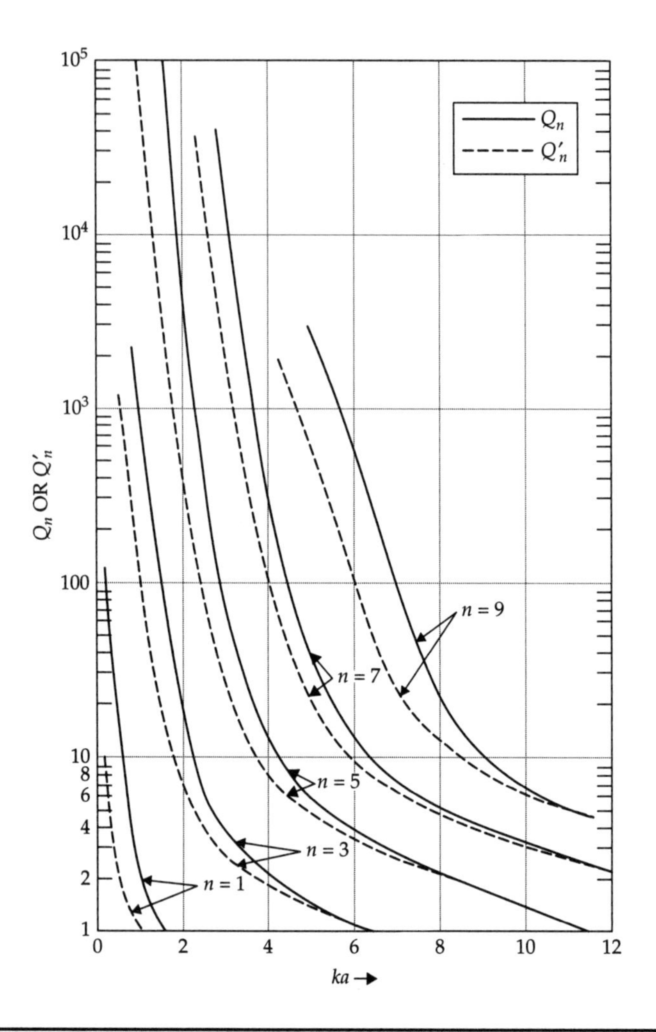

FIGURE 1.25 Quality factor components Q_n (solid) and Q'_n (dashed) for different mode indices n (*After Fante © IEEE, 1969* [19].)

and Eq. (1.76) reduces to

$$Q = \frac{\sum_{n=1}^{\infty} a_n^2 \hat{Q}_n}{\sum_{n=1}^{\infty} a_n^2} \tag{1.81}$$

with

$$\hat{Q}_n = \frac{Q_n + Q'_n}{2} \tag{1.82}$$

ka	$Q_1/2$	$\hat{Q}_{\min}$
0.3	20.00	21.600
0.6	3.16	3.990
1.0	1.00	1.505

TABLE 1.3 $\hat{Q}_{\min}$ and $Q_1/2$ for the *Chu Antenna* (See Fante [19])

From Eq. (1.81), Fante concluded that the minimum possible Q occurs for a *Chu antenna* exciting equal parts TM_{1m} and TE_{1m}

$$\hat{Q}_{\min} = Q_{\text{TM10+TE10,Chu}} = \frac{Q_1 + Q_1'}{2} \tag{1.83}$$

$$\hat{Q}_{\min} \approx \frac{1}{2} Q_1 = \frac{1}{2} Q_{\text{TE10,Chu}} = \frac{1}{2} Q_{\text{TM10,Chu}} \quad \text{for} \quad ka \ll 1 \tag{1.84}$$

Clearly, Eq. (1.84) is the same result as that derived by Harrington [36]. Also, Fig. 1.25 and Eq. (1.83) demonstrate an important issue alluded by Chu [11] and Harrington [4] (though never fully quantified). That is, the Q for a *Chu antenna* exciting equal parts of TM_{1m} and TE_{1m} modes is slightly greater than half the Q associated with either TE or TM modes. Indeed, Fig. 1.25 shows that for small ka, Q_n remains approximately an order of magnitude greater than Q_n'. Fante remarks that Q_n is always greater than Q_n'. Thus, from Eq. (1.83) the Q corresponding to equally excited TM_{1m} and TE_{1m} modes is approximately equal to $Q/2$ for the *Chu antenna*. However, as ka approaches unity, Q_n' can no longer be ignored, and the approximation Eq. (1.84) breaks down. Table 1.3 demonstrates this principle by comparing the actual $\hat{Q}_{\min}$ Eq. (1.83) to the $Q_1/2$ approximation.

1.3.4.3 Relationship Between *Q* and Fractional Bandwidth *B*

In [19] Fante gives the first direct relationship between Q and the 3 dB fractional bandwidth B for small antennas. To do so, Fante begins by assuming a high-Q, perfectly efficient antenna having an input impedance

$$Z_{\text{in}} = R_{\text{rad}} + jX_A(\omega) = \frac{2}{|I|^2}\left[P_{\text{rad}} + j2\omega(W_M - W_E)\right] \tag{1.85}$$

At resonance ($\omega = \omega_0$), Z_{in} can be approximated by the first two terms of its Taylor series expansion about ω_0. Specifically, we have

$$Z_{\text{in}} = R_{\text{rad}} + j(\omega - \omega_0)\left(\frac{dX_A(\omega)}{d\omega}\right)\bigg|_{\omega_0} \tag{1.86}$$

For this, the 3 dB return loss point occurs when $(\omega - \omega_0) d X_A \omega_0)/d\omega = R_{\text{rad}}$. Therefore, the 3 dB fractional bandwidth B can be identified as

$$B \approx \frac{2R_{\text{rad}}}{\omega_0 \left(\dfrac{dX_A(\omega)}{d\omega} \right) \Bigg|_{\omega_0}} = \frac{4P_{\text{rad}}}{\omega_0 \, |I|^2 \left(\dfrac{dX_A(\omega)}{d\omega} \right) \Bigg|_{\omega_0}} \tag{1.87}$$

To find $X'_A(\omega_0)$, Fante begins by considering a volume Ω bounded by a surface S coinciding with the entire antenna structure (including the feed port A) and the spherical surface S_∞ at infinity (see Fig. 1.26). Fante starts with the identity [36, pp. 394–396]

$$\iint_S \left(\frac{\partial \boldsymbol{E}}{\partial \omega} \times \boldsymbol{H}^* - \frac{\partial \boldsymbol{H}}{\partial \omega} \times \boldsymbol{E}^* \right) \bullet d\boldsymbol{S} + \iint_{S_\infty} \left(\frac{\partial \boldsymbol{E}}{\partial \omega} \times \boldsymbol{H}^* - \frac{\partial \boldsymbol{H}}{\partial \omega} \times \boldsymbol{E}^* \right) \bullet d\boldsymbol{S} = -j \iiint_\Omega \left(\mu_0 \, |\boldsymbol{H}|^2 + \varepsilon_0 \, |\boldsymbol{E}|^2 \right) dv$$
$$= -j4 \left(W_{M,\text{total}} + W_{E,\text{total}} \right) \tag{1.88}$$

where $W_{M,\text{total}}$ and $W_{E,\text{total}}$ represent the total magnetic and electric energy in Ω. For an antenna fabricated of perfect conductor, the

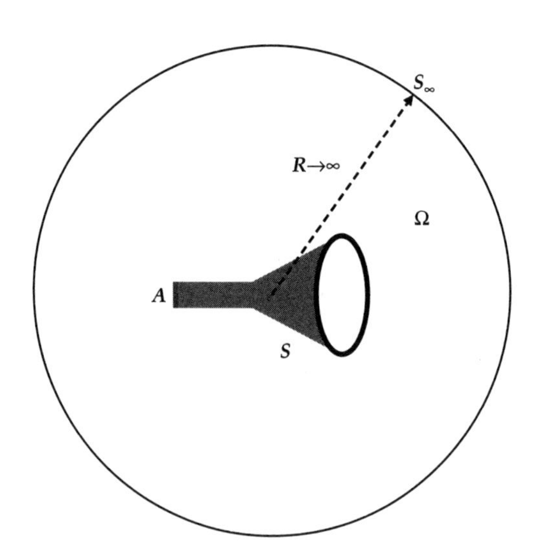

FIGURE 1.26 Volume Ω bounded by surface S coinciding with the entire antenna structure (including the feed port A) and far-field sphere S_∞.

boundary conditions can be applied

$$d\boldsymbol{S} \bullet \left[\frac{\partial \boldsymbol{E}}{\partial \omega} \times \boldsymbol{H}^* \right] = 0$$

$$d\boldsymbol{S} \bullet \left[\frac{\partial \boldsymbol{H}}{\partial \omega} \times \boldsymbol{E}^* \right] = 0 \qquad \text{on } S\text{, excluding } A$$

to rewrite Eq. (1.88) as

$$\iint\limits_{S_\infty} \left(\frac{\partial \boldsymbol{E}}{\partial \omega} \times \boldsymbol{H}^* - \frac{\partial \boldsymbol{H}}{\partial \omega} \times \boldsymbol{E}^* \right) \bullet d\boldsymbol{S} - \left(I^* \frac{\partial V}{\partial \omega} + V^* \frac{\partial I}{\partial \omega} \right)$$
$$= -j4(W_{M,\text{total}} + W_{E,\text{total}}) \tag{1.89}$$

where the second term on the left hand side is associated with the feed port equivalent voltage V and equivalent current I. Next, introducing the far zone field expressions valid on S_∞, we get

$$\boldsymbol{E}(r \to \infty) = \boldsymbol{E}_\infty(\omega) \frac{e^{-jkr}}{r} \tag{1.90a}$$

$$\boldsymbol{H}(r \to \infty) = \boldsymbol{H}_\infty(\omega) \frac{e^{-jkr}}{r} \tag{1.90b}$$

Using Eqs. (1.90a) and (1.90b) along with the chain rule for derivatives (with respect to frequency), Eq. (1.89) can be simplified to

$$\left(I^* \frac{\partial V}{\partial \omega} + V^* \frac{\partial I}{\partial \omega} \right) = \iint\limits_{S_\infty} \frac{1}{r^2} \left(\frac{\partial \boldsymbol{E}_\infty}{\partial \omega} \times \boldsymbol{H}_\infty^* - \frac{\partial \boldsymbol{H}_\infty}{\partial \omega} \times \boldsymbol{E}_\infty^* \right) \bullet d\boldsymbol{S}$$
$$+ j4 \left[W_{M,\text{total}} + W_{E,\text{total}} - \frac{r}{2c} \text{Re} \iint\limits_{S_\infty} (\boldsymbol{E} \times \boldsymbol{H}^*) \bullet d\boldsymbol{S} \right] \tag{1.91}$$

As $r \to \infty$ in Eq. (1.91), we immediately identify the bracketed term in the imaginary part as the total stored electric and magnetic energy $W_E + W_M$. Subtracting the conjugate of Eq. (1.91) from itself and utilizing vector identities inside the integrand, Fante derived an equation for the frequency derivative of the reactance $X^{A'}(\omega_0)$ as

$$\frac{\partial X_A}{\partial \omega} = \frac{4(W_E + W_M)}{|I|^2} - \frac{2}{Z_0 |I|^2} \text{Im} \iint\limits_{S_\infty} \left(\boldsymbol{E}_\infty \bullet \frac{\partial \boldsymbol{E}_\infty^*}{\partial \omega} \right) dS \tag{1.92}$$

where it was assumed the current was held constant with frequency. Substituting this into the fractional bandwidth definition of Eq. (1.87) gives

$$B \approx \left[\frac{\omega_0(W_E + W_M)}{P_{\text{rad}}} + P(\omega_0)\right]^{-1} = \left[\frac{2\omega_0 W_E}{P_{\text{rad}}} + P(\omega_0)\right]^{-1}$$
$$= \left[\frac{2\omega_0 W_M}{P_{\text{rad}}} + P(\omega_0)\right]^{-1} \tag{1.93}$$

From Eq. (1.93) it is clear that in the case where $P(\omega_0)$ is small, this expression for B results in the well known relation of $Q \approx 1/B$. To quantify $P(\omega_0)$ in relation to Q, Fante derived the following inequality

$$P(\omega_0) \leq \frac{2}{B}\left[\frac{\sum_{n,m} |\Delta C_{\text{nm}}|^2}{\sum_{n,m} |C_{\text{nm}}|^2}\right]^{1/2} \tag{1.94}$$

where C_{nm} are proportional to the TE and TM mode excitation coefficients present in the far-field, and ΔC_{nm} is the total change in C_{nm} over approximately half the antenna's bandwidth. For this derivation, $R_{\text{rad}}{}'(\omega_0)$ was assumed nearly zero, implying that the derivative of the radiated power about the resonant frequency is approximately zero. Consequently, there is little change in the C_{nm} coefficients as these are proportional to the TE and TM mode coefficients present in the far-field, and the square root term in Eq. (1.94) is very small with respect to unity.

1.3.4.4 Comments

Collin and Rothschild developed a field-based method for determining the Q of a *Chu antenna* exciting either TM_{nm} or TE_{nm} modes. Their results were shown to be consistent with the results derived by the Chu equivalent circuits. Collin and Rothschild found that the Q associated with TM_{1m} or TE_{1m} modes represents the absolute lower bound on Q for a small antenna radiating only TE or TM modes, given below as

$$Q_{\text{min}} = \frac{1}{(ka)^3} + \frac{1}{ka} \tag{1.95}$$

We do note that compared to Wheeler's result, Eq. (1.95) has the extra term $1/ka$. This extra term is predictably due to the TM_{1m} mode having a small inductive energy component. Likewise, the TE_{1m} mode will have a small capacitive energy component. Nevertheless, both Eqs. (1.42) and (1.95) are very close for $ka \ll 1$.

Fante used the same procedure as Collin and Rothschild, and affirmed the statements made by Harrington and Chu that equal excitations of TM_{1m} and TE_{1m} modes lead to the lowest possible Q. This Q in Eq. (1.83) is half that in Eq. (1.95) for $ka \ll 1$, and slightly greater than in Eq. (1.95) for $ka \approx 1$. It will be discussed later that an exact expression for this lowest possible Q will be derived by McLean [37] in Sec. 1.3.6.

We remark that Fante was the first to give a direct relationship between Q and the 3 dB fractional bandwidth B. He showed that the approximation $Q \approx 1/B$ remains valid for large Q and assumes that the input resistance does not vary rapidly near the resonant frequency. Another derivation relating Q and fractional bandwidth will be given later in Sec. 1.3.10.

1.3.5 Work of Hansen (1981–2006)

Hansen [2] sought a closed form expression for the minimum Q based on the results of Chu's work. He begins by representing the radiated field as a superposition of spherical vector waves, and notes that all excited modes (TE_{nm} and TM_{nm}) have an associated stored electric and magnetic energy. However, only the propagating modes contribute to radiation. Hansen reiterates that Q_n rises rapidly for $ka < n$.

Hansen [38] has reviewed the performance of many practical small antennas including loaded dipoles and loops, dielectric resonator antennas (DRA), small patches, and partial sleeves. He also analyzed several published small antenna designs which have poor radiation characteristics when compared to traditional designs, as well as a detailed criticism of antennas that claim to beat McLean's [37] Q limits [38].

1.3.5.1 Closed Form *Q* for TM or TE Omni-directional Chu Antenna Using Chu RLC Approximation

Hansen [2] used the series RLC approximation employed by Chu for the TM mode network in Fig. 1.19 to determine a closed form expression for the minimum Q of a *Chu antenna*. Chu's RLC series approximation of the networks in Fig. 1.19 is based on equating the input resistance, reactance, and frequency derivative of the reactance for each mode TM_{nm} to that of an equivalent series RLC circuit. Using Eq. (1.3.2.2), he cites the input resistance and reactance for each mode (as also obtained by Chu) as

$$R_n = \frac{1}{\left|ka h_n^{(2)}(ka)\right|^2} \qquad X_n = \text{Re}\left\{\frac{\left(ka h_n^{(2)}(ka)\right)'}{ka h_n^{(2)}(ka)}\right\} \tag{1.96}$$

where the primes indicate derivatives with respect to ka. Using the series RLC approximation, Q_n for each TE_{nm} and TM_{nm} mode can be found using Eqs. (1.96), (1.3), and the duality principles in Table 1.2 as

$$Q_n = \frac{2\omega_0 \max(W_{En}, W_{Mn})}{P_{a,n}} = \frac{ka\left|ka h_n^{(2)}(ka)\right|^2}{2}(ka\, X_n)' \tag{1.97}$$

The total Q for a pure TE or pure TM antenna system can then be obtained from Eq. (1.76). For $ka < 1$, Hansen states that based on the Chu equivalent circuits of Figs. 1.19 and 1.22, modes with $n > 1$ can be considered evanescent. Accordingly, setting $a_n^2 = 0$ for $n \neq 1$, we get the closed form solution using the series RLC approximation as

$$Q_{\mathrm{TM}1m,\mathrm{Chu}} = Q_{\mathrm{TE}1m,\mathrm{Chu}} \approx \frac{1 + 2(ka)^2}{(ka)^3\left[1 + (ka)^2\right]} \quad \text{for } ka < 1 \qquad (1.98)$$

As mentioned previously, Eq. (1.98) is an approximate minimum possible Q for a TM or TE antenna circumscribed by a *Chu sphere* of radius a. We note that Eq. (1.98) is the corrected result derived by McLean based on Chu's RLC circuit approximation, as the original result [2] has an algebra mistake [3]. Figure 1.27 shows the Chu approximate result Eq. (1.98) alongside the approximate limit derived by Wheeler in Eq. (1.42) and exact limit derived by Collin and Rothschild in Eq. (1.95).

1.3.6 Work of McLean (1996)

In 1996, McLean [3] presented another rigorous method to determine Q for a *Chu antenna* supporting TE_{10}, TM_{10}, or equally excited TE_{10} and TM_{10} modes. McLean focused on this class of modes as they represent the radiated fields by small antennas. In his approach, McLean

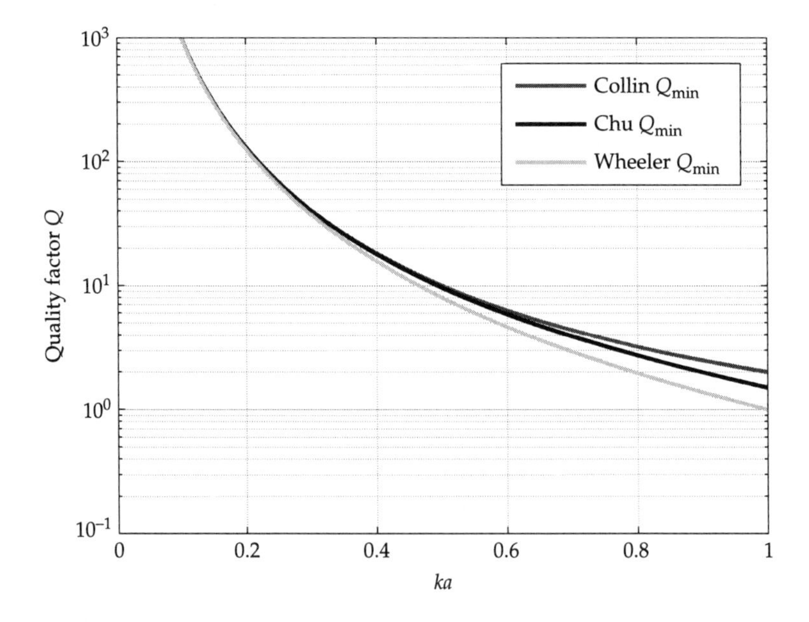

FIGURE 1.27 $Q_{\min}$ comparison for TM or TE antenna circumscribed by a *Chu sphere* of radius a.

challenged the approximate Q limits derived by Wheeler [1], Chu [11], Harrington [3], and Hansen [2].

1.3.6.1 McLean's Exact Q for Chu Antenna Radiating TM_{10} or TE_{10}

McLean starts by using (1.51a) to (1.51c) to compute the near-zone TM_{10} mode radiated fields

$$H_\phi = \sin\theta\, e^{-jkr}\left(\frac{j}{kr^2} - \frac{1}{r}\right) \tag{1.99a}$$

$$E_\theta = \frac{1}{j\omega\varepsilon}\sin\theta\, e^{-jkr}\left(-\frac{1}{r^2} - \frac{jk}{r} + \frac{j}{kr^3}\right) \tag{1.99b}$$

$$E_r = \frac{2}{\omega\varepsilon}\cos\theta\, e^{-jkr}\left(\frac{1}{kr^3} + \frac{j}{r^2}\right) \tag{1.99c}$$

and become

$$H_\phi^{\text{rad}} = -\sin\theta\,\frac{e^{-jkr}}{r} \tag{1.100a}$$

$$E_\theta^{\text{rad}} = -Z_0\sin\theta\,\frac{e^{-jkr}}{r} \tag{1.100b}$$

as $r \to \infty$. As the TM_{10} mode has capacitive energy, McLean used only the electric energy density w_e in his Q computations. It is given by

$$w_e = \frac{1}{4}\varepsilon\,|E|^2 \tag{1.101}$$

which reduces to the radiated energy density

$$w_e^{\text{rad}} = \frac{Z_0^2}{r^2}\sin^2(\theta) \tag{1.102}$$

for $r \to \infty$. Using a method similar to Collin and Rothschild, McLean subtracts the electric energy density Eq. (1.102) from the total electric energy density Eq. (1.101) to obtain the stored electric energy density w_e^s. Integrating w_e^s over all space outside the *Chu sphere* of radius a then gives the stored electric energy W_E to be used in Eq. (1.3) to compute Q. The associated radiated power P_{rad} is readily obtained using the far zone fields Eqs. (1.100a) and (1.100b) and the Poynting theorem. Using P_{rad} and W_E in Eq. (1.3), we have

$$Q_{\text{TM10,Chu}} = Q_{\min} = \frac{2\omega_0 W_E}{P_{\text{rad}}} = \frac{1}{ka} + \frac{1}{(ka)^3} \tag{1.103}$$

Equation (1.103) is the same as that derived by Collin and Rothschild [18] and the same as the TE_{10} or TM_{10} mode Q for the *Chu antenna*.

1.3.6.2 *Q* for Radiating Equally Excited TM_{10} and TE_{10}

For equally excited TM_{10} and TE_{10} modes within the *Chu antenna*, McLean invoked duality to represent the TE_{10} fields (see Table 1.2). He excited the modes to achieve CP polarization by scaling the TE_{10} coefficients with jZ_0. In computing the Q, he used the same procedure to evaluate the W_E and P_{rad}. Doing so, McLean concluded that

$$Q_{TM10-TE10,Chu} = \hat{Q}_{min} = \frac{1}{2}\left(\frac{2}{ka} + \frac{1}{(ka)^3}\right) \tag{1.104}$$

We note that Eqs. (1.103) and (1.104) are exact and valid for all *ka*. Also, Eq. (1.104) is the absolute minimum Q, and is approximately half of Eq. (1.103) for $ka \ll 1$. For such small sizes, the TM_{10} contributes primarily all W_E and TE_{10} does the same for W_H. Meanwhile, P_{rad} is doubled when both TE_{10} and TM_{10} are excited, leading to the half factor. However, this approximation breaks down as *ka* approaches unity.

Figure 1.28 compares the Q for TM_{10} and $TM_{10} + TE_{10}$ *Chu antennas* using Chu's RLC approximation [12] versus the exact results in Eqs. (1.103) and (1.104). It is evident that for $ka < 0.5$, the Chu RLC approximation remains accurate.

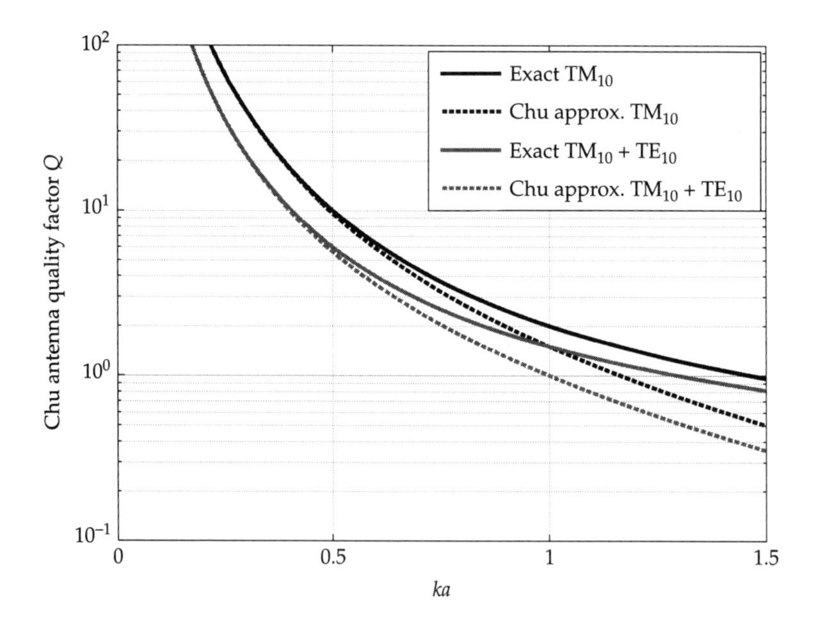

FIGURE 1.28 Comparison between exact and Chu RLC approximated *Chu antenna Q*.

1.3.6.3 Exact *Q* for Chu Antenna Derived from Mode Circuits

The Q values in Eqs. (1.103) and (1.104) can be verified using the equivalent impedance and admittance circuits in Figs. 1.19 and 1.22. For the TM_{10} mode, the circuit is given in Fig. 1.29 having the corresponding stored electric and radiated power given by

$$W_E = \frac{1}{4}C\,|V_C|^2 = \frac{1}{2\omega(ka)} \tag{1.105}$$

$$P_{\text{rad}} = \frac{1}{2}\,|I_r|^2\,R = \frac{(ka)^2}{1+(ka)^2} \tag{1.106}$$

Using these in the definition of Q in Eq. (1.3) we get

$$Q_{\text{TM10,Chu}} = Q_{\min} = \frac{2\omega_0 W_E}{P_{\text{rad}}} = \frac{1}{ka} + \frac{1}{(ka)^3} \tag{1.107}$$

Obviously, this is identical to Eq. (1.103) obtained via full-wave analysis. Correspondingly, using the TM_{10} and TE_{10} equivalent circuits, equally excited TM_{10} and TE_{10} modes radiating CP fields lead to the result in Eq. (1.104). The relation $\hat{Q}_{\min} \approx 1/2\ Q_{\min}$ for $ka \ll 1$ can be easily seen through this circuit method. Specifically, for low frequencies the circuits in Figs. 1.19 and 1.22 ($n = 1$) are dominated by the capacitive and inductive elements, respectively. However, as the frequency increases and ka approaches unity, the contribution of the inductive (Fig. 1.19) and capacitive (Fig. 1.22) elements become more prominent, and the half factor relation begins to break down.

Using Figs. 1.19 and 1.22 for $n = 1$, we further note that mode circuits also demonstrate the breakdown of the often cited Q

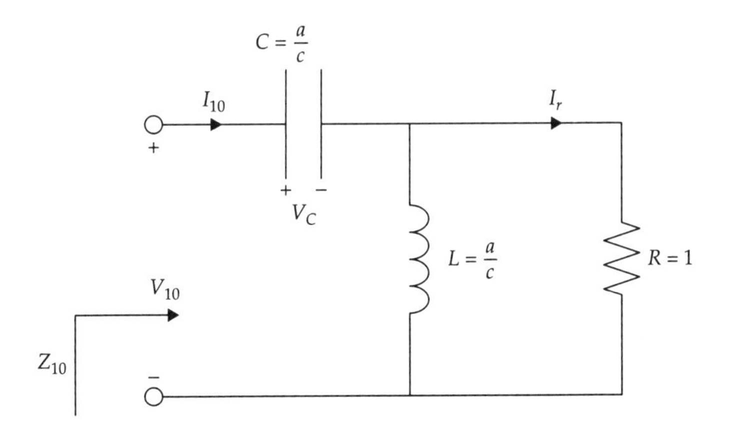

FIGURE 1.29 Equivalent RLC circuit for the TM_{10} mode.

approximation [39] (for small ka) given by

$$Q \approx \frac{|X_A|}{R_A} \tag{1.108}$$

For $ka \ll 1$ the net stored energy $|W_M - W_E| \approx |W_E|$, due to negligible inductor reactance. However, as ka approaches 1, the inductor's reactance becomes larger and $|W_M - W_E|$ can no longer be approximated by $|W_E|$. Consequently, Eq. (1.108) is no longer a valid approximation of Eq. (1.3). This is depicted in Fig. 1.30, where Yaghjian and Best's computed Q [10] [also given in Eq. (1.153)] is compared to Eq. (1.108) for a small dipole and small loop. We can state that Eq. (1.108) remains accurate for a small dipole with $ka < 0.5$, and for a small loop with $ka < 0.3$.

1.3.6.4 Comments

McLean demonstrated two procedures (using fields and circuits) to determine the exact minimum Q limits for an antenna circumscribed by a *Chu sphere*. For an antenna exciting only TE modes or TM modes, the fundamental Q limit is given by Eq. (1.103) and valid for all ka. When both TE and TM modes are excited, the fundamental limit is given by Eq. (1.104) and valid for all ka.

In Secs. 1.3.2 through 1.3.6, the minimum Q was derived using a *Chu antenna*—an antenna which is enclosed by a *Chu sphere* and has zero stored energy inside the sphere. Though the Q given by McLean represents an absolute lower bound, it remains much lower than those of practical antennas. It has been noted that this is due to the nonzero energy stored within the *Chu sphere*. This additional energy component was used by Lopez [31] to define the quality factor ratio (QR) for a small antenna

$$QR = \frac{Q}{Q_{\text{Chu}}} \tag{1.109}$$

where Q is the actual value and Q_{Chu} is that given by Eqs. (1.103) and (1.104). In essence, QR gives the ratio of the total stored energy to that stored external to the *Chu sphere*. If only the TM_{10} or TE_{10} mode is excited, then Q_{Chu} is given as Eq. (1.103). Conversely, if both TM_{10} and TE_{10} modes are excited, Q_{Chu} is given as Eq. (1.104). Lopez gives the QR for several practical antennas in Table 1.4.

1.3.7 Work of Foltz and McLean (1999)

Foltz and McLean [21] recognized that the derived fundamental Q limit is not close to the verifiable values for many practical antennas (see Table 1.4). Wheeler [9] and Hansen [2] had also noted that dipole antennas deviate from the optimal Q as they do not utilize the

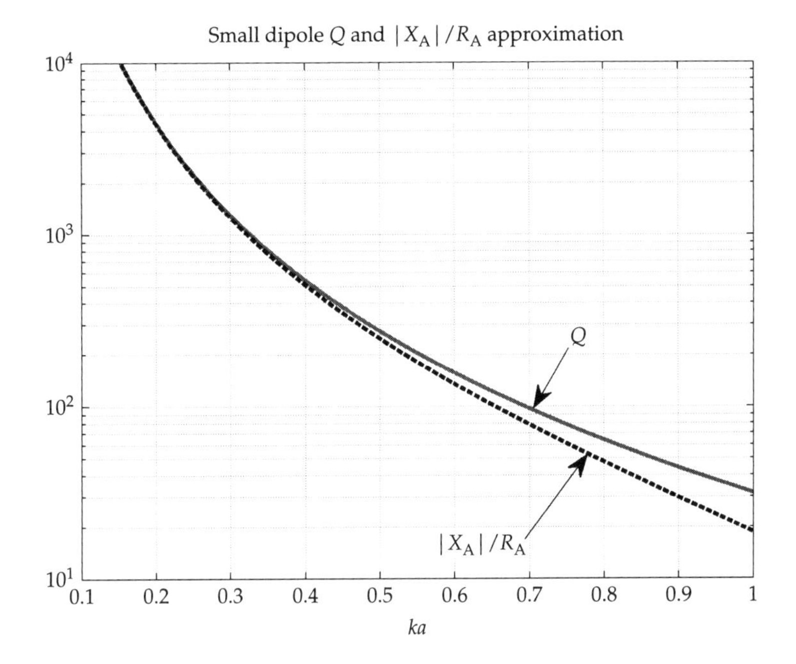

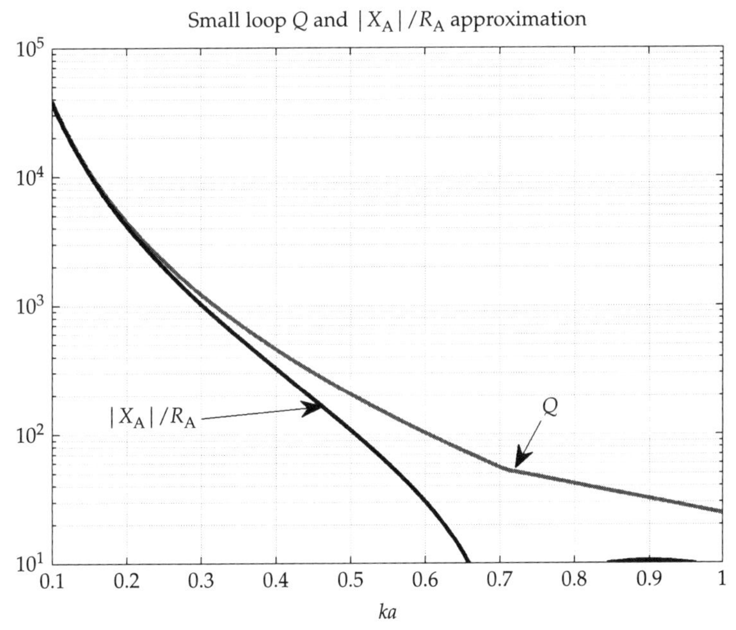

FIGURE 1.30 Yaghjian and Best's [10] computed Q compared to $|X_A|/R_A$ approximation for a small dipole and small loop.

Antenna Type	QR
Spherical coil (Fig. 1.15), $\mu_r = \infty$	1
Spherical coil (Fig. 1.15), $\mu_r = 1$	3
L-type cylindrical antenna (Fig. 1.13), $\mu_r = 1$, $D/b = 2.24$	4.4
Disc dipole (Fig. 1.31), $D/b = 0.84$	2.4
Spherical-cap dipole (Fig. 1.32)	1.75

TABLE 1.4 Quality Factor Ratio QR for Several Common Small Antennas (*See Lopez* [31])

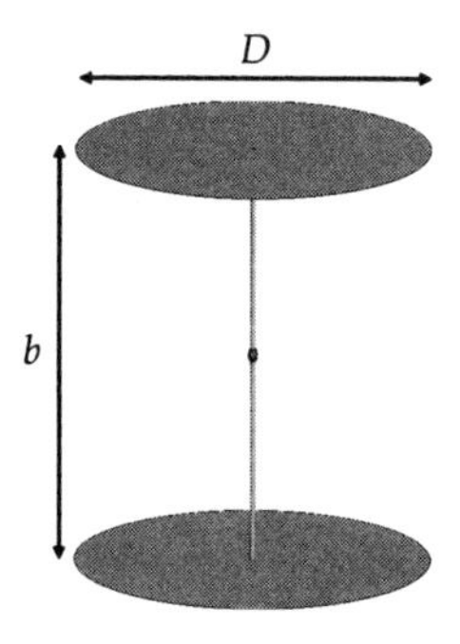

FIGURE 1.31 Disc dipole.

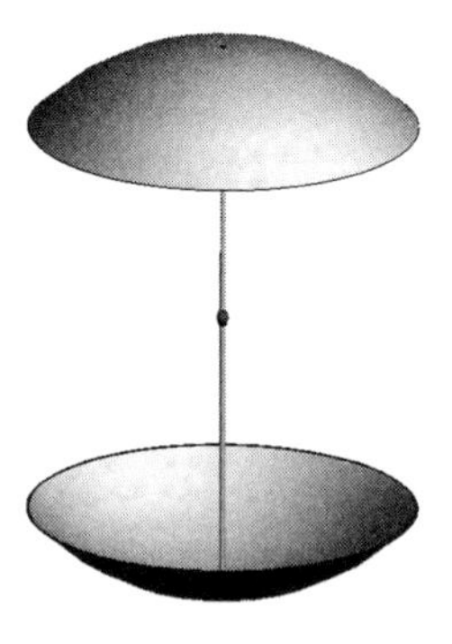

FIGURE 1.32 Spherical-cap dipole.

Chu sphere effectively. To examine this hypothesis, Foltz and McLean considered the Q of an enclosure having the surface of a prolate spheroid (see Fig. 1.33) with zero interior energy. We will refer to this as the *prolate Chu antenna*. Foltz and McLean assume the *prolate Chu antenna* has azimuthal symmetry. This is analogous to setting $m = 0$ for the spherical case.

The minimum Q associated with the *prolate Chu antenna* gives a more restrictive class of limitations. The analysis reported [20] is summarized in the preceding sections.

1.3.7.1 Minimum *Q* for the TM Prolate Chu Antenna

Foltz and McLean represented the geometry of Fig. 1.33 in a prolate spheroidal coordinate system (ξ, η, ϕ), where the surface of the prolate spheroid lies on $\xi = \xi_0$. The relation between ξ_0 and the prolate dimensions are

$$a = \xi_0 f$$
$$b = \left(\xi_0^2 - 1\right)^{1/2} f$$

where $2f$ is the foci distance. Foltz and McLean expanded the fields generated by the TM *prolate Chu antenna* as a superposition of azimuthally symmetric TM (to ξ) spheroidal vector wavefunctions.

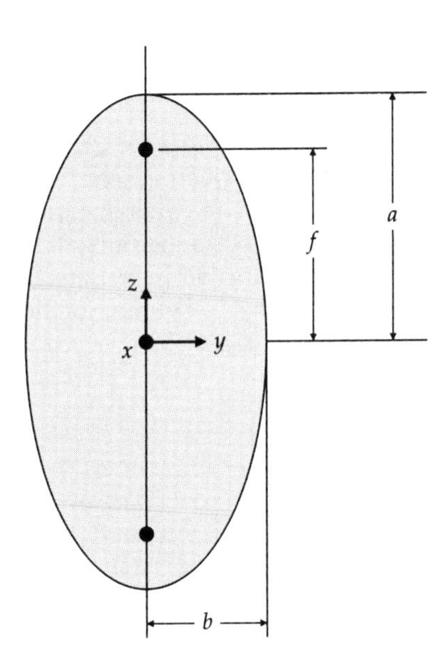

FIGURE 1.33 Cross-section of a prolate spheroid volume and coordinates, used in [20].

The wave admittance of these modes looking out from the surface ξ_0 was computed to be

$$Y_{n+1} = -jc_x\omega\varepsilon \frac{\sqrt{\xi_0^2 - \eta^2} R_{n,n+1}^{(4)}(\xi_0)}{\frac{d}{d\xi}\left(\sqrt{\xi^2 - 1} R_{n,n+1}^{(4)}(\xi)\right)\Big|_{\xi_0}} \tag{1.110}$$

where $c_x = fk$, and $R_{n,n+1}^{(4)}$ is the (outward) radial spheroidal function of the fourth kind. This is analogous to the second kind spherical Hankel function used in spherical wavefunction analysis. The transition from a prolate spheroid to an oblate spheroid can be achieved by simply replacing c_x with $\pm jc_x$. The reader is referred to Stratton [40] and Flammer [41] for more details on spheroidal wavefunctions.

Foltz and McLean state that Eq. (1.110) cannot be easily represented as an exact lumped network as was done by Chu [11]. Instead, for $ka < 0.5$ they fitted Eq. (1.110) to the numerical representation

$$Y_{n+1} \approx \frac{(j\omega)a_1 + (j\omega)^2 a_2 + \cdots}{1 + (j\omega)b_1 + (j\omega)^2 b_2 + \cdots} \tag{1.111}$$

Foltz and McLean derived an equivalent high-pass network for Eq. (1.110) based on Eq. (1.111). They also noted that the minimum Q for the TM *prolate Chu antenna* is associated with the first order TM (to ξ) mode $n = 1$, which we denote as $Q_{\text{min,prolate}}$.

1.3.7.2 Comments

Foltz and McLean plotted their results for $Q_{\text{min,prolate}}$ versus ka for varying a/b ratios of the prolate spheroid (see Fig. 1.34). The longest dimension for this spheroid is $2a$. In general, $Q_{\text{min,prolate}}$ increases as the length/diameter ratio increases (spheroid gets thinner). However, as the length/diameter ratio is further increased, $Q_{\text{min,prolate}}$ changes much less drastically than the physical volume of the prolate spheroid. This relationship is depicted in Fig. 1.35.

1.3.8 Work of Thiele (2003)

Thiele and his associates Detweiler and Penno [7] also recognized that the theoretical lower limits on Q described in previous sections are far from the results attained for actual antennas. Thiele also addresses an ambiguity in Foltz and McLean [20], stating that multiple prolate spheroid shapes can qualify as minimally enclosing a given antenna structure, resulting in various minimum Q limits.

In an attempt to find the Q for more practical small antennas, Thiele et al. take a much different approach to determine the Q of an ESA based purely on the far-fields. Central to their method is the concept of

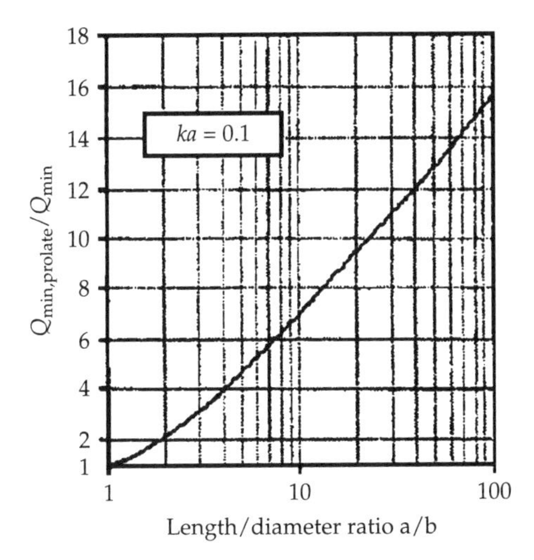

FIGURE 1.34 $Q_{min,prolate}$ versus *ka* for various length/diameter ratios for the prolate *Chu antenna*. (*After Foltz and McLean ©IEEE, 1999* [20].)

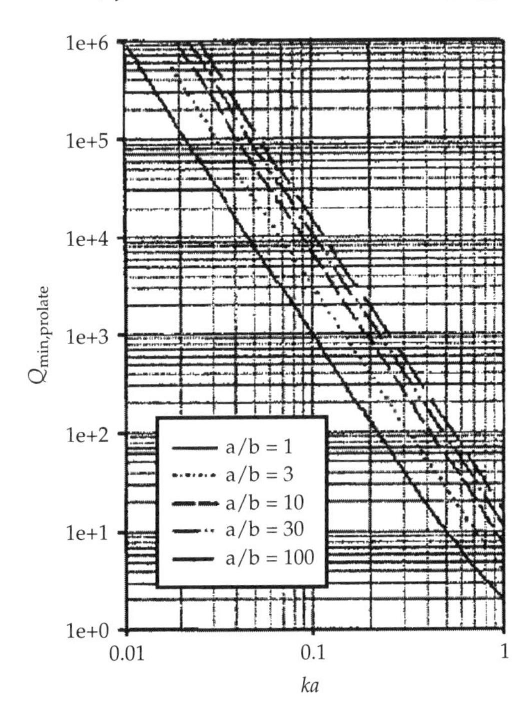

FIGURE 1.35 $Q_{min,prolate}/Q_{min}$ length/diameter ratio with Q_{min} given in Eq. (1.103) and $ka = 0.1$. (*After Foltz and McLean ©IEEE, 1999* [20].)

superdirectivity, stated by many authors [42–45] to be directly related to the antenna Q. Thiele et al. computed the Q of a dipole supporting a sinusoidal current, and compared their results to the minimum Q in Eq. (1.103) as derived by McLean.

1.3.8.1 Superdirectivity and *Q*

To determine Q using [7], we first review the superdirective ratio R_{SD}. We begin by introducing the variable transformation

$$u = \frac{kd}{2}\cos(\theta) \tag{1.112}$$

where

$$\frac{\pi d}{\lambda} \leq u \leq \frac{\pi d}{\lambda} \quad \text{is the visible region}$$

and

$$\begin{bmatrix} -\infty \leq u \leq -\dfrac{\pi d}{\lambda} \\ \dfrac{\pi d}{\lambda} \leq u \leq \infty \end{bmatrix} \quad \text{is the invisible region}$$

For an antenna array (see Fig. 1.36), the array factor is then given by

$$f(u) = \frac{\sin\left(\dfrac{Nu}{2}\right)}{N\sin\left(\dfrac{u}{2}\right)} \tag{1.113}$$

and R_{SD} is given by

$$R_{SD} = \frac{\displaystyle\int_{-\infty}^{\infty} |f(u)|^2\,du}{\displaystyle\int_{-\pi d/\lambda}^{\pi d/\lambda} |f(u)|^2\,du} \tag{1.114}$$

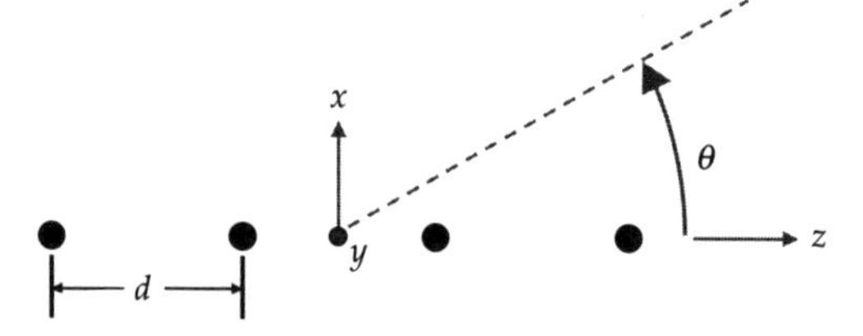

FIGURE 1.36 Linear array along z-axis with element spacing d.

This superdirective ratio has also been used to measure the realizability of an antenna. Superdirective antennas are antennas that exhibit higher directivity than nominal, and an "electrically small antenna (ESA)" can be classified as superdirective since its directivity stays approximately constant as $ka \to 0$. Thiele et al. noted that Rhodes [45] compared R_{SD} to "somewhat similar to $Q + 1$." If R_{SD} is assumed to be equal to $Q + 1$ as in [42–44], then Q can be represented as

$$Q = \frac{\displaystyle\int_{-\infty}^{-\pi d/\lambda} |f(u)|^2\,du + \int_{\pi d/\lambda}^{\infty} |f(u)|^2\,du}{\displaystyle\int_{-\pi d/\lambda}^{\pi d/\lambda} |f(u)|^2\,du} \tag{1.115}$$

We may also replace $f(u)$ in Eq. (1.115) by the array's normalized electric field $E(u)$

$$|E(u)| = |g(u)\,f(u)| \tag{1.116}$$

where $g(u)$ is the element pattern. Using this in Eq. (1.115) gives

$$Q = \frac{\displaystyle\int_{-\infty}^{-\pi d/\lambda} |E(u)|^2\,du + \int_{\pi d/\lambda}^{\infty} |E(u)|^2\,du}{\displaystyle\int_{-\pi d/\lambda}^{\pi d/\lambda} |E(u)|^2\,du} \tag{1.117}$$

In Fig. 1.37, Thiele et al. used Eq. (1.117) for a dipole element pattern (of sinusoidal distribution) and compared it to Eq. (1.103). Unlike Eq. (1.103), Eq. (1.117) does not assume TM mode propagation, but is instead based on the far fields of the actual antenna. The computed Q for a bowtie and end-loaded dipole are also given in Fig. 1.37 (see Figs. 1.38 and 1.39 for the antenna geometries), as well as the Q associated with dipoles of radius $a_0 = 0.0005\lambda$ and 0.001λ.

Of importance in Fig. 1.37 is that the Thiele et al. Q values for the bowtie and end-loaded dipole (enclosed in a *Chu sphere*) are closer to the far-field Q in Eq. (1.117). In contrast, McLean's minimum Q Eq. (1.103) is often an order of magnitude smaller. We note that Thiele makes a significant approximation (see Fig. 1.30) in using $Q \approx |X_A|/R_A$ to compute Q for the two dipole curves ($a_0 = 0.0005\lambda$ and 0.001λ), the bowtie antenna, and the end-loaded dipole antenna. Thiele remarks that the patterns of the sinusoidally distributed dipoles are slightly narrower than that of a Hertzian (constant current) dipole (see Fig. 1.2). Thus, from Eq. (1.117) the sinusoidal dipoles are more

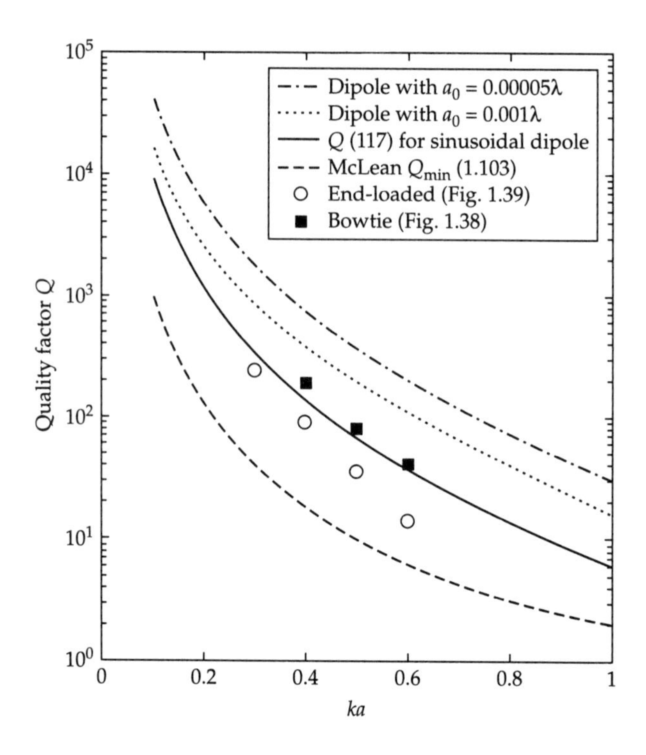

FIGURE 1.37 Computed bowtie, end-loaded, and sinusoidal dipole Q with Eqs. (1.103) and (1.117). (*After Thiele et al. ©IEEE, 2003* [7].)

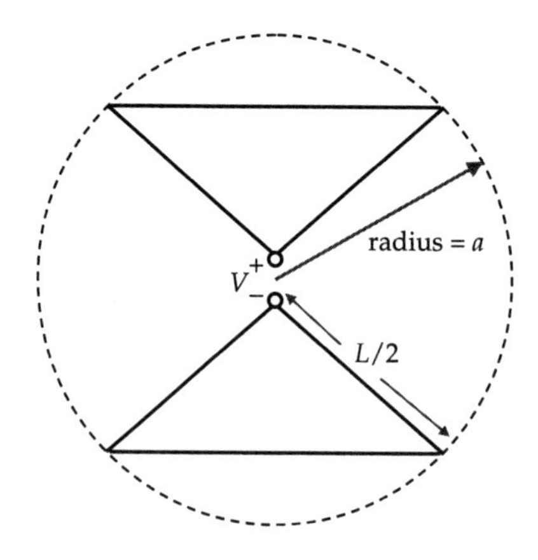

FIGURE 1.38 Bowtie dipole circumscribed by *Chu sphere*.

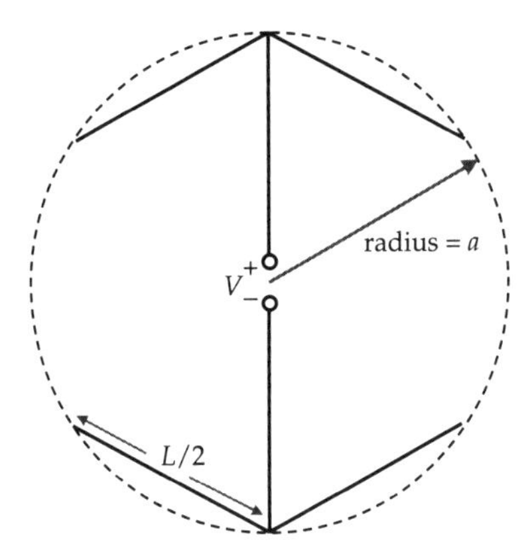

FIGURE 1.39 End-loaded dipole circumscribed by *Chu sphere*.

superdirective and as a result have a higher Q. This is also seen in the current distributions of the bowtie and end-loaded dipoles [7]. From Fig. 1.37 we also observe that as the dipole thickness decreases, the Q value increases. This effect was also predicted in Fig. 1.34 by Foltz and McLean.

1.3.8.2 Comments

The approach used by Thiele et al. [7] is solely based on the antenna or array far field, regardless of its shape and form. Therefore, it can be readily used to compute the Q for any other antenna. However, it is important to examine Eq. (1.117) and verify that it agrees with standard definition of Q Eq. (1.3). Indeed, the denominator of Eq. (1.117) involves integration of the normalized field pattern over the visible region. Thus, it is proportional to the radiated power. So, for Eq. (1.117) to be equivalent to Eq. (1.3), the numerator of Eq. (1.117) must represent twice the reactive electric power in the space surrounding the antenna (assuming the proportionality factor is the same as in the numerator). For that we refer to Rhodes [46]. Rhodes determined that the stored electric and magnetic energies generated by a planar aperture can be found using the field pattern given in Eq. (1.116). His result is indeed twice the reactive electric power in [46] making it consistent with the numerator of Eq. (1.117).

Rhodes [46] also examined the fields of a planar dipole (of sinusoidal distribution) to validate his method. He remarked that the input reactance of the planar dipole using his method closely matches that derived by King [12]. He later verified the approximate relationship between Q and the 3 dB fractional bandwidth B for this planar dipole using his far-field integrals over the visible and invisible regions to determine real and reactive electric power, respectively (see Fig. 1.40).

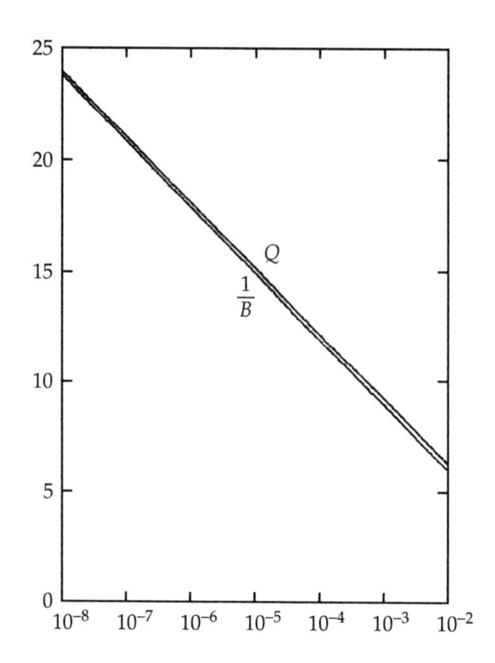

FIGURE 1.40 Q and $1/B$ for a sinusoidal current planar dipole of width "W." (*After Rhodes ©IEEE, 1966* [46].)

1.3.9 Work of Geyi (2003)

Geyi [22] stated that direct calculation of the exact Q using the Collin and Rothschild method [18] is difficult in practice, as it involves numerical integrations over an infinite spatial domain. To overcome this, Geyi formulated an approximate field-based method applicable for the small antenna range ($ka < 0.5$) using the Poynting theorem in both the time and frequency domains. Specifically, he determined Q by integrating the current distribution of the antenna structure. This simplifies numerical integrations significantly as the domain of integration is on the antenna only. Geyi derived the Q for several simple antennas and compared his results to other independent methods. He further reconsidered the fundamental Q limits for omni-directional and directional small antennas in a straightforward manner using spherical wavefunctions [21].

Geyi, along with Jarmuszewski and Qi [47], calculated Q as a function of input impedance via Maxwell's equations extended to the complex frequency domain. Essential to Geyi's results is the assertion that the Foster reactance theorem is valid for general antenna systems. However, several aspects of this derivation has been criticized by Best [48] and Andersen [49] under the premises that the frequency derivatives of the antenna input reactance $X'_A(\omega)$ and input susceptance

$B'_A(\omega)$ at resonance and antiresonance are not always positive. Additionally, Andersen [49] notes that the Cauchy-Reimann (C-R) equation is improperly applied in the derivation of Q, as the complex function for which the C-R theorem is used is not analytic. Due to the controversy surrounding [47], we choose to omit this derivation of Q as a function of input impedance and only state the result used by Geyi et al. [47] as

$$Q = \frac{\omega \left[\dfrac{\partial X_A}{\partial \omega} \pm \dfrac{X_A}{\omega} \right]}{2R_A} \tag{1.118}$$

where $\pm$ is chosen to yield the larger Q. In a later section, we will provide the derivation of the Q in terms of input impedance as given by Yaghjian and Best [10]. This Q will be shown to be

$$\begin{aligned} Q &= \frac{\omega}{2R_A(\omega)} \left| \frac{dZ_{\text{in}}(\omega)}{d\omega} \right| \\ &= \frac{\omega}{2R_A(\omega)} \left[\left(\frac{dR_A(\omega)}{d\omega} \right)^2 + \left(\frac{\partial X_A}{\partial \omega} + \frac{|X_A|}{\omega} \right)^2 \right]^{1/2} \end{aligned} \tag{1.119}$$

We remark that if the derivative for the input resistance in Eq. (1.119) is negligible, then Eq. (1.119) can be reduced to Eq. (1.118) as given by Geyi [47]. Indeed, in practice, the frequency derivative of the reactance dominates for small antennas operating away from their natural antiresonance regions [10].

1.3.9.1 Field-Based Evaluation of Antenna *Q*

To find a practical method of computing Q for $ka < 0.5$, Geyi [22] begins by assuming a perfectly conducting small antenna with volume V_0. For his derivations, he approximates the near-zone-antenna fields by introducing a power series expansion for the exponential $\exp(-jk|\boldsymbol{r}-\boldsymbol{r}'|)$ term in the frequency domain scalar and vector potentials. This low frequency approximation for $\exp(-jk|\boldsymbol{r}-\boldsymbol{r}'|)$ remains accurate under the condition that $ka < 0.5$. The Poynting theorem (in the frequency domain) is then used to find the radiated power (P_{rad}) and the difference of the average electric and magnetic energies ($W_E - W_M$). To represent the near-zone fields in the time domain, the charge and current density sources, $\rho(\boldsymbol{r}', T)$ and $\boldsymbol{J}(\boldsymbol{r}', T)$, can be approximated in a power series with respect to retarded time $T = t - |\boldsymbol{r} - \boldsymbol{r}'|/c$ about the point t. With this, the time domain scalar and vector potentials can then be found, and subsequently the Poynting theorem (in the time domain) is used to compute the sum of the average electric and magnetic energies ($W_E + W_M$).

The actual expressions for W_E, W_M, and P_{rad} used by Geyi [22] are

$$W_E \approx \frac{cZ_0}{16\pi} \int_{V_0} \int_{V_0} \frac{1}{R} \left[\rho(\boldsymbol{r})\rho^*(\boldsymbol{r}')\right] dv(\boldsymbol{r})dv(\boldsymbol{r}') \tag{1.120}$$

$$W_M \approx \frac{cZ_0}{16\pi} \left[\frac{1}{c^2} \int_{V_0} \int_{V_0} \frac{\boldsymbol{J}(\boldsymbol{r}) \bullet \boldsymbol{J}^*(\boldsymbol{r})}{R} dv(\boldsymbol{r})dv(\boldsymbol{r}') + \frac{k^2}{2} \int_{V_0} \int_{V_0} R \left[\rho(\boldsymbol{r})\rho^*(\boldsymbol{r}')\right] dv(\boldsymbol{r})dv(\boldsymbol{r}') \right] \tag{1.121}$$

$$P_{\text{rad}} \approx \frac{Z_0 k^4}{12\pi} \left[c^2 \left|\boldsymbol{p}\right|^2 + \left|\boldsymbol{m}\right|^2 \right] \tag{1.122}$$

with the electric and magnetic dipole moments $\boldsymbol{p}$ and $\boldsymbol{m}$ defined as

$$\boldsymbol{p} = \int_{V_0} \boldsymbol{r}' \rho(\boldsymbol{r}') dv(\boldsymbol{r}') \tag{1.123}$$

$$\boldsymbol{m} = \int_{V_0} \frac{\boldsymbol{r} \times \boldsymbol{J}(\boldsymbol{r}) dv(\boldsymbol{r})}{2} \tag{1.124}$$

Using the above in Eq. (1.3), Geyi computed the following Q expressions for the three antennas in Fig. 1.41

Dipole antenna
$$Q = \frac{2\omega W_E}{P_a} \approx \frac{6\left[\ln\left(\frac{a}{a_0}\right) - 1\right]}{(ka)^3} \tag{1.125}$$

Loop antenna
$$Q = \frac{2\omega W_M}{P_a} \approx \frac{6\ln\left(\frac{a}{a_0}\right)}{\pi (ka)^3} \tag{1.126}$$

Inverted-L antenna
$$Q = \frac{2\omega W_E}{P_a} = \frac{6\left\{\left[\ln\left(\frac{h}{a_0}\right) - 1\right] + b\left[\ln\left(\frac{2b}{a_0}\right) - 1\right]\right\}}{k^3 h^2 (h+b)^2} \tag{1.127}$$

1.3.9.2 Reinvestigation of Small Antenna Gain and Q Limitations

Geyi [21] reconsidered the small antenna physical limits. To find the minimum Q for a *Chu antenna*, Geyi directly minimized Eq. (1.76)

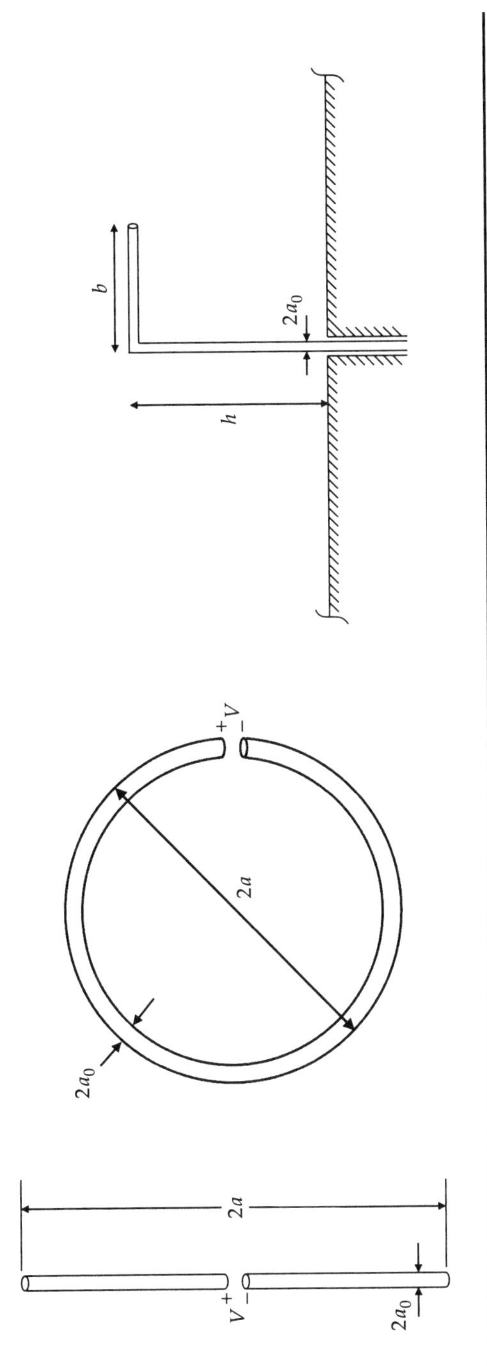

FIGURE 1.41 Small dipole, loop, and inverted L antennas.

in terms of spherical vector wavefunctions. His results conclude that the minimum Qs for TM, TE, or TM + TE mode antennas are exactly those derived previously by McLean [3] in Eqs. (1.103) and (1.104), and correspond to only first order mode ($n = 1$) excitations. Furthermore, he notes that the minimum Q TE + TM mode antenna is a *Chu antenna* with equally excited TM_{1m} and TE_{1m} modes.

Geyi believed that Fante's treatment [19] on the maximum G/Q for a directional antenna was incorrect, as Fante ignored a constraint on the mode coefficients necessary to maintain consistency with W_M or W_E in Eq. (1.3). Geyi then examines the maximum G/Q ratio by using the general form of Q given in Eq. (1.76) along with the definition of directivity written in terms of the spherical vector wavefunctions. Through a maximization process of this G/Q ratio, Geyi confirms that for both the omni-directional and directional case, the G/Q ratio is maximized when the TE and TM modes of a *Chu antenna* are equally excited (as already noted by several authors). For $ka \ll 1$, Geyi's formulas indicate that contributions from higher order modes can be ignored, and they are given by

$$\max\left(\frac{G}{Q}\right)\Bigg|_{\text{directional}, ka<<1} \approx \frac{6(ka)^3}{2(ka)^2+1} \tag{1.128a}$$

$$G^{\max}_{\text{directional}}\Big|_{ka<<1} \approx 3 \tag{1.128b}$$

$$Q^{\min}_{\text{directional}}\Big|_{ka<<1} \approx \frac{1}{ka} + \frac{1}{2(ka)^3} \tag{1.128c}$$

and

$$\max\left(\frac{G}{Q}\right)\Bigg|_{\text{omni}, ka<<1} \approx \frac{3(ka)^3}{2(ka)^2+1} \tag{1.129a}$$

$$G^{\max}_{\text{omni}}\Big|_{ka<<1} \approx 1.5 \tag{1.129b}$$

$$Q^{\min}_{\text{omni}}\Big|_{ka<<1} \approx \frac{1}{ka} + \frac{1}{2(ka)^3} \tag{1.129c}$$

In these equations, the superscript *max* indicates that gain is maximized under the constraint that antenna Q is at its minimum value. The superscript *min* similarly indicates that Q is minimized under the constraint that G has its maximum value. From Eqs. (1.128) and (1.129), Geyi concludes that in theory a small antenna can have maximum gain and minimum Q simultaneously [21]. We remark that this claim will later be challenged by Thal [29] in Sec. 1.3.12. Figs. 1.42 to 1.44 plot $Q_{\min}$, $G_{\max}$, and $(G/Q)_{\max}$, for the omni-directional and directional antennas as determined through Geyi's optimization procedure. We note that the "min Q" in Fig. 1.42 is that in Eq. (1.104)

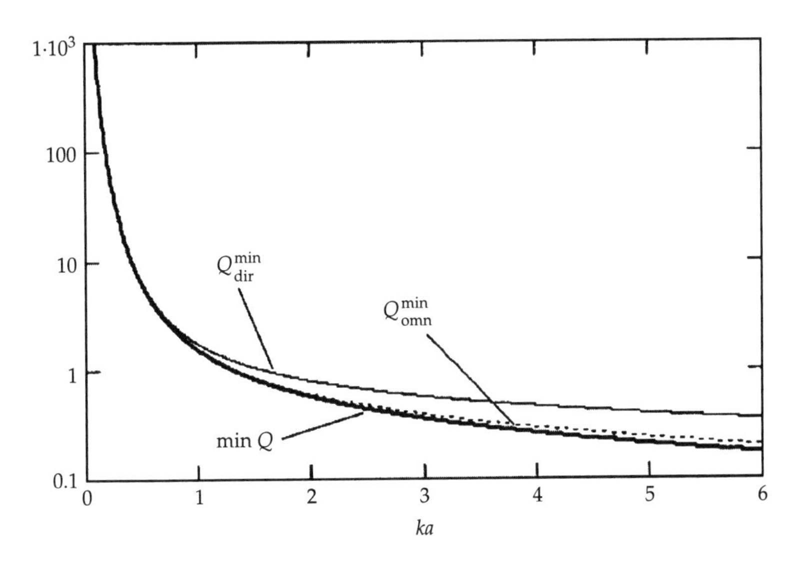

FIGURE 1.42 Minimum possible Q for directional and omni-directional antennas subject to maximum gain constraint; the min Q curve refers to Eq. (1.103) given by McLean [3]. (*After Geyi ©IEEE, 2003* [21].)

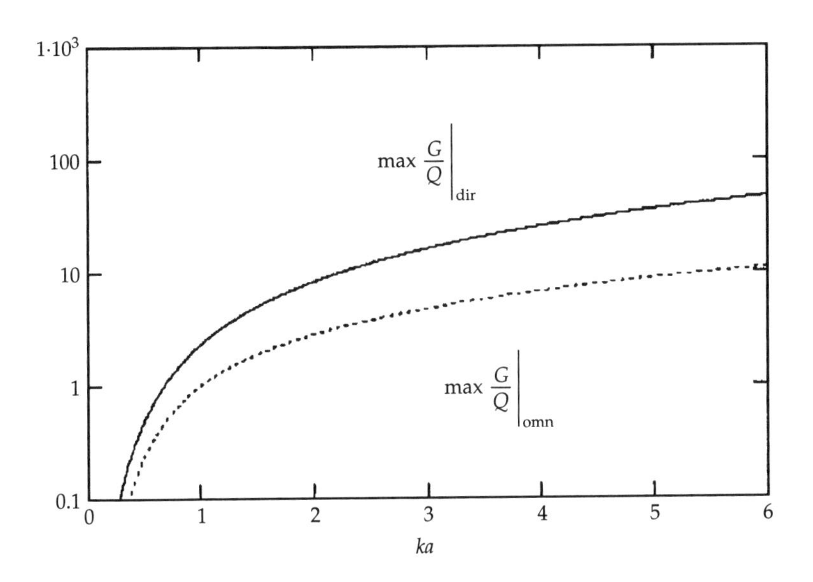

FIGURE 1.43 Maximum G/Q for directional and omni-directional antennas. (*After Geyi ©IEEE, 2003* [21].)

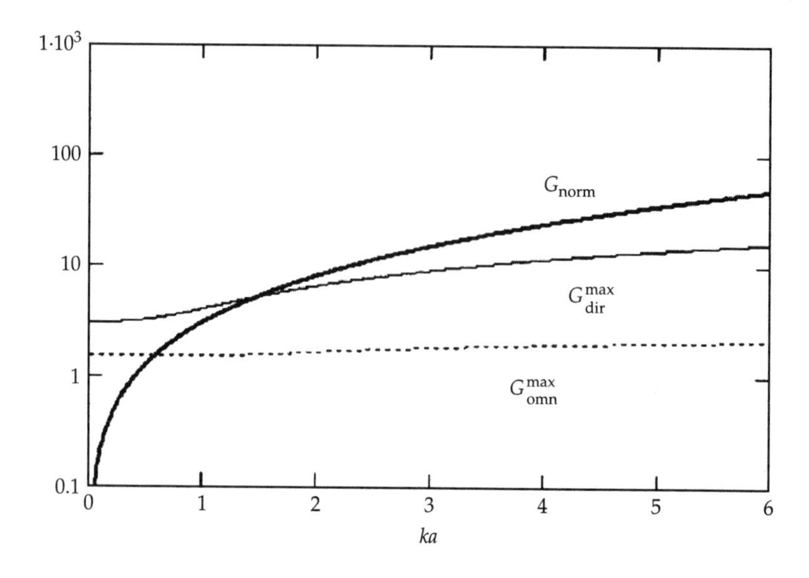

FIGURE 1.44 Maximum possible gain under minimized Q constraint. (*After Geyi ©IEEE, 2003* [21].)

derived by McLean [3]. Figure 1.44 plots the maximum G versus *ka* (Harrington [4] states that any antenna with a gain higher than the normal gain G_{norm} can be classified as a supergain antenna; see also Chu [11]).

1.3.10 Work of Best (2003-2008)

Yaghjian [10] and Best [10,13,23–26] reconsidered past developments on small antenna theory. A key contribution by Yaghjian and Best [10] is the new expression of Q in terms of input impedance, valid at all frequencies. They also formulated a relationship between the antenna Q and the matched voltage standing wave ratio (VSWR) bandwidth, allowing for antenna bandwidth to be defined in a more flexible manner. Yaghjian and Best first derived the approximate expression for the matched VSWR fractional bandwidth (B_V) of a tuned antenna in terms of its input impedance in resonance and antiresonance ranges. They detail the advantages in using matched VSWR bandwidth over conductance bandwidth, demonstrating that the conductance bandwidth definition breaks down as the impedance approaches antiresonance. Best [23] also gave the limiting relationships between B_V and antenna Q. The validity and accuracy of the derived expressions were confirmed with numerical data for several antennas, including lossy and lossless tuned antennas over a wide range of frequencies.

Best also considered several wire antennas and their Q limits. He showed that a spherical helix antenna [25] has the lowest realized Q to date. Best also explored the folding of small antennas to minimize their Q [14], the relationship between wire antennas and their resonant properties [14,25–26], and antenna volume utilization issues [14].

1.3.10.1 B_V and Antenna Input Impedance

With respect to the transmitting antenna system in Fig. 1.45, we define the following parameters:

- V_P - Shielded power supply and waveguide volume consisting of perfectly conducting walls
- V_A - Antenna volume, including the volume of the tuning reactance bringing the antenna to resonance
- V_0 - Entire volume outside the shielded power supply, including V_A
- S_0 - Antenna input feed port
- X_S - Series tuning reactance bringing the antenna to resonance
- Z_{in} - Input impedance looking into the feed port S_0
- Γ - Input reflection coefficient

Yaghjian and Best [10] begin their study by noting that antennas can be tuned to resonance (zero reactive impedance) by including a series reactance (see Fig. 1.45). Thus, we can represent the input impedance

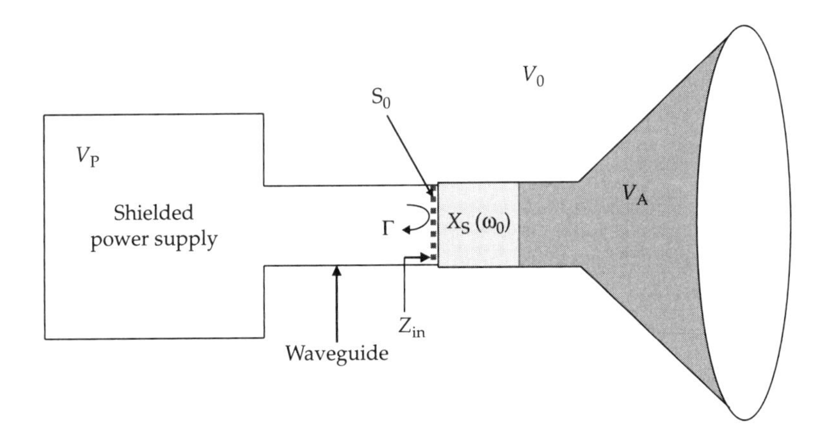

FIGURE 1.45 General transmitting antenna system. (*See Yaghjian and Best* [10].)

at the feed port as

$$Z_{\text{in}}(\omega) = R_A(\omega) + j\,[X_A(\omega) + X_S(\omega)] = R_A(\omega) + j\,X_{\text{in}}(\omega) \qquad (1.130)$$

Here, the antenna input impedance is $Z_A(\omega) = R_A(\omega) + j\,X_A(\omega)$, and the series tuning reactance $X_S(\omega)$ is

$$X_S(\omega) = \begin{cases} \omega L_S & \text{for} \quad X_A(\omega_0) < 0 \\ -1/(\omega C_S) & \text{for} \quad X_A(\omega_0) > 0 \end{cases} \qquad (1.131)$$

in which L_S and C_S correspond to the associated inductance and capacitance values. For resonance ($\omega = \omega_0$)

$$X_{\text{in}}(\omega_0) = X_A(\omega_0) + X_S(\omega_0) = 0 \qquad (1.132)$$

The frequency derivative of the tuned impedance at resonance can then be written as

$$X'_{\text{in}}(\omega_0) = X'_A(\omega_0) + \frac{|X_A(\omega_0)|}{\omega_0} \qquad (1.133)$$

This derivative will be used later in computing matched VSWR fractional bandwidth B_V.

The matched VSWR fractional bandwidth (B_V) is defined as the band between frequencies ω_+ and ω_- having a prescribed VSWR, given $Z_C = R_A(\omega_0)$, where Z_C is the feed line characteristic impedance. A parameter associated with the reflection coefficient Γ (see Fig. 1.45) is given by

$$\Gamma_{\max} = \Gamma(\omega_+) = \Gamma(\omega_-) = \left[\frac{VSWR - 1}{VSWR + 1}\right]^2 \qquad (1.134)$$

Using Eq. (1.130) and $Z_C = R_A(\omega_0)$, we have

$$|\Gamma(\omega)|^2 = \frac{X^2_{\text{in}}(\omega) + [R_A(\omega) - R_A(\omega_0)]^2}{X^2_{\text{in}}(\omega) + [R_A(\omega) + R_A(\omega_0)]^2} \qquad (1.135)$$

Since Eq. (1.135) is valid at any resonance location, Eq. (1.135) is valid for both resonant and antiresonant frequency ranges of the antenna. Following an algebraic manipulation of Eq. (1.135) and a Taylor series expansion of $X^2_{\text{in}}(\omega) + [R_A(\omega) - R_A(\omega_0)]^2$ about ω_0, Yaghjian and Best obtain the equation

$$\left|Z'_{\text{in}}(\omega_0)\right|^2 (\Delta\omega_+)^2 = \left|Z'_{\text{in}}(\omega_0)\right|^2 (\Delta\omega_-)^2 \approx 4\beta R^2_A(\omega_0) \qquad (1.136)$$

where

$$\sqrt{\beta} = \frac{|\Gamma_{\max}|^2}{1 - |\Gamma_{\max}|^2} = \frac{VSWR - 1}{2\sqrt{VSWR}} \qquad (1.137)$$

and

$$\Delta\omega = \omega_+ - \omega_- \approx \pm \frac{4\beta R_A(\omega_0)}{\omega_0 \left| Z_0'(\omega_0) \right|} \tag{1.138}$$

Using these values in the expression for the fractional bandwidth, we have

$$B_V(\omega_0) = \frac{\omega_+ - \omega_-}{\omega_0} \approx \frac{4\beta R_A(\omega_0)}{\omega_0 \left| Z_0'(\omega_0) \right|} \tag{1.139}$$

Yaghjian and Best remark that Eq. (1.139) remains valid when $B_V(\omega_0) \ll 1$, or equivalently β small enough (stated as $\beta \leq 1$) [10].

Yaghjian and Best stressed that the value of B_V is that it uses impedances as opposed to conductances. The conductance bandwidth at a resonant frequency ω_0 is defined as the difference between the two frequencies about ω_0 such that the power delivered to the antenna is a chosen fraction of the power delivered at resonance. Yaghjian and Best noted that the conductance will only reach its maximum at resonance when $R_A'(\omega)$ is zero [10]. But this is not generally true for small antennas. They also recognized that away from resonance there may be a conductance maximum. Thus the conductance bandwidth cannot be properly defined.

1.3.10.2 Exact *Q* Derived from Maxwell's Equations

Assuming a feed port impedance described by Eq. (1.130), the computation of Q requires $P_A = 1/2|I_{\text{in}}|^2 R_A$ and the stored energy. A convenient form for the stored energy can be found using the frequency derivative of the input reactance $X'_{\text{in}}(\omega_0)$. By coupling Maxwell's equations to the frequency derivative of Maxwell's equations for the antenna system in Fig. 1.45, Yaghjian and Best derive (see [10, App. A])

$$|I_{\text{in}}|^2\, X_{\text{in}}'(\omega_0) = \lim_{r\to\infty} \left[\text{Re} \int_{V_0} (\boldsymbol{B}^* \bullet \boldsymbol{H} + \boldsymbol{D}^* \bullet \boldsymbol{E})dv \right.$$

$$\left. - \; 2\varepsilon_0 r \int_{S_\infty} |\boldsymbol{E}_\infty|^2\, dS \right] + \omega_0 \text{Re} \int_{V_A} \left((\boldsymbol{B}')^* \bullet \boldsymbol{H} - \boldsymbol{B}^* \bullet \boldsymbol{H}' + (\boldsymbol{D}')^* \bullet \boldsymbol{E} \right.$$

$$\left. - \; \boldsymbol{D}^* \bullet \boldsymbol{E}' \right) dv + \frac{2}{Z_0} \text{Im} \int_{S_\infty} \boldsymbol{E}'_\infty \bullet \boldsymbol{E}^*_\infty dS \tag{1.140}$$

where primes indicate differentiation with respect to frequency, S_∞ is the far field sphere, and $\boldsymbol{E}_\infty$ is defined as

$$\lim_{r\to\infty} \boldsymbol{E}(\boldsymbol{r}) = \boldsymbol{E}_\infty \frac{e^{-jkr}}{r} \tag{1.141}$$

Assuming a linear, isotropic antenna radiating into free space, we can use the constitutive relations

$$B = \mu H \qquad D = \varepsilon E \tag{1.142}$$

and

$$\mu = \mu_r - j\mu_i \qquad \varepsilon = \varepsilon_r - j\varepsilon_i \tag{1.143}$$

Equation (1.140) can then be written as

$$|I_{\text{in}}|^2 X'_{\text{in}}(\omega_0) = 4\left[W_E(\omega_0) + W_M(\omega_0) + W_L(\omega_0) + W_R(\omega_0)\right] \tag{1.144}$$

with

$$W_E(\omega_0) = \frac{1}{4}\lim_{r\to\infty}\left[\int_{V_0}(\omega_0\varepsilon_r)'\,|\boldsymbol{E}|^2\,dV - \varepsilon_0 r\int_{S_\infty}|\boldsymbol{E}_\infty|^2\,dS\right] \tag{1.145}$$

$$W_M(\omega_0) = \frac{1}{4}\lim_{r\to\infty}\left[\int_{V_0}(\omega_0\mu_r)'\,|\boldsymbol{H}|^2\,dV - \varepsilon_0 r\int_{S_\infty}|\boldsymbol{E}_\infty|^2\,dS\right] \tag{1.146}$$

$$W_L(\omega_0) = \frac{\omega_0}{2}\,\text{Im}\int_{V_A}(\mu_i \boldsymbol{H}' \bullet \boldsymbol{H}^* + \varepsilon_i \boldsymbol{E}' \bullet \boldsymbol{E}^*)dv \tag{1.147}$$

$$W_R(\omega_0) = \frac{1}{2Z_0}\,\text{Im}\int_{S_\infty}\boldsymbol{E}'_\infty \bullet \boldsymbol{E}^*_\infty\,dS \tag{1.148}$$

$$\lim_{r\to\infty}\boldsymbol{E}(\boldsymbol{r}) = \boldsymbol{E}_\infty\frac{e^{-jkr}}{r} \tag{1.149}$$

where primes indicate differentiation with respect to the resonant frequency. W_E and W_M are as defined before, and the second terms in Eqs. (1.145) and (1.146) are due to the fact that the electric and magnetic energy densities of the radiated fields must be equivalent. This was an essential aspect of McLean's [3] analysis. The dispersive energies $W_L(\omega_0)$ and $W_R(\omega_0)$ are associated with the total dissipated and radiated power in the antenna structure, respectively. However, $W_L(\omega_0)$ and $W_R(\omega_0)$ do not represent the actual energy dissipated or radiated in the antenna structure itself. Furthermore, their sum can be either positive or negative and $X'_0(\omega_0)$ can take different signs. Due to this, Yaghjian and Best state that Eq. (1.144) proves the Foster reactance theorem does not hold for antennas.

Using Eq. (1.144) and the aforementioned P_A expression, we get

$$Q(\omega_0) = \frac{\omega_0 \left[W_M(\omega_0) + W_E(\omega_0)\right]}{P_A} = \frac{\omega_0}{2R_A(\omega_0)} X'_{\text{in}}(\omega_0) - \frac{2\omega_0}{|I_{\text{in}}|^2 R_A(\omega_0)} \left[W_L(\omega_0) + W_R(\omega_0)\right] \quad (1.150)$$

Note that the dispersive energy term $W_L + W_R$ is not present in Geyi's controversial expression (1.118).

1.3.10.3 Representing *Q* as a Function of Antenna Input Impedance

To obtain a useful representation of Eq. (1.150) as a function of the input impedance $Z_{\text{in}}(\omega_0)$ and bandwidth, the dispersion energies associated with dissipation W_L and radiation W_R must be evaluated. Yaghjian and Best choose to model these dispersion energies in terms of a series RLC circuit away from antiresonant frequencies, and as a parallel RLC circuit near the antiresonant frequency ranges. In both cases, the resistance at resonance is assumed to be frequency dependent. Using the models derived in the appendix of [10], they conclude that away from antiresonance regions $|R'_A(\omega_0)| \ll X'_{\text{in}}(\omega_0)$. Thus,

$$Q(\omega_0) \approx \frac{\omega_0}{2R_A(\omega_0)} X'_{\text{in}}(\omega_0) \approx \frac{\omega_0}{2R_A(\omega_0)} \left|Z'_{\text{in}}(\omega_0)\right| \quad (1.151)$$

However, when $X'_{\text{in}}(\omega_0) < 0$, the parallel RLC model approximates the dispersion energies as

$$X'_{\text{in}}(\omega_0) - \frac{4}{|I_{\text{in}}|^2} \left[W_L(\omega_0) + W_R(\omega_0)\right] \approx \sqrt{\left(X'_{\text{in}}(\omega_0)\right)^2 + \left(R'_A(\omega_0)\right)^2} = \sqrt{\left(X'_A(\omega_0) + \frac{|X_A(\omega_0)|}{\omega_0}\right)^2 + \left(R'_A(\omega_0)\right)^2} \quad (1.152)$$

Using Eqs. (1.151) and (1.152), Yagjian and Best [10] note that Q can be accurately represented in both the resonance and antiresonance frequency ranges as

$$Q(\omega_0) \approx \frac{\omega_0}{2R_A(\omega_0)} \left|Z'_{\text{in}}(\omega_0)\right| = \frac{\omega_0}{2R_A(\omega_0)} \sqrt{\left(R'_A(\omega_0)\right)^2 + \left(X'_A(\omega_0) + \frac{|X_A(\omega_0)|}{\omega_0}\right)^2} \quad (1.153)$$

Comparing Eqs. (1.153) and (1.139) gives the relationship between Q and B_V

$$Q(\omega_0) \approx \frac{2\sqrt{\beta}}{B_V(\omega_0)} \quad (1.154)$$

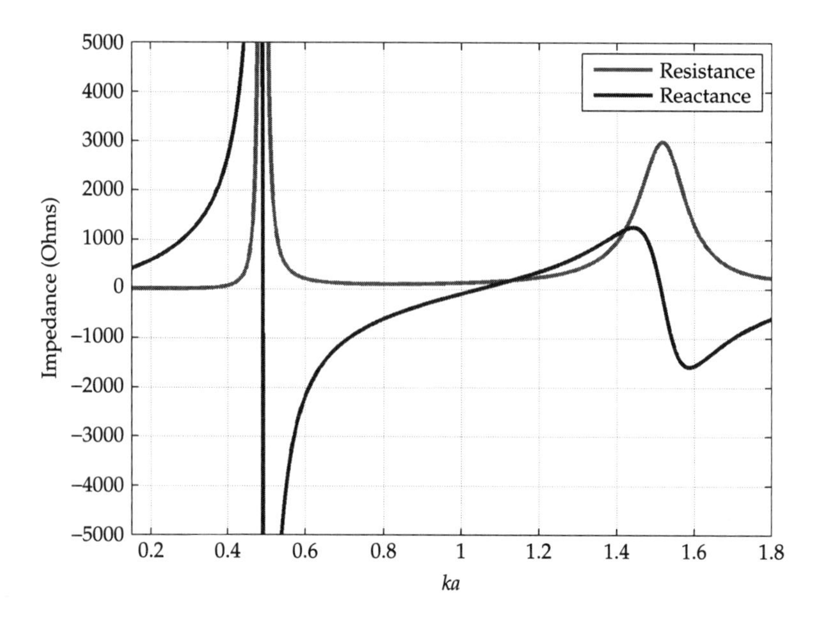

FIGURE 1.46 Impedance for a lossless circular loop of diameter 70 cm and wire diameter 1 mm. (*See Yaghjian and Best* [10].)

Equation (1.154) remains a good approximation when Q is greater than 4 [10]. In [10], method of moments simulations were used to determine the Q for a straight-wire antenna, a lossless or lossy circular loop, a lossless yagi, and a straight-wire embedded in a lossy dispersive dielectric. In each case the approximated Q from Eq. (1.153) remains highly accurate as compared to the rigorous Q values in Eq. (1.150) for all frequencies. However, Geyi's Eq. (1.118) breaks down near the antiresonance regions. Figures 1.46 and 1.47 show the impedance and Q for a lossless circular loop of diameter 70 cm with 1 mm wire diameter. The breakdown of Eq. (1.118) is clearly seen in the bands where $X'_{in}(\omega_0) < 0$ (antiresonance band).

1.3.10.4 Fundamental Limitations on B_V

Best [23] provided an upper bound for the matched VSWR fractional bandwidth B_V. This is similar to the minimum achievable value for Q. To do so, we begin by the lower bound on Q for an antenna radiating only TE or TM modes including the radiation efficiency η

$$Q_{\min} = \eta\left(\frac{1}{(ka)^3} + \frac{1}{ka}\right) \tag{1.155}$$

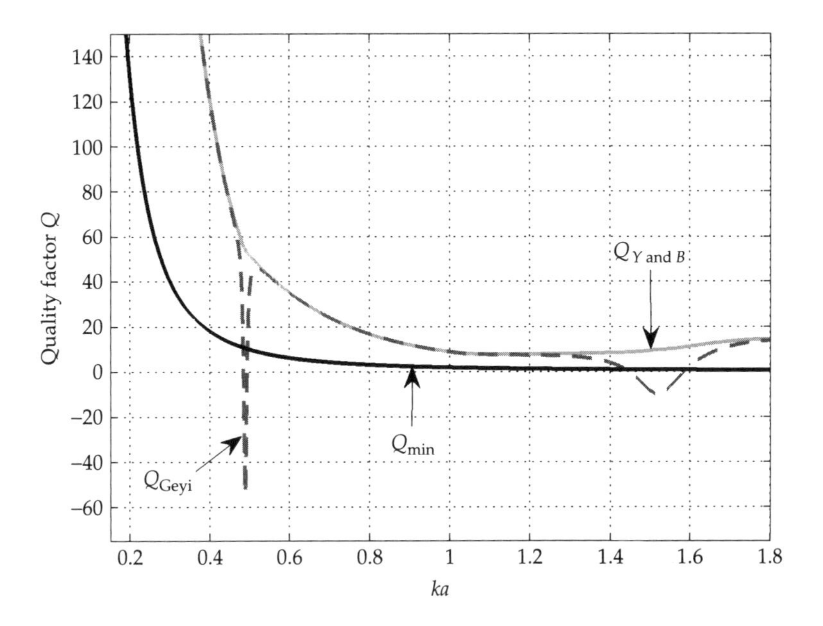

FIGURE 1.47 Q for the lossless circular loop using Geyi's method Q_{Geyi} in Eq. (1.118), Yaghjian and Best's method $Q_{Y \text{ and } B}$ in Eq. (1.153), along with McLean's $Q_{\min}$ in Eq. (1.103). (*See Yaghjian and Best* [10].)

This is a generalization of McLean's expression in Eq. (1.103). Using Eqs. (1.154) and (1.155), and assuming 100% radiation efficiency, the upper bound of B_V is

$$B_{V,\,\text{UB}} = \frac{(ka)^3}{1 + (ka)^2} \frac{VSWR - 1}{\sqrt{VSWR}} \tag{1.156}$$

Best [23] mentioned two methods for maximizing the bandwidth of a small antenna: (1) increasing the number of tuning circuits to approach the Bode-Fano limits (see Chap. 2), and (2) designing for multiple antenna resonances. He first considered maximizing B_V for a case where the small antenna is matched to a load using a number of lossless tuned circuits. From Sec. 1.2.3, it is known that the Q of the antenna itself does not change (as it is an intrinsic quantity). However, the operating bandwidth can change depending on the external Q. Best showed that by applying Eq. (1.156) to the Bode-Fano upper limit relationship in Eq. (1.20), the maximum B_V using an infinite number of lossless tuning circuits gives

$$B_{V,\,\mathrm{UB,\,Bode}} = \frac{\pi}{Q \ln\left(\dfrac{VSWR+1}{VSWR-1}\right)} = \frac{(ka)^3}{1+(ka)^2} \frac{\pi}{\ln\left(\dfrac{VSWR+1}{VSWR-1}\right)} \tag{1.157}$$

Best then examined the second method (viz. using multiband with closely spaced resonances or wideband antennas) for maximizing B_V, and the validity of Eq. (1.156) and the minimum Q limit in Eq. (1.103) for such antennas. From the numerical computations of a wideband disk-loaded antenna, Best found that Eq. (1.103) was never violated through the entire frequency band and the measured B_V falls below Eq. (1.157) when evaluating at the center frequency.

1.3.10.5 The Spherical Helix Antenna

Apart from the minimum Q and maximum B_V limits discussed earlier, Best [13] also pursued Wheeler's suggestion to better utilize the volume of the antenna's *Chu sphere* to minimize Q. He showed that the resonant properties of small wire antennas are a much heavier function of utilized *Chu sphere* volume rather than conductor arrangement, and that the resonant frequency can be easily decreased by increasing the effective inductance or capacitance seen at the feed point. The latter can be accomplished by techniques such as adding longer wire lengths to fill the *Chu sphere* (increases inductance) or top-loading (increases capacitance). Doing so, the resonance ($X_{\mathrm{in}}(\omega_0) = 0$) shifts, but the resistance curve remains relatively unchanged (as predicted by $R \propto h^2/\lambda^2$ for a monopole height h). This holds as long as the small antenna is far enough away from its initial, first natural antiresonant frequency [14]. Consequently, a great deal of flexibility is available to the designer for choosing the resonant properties of a small antenna.

One monopole design that fills well the *Chu sphere* volume is that of the four-arm spherical helix [24], shown in Fig. 1.48. This is a realization of the coupled $\mathrm{TM}_{10} - \mathrm{TE}_{20}$ mode equivalent circuit given in Fig. 1.57. This four-arm design by Best [24] in Fig. 1.48 is an extension of the single arm case, where four folded arms are used to increase the resonant resistance to values approaching practical transmission line impedances. Using the usual resistance for wire loop antennas, the resistance of the geometry in Fig. 1.48 can be approximated as

$$R_{N\,\mathrm{arm}} \approx N^2 R_{1\,\mathrm{arm}} \rightarrow R_{4\,\mathrm{arm}} \approx 16 R_{1\,\mathrm{arm}} \tag{1.158}$$

where N refers to the number of arms making up the helix in Fig. 1.48. It is important to note that the four-arm antennas in Fig. 1.48 have efficiencies greater than 97% when copper wires are used [24].

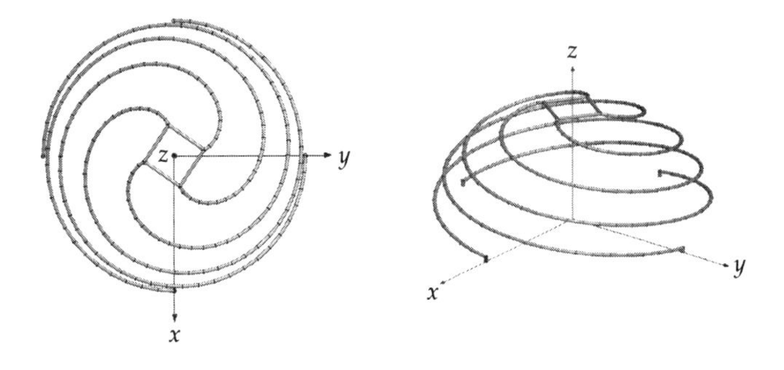

FIGURE 1.48 Four-arm spherical helix. (*See Best* [24].)

Apart from the four-arm spherical helix by Best [24], other shapes can be considered to fill up the *Chu sphere* volume. Examples are shown in Figs. 1.49 and 1.50. These are staircase spherical helices (SSH) and were considered due to their more practical realization. To take advantage of Best's [24] four-arm properties, the design in Fig. 1.49 was constructed using the same design equations given in [24, Eqs. 2–5], with eight vertical jumps of height 2/3 cm between planar layers at the angles

$$\phi_{\text{step}} = 45i\,N(\text{degrees}) \quad \text{for} \quad i = 1, 2, \ldots, 8 \tag{1.159}$$

(ϕ is measured from the x-axis, see Fig. 1.49). As can be understood, the angles and vertical jump heights were chosen so that the SSH would fully utilize the *Chu sphere* using eight layers. Without any dielectric substrate, the SSH was simulated using HFSS [50] for various values of turns N.

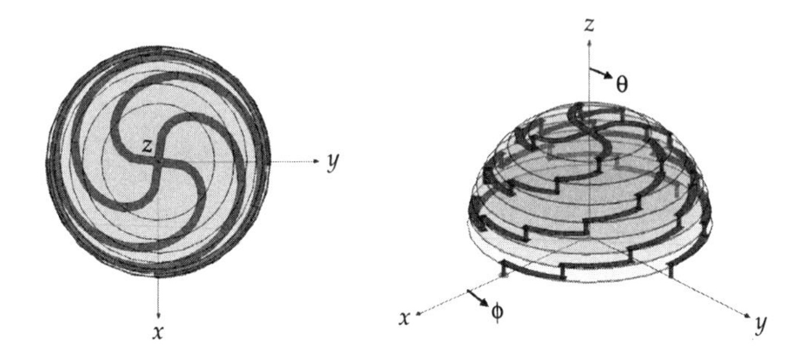

FIGURE 1.49 Top and isometric views of the $N = 0.75$ turn spherical staircase helix (SSH) antenna.

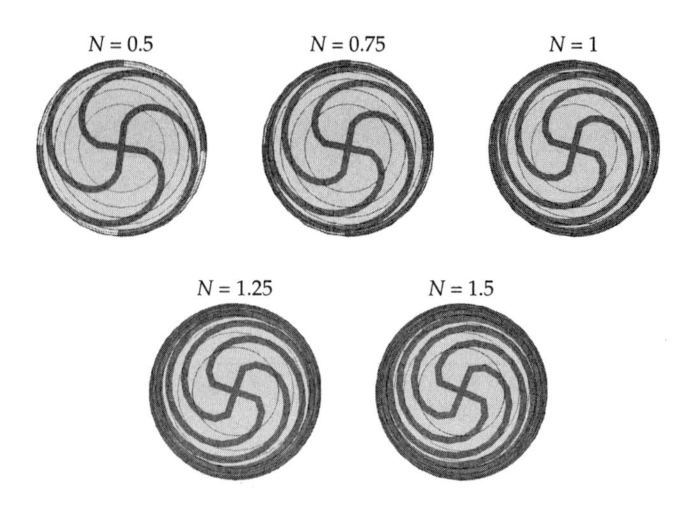

FIGURE 1.50 Top view of SSH antenna with varying turns.

Figures 1.51 and 1.52 give the quality factor and resonant resistance for the SSH for various values of turns N. As depicted, the efficiency for all SSH antennas using copper conductors is greater than 94%.

As seen from Figs. 1.51 and 1.52, the Q of the SSH closely follows that of the two-arm case [24]. However, the resonant resistance follows the four-arm [24] values. The differences between the four-arm Q values in [24] and those of the SSH are likely due to increased stored energy within the SSH *Chu sphere*. We do note, that unlike the four-arm antenna by Best [24], the SSH antenna does not utilize the entire *Chu sphere* surface (see Fig. 1.49). Concurrently, the SSH radiates much better than the two-arm and is closer to the four-arm radiation level.

It is important to note that since the SSH radiates a TM_{10} mode, the stored energy is primarily electric and the addition of dielectric inside the *Chu sphere* would increase the stored electric energy and the Q value. Therefore for the SSH antenna, it is important that the dielectric constant remains low, and the substrate is thin enough for improved radiation. This was also observed in [28] where Thal notes that filling the *Chu sphere* with a dielectric of permittivity ε_r corresponds to multiplying the internal region capacitances by ε_r (see Sec. 1.3.12).

1.3.11 Work of Kwon and Pozar (2005–2009)

Recent work by Kwon [4,5] and Pozar [6] aimed to consolidate the published results about small antenna gain and Q. As described in McLean [37] and Chu [11], a circularly polarized *Chu antenna* radiating equally excited TE_{10} and TM_{10} modes has a Q which is slightly greater than half

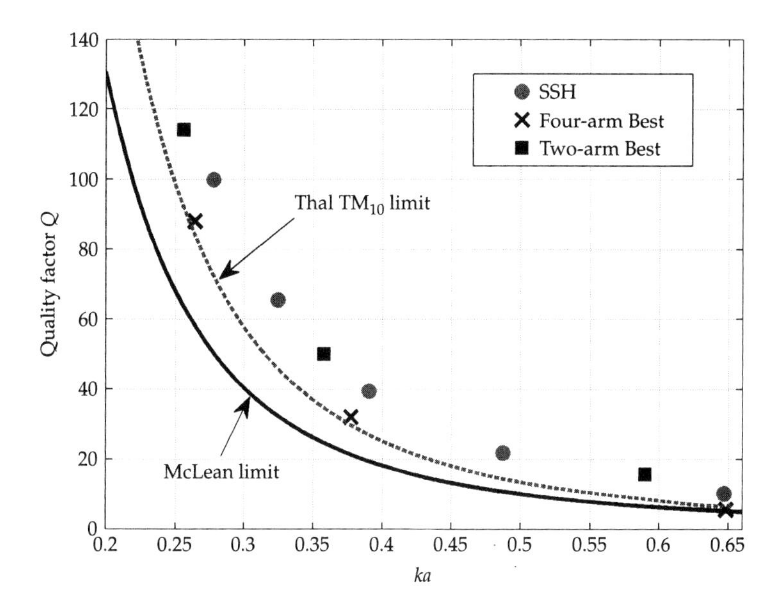

FIGURE 1.51 Quality factor for the four-arm SSH [51], four-arm [24], and two-arm [24] small antennas with the McLean limit in Eq. (1.103) and Thal TM_{10} limit in Eq. (1.170).

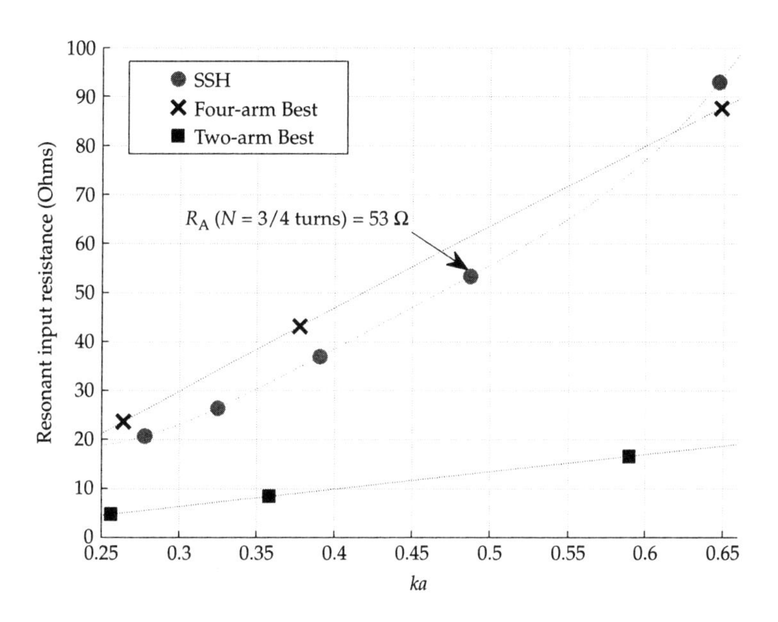

FIGURE 1.52 Resonant resistance for the four-arm SSH [51], four-arm [24], and two-arm [24] small antennas.

that of the TE_{10} or TM_{10} only case, and is given as $\hat{Q}_{min}$ in Eq. (1.104). However, this derivation does not specify if circular polarization is a necessary condition to obtain $\hat{Q}_{min}$ given in Eq. (1.104). Harrington [4] found that a unidirectional linearly polarized *Chu antenna* can achieve a gain of 3 and have $\hat{Q}_{min}$ in Eq. (1.104). Also, Hansen [2] does not mention the excitation levels between the TE and TM modes for achieving Eq. (1.104). However, Fante [19] showed that $\hat{Q}_{min}$ is possible for a *Chu antenna* with equally excited TE_{1m} and TM_{1m} modes, but did not mention the associated gain of such an antenna. On the other hand, although Geyi [21] categorized the limits of omni-directional and directional antennas, he did not specify the excitation.

To clarify the issue of excitation and polarization for the *Chu antenna*, Kwon [4,5] explored various crossed electric and magnetic dipole excitations. Later, Pozar [6] presented a summary of the gain and minimum Q values for small antennas along with the corresponding excitations. In the following sections, we provide a summary of the work by Kwon and Pozar.

1.3.11.1 Gain and *Q* for a Crossed-Dipole Chu Antenna

Kwon explored the gain and Q for a *Chu antenna* generating fields corresponding to those radiated by a pair of ideal, crossed electric and magnetic dipoles [4]. These dipoles are depicted in Fig. 1.53, with their electric and magnetic dipole moments given by

$$\mathbf{p}_e = \hat{z}p_e \tag{1.160}$$

$$\mathbf{p}_m = \hat{z}p_m \cos\theta_m + \hat{y}p_m \sin\theta_m \tag{1.161}$$

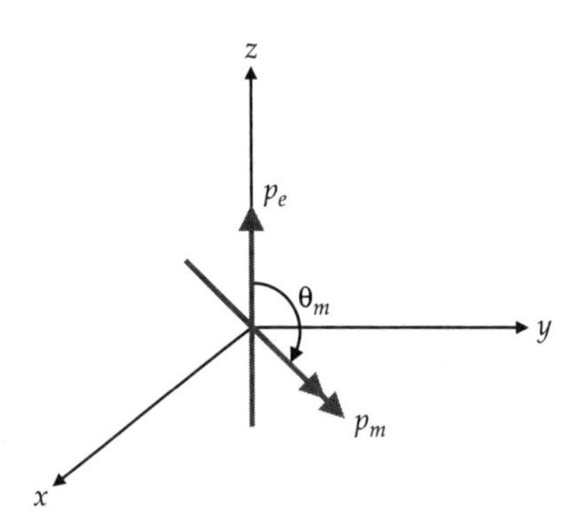

FIGURE 1.53 Set of crossed electric and magnetic dipoles about the origin, radiating the same fields as a CDCA outside the *Chu sphere*. (*After Kwon* [4].)

We will refer to this type of antenna as a *crossed-dipole Chu antenna (CDCA)*. To compute the TE_{10} and TM_{10} mode excitations corresponding to a CDCA, Kwon equated the fields due to the pair in Fig. 1.53 to the spherical mode representations. Invoking the usual mode orthogonalities of spherical wavefunctions representing the radiated field, Kwon found that the only nonzero mode coefficients excited by the crossed dipole arrangement of Fig. 1.53 are those of the TM_{10}, TE_{10}, and TE_{11} spherical modes. With the nonzero spherical mode coefficients known, the following relation between Q and polarization was obtained using Eq. (1.76)

$$Q_{\mathrm{CDCA}} = \frac{1}{ka} + \frac{1}{(ka)^3} \cdot \frac{1}{1 + \dfrac{|\boldsymbol{p}_m|^2}{Z_0^2 |\boldsymbol{p}_e|^2}} \tag{1.162}$$

assuming $Z_0|\boldsymbol{p}_e| \geq |\boldsymbol{p}_m|$. Note the similarity to $Q_{\min}$ given in Eq. (1.103) with the one in Eq. (1.162), though Q_{CDCA} contains an extra multiplying factor. Kwon's results [4] find that:

1. Q is independent of polarization: – Both $\boldsymbol{p}_e$ and $\boldsymbol{p}_m$ are present in Q_{CDCA} as magnitude squared terms, so phase does not affect Q. That is linear, circular, and elliptical polarizations can all achieve the same Q. Also, omni-directional and directional antennas can have the same Q.
2. Equal electric and magnetic source strengths give the minimum possible Q for *Chu antenna* – That is, $Z_0|\boldsymbol{p}_e| = |\boldsymbol{p}_m|$ reduces $Q_{\mathrm{CDCA}} = \hat{Q}_{\min}$ in Eq. (1.104).

Of course, duality also implies that interchanging magnetic and electric polarizability strengths does not impact these conclusions.

Of interest is the maximum possible gain for the arrangement in Fig. 1.53. To proceed, without loss of generality, we may choose to relate p_e and p_m from Eqs. (1.160) and (1.161) through the phase and amplitude parameters α and β, respectively. That is, we set

$$p_m = (\alpha e^{j\beta}) Z_0 p_e \tag{1.163}$$

With this choice of p_e and p_m, Kwon [4, Eq. 27] maximized the gain expression for the CDCA antenna to obtain

$$G_{\max,\mathrm{CDCA}} = \frac{3}{2}\left(1 + \frac{2\alpha\,|\sin\theta_m \cos\beta|}{1+\alpha^2}\right) \tag{1.164}$$

with a front-to-back ratio

$$FBR = \frac{G_{\max}}{3 - G_{\max}} \tag{1.165}$$

From Eq. (1.164), it is seen that the maximum possible gain for the CDCA is 3, with $\theta_m = \pi/2$, $\beta = \pm\pi$, and $\alpha = 1$, corresponding to linear polarization. That is, the maximum (directional) gain is the sum of the gains from the ideal electric and magnetic dipoles. We also note that these excitation dipoles must be normal to each other. Further, the gain is maximum in the direction normal to the plane containing the sources.

1.3.11.2 Gain and *Q* for Dual-Set Chu Antenna (DSCA)

Kwon extended the analysis in [4] by examining the properties of a *Chu antenna* whose radiating fields correspond to more general sources. Specifically, he chose the sources $(\boldsymbol{J}_a, \boldsymbol{M}_a)$ to represent the radiating TM modes and the sources $(\boldsymbol{J}_b, \boldsymbol{M}_b)$ to radiate the spherical TE modes [5]. As usual, $\boldsymbol{J}_{a,b}$ represents an electric dipole source and $\boldsymbol{M}_{a,b}$ refers to a magnetic dipole source. To have equal power radiated from the electric and magnetic sources, Kwon [5] followed the relations

$$\boldsymbol{J}_b = -\frac{1}{Z_0} e^{j\beta} \boldsymbol{M}_a \tag{1.166}$$

$$\boldsymbol{M}_b = Z_0 e^{j\beta} \boldsymbol{J}_a \tag{1.167}$$

That is, $(\boldsymbol{J}_a, \boldsymbol{M}_a)$ are the dual of $(\boldsymbol{J}_b, \boldsymbol{M}_b)$; each set radiating equal power. We will refer to this type of antenna as the *dual-set Chu antenna (DSCA)*. We do note the DSCA is simply a linear combination of two CDCA antennas. Thus, the DSCA has the same Q as the CDCA antenna. In essence the DSCA provides more degrees of freedom in choosing the excitation sources, but the radiated mode choices are still the same. Consequently, Kwon finds the extra degrees of freedom allow for all types of polarization to be realized, simultaneously with a gain of 3 and minimum Q [5].

Table 1.5 summarizes the results derived by Kwon [4,5] for the CDCA and DSCA antennas, where $Q_{\min}$ is given in Eq. (1.103) and $\hat{Q}_{\min}$ is given in Eq. (1.104). As depicted, for all possible combinations of sources, $\hat{Q}_{\min}$ is achieved only when equally excited first order TM and TE modes are present. However, maximum gain is attained only when $\boldsymbol{J}$ and $\boldsymbol{M}$ are present. For linear polarization, maximum gain is attained when $\boldsymbol{J}$ and $\boldsymbol{M}$ are in phase but normal to each other. In the case of CP radiation, the sources must be individually CP. That is, $\boldsymbol{J} = \hat{z}J_z$ and $\boldsymbol{M} = j\hat{y}M_y$ will not give maximum gain. Finally, we remark that the results in Table 1.5 and the derivations by Kwon [4,5] have recently been contested by Thal as unrealizable [29]. This will be examined in Sec. 1.3.12.

Equivalent Hertzian Dipole Sources for CDCA/DSCA	Polarization/Pattern	Gain	Q
$\hat{z}J_z$(or $\hat{x}J_x$ or $\hat{y}J_y$)	Linear/Omni	1.5	Q_{min}
$\hat{z}M_z$(or $\hat{x}M_x$ or $\hat{y}M_y$)	Linear/Omni	1.5	Q_{min}
$\hat{z}J_z \pm \hat{z}M_z$	Linear/Omni	1.5	$\hat{Q}_{\text{min}}$
$\hat{z}J_z \pm \hat{y}M_y$	Linear/Directional	3	$\hat{Q}_{\text{min}}$
$\hat{z}J_z \pm j\hat{z}M_z$	Circular/Omni	1.5	$\hat{Q}_{\text{min}}$
$\hat{z}J_z \pm j\hat{y}M_y$	Linear/Bidirectional	1.5	$\hat{Q}_{\text{min}}$
$\hat{x}J_x \pm j\hat{y}J_y$	Circular/Bidirectional	1.5	Q_{min}
$\hat{x}M_x \pm j\hat{y}M_y$	Circular/Bidirectional	1.5	Q_{min}
$\hat{x}J_x \pm j\hat{y}J_y$ and $\hat{x}M_x \pm j\hat{y}M_y$	Circular/Directional	3	$\hat{Q}_{\text{min}}$

TABLE 1.5 Summary of CDCA and DSCA *Chu Antenna* Performance. (*See Pozar* [6])

1.3.12 Work of Thal (2006–2009)

Similar to Foltz and McLean in Sec. 1.3.7, Thal set out to find a stricter Q limit to more accurately represent real antennas. Thal [28] considered the antenna geometry in Fig. 1.54, that is, an antenna with its current distribution on the *Chu sphere* surface. Thal's analysis of the antenna in Fig. 1.54 closely matches the approach taken by Chu [11]. That is, he represented each propagating mode radiated outside the Chu surface with an equivalent mode circuit. However, Thal also included in his mode representation an equivalent circuit for the mode extending inside the *Chu sphere* to account for the mode energy stored interior

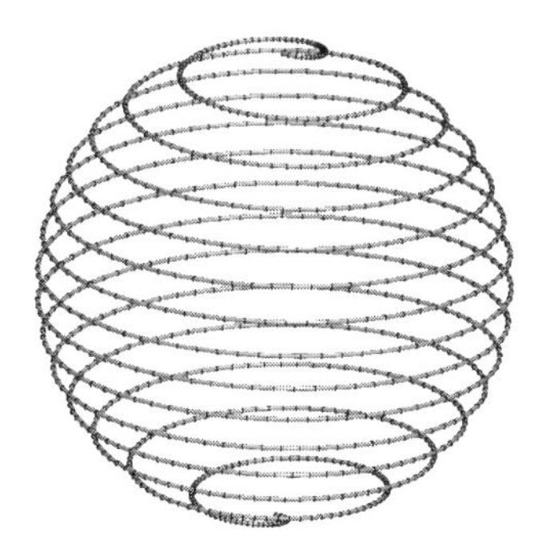

FIGURE 1.54 Possible antenna geometry considered by Thal [28] where the wiring lies on the *Chu sphere* surface.

to the sphere. We remark, that contrary to the *Chu antenna*, in Thal's analysis Q between the TM and TE modes differ significantly.

A recent publication [29] by Thal takes another approach to the problem of quantifying the energy internal to the *Chu sphere*. In [29], Thal examines the effect of TM and TE mode phases on the small antenna gain and Q limits by analyzing the time-varying component of the power in Chu's mode circuits. His results conclude, contrary to Geyi [21], Kwon [4,5], and Pozar [6], that a small antenna cannot simultaneously achieve a gain of 3 and a Q given in Eq. (1.104).

1.3.12.1 Spherical Mode Circuits at the Chu Sphere Surface

The TM and TE spherical mode wave impedances can be formulated from the orthogonal field components of Eqs. (1.51a) to (1.51c) and their duals. Table 1.6 gives the TM and TE normalized wave impedances looking outside and inside the *Chu sphere* surface [36]. As would be expected, these normalized wave impedances are independent of the azimuthal index m. We also note that for m greater than n, the associated Legendre polynomials in Eqs. (1.51a) to (1.51c) are zero. As a result, no modes for $m > n$ exist.

To construct the equivalent circuits for an antenna with its current over the *Chu sphere* surface (see Fig. 1.54), we note the recurrence relation for spherical Bessel functions $b_n(ka)$

$$\left[\frac{ka\,b_{n+1}(ka)}{j^{n+2}}\right] = \left[\frac{ka\,b_{n-1}(ka)}{j^{n}}\right] + \frac{2n+1}{jka}\left[\frac{ka\,b_{n}(ka)}{j^{n+1}}\right] \tag{1.168}$$

$$j\left[\frac{ka\,b_{n}(ka)}{j^{n+1}}\right]' = \left[\frac{ka\,b_{n-1}(ka)}{j^{n}}\right] + \frac{n}{jka}\left[\frac{ka\,b_{n}(ka)}{j^{n+1}}\right] \tag{1.169}$$

Using these, Thal then constructed equivalent mode networks to represent the impedances in Table 1.6 using Chu's procedure in Sec. 1.3.2 [52]. That is, he related the Bessel functions to analogous voltage and current quantities. In essence, these equivalent circuit networks are "spherical Bessel function generators" for each index n.

	$\mathbf{TM_{nm}}$	$\mathbf{TE_{nm}}$
Normalized wave impedance looking out of the sphere	$\dfrac{j\left[h_n^{(2)}(ka)\right]'}{\left[h_n^{(2)}(ka)\right]}$	$\dfrac{-j\left[h_n^{(2)}(ka)\right]}{\left[h_n^{(2)}(ka)\right]'}$
Normalized wave impedance looking into the sphere	$\dfrac{-j\,[j_n(ka)]'}{[j_n(ka)]}$	$\dfrac{j\,[j_n(ka)]}{[j_n(ka)]'}$

TABLE 1.6 Normalized Wave Impedances at Sphere $r = a$ for Spherical Mode n, with Primes Indicating Differentiation With Respect to ka

Figure 1.55 shows the two equivalent spherical mode networks for all TE_{nm} and TM_{nm} modes, with each terminal corresponding a certain TE_{nm} or TM_{nm} mode. As was in Chu [12], a is the radius of the *Chu sphere* and c is the speed of light. At a terminal, the impedance looking to the left corresponds to the wave impedance looking out of the sphere, and the impedance looking to the right corresponds to the wave impedance looking into the sphere. This is indicated in Fig. 1.55 by the exterior and interior region arrows. We note that c may be different in the interior and exterior regions if the sphere is filled with a material other than free space. We assume c is that of free space unless specified otherwise.

Given the orthogonality of the spherical modes, each antenna mode radiated by the surface current can be represented by an independent circuit of Fig. 1.55. The Q for a given mode is, of course, determined by the standard definition in Eq. (1.3). Specifically, W_E is found by summing the energies stored in all capacitors, and likewise W_M is found by summing the energies stored in all inductors. The radiated power is extracted from the termination resistance. We lastly observe that for odd n, the exterior region circuits are equivalent to Chu's circuits in Figs. 1.19 and 1.22 (Chu only found circuits for odd n). As an example, the mode circuit for a TM_{10} mode is depicted in Fig. 1.56, where the *Chu sphere* surface current exciting the TM_{10} mode is represented by a current source I_{TM10}.

1.3.12.2 *Q* Value for the TM_{1m} and TE_{1m} Modes

Thal [28] tabulated the Q values obtained using the mode circuits in Fig. 1.55 for the TM_{1m} and TE_{1m} modes. His results are shown in Table 1.7 where Q_{min} is the minimum possible Q for TM or TE operation given in Eq. (1.103) (McLean [3]). $Q_{TM1m,Thal}$ and $Q_{TE1m,Thal}$ are the Q values for TM_{1m} and TE_{1m} mode excitation in Fig. 1.55, respectively.

From Table 1.7, we can conclude that for $ka < 0.5$

$$Q_{TM1m,Thal} \approx 1.5Q_{min} \tag{1.170}$$

$$Q_{TE1m,Thal} \approx 3Q_{min} \tag{1.171}$$

It is observed that for vanishing ka, the stored energy is in the TM_{1m} mode and is dominated by the capacitors next to the current source excitation. That is, the inductors store very little magnetic energy. We can also conclude that the energy stored in the internal capacitor is half that of the external one. Therefore, the Q computed by Thal [28] is 1.5 times the McLean limit in Eq. (1.103).

Similarly, if we place a current source at the TE_{1m} port as ka decreases, the stored energy will then be dominated by the shunt inductors next to the current source since the nearby capacitors store very little electric energy. Thus, we can ignore the capacitors and conclude

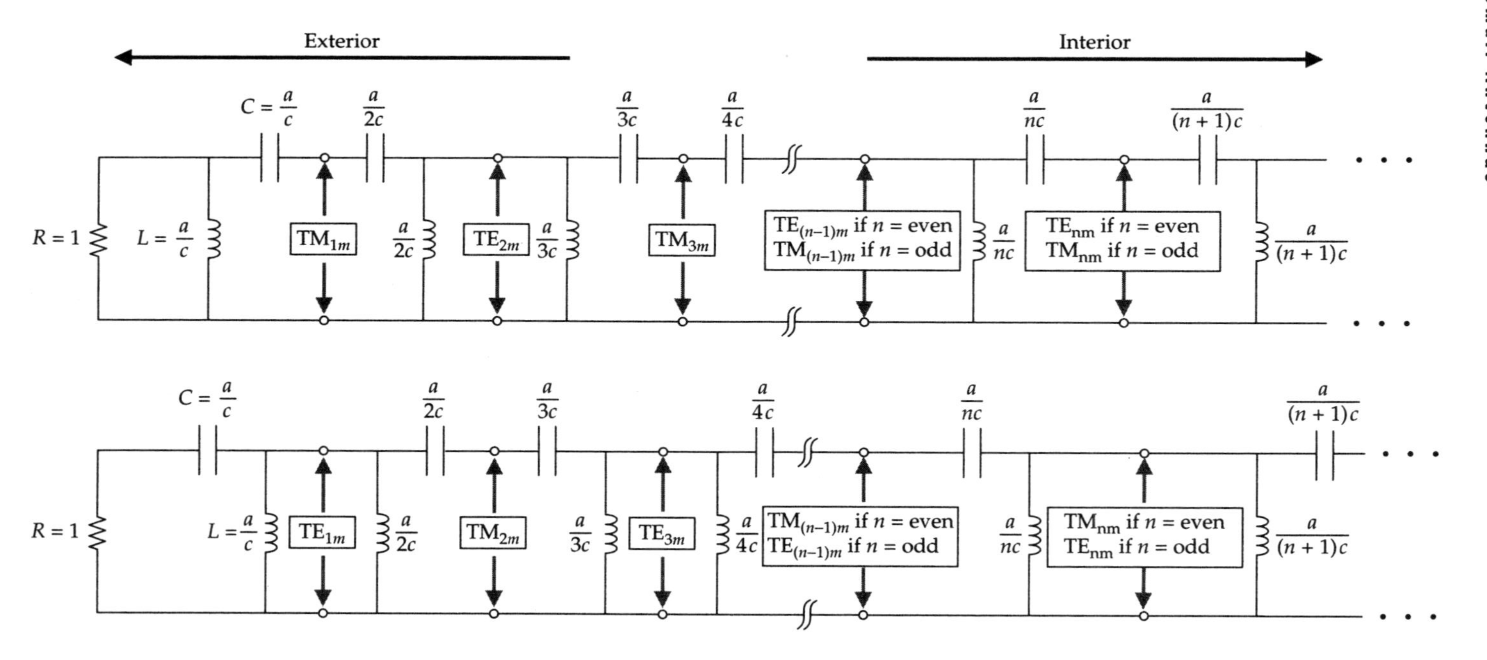

FIGURE 1.55 Spherical mode circuits at the *Chu sphere* surface. (*See Thal* [28].)

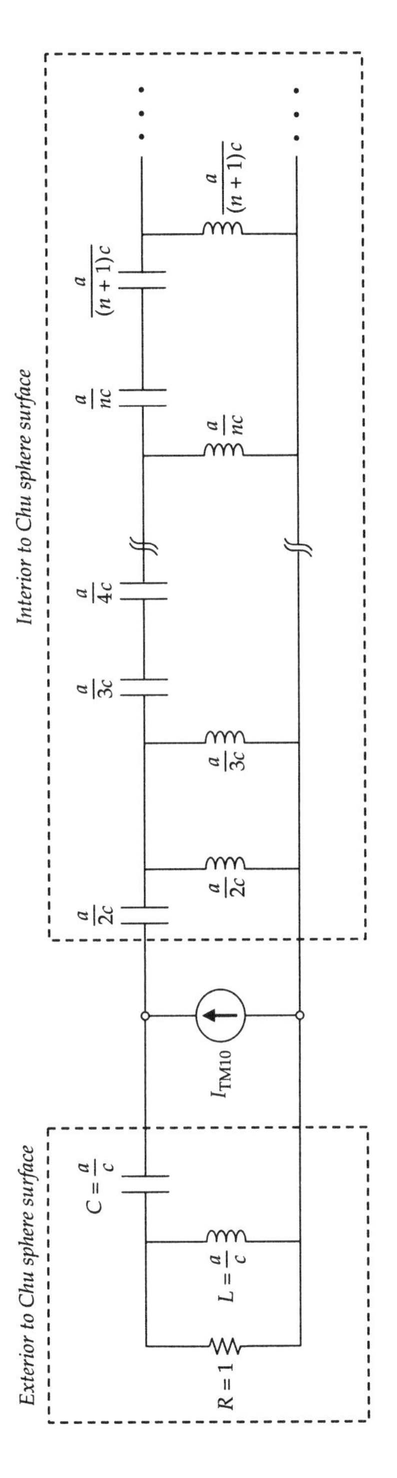

FIGURE 1.56 Spherical mode circuit for TM_{10} mode excitation at *Chu sphere* surface.

ka	$Q_{\min}$ (1.103)	$Q_{\mathrm{TM1}m,\mathrm{Thal}}$	$Q_{\mathrm{TE1}m,\mathrm{Thal}}$	$Q_{\mathrm{TM1}m,\mathrm{Thal}}/Q_{\min}$	$Q_{\mathrm{TE1}m,\mathrm{Thal}}/Q_{\min}$
0.050	8020.0	12012.1	24062.0	1.50	3.00
0.100	1010.0	1506.0	3029.9	1.49	3.00
0.200	130.0	190.6	390.0	1.47	3.00
0.300	40.4	57.7	121.1	1.43	3.00
0.400	18.1	25.1	54.4	1.39	3.00
0.500	10.0	13.4	30.0	1.34	3.00
0.600	6.3	8.2	18.9	1.30	3.00

TABLE 1.7 Comparison of McLean's $Q_{\min}$ (1.103) and the ones derived by Thal for the TE_{1m} and TM_{1m} modes in Fig. 1.55. (*See Thal* [28])

that the energy stored by the internal circuit inductor is twice that of the external inductor. Thus, the Q value using Thal's analysis is predictably three times that of McLean's limit in Eq. (1.103). We also note that there is very little variation in $Q_{\mathrm{TE1}m,\mathrm{Thal}}/Q_{\min}$ as ka increases, indicating that the ratio of magnetic stored energy to radiated power remains fairly constant in the small antenna limit. This is in contrast to the TM_{1m} case, where we see a decrease in the $Q_{\mathrm{TM1}m,\mathrm{Thal}}/Q_{\min}$ as ka increases, indicating electric stored energy to radiated power decreases.

Recently, Hansen and Collin [53] verified the accuracy of Thal's mode circuits for an antenna with its current distribution on the *Chu sphere* surface (see Fig. 1.54). After using Collin's results in [18] (see Sec. 1.3.4) to account for the exterior region energy, the internal stored energy can be found by representing the fields as spherical Bessel functions of the first kind and enforcing the tangential electric field continuity at the *Chu sphere* surface. After a curve-fitting procedure, Hansen and Collin state that Eq. (1.170) can be more accurately represented for $ka < 0.5$ as

$$Q_{\mathrm{TM1}m,\mathrm{Thal}} \approx \frac{1}{\sqrt{2}ka} + \frac{3}{2(ka)^3} \tag{1.172}$$

1.3.12.3 Self-Resonant Mode Configurations

The Q values in Table 1.7 are found using the standard Q formula of Eq. (1.3) with the assumption that the antenna is tuned to resonance using a lossless reactive element. However, the circuits of Fig. 1.55 can be used to determine the combinations of modes necessary for a self-resonant antenna described by Fig. 1.54.

TM_{10} – TE_{20} Self-Resonant Antenna Thal [28] gives two examples of self-resonant small antennas of the type described in Fig. 1.54. For a TM_{10} mode, an inductive reactance is needed to tune the antenna to resonance. Figure 1.57 shows one possible tuning solution, where the

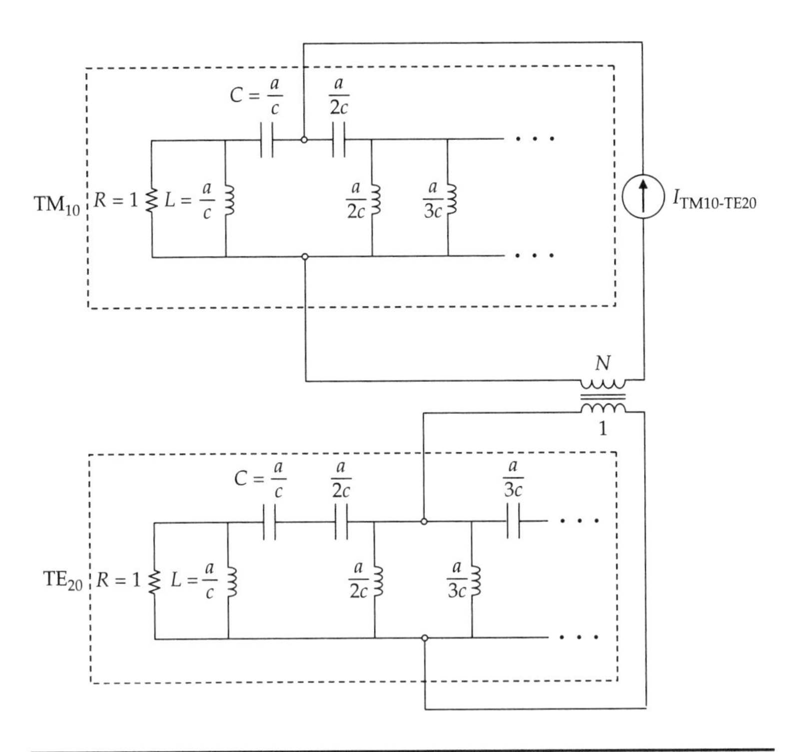

FIGURE 1.57 $TM_{10} - TE_{20}$ mode circuit for a small antenna. (*See Thal* [28].)

surface current distribution radiates both a TM_{10} and a TE_{20} mode. The TM_{10} mode remains the primary mode of radiation, with the TE_{20} mode excited through a transformer of ratio N:1 (N is adjusted to cancel the capacitive reactance due to the TM_{10} mode). The same numerical procedure noted in [52, (Fig. 4*c*)] is used to determine the Q values using the circuits in Fig. 1.57. The results of Table 1.8 show

ka	Q_{min} (1.103)	$Q_{TM10,Thal}$	$Q_{TM10-TE20,Thal}$	TE_{20}/TM_{10} (dB)
0.050	8020.0	12012.1	12013.1	−40.80
0.100	1010.0	1506.0	1506.5	−34.80
0.200	130.0	190.6	190.8	−28.85
0.400	18.1	25.1	25.4	−23.08
0.500	10.0	13.4	13.6	−21.31
0.600	6.3	8.2	8.4	−19.90

TABLE 1.8 Comparison of McLean's Q_{min} (1.103) with the Q Derived by Thal [28] Using the $TM_{10} - TE_{20}$ Self-Resonant Circuit in Fig. 1.57. (*See Thal* [28])

$Q_{\mathrm{TM10,Thal}}$ (identical to that of Table 1.7) and $Q_{\mathrm{TM10-TE20,Thal}}$ when the antenna in Fig. 1.57 is tuned to resonance by the appropriate value of N. $\mathrm{TE}_{20}/\mathrm{TM}_{10}$ is the ratio of the power radiated by the TE_{20} mode to the power radiated by the TM_{10} mode.

From Table 1.8 it is evident that

$$Q_{\mathrm{TM10-TE20,Thal}} \approx Q_{\mathrm{TM10,Thal}} \tag{1.173}$$

However, $Q_{\mathrm{TM10-TE20,Thal}}$ remains slightly larger due to that the stored energy of the TE_{20} mode is not completely inductive. Also, the $\mathrm{TE}_{20}/\mathrm{TM}_{10}$ power ratio verifies that the familiar dipole pattern for small antennas remains undisturbed using this tuning technique (as the TE_{20} mode is evanescent). Of course, the TM_{10} and TE_{20} mode field patterns are given by:

$$\mathrm{TM}_{10} \sim \sin\theta,\ E_\theta \text{ polarization (vertical)}$$

$$\mathrm{TE}_{20} \sim \sin 2\theta,\ E_\phi \text{ polarization (horizontal)}$$

That is, the TE_{20} mode can be used to tune small antennas placed over ground planes. This is because the E_ϕ component of the electric field vanishes to zero for $\theta = 90°$, automatically satisfying the PEC boundary conditions. One implementation of the $\mathrm{TM}_{10} - \mathrm{TE}_{20}$ in Fig. 1.57 is to use the four-arm spherical helix given by Best [24] and shown in Fig. 1.48. This antenna exhibits Q values that closely match the numerically computed $Q_{\mathrm{TM10-TE20,Thal}}$. Additionally, the $\mathrm{TE}_{20}/\mathrm{TM}_{10}$ power ratio of the four-arm antenna in [24] is consistent with the values obtained in Table 1.8, as evidenced by the vertical and horizontal radiation patterns given in Fig. 1.58.

TM_{1m} - TE_{1m} Self-Resonant Antenna It was shown above that if a small antenna radiates a TM_{10} mode along with a TE_{10} mode, the resulting Q can be smaller than the value obtained if the antenna radiated only the TM_{10} or TE_{10} mode. Furthermore, for a *Chu antenna*, when the radiated powers by TM_{10} and TE_{10} modes are same, the Chu limit becomes slightly greater than half the limit for the single TM_{10} mode (or TE_{10} mode) [3], as given in Eq. (1.104).

Figure 1.59 depicts the equivalent circuit for an antenna of radius a radiating both TM_{1m} and TE_{1m} modes using Thal's approach. The TM_{1m} mode remains the primary mode of radiation. As before (see Fig. 1.57), the current source representing the magnetic field discontinuity at the surface of the *Chu sphere* is coupled to the TE_{1m} mode through a transformer of ratio N:1. This ratio can be adjusted to cancel the capacitive reactance of the TM_{1m} circuit.

Using the same numerical procedure as in [52, Fig. 4c], Thal computed the Q values corresponding to the circuit in Fig. 1.59. The results of this procedure are given in Table 1.9 where $Q_{\mathrm{TM1}m\mathrm{-TE1}m\mathrm{,Thal}}$ is the

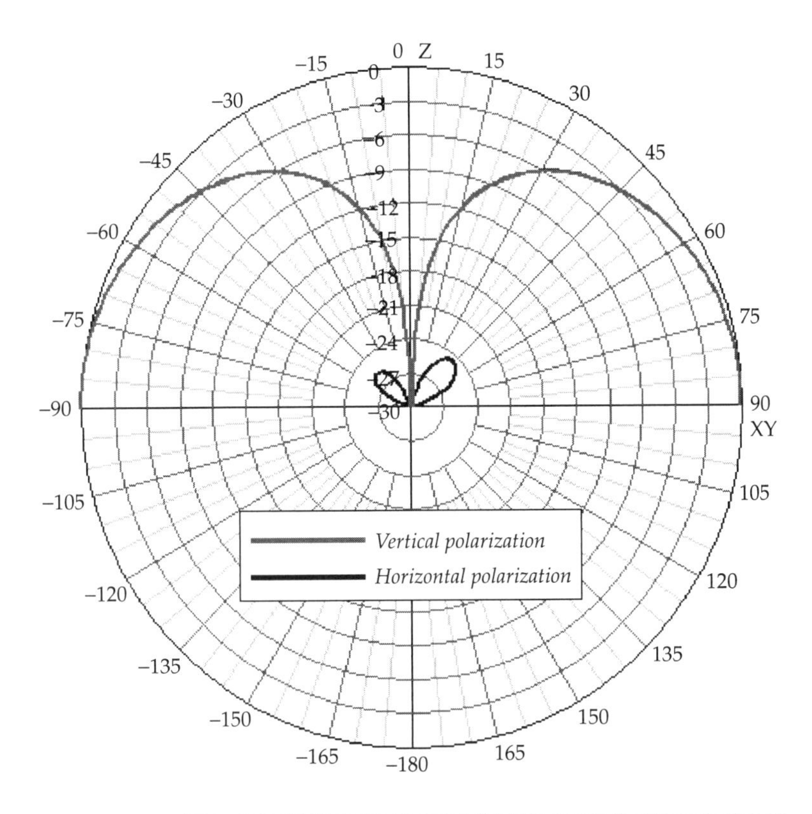

FIGURE 1.58 Vertical and horizontal patterns for the $N = 1$ turn four-arm Best antenna at resonance. (*See Best* [24].)

Q for the TM_{1m}–TE_{1m} mode circuit in Fig. 1.59 tuned to resonance by the appropriate value of N. The last column in Table 1.9 is the ratio of the power radiated by the TE_{1m} mode to the power radiated by the TM_{1m} mode.

Table 1.9 shows that the minimum Q for the TM_{1m}–TE_{1m} self-resonant antenna is approximately twice the McLean limit $\hat{Q}_{\min}$ given in Eq. (1.104)

$$Q_{TM1m-TE1m,\text{Thal}} \approx 2\hat{Q}_{\min} \approx Q_{\min} \tag{1.174}$$

Since it was shown in Table 1.7 that $Q_{TM1m,\text{Thal}} < Q_{TE1m,\text{Thal}}$, we can conclude that $Q_{TM1m-TE1m,\text{Thal}}$ represents the minimum possible Q for a small air-filled antenna of the type in Fig. 1.54. We should note the $\hat{Q}_{\min}$ was derived assuming a self-resonant *Chu antenna* with equally radiating TM_{1m} and TE_{1m} modes. As already noted, the results of Table 1.9 can be easily predicted by considering only the energy stored

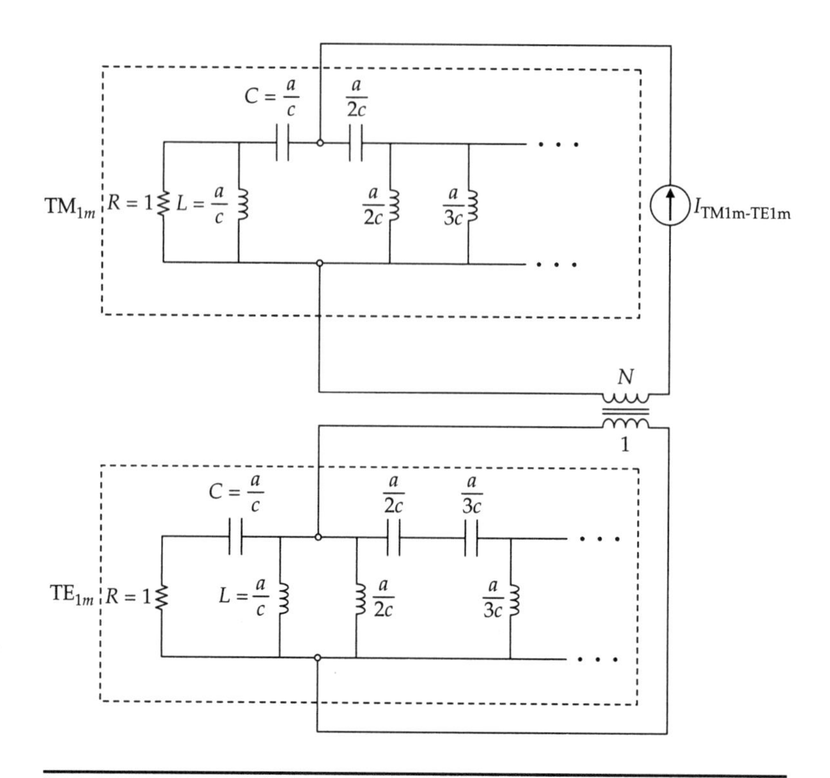

FIGURE 1.59 $TM_{1m} - TE_{1m}$ mode circuit for a small antenna using Thal's procedure. (*See Thal* [28].)

ka	Q_{min} (1.103)	$\hat{Q}_{min}$ (1.104)	$Q_{TM1m-TE1m,\,Thal}$	TE_{1m}/TM_{1m}
0.050	8020.0	4020.0	8022.5	−3.02
0.100	1010.0	510.0	1011.0	−3.05
0.200	130.0	67.5	130.5	−3.16
0.300	40.4	21.9	40.7	−3.34
0.400	18.1	10.3	18.4	−3.58
0.500	10.0	6.0	10.2	−3.85
0.600	6.3	4.0	6.5	−4.13

TABLE 1.9 Comparison of the Q_{min} (1.103) and $\hat{Q}_{min}$ (1.104) computed by McLean [3] as compared to the Q obtained by Thal [28] for the circuit in Fig. 1.59. (*See Thal* [28].)

Radiating Mode	Q/Q_{min}	Axial Ratio
TM_{10}	1.5	-
TE_{10}	3	-
$TM_{10} - TE_{10}$	1	3 dB (elliptical polarization)
$TM_{10} - TE_{20}$	1.5	≈40 dB

TABLE 1.10 Q Values Given by Thal for $ka \ll 1$ [Q_{min} is that Given by McLean in Eq. (1.103)]

by elements adjacent to the source, and by assuming all other capacitors open and all other inductors shorted. Another comment from Table 1.9 is that the power ratio TE_{1m}/TM_{1m} is approximately the inverse of $Q_{TE10,Thal}/Q_{TM10,Thal}$. Also, the power ratio TE_{1m}/TM_{1m} is no longer unity, as it was in the case for McLean's calculations with the minimum possible quality factor $\hat{Q}_{min}$. As a result, Thal's approach predicts elliptical polarization [28].

The Q calculations by Thal [28] are summarized in Table 1.10 as compared to the Q_{min} given by McLean in Eq. (1.103). Also, Table 1.11 gives the corresponding surface current distributions necessary to excite the TM_{10}, TE_{10}, self-resonant $TM_{10} - TE_{10}$, and self-resonant $TM_{10} - TE_{20}$ modes [28] listed in Table 1.10. The necessary transformer turns N to make the resonant modes are listed in Table 1.12 [28].

1.3.12.4 Thal's Energy Lower Bound on the Mode Coupling Network

In [29], Thal seeks a relationship between small antenna gain and Q, and reexamines previous claims made on the topic. He notes that previous authors [4,6,21] who have examined the gain and Q relationship ignored any conditions on the energy inside the *Chu sphere*, and simply assumed zero internal stored energy. Consequently, it is found in [4,6,21] that it is theoretically possible for a small antenna to realize a gain of 3 and a Q given in Eq. (1.104) by equally exciting first order TE and TM modes. To find stricter limits on Q and gain, Thal begins by finding a lower bound on the energy inside the *Chu sphere*. As was in Chu [11], Thal represents the antenna system as the equivalent circuit shown in Fig. 1.60, where the input port and mode coupling network

Radiating Mode	Surface Current Distribution
TM_{10}	$J = \pm\hat{\theta}\,[A\sin\theta]$
TE_{10}	$J = \pm\hat{\phi}\,[A\sin\theta]$
$TM_{10} - TE_{10}$	$J = -\hat{\theta}\,[A\sin\theta] \pm \hat{\phi}\,[NA\sin\theta]$
$TM_{10} - TE_{20}$	$J = -\hat{\theta}\,[A\sin\theta] \pm \hat{\phi}[\sqrt{1.25}NA\sin 2\theta]$

TABLE 1.11 Current Distributions Over *Chu Sphere* Surface to Excite of Various Mode Configurations, with N given in Table 1.12. (*See Thal* [28])

ka	N for TM_{1m}-TE_{1m}	N for TM_{10}-TE_{20}
0.050	28.25	36.48
0.100	14.06	18.19
0.200	6.92	8.99
0.300	4.50	5.88
0.400	3.26	4.30
0.500	2.50	3.33
0.600	2.00	2.68

TABLE 1.12 Transformer Turns N to Achieve Self-Resonance for the Circuits Given in Figs. 1.57 and 1.59. (*See Thal* [28])

represent the antenna of *Chu sphere* radius a, and the radiated modes outside the *Chu sphere* are represented by the mode circuits given in Fig. 1.55 (using the exterior region circuits only). Thal states that the instantaneous power at each port is the sum of the average and time-varying powers.

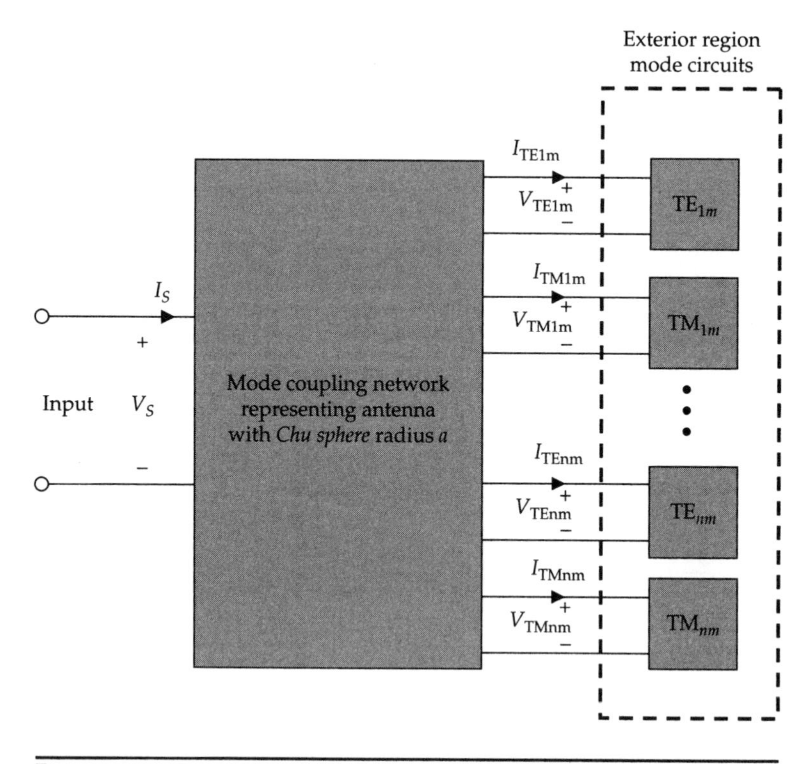

FIGURE 1.60 Equivalent network used in Thal's [29] time-varying circuit analysis to represent an arbitrary antenna with *Chu sphere* radius a.

Assuming a resonant antenna with no ohmic losses (zero average power in the coupling network), Thal begins his derivation of the coupling network energy bound by stating the time-varying power balance relation between the ports. Using Fig. 1.60 and assuming cosine referenced phasors for voltages and currents, Thal [28] expresses the time-varying power at the mode coupling network in terms of the voltages and currents at the ports through

$$\mathrm{Re}\left\{P_{C,\mathrm{tv}}e^{j2\omega t}\right\} = \mathrm{Re}\left\{\left(V_S I_S - \sum_{n=1}^{N} V_n I_n\right) e^{j2\omega t}\right\} \tag{1.175}$$

where $P_{C,\mathrm{tv}}$ is the time-varying power in the coupling network, V_S and I_S are the complex amplitudes of the time-varying voltage and current of the source, V_n and I_n are the complex amplitudes of the time-varying voltage and current at each of the mode circuit inputs, and N is the number of radiating modes. Thal remarks that for a resonant antenna, the instantaneous voltage and current at the source are in phase. Consequently, the time-varying power $V_S I_S$ must be the radiated power with an additional phase term $e^{j\xi}$. Suppressing the Re{ } operator and the $e^{j2\omega t}$ common to each term, Eq. (1.175) becomes

$$P_{C,\mathrm{tv}} = P_{\mathrm{rad}}e^{j\xi} - \sum_{n=1}^{N} V_n I_n \tag{1.176}$$

P_C is minimized when the phases of $P_{\mathrm{rad}}e^{j\xi}$ and $\sum_{n=1}^{N} V_n I_n$ are equal. Using Eq. (1.176) Thal forms the inequality

$$\left|P_{C,\mathrm{tv}}\right| \geq \left|P_{\mathrm{rad}} - \left|\sum_{n=1}^{N} V_n I_n\right|\right| \tag{1.177}$$

Due to the $e^{j2\omega t}$ time dependence of the time-varying components, the time-varying component of the energy in the coupling network is

$$W_{C,\mathrm{tv}} = \int P_{C,\mathrm{tv}} dt = \frac{P_{C,\mathrm{tv}}}{j2\omega} \tag{1.178}$$

Inserting Eq. (1.178) in (1.177) and multiplying both sides by ω, Eq. (1.177) can be written as

$$\frac{\omega\left|W_{C,\mathrm{tv}}\right|}{P_{\mathrm{rad}}} \geq \frac{1}{2}\left|1-\frac{\left|\sum_{n=1}^{N} V_n I_n\right|}{P_{\mathrm{rad}}}\right| \tag{1.179}$$

Thal then notes that the instantaneous energy in the coupling network can never be less than zero, thus the time-varying component of the energy in the coupling network $W_{C,\mathrm{tv}}$ can never exceed the average energy in the coupling network $W_{C,\mathrm{avg}}$. As a result, a lower bound on the coupling network energy can be found as

$$\frac{\omega_0\left|W_{C,\mathrm{avg}}\right|}{P_{\mathrm{rad}}} \geq \frac{1}{2}\left|1-\frac{\left|\sum_{n=1}^{N} V_n I_n\right|}{P_{\mathrm{rad}}}\right| \tag{1.180}$$

where ω has been replaced with ω_0 to stress that the antenna is resonant. Thal states that using the minimum energy bound in Eq. (1.180), a stricter Q limit can be enforced as

$$Q = Q_{\mathrm{exterior}} + \frac{\omega_0\left|W_{C,\mathrm{avg}}\right|}{P_{\mathrm{rad}}} \tag{1.181}$$

where Q_{exterior} is the quality factor of the antenna, assuming zero energy inside its *Chu sphere* (thus a *Chu antenna*). Q_{exterior} can be found simply by applying Eq. (1.3) to the exterior region mode circuits of Fig. 1.55 (omitting the internal region), in conjunction with knowledge of the modal excitation levels.

1.3.12.5 Applications of the Energy Lower Bound

Using the minimum energy bound Eq. (1.180) and the sharpened Q limit Eq. (1.181), Thal examines the often cited [4,6,21] result that it is theoretically possible for a small antenna to achieve both a gain of 3 and Q given in Eq. (1.104) using first order TE and TM modes. Thal considers the case where the TE_{1m} and TM_{1m} modes radiate equal power, with the phase of the TE pattern advanced by a phase angle $\Psi/2$ and the phase of the TM pattern delayed by $-\Psi/2$. This corresponds to the currents I_R in the unit terminating resistances of the exterior circuits in Fig. 1.55 having the form of

$$I_{R,\mathrm{TE}1m} = I_0 e^{j\Psi/2} \qquad I_{R,\mathrm{TM}1m} = I_0 e^{-j\Psi/2} \tag{1.182}$$

The power radiated by each mode is found as the power dissipated in the terminating resistances. With input currents and voltages at the TM_{1m} and TE_{1m} mode ports written in terms of the terminating resistance currents in Eq. (1.182), the coupling network energy bound Eq. (1.180) becomes [29]

$$\frac{\omega_0 \left| W_{C,\text{avg}} \right|}{P_{\text{rad}}} \geq \frac{1}{2} \left| 1 - \cos\Psi \sqrt{1 + 1/(ka)^6} \right| \tag{1.183}$$

For the case where the TM_{1m} and TE_{1m} modes correspond to the fields radiated by a z-polarized electric Hertzian dipole (TM) and a y-polarized magnetic Hertzian dipole (TE), respectively, the gain can be written as [29]

$$G = 3\cos^2(\Psi/2) \tag{1.184}$$

To achieve minimum possible Q, the coupling mode energy bound in Eq. (1.183) must go to zero. Under such conditions

$$\cos\Psi = (1 + 1/(\text{ka})^6)^{-1/2} \tag{1.185}$$

and from Eq. (1.184), the corresponding gain is

$$G_{\text{Qmin}} = 1.5 \left(1 + \frac{(ka)^3}{\sqrt{1 + (ka)^6}} \right) \tag{1.186}$$

with the corresponding Q given in Eq. (1.104) [29]. For a gain of 3, $\Psi = 0$ and Eq. (1.183) becomes

$$\left. \frac{\omega_0 \left| W_{C,\text{avg}} \right|}{P_{\text{rad}}} \right|_{\Psi=0} \geq \frac{1}{2} \left[\frac{\sqrt{1 + (ka)^6}}{(ka)^3} - 1 \right] \tag{1.187}$$

Using Eqs. (1.181) and (1.187), the corresponding minimum possible Q in the small antenna region is

$$Q_{\text{max(G)}} \approx \frac{1}{ka} + \frac{1}{(ka)^3} - \frac{1}{2} \tag{1.188}$$

Equation (1.188) is nearly identical to the single mode minimum Q in Eq. (1.103). Thal concludes that the statements made in [4,6,21] proclaiming it is theoretically possible for a small antenna to achieve a gain of 3 and a Q of Eq. (1.104) are erroneous [29].

1.3.13 Work of Gustafsson (2007)

So far, spherical wave functions were used to represent the radiation outside the *Chu sphere*. In contrast, recent publications [30,53–54] considered a different approach to analyzing small antennas. As already noted by Thiele [7], Thal [28], and Foltz and McLean [20], Chu's approach may not be suitable for practical antennas. Gustafsson et al. [30] proceeded to instead use the scattering properties of small particles (i.e., their polarizability dyads, $\bar{\bar{\gamma}}_e$ and $\bar{\bar{\gamma}}_m$) to extract the minimum Q, gain, and bandwidth of small antennas. His approach is evocative of Green's [54] work that related antenna gain to its radar cross-section.

To describe the approach in [30], it is important that we first introduce the small particle or low frequency scattering parameters found in many books and papers [57–60]. With the aid of Fig. 1.61, they are (see [55] and [56]):

- $\bar{\bar{\chi}}_e$: electric susceptibility dyad
- $\bar{\bar{\chi}}_m$: magnetic susceptibility dyad
- $\bar{\bar{\gamma}}_e$: electric polarizability dyad
- $\bar{\bar{\gamma}}_m$: magnetic polarizability dyad
- $\bar{\bar{\gamma}}_\infty$: high contrast polarizability; limiting value of $\bar{\bar{\gamma}}_e$ and $\bar{\bar{\gamma}}_m$ when the small scatterer is perfectly (electric or magnetic) conducting. It follows the $\bar{\bar{\gamma}}_\infty$ is symmetric 3×3 dyadic. It is therefore diagonalizable with its eigenvalues being $\gamma_{1,2,3}$ and $\gamma_1 > \gamma_2 > \gamma_3$.

1.3.13.1 Limitations on Small Antenna Gain-Bandwidth Product and *D*/*Q* Ratio

Gustafsson et al. considered the limitations on the gain-bandwidth product and D/Q ratio for a single resonance small antenna. A key

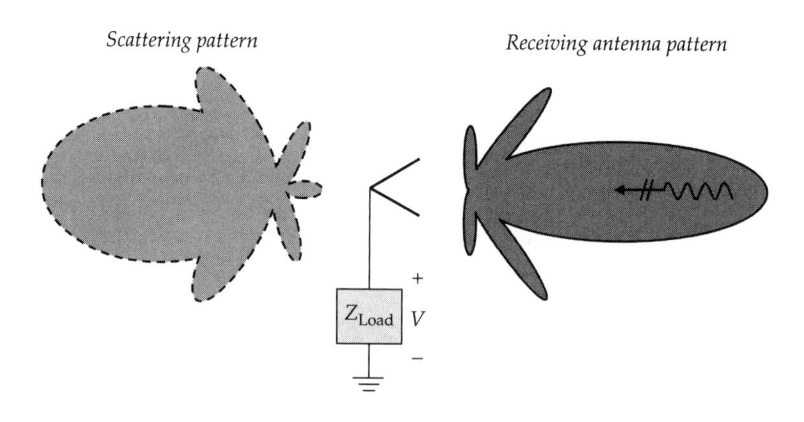

FIGURE 1.61 Antenna receiving and scattering patterns.

starting expression for the analysis is given in [30] as

$$\int_0^\infty \sigma_{\text{ext}}(\lambda)d\lambda = \pi^2 \left(\hat{p}_e^* \bullet \bar{\bar{\gamma}}_e \bullet \hat{p}_e + \hat{p}_m^* \bullet \bar{\bar{\gamma}}_m \bullet \hat{p}_m\right) \tag{1.189}$$

In this, $\hat{p}_e = E_0/|E_0|$ is the polarization of the incident field to the small scatter (small antenna), and $\hat{p}_m = \hat{k} \times \hat{p}_e$ where $\hat{k}$ is the direction of propagation. It is shown in [30] that

$$\begin{aligned}\int_0^\infty \sigma_{\text{ext}}(\lambda)d\lambda &\geq \frac{1}{\tilde{\eta}}\int_{\lambda_1}^{\lambda_2} \sigma_a(\lambda)d\lambda \geq \frac{1}{\tilde{\eta}}\int_{\lambda_1}^{\lambda_2} \left(1-|\Gamma|^2\right)\lambda^2 G(\lambda)d\lambda \\ &\geq \frac{1}{\tilde{\eta}}\min\left\{\left(1-|\Gamma|^2\right)G\right\}\cdot\left(\int_{\lambda_1}^{\lambda_2}\lambda^2 d\lambda\right) \\ &\geq \frac{1}{\tilde{\eta}}\lambda_0^3 \min\left\{\left(1-|\Gamma|^2\right)G\right\}\cdot B\end{aligned} \tag{1.190}$$

where $\lambda_1 = 3.10^8 f_1$ = lower operational frequency wavelength
$\lambda_2 = 3.10^8 f_2$ = upper operational frequency wavelength
$\lambda_0 = 2\pi/k_0 = (\lambda_1 + \lambda_2)/2$ = average wavelength

$$\begin{aligned}B &= 2\left(\frac{\lambda_2+\lambda_1}{\lambda_2-\lambda_1}\right) = 2\left(\frac{k_1-k_2}{k_2+k_1}\right) = 2\left(\frac{f_2-f_1}{f_2+f_1}\right) \\ &= \text{fractional bandwidth}\end{aligned} \tag{1.191}$$

G = antenna gain
Γ = reflection coefficient at the feed port
$\tilde{\eta}$ = maximum value of the absorption efficiency over the wavelength interval $[\lambda_1, \lambda_2]$

It is also implied that $\min\left\{\left(1-|\Gamma|^2\right)G\right\}$ is taken over the wavelength interval $[\lambda_1, \lambda_2]$. We note that in deriving Eq. (1.190) we used the approximation $\int_{\lambda_1}^{\lambda_2}\lambda^2 d\lambda = \lambda_0^3(1 + B^2/12) \approx \lambda_0^3$. It follows from Eqs. (1.189) and (1.190) that

$$\min\left\{\left(1-|\Gamma|^2\right)G\right\}\cdot B \leq \tilde{\eta}\frac{\pi^2}{\lambda_0^3}\left(\hat{p}_e^* \bullet \bar{\bar{\gamma}}_e \bullet \hat{p}_e + \hat{p}_m^* \bullet \bar{\bar{\gamma}}_m \bullet \hat{p}_m\right) \tag{1.192}$$

and for a perfectly (electric or magnetic) conducting antenna structure, we have

$$\min\left\{\left(1-|\Gamma|^2\right)G\right\}\cdot B \leq \tilde{\eta}\frac{\pi^2}{\lambda_0^3}\left(\hat{p}_e^* \bullet \bar{\bar{\gamma}}_\infty \bullet \hat{p}_e + \hat{p}_m^* \bullet \bar{\bar{\gamma}}_\infty \bullet \hat{p}_m\right) \tag{1.193}$$

with, of course, $\hat{p}_e \bullet \hat{p}_m = 0$. From the relations of $\bar{\bar{\gamma}}_{e,m,\infty}$, we can conclude that

$$\min\left\{\left(1 - |\Gamma|^2\right) G\right\} \cdot B \leq \tilde{\eta}\frac{4\pi^3}{\lambda_0^3}(\gamma_1 + \gamma_2) \tag{1.194}$$

where $\gamma_{1,2}$ are the largest two eigenvalues of $\bar{\bar{\gamma}}_\infty$.

Gustafsson et al. proceeded to find a lower limit for the Q of a single resonance, lossless small antenna matched at the resonant frequency. From a single resonance model for the absorption cross-section, Gustafsson et al. argue that the antenna Q factor can be extracted [30]. Using the absorption cross-section model and Eq. (1.189), he finds that the directivity and Q for a small antenna are related through [30]

$$\frac{D}{Q} \leq \tilde{\eta}\frac{k_0^3}{2\pi}(\gamma_1 + \gamma_2) \tag{1.195}$$

We note the similarity of Eqs. (1.195) to (1.194), and we also remark the expected behavior that Q is inversely proportional to the antenna volume (since $\gamma_{1,2}$ are proportional to the antenna volume).

1.3.13.2 Applications of the Gustafsson Limits

Considering now an antenna being perfectly electric conducting (PEC), it follows that $\bar{\bar{\gamma}}_m$ and thus $\gamma_2 = 0$ in Eqs. (1.194) and (1.195) [30].

Also, $\gamma_1 = 4\pi a^3 \gamma_1^{\text{norm}}$ [30], where a is the radius of the *Chu sphere* and γ_1^{norm} is given in Fig. 1.62 for several antenna geometries (scatterers). Further, since the directivity of single mode radiating small antennas (like the Hertzian dipole) is $D = 1.5$, it follows from Eq. (1.195) that

$$Q_{\min} = \frac{1.5}{(ka)^3 \gamma_1^{\text{norm}}} \tag{1.196}$$

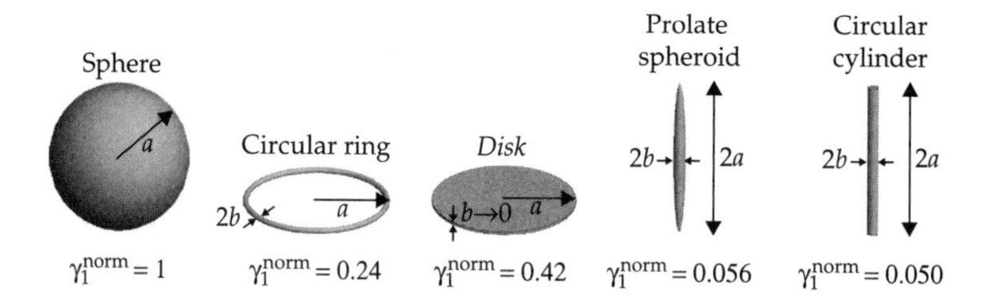

FIGURE 1.62 Normalized γ_1^{norm} eigenvalues for several small antenna geometries, with $b/a = 10^{-3}$ for the circular ring, prolate spheroid, and circular cylinder. (*See Gustafsson* [30].)

with γ_1^{norm} given in Fig. 1.62. Gustafsson also provides closed-form $\gamma_{1,2}$ expressions for prolate and oblate spheroids in [30] and shows that the oblate spheroid geometry has lower Q_{min}.

For a spherical volume (*Chu sphere*), with $D = 1.5$ and $\tilde{\eta} = 0.5$, since $\gamma_1^{norm} = 1$, Eq. (1.196) gives

$$Q_{\text{Spherical, Gustafsson}} = \frac{1.5}{(ka)^3} \tag{1.197}$$

This is identical to Thal's antenna supporting a single TM_{10} mode. We also note that Gustafsson's (1.197) expression is identical to the Q values obtained by Best [24] for a N-turn four-arm spherical helix.

Another comparison to Gustafsson's analytic Q_{min} values are shown for a prolated spheroid (see Fig. 1.33). For numerical computations, the prolate spheroid (of height h and width $2a/3$) was modeled as a four-arm elongated helix antenna (see Fig. 1.63). Using the TM_{10} – TE_{20} tuning technique describe in Thal [28] (varying the number of turns N in the four-arm elongated helix) the resonance was adjusted to generate the data in Fig. 1.64. The calculated values using the radiation data from the NEC code plotted in Fig. 1.64 and compared to the analytic Q in [30] with $\gamma_2 = 0$, $\tilde{\eta} = 0.5$, and $D = 1.5$. Indeed, though D/Q was derived assuming $|k| \to 0$, the Gustafsson limit and measured data of Fig. 1.64 are in close agreement even for $ka = 0.65$. From Fig. 1.64, we can conclude that unlike the limits derived using the fictitious *Chu antenna* with assumed mode excitation, Eqs. (1.194) and (1.195) allow for stricter upper bounds if the antenna is composed of purely electric material. Knowledge of the antenna's absorption characteristics are, however, necessary to use the Gustafsson limits.

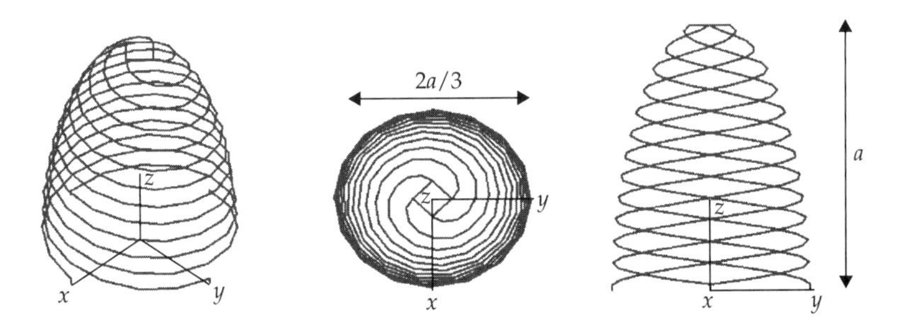

FIGURE 1.63 Three-turn four-arm elongated helix antenna with height/width = 2/3.

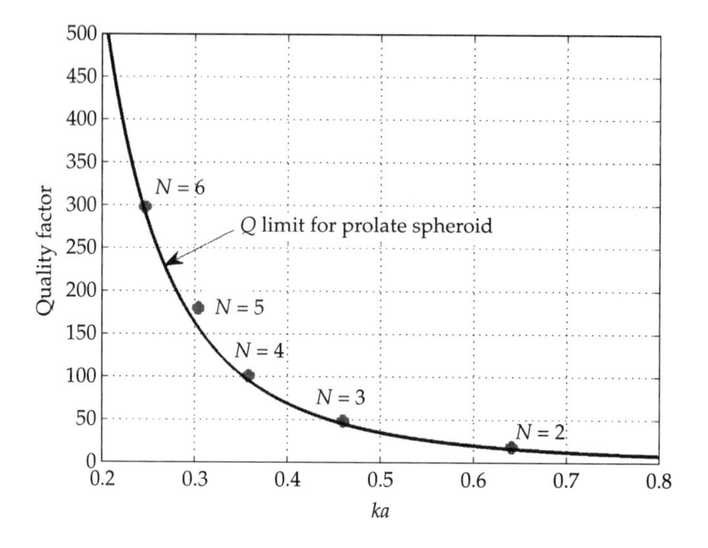

FIGURE 1.64 Gustafsson Q limit for prolate spheroid with width to height ratio 2/3 along with 4NEC2 measured N-turn four-arm elongated helix Q. *(See Fig. 1.63.)*

Comments

Though the methods used by different authors to evaluate small antenna properties have become more sophisticated, the basic principles relating antenna size and Q presented by Wheeler and Chu remain valid. The circuit approximations for the supported mode are likely most valuable in understanding small antenna parameters and their properties. Recent work has shown that the relationship between gain and Q remains a controversial topic. As seen by Chu [11], and recently by Thal [29], Gain and Q cannot be considered independent quantities in practice.

The work of Yaghjian and Best [10] provided a thorough evaluation of Q computation methods. While Q can be computed exactly through Maxwell's equations, in practical applications it is necessary to determine Q based on the antenna input impedance. Q can be easily computed from the antenna input impedance $Z_A(\omega) = R_A(\omega) + jX_A(\omega)$ as [10]

$$Q(\omega) \approx \frac{\omega}{2R_A(\omega)} \sqrt{\left(R'_A(\omega)\right)^2 + \left(X'_A(\omega) + \frac{|X_A(\omega)|}{\omega}\right)^2} \tag{1.198}$$

Equation (1.198) remains valid for all frequency ranges and antenna sizes [10], and as a result is the most robust method for evaluating Q based on input impedance.

The minimum possible Q for a small antenna circumscribed by a *Chu sphere* (see Fig. 1.1) is a topic of significant practical relevance, as it provides a valuable reference as to how well a small antenna's bandwidth compares to an ideal case. We summarize several minimum Q formulas in Table 1.13. Notes on each formula are also provided, as it is critical to understand the underlying assumptions in each equation in order to make an appropriate comparison with the small antenna.

Recent works by Gustafsson, Best, and Thal did highlight the uncertainties in characterizing small antennas. So questions such as

- What antenna current gives the lowest Q?
- How can small antenna efficiency be increased?
- Can shape play a more important role in small antenna theory and parameter control?

although not fully addressed, are now much better understood.

Minimum Q	Reference	Notes
$\frac{1}{ka}+\frac{1}{(ka)^3}$	McLean [37]	Assumes TM or TE mode radiation only
$\frac{1}{2}\left(\frac{2}{ka}+\frac{1}{(ka)^3}\right)$	McLean [37]	Assumes TM and TE mode radiation
$\frac{1.5}{(ka)^3}$	Thal [28]	Assumes a surface current distribution over the *Chu sphere* surface radiating a TM mode
$\frac{3}{(ka)^3}$	Thal [28]	Assumes a surface current distribution over the *Chu sphere* surface radiating a TE mode
$\frac{1}{(ka)^3}$	Thal [28]	Assumes a surface current distribution over the *Chu sphere* surface radiating TM and TE modes
$\frac{G}{\tilde{\eta}}\frac{1}{2(ka)^3}$	Gustafsson et al. [30]	G = antenna gain $\tilde{\eta}$ = antenna absorption efficiency Assumes antenna composed of PEC material

TABLE 1.13 Summary of Small Antenna Q Limits for *Chu Sphere* of radius = a

References

1. H. A. Wheeler, "Fundamental limitations of small antennas," *Proceedings of the IRE*, vol. 35, December 1947, pp. 1479–1484.
2. R. C. Hansen, "Fundamental limitations in antennas," *Proceedings of the IEEE*, vol. 69, no. 2, February 1981, pp. 170–182.
3. R. F. Harrington, "Effect of antenna size on gain, bandwidth, and efficiency," *Journal of Research of the National Bureau of Standards*, vol. 64D, January–February 1960, pp. 1–12.
4. D. H. Kwon, "On the radiation *Q* and the gain of crossed electric and magnetic dipole moments," *IEEE Transactions on Antennas and Propagation*, vol. AP-53, May 2005, pp. 1681–1687.
5. D. H. Kwon, "Radiation *Q* and gain of TM and TE sources in phase-delayed rotated configurations," *IEEE Transactions on Antennas and Propagation*, vol. AP-56, August 2008, pp. 2783–2786.
6. D. M. Pozar, "New results for minimum *Q*, maximum gain, and polarization properties of electrically small arbitrary antennas," EuCAP 2009, Berlin, Germany, March 2009, pp. 23–27.
7. G. A. Thiele, P. L. Detweiler, and R. P. Penno, "On the lower bound of the radiation Q for electrically small antennas," *IEEE Transactions on Antennas and Propagation*, vol. AP-51, June 2003, pp. 1263–1269.
8. H. A. Wheeler, "Small antennas," *IEEE Transactions on Antennas and Propagation*, vol. 23, July 1975, pp. 462–469.
9. Y. Huang, R. M. Narayanan, G. R. Kadambi, "On Wheeler's method for efficiency measurement of small antennas," *IEEE International Symposium on Antennas and Propagation*, 2001, pp. 346–349.
10. A. D. Yaghjian and S. R. Best, "Impedance, bandwidth, and Q of antennas," *IEEE Transactions on Antennas and Propagation*, vol. AP-53, April 2005, pp. 1298–1324.
11. L. J. Chu, "Physical limitations on omni-directional antennas," *Journal of Applied Physics*, vol. 19, December 1948, pp. 1163–1175.
12. R. W. P. King, *The Theory of Linear Antennas*, Harvard University Press, Cambridge, Mass., 1956.
13. S. R. Best, "The performance properties of electrically small resonant multiple-arm folded wire antennas," *IEEE Antennas and Propagation Magazine*, vol. 47, no. 4, August 2005, pp. 13–27.
14. D. M. Pozar, *Microwave Engineering*, 3d ed., John Wiley & Sons, Hoboken, N.J., 2005.
15. R. E. Collin, *Foundations for Microwave Engineering*, 2d ed. McGraw-Hill, New York, 1992.
16. P. A. Rizzi, *Microwave Engineering: Passive Circuits*, Prentice Hall, Englewood Cliffs, N.J., 1987.
17. R. M. Fano, "Theoretical limitations on the broadband matching of arbitrary impedances," *Journal of the Franklin Institute*, vol. 249, Jan. 1950.
18. R. E. Collin and S. Rothschild, "Evaluation of antenna Q," *IEEE Transactions on Antennas and Propagation*, vol. AP-12, January 1964, pp. 23–27.
19. R. L. Fante, "Quality factor of general ideal antennas," *IEEE Transactions on Antennas and Propagation*, vol. AP-17, March 1969, pp. 151–155.
20. H. D. Foltz and J. S. McLean, "Limits on the radiation Q of electrically small antennas restricted to oblong bounding regions," *Proceedings of the IEEE AP-S International Symposium*, vol. 4, July 11–16, 1999, pp. 2702–2705.
21. Y. Geyi, "Physical limitation of antenna," *IEEE Transactions on Antennas and Propagation*, vol. AP-51, August 2003, pp. 2116–2123.
22. Y. Geyi, "A method for the evaluation of small antenna Q," *IEEE Transactions on Antennas and Propagation*, vol. AP-51, August 2003, pp. 2124–2129.
23. S. R. Best, "Bandwidth and the lower bound on Q for small wideband antennas," *IEEE International Symposium on Antennas and Propagation*, 2006, pp. 647–650.

24. S. R. Best, "The radiation properties of electrically small folded spherical helix antennas," *IEEE Transactions on Antennas and Propagation*, vol. AP-52, no. 4, April 2004, pp. 953–960.
25. S. R. Best, "Low Q electrically small linear and elliptical polarized spherical dipole antennas," *IEEE Transactions on Antennas and Propagation*, vol. AP-53, March 2005, pp. 1047–1053.
26. S. R. Best and J. D. Morrow, "On the significance of current vector alignment in establishing the resonant frequency of small space-filling wire antennas," *IEEE Antennas and Wireless Propagation Letters*, vol. 2, 2003, pp. 201–204.
27. H. A. Wheeler, "The spherical coil as an inductor, shield, or antenna," *Proceedings of the IRE*, vol. 46, September 1958, pp. 1595–1602.
28. H. L. Thal, "New radiation *Q* limits for spherical wire antennas," *IEEE Transactions on Antennas and Propagation*, vol. AP-54, October 2006, pp. 2757–2763.
29. H. L. Thal, "Gain and *Q* bounds for coupled TM-TE modes," *IEEE Transactions on Antennas and Propagation*, vol. AP-57, no. 7, July 2009, pp. 1879–1885.
30. M. Gustafsson, C. Sohl, and G. Kristensson, "Physical limitations on antennas of arbitrary shape," *Proceedings of the Royal Society A: Mathematical, Physical and Engineering Sciences*, vol. 463, issue 2086, 2007, pp. 2589–2607.
31. A. R. Lopez, "Fundamental limitations of small antennas: validation of Wheeler's formulas," *IEEE Antennas and Propagation Magazine*, vol. 48, no. 4, August 2006, pp. 28–36.
32. H. A. Wheeler, "The radiansphere around a small antenna," *Proceedings of the IRE*, vol. 47, August 1959, pp. 1325–1331.
33. T. Simpson, J. Cahill, "The electrically small elliptical loop with an oblate spheroidal core," *IEEE Antennas and Propagation Magazine*, vol. 49, no. 5, October 2007, pp. 83–92.
34. T. Simpson, Y. Zhu, "The electrically small multi-turn loop antenna with a spheroidal core," *IEEE Antennas and Propagation Magazine*, vol. 48, no. 5, October 2006, pp. 54–66.
35. C. A. Balanis, *Advanced Engineering Electromagnetics*, John Wiley & Sons, New York, 1989.
36. R. F. Harrington, *Time-Harmonic Electromagnetic Fields*, McGraw-Hill, New York, 1961.
37. J. S. McLean, "A re-examination of the fundamental limits on the radiation Q of electrically small antennas," *IEEE Transactions on Antennas and Propagation*, vol. AP-44, May 1996, pp. 672–675.
38. R. C. Hansen, *Electrically Small, Superdirective, and Superconducting Antennas*, Wiley, Hoboken, N.J., 2006.
39. S. R. Best, "A discussion on power factor, quality factor, and efficiency of small antennas," *IEEE International Symposium on Antennas and Propagation*, June 2007, pp. 2269–2272.
40. J. A. Stratton et al., *Spheroidal Wave Functions*, Technology Press of M.I.T. and John Wiley & Sons, New York, 1956.
41. C. Flammer, *Spheroidal Wave Functions*, Stanford Univ. Press, Stanford, Calif., 1957.
42. R. C. Hansen, *Microwave Scanning Antennas*, Peninsula, Los Altos, Calif., 1985.
43. C. A. Walter, *Traveling Wave Antennas*, 2d ed. Peninsula, Los Altos, Calif., 1990.
44. W. L. Stutzman and G. A. Thiele, *Antenna Theory & Design*, 2d ed. Wiley, New York, 1998.
45. D. R. Rhodes, *Synthesis of Planar Antenna Sources*. Clarendon, Oxford, U.K., 1974.
46. D. R. Rhodes, "On the stored energy of planar aperatures," *IEEE Transactions on Antennas and Propagation*, vol. AP-14, November 1966, pp. 676–683.
47. Y. Geyi, P. Jarmuszewski, and Y. Qi, "The Foster reactance theorem for antennas and radiation Q," *IEEE Transactions on Antennas and Propagation*, vol. AP-48, March 2000, pp. 401–408.
48. S. R. Best, "The Foster reactance theorem and quality factor for antennas," *IEEE Antennas and Wireless Propagation Letters*, vol. 3, 2004, pp. 306–309.

49. J. A. Andersen and S. Berntsen, "Comments on 'the Foster reactance theorem for antennas and radiation Q," *IEEE Transactions on Antennas and Propagation*, vol. AP-55, March 2007, pp. 1013–1014.
50. HFSS. ver. 11, Ansoft Corporation. Pittsbugh, PA, 2008.
51. J. Chalas and K. Sertel, The Ohio State University ElectroScience Lab, Personal Communication.
52. H. L. Thal, "Exact circuit analysis of spherical waves," *IEEE Transactions on Antennas and Propagation*, vol. AP-26, March 1978, pp. 282–287.
53. R. C. Hansen and R. E. Collin, "A new Chu formula for Q," *IEEE Antennas and Propagation Magazine*, vol. 51, no. 5, October 2009, pp. 38–41.
54. R. B. Green, "The general theory of antenna scattering," Ph.D thesis, The Ohio State University, November 1963.
55. C. Sohl, M. Gustafsson, and G. Kristensson, "Physical limitations on broadband scattering by heterogeneous obstacles," *Journal of Physics A: Mathematical & Theoretical*, September 2007.
56. C. Sohl, M. Gustafsson, and G. Kristensson, "Physical limitations on metamaterials: restrictions on scattering and absorption over a frequency interval," Technical Report LUTEDX/(TEAT-7154)/1-10/(2007), Lund University, Department of Electrical and Information Technology, Lund, Sweden.
57. J. B. Andersen and A. Frandsen, "Absorption efficiency of receiving antennas," *IEEE Transactions on Antennas and Propagation*, vol. AP-53, September 2005, pp. 2843–2849.
58. R. G. Newton, *Scattering Theory of Waves and Particles*, 2d ed., Springer-Verlag, New York, 1982.
59. J. R. Taylor, *Scattering theory: The Quantum Theory of Nonrelativistic Collisions*, Robert E. Krieger Publishing Company, Malabar, Fa., 1983.
60. R. Kleinman and T. Senior, "Low frequency scattering by space objects," *IEEE Transactions on Aerospace and Electronic Systems*, vol. AES-11, pp. 672–675.

CHAPTER 2

Fundamental Limits and Design Guidelines for Miniaturizing Ultra-Wideband Antennas

B. A. Kramer and John L. Volakis

2.1 Introduction

Chapter 1 presented the fundamental limits on Q, directivity, and limits on efficiency for electrically small antennas. In this chapter, we present the Fano-Bode theory [1–3] which is used to determine the maximum impedance bandwidth that can be obtained using a passive lossless matching network. By applying the Fano-Bode theory to a minimum Q antenna (i.e., using the equivalent circuit of the lowest order TE or TM mode as the load impedance), we can demonstrate how the electrical size impacts impedance matching for narrowband (band-pass response) and UWB antennas (high-pass response). In doing so, we will derive a set of antenna limits to characterize UWB antenna performance.

For the remainder of this chapter, we discuss how to approach these UWB limits by means of antenna miniaturization. Specifically, we discuss the concept of antenna miniaturization and how it can be used to enhance the performance of electrically small antennas. We remark that this topic has been previously addressed in [4]. Specifically, [4] discusses several practical techniques to reduce size and improve impedance matching for specific antennas. In contrast, this chapter focuses on the theory that defines the limits of miniaturization for wideband antennas. We also discuss trade-offs associated with broadband antenna miniaturization and provide basic guidelines for size reduction using a spiral antenna as an illustrative example. Details for implementing the various miniaturization techniques (e.g., dielectric loading or inductive loading) are discussed in Chaps. 3 and 4. Here, we focus only on miniaturization limits for UWB antenna without reference to the way such miniaturizations will be implemented.

2.2 Overview of Fano-Bode Theory

The classic problem of designing a passive reactive network to maximize impedance bandwidth between an arbitrary load and a resistive generator was initiated by Bode [1]. Bode considered a two-element RC or RL load and determined the maximum possible bandwidth for a given maximum tolerable reflection coefficient Γ_0 within the passband as illustrated in Fig. 2.1. It is important to note that Bode only considered a matching network with an infinite number of stages. Later, Fano generalized Bode's work by extending it to include arbitrary loads and an arbitrary complex matching network [2,3]. In both Bode's and Fano's work, the system response is arbitrary. However, in the literature (especially in antenna theory) their work is presented

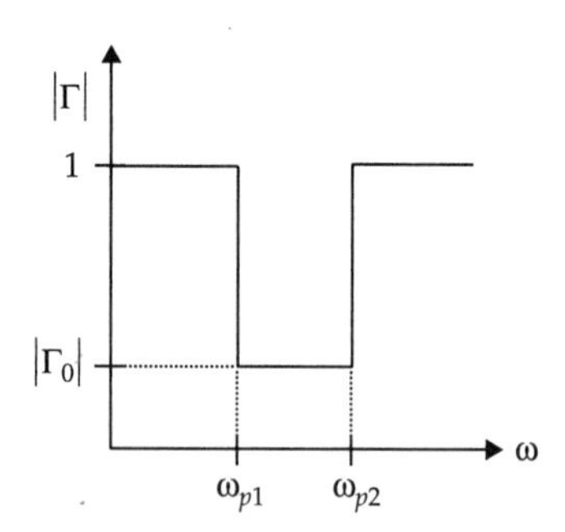

FIGURE 2.1 A possible band-pass response for the reflection coefficient that illustrates Fano-Bode criterion.

for the band-pass response [5–7], which is suitable for narrowband antennas. Here, we discuss impedance matching limitations for both the narrowband (band-pass) and broadband (high-pass) cases. We begin by discussing Fano's method for determining impedance matching limitations for an arbitrary load. Due to the complexity of Fano's method, we will restrict our discussion to a high level overview of his approach.

Consider the problem of designing an optimum lossless passive network to match an arbitrary load impedance to a resistive generator as illustrated in Fig. 2.2. Fano solved this problem in general for any load impedance that could be represented using a finite number of passive circuit elements. The basic idea of Fano's approach was to tolerate a certain amount of mismatch between the load and generator such that the bandwidth (or matching area [8]) is maximized. The first step in Fano's method was to simplify the problem by replacing the load impedance with its Darlington equivalent circuit as shown in Fig. 2.3. He then normalized the impedance making the resistances equal to unity. The resulting network could then be described by two reactive networks connected in cascade to form a single two-port network N terminated at each port with a 1-Ω resistor. Since the reflection coefficient Γ of a lumped element reactive network must be a ratio of two real polynomials in the complex frequency variable $s = \sigma + j\omega$ [2], Γ must have the following form

$$\Gamma = K\frac{(s - s_{01})(s - s_{02})\cdots(s - s_{0n})}{(s - s_{p1})(s - s_{p2})\cdots(s - s_{pn})} \tag{2.1}$$

where K is a real number, s_{0m} are the zeros and s_{pm} are the poles. Knowing the form of Γ, Fano proceeded to determine the zeros and poles which would maximize the bandwidth subject to the tolerance of match Γ_0. However, the selection of the zeros and poles had to satisfy various constraints (see [2] for details). For instance, Γ has to satisfy constraints based on the behavior of the load impedance. Also, the

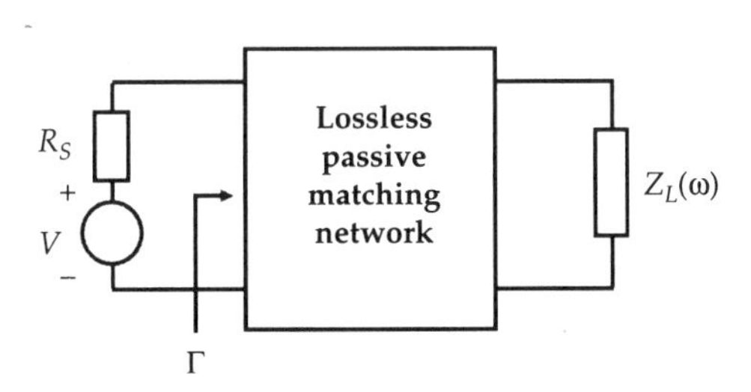

FIGURE 2.2 Matching network for an arbitrary load impedance [3].

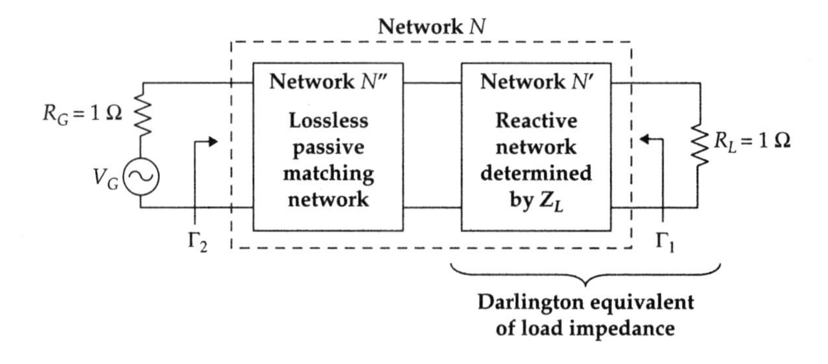

FIGURE 2.3 Matching network and Darlington equivalent of the load impedance in cascade [3].

network N and therefore Γ must be physically realizable. Essentially, these constraints place limits on how well the load impedance can be matched. To satisfy these constraints, Fano developed a series of integral equations involving the logarithm of the magnitude of the reflection coefficient. The form of these equations depends upon the type of response (i.e., band-pass, high-pass, low-pass, etc.) for the overall system and the response of the load.

For an overall high-pass system response, the equations given by Fano involve an integral of the form

$$\int_0^\infty \omega^{-2(k+1)} \ln\left(\frac{1}{|\Gamma_1|}\right) d\omega \tag{2.2}$$

for $k = 0, 1, \ldots, N-1$ with N representing the multiplicity of the transmission zeros associated with the load.* Each integral is evaluated over the frequency spectrum of a return loss function, $\ln(1/|\Gamma_1|)$, which is used instead of $|\Gamma_1|$ for purely mathematical reasons. By using the calculus of residues, Fano evaluated Eq. (2.2) to obtain the following expression for a high-pass response

$$\int_0^\infty \omega^{-2(k+1)} \ln\left(\frac{1}{|\Gamma_1|}\right) d\omega = (-1)^k \frac{\pi}{2}\left[A^0_{2k+1} - \frac{2}{2k+1}\sum_i s_{ri}^{-(2k+1)}\right] \tag{2.3}$$

The resulting expression relates the return loss function (left-hand side) to the zeros and poles of the load and matching network

*The integral equations can be written in terms of either Γ_1 or Γ_2.

(right-hand side). Specifically, the A^0_{2k+1} coefficients come from the Taylor series expansion of the return loss function $\ln(1/|\Gamma|)$ and depend only on the load [2]. They are given by

$$A^0_{2k+1} = \frac{1}{2k+1}\left(\sum_i s_{oi}^{-(2k+1)} - \sum_i s_{pi}^{-(2k+1)}\right) \tag{2.4}$$

where s_{oi} and s_{pi} are the zeros and poles of the load impedance respectively. The last term on the right-hand side of Eq. (2.3) are the unknown zeros s_{ri} of the matching network that lie in the right half plane.

The optimum tolerance of match is found by determining the function $|\Gamma_1|$ which simultaneously maximizes the integral relations. In other words, $|\Gamma_1|$ must maximize the "matching area" given by the integral relations. For a given load, this is accomplished by proper selection of the zeros s_{ri} of the matching network. Ideally, the matching area is maximized when $|\Gamma_1|$ has a rectangular-shaped response (see Fig. 2.1 or Fig. 2.4). For the rectangular-shaped response, the left-hand side of Eq. (2.3) can be evaluated analytically making a closed form solution possible. Of course, the ideal rectangular-shaped response can only be obtained if the matching network has an infinite number of stages. When the matching network has finite complexity, the optimum $|\Gamma_1|$ is found by approximating the ideal rectangular response with a set of basis functions. In [2], Fano used Tchebysheff polynomials to represent $|\Gamma_1|$ and to obtain a solution for an RL (or RC) load with a finite number of matching stages. Because this solution uses a simple RL load, it is only accurate for antennas with an electrical size ka less than about 0.3. We now briefly review this solution which can be used to determine the maximum possible bandwidth for a narrowband antenna.

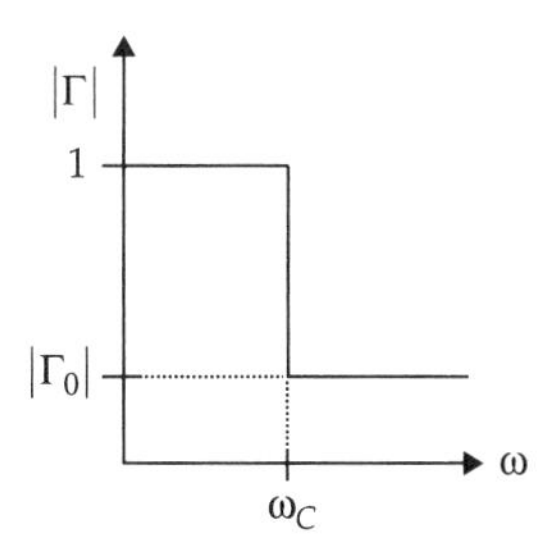

FIGURE 2.4 The ideal high-pass reflection coefficient response.

2.3 Fano-Bode Limit for the Band-Pass Response

To begin, the fractional bandwidth for the band-pass response is defined as

$$B_n = \frac{\omega_{p2} - \omega_{p1}}{\sqrt{\omega_{p2}\omega_{p1}}} \tag{2.5}$$

where ω_{p2} and ω_{p1} are the high and low edge-band frequencies for which $|\Gamma| \leq |\Gamma_0|$ (see Fig. 2.1). Additionally, the center frequency of the band-pass response is defined to be $\sqrt{\omega_{p2}\omega_{p1}}$. Fano's solution for the band-pass response is in terms of a set of equations [2]

$$QB_n = \frac{2\sin\left(\frac{\pi}{2n}\right)}{\sinh(a) - \sinh(b)} \tag{2.6}$$

$$\frac{\tanh(na)}{\cosh(a)} = \frac{\tanh(nb)}{\cosh(b)} \tag{2.7}$$

$$\Gamma_0 = \frac{\cosh(nb)}{\cosh(na)} \tag{2.8}$$

where Q is the quality factor associated with the load, n is the number of matching stages* and the coefficients a and b are unknowns. Matthaei, Young, and Jones were among the first to publish solutions to Fano's equations in the form of tables for the coefficients a and b [9]. However, their coefficients were dependent upon the choice of $|\Gamma_0|$. Lopez [5], using a methodology developed by Wheeler [10], provided a closed form solution in which his coefficients were independent of $|\Gamma_0|$. Lopez's equation is as follows

$$B_n = \frac{1}{Q}\frac{1}{b_n \sinh\left[\frac{1}{a_n}\ln\left(\frac{1}{\Gamma_0}\right)\right] + \frac{1-b_n}{a_n}\ln\left(\frac{1}{\Gamma_0}\right)} \tag{2.9}$$

where the (a_n, b_n) coefficients are given in [5] and depend only on the number of matching stages. From Eq. (2.9), it is evident that for a given n and $|\Gamma_0|$, B_n is inversely proportional to Q. Therefore, for

*The $n = 1$ case corresponds to the antenna or load being connected directly to the generator (no matching network).

a band-pass type antenna, the upper bound on B_n is obtained upon substitution of the minimum Q.

Before we proceed to discuss broadband (high-pass) impedance matching, it is useful to note an important difference between the band-pass and high-pass cases. For a given matching network complexity n and load impedance, there is a fundamental limitation or trade-off between B_n and $|\Gamma_0|$ for the band-pass case. It is important to note that the center frequency is not restricted in any way. That is, regardless of antenna size, any $|\Gamma_0|$ is achievable as long as one is willing to reduce B_n. However, for the high-pass response, we will show that the electrical size will determine the lowest possible $|\Gamma_0|$. That is, for the high-pass response the fundamental trade-off is between $|\Gamma_0|$ and the cutoff frequency ω_c or "cutoff size" $(ka)_c$.

2.4 Fano-Bode Limit for the High-Pass Response

For antennas having a continuous high-pass response, such as a spiral or some other frequency independent antenna, the impedance matching limitations for the band-pass case are not relevant. Here, Fano's work is adapted to the high-pass matching case which is more applicable to broadband antennas. Similar to the band-pass limitations, the impedance matching limitations for the high-pass case are obtained using the equivalent circuits of the lowest order spherical modes (TE_{01} and TM_{01}) as the load impedance. Using Fano's method, a relationship between the cutoff frequency ω_c and maximum tolerable reflection coefficient $|\Gamma_0|$ is obtained for three specific cases. Note that the same approach was used in [11] to determine matching limits on UWB antennas. However, the cases considered in [11] are different than those considered here and involve higher order spherical modes. All of the cases considered here involve only the lowest order modes.

For the first case, we consider the matching limitations obtained by connecting the equivalent circuit of the TE_{01} or TM_{01} mode directly to the generator (no matching network, $n = 1$). In the second case, we consider the matching limitations obtained when both TE_{01} and TM_{01} modes are excited equally (lowest possible antenna Q). As in [12], it is assumed that the same current distribution equally excites both modes resulting in the series combination of their respective equivalent circuits. Using this series combination, the matching limitations are determined by connecting the load directly to the generator. For the third case, we demonstrate the potential improvement that can be obtained by including an infinitely complex matching network ($n = \infty$) in the first case.

We remark that for the first two cases ($n = 1$) it is not necessary to use Fano's integral relations to determine the impedance matching limits, since the form of the reflection coefficient is known and there is no matching network. For these cases, we only need to select the generator resistance R_s to calculate the reflection coefficient looking into the load. For the first case, R_s is set equal to the load resistance of the equivalent circuit which happens to be the impedance of free-space η_0 (see Fig. 2.1). For the second case, the series combination of the equivalent circuits results in a load resistance equal to $2\eta_0$ [12]. Therefore, R_s is set equal to $2\eta_0$ in the second case. For the first two cases, the fundamental relation between ω_c and $|\Gamma_0|$ can now be determined by calculating the reflection coefficient with respect to R_s looking into the load. The results are shown in Fig. 2.5.

To demonstrate the impact of an infinite stage matching network (third case), we must solve the set of equations given by Eq. (2.3) using the equivalent circuit of the TM_{01} or TE_{01} mode as the load. For this load there are two transmission zeros at zero, implying a multiplicity of two ($N = 2$). For $N = 2$, the integral relations obtained from Eq. (2.3)

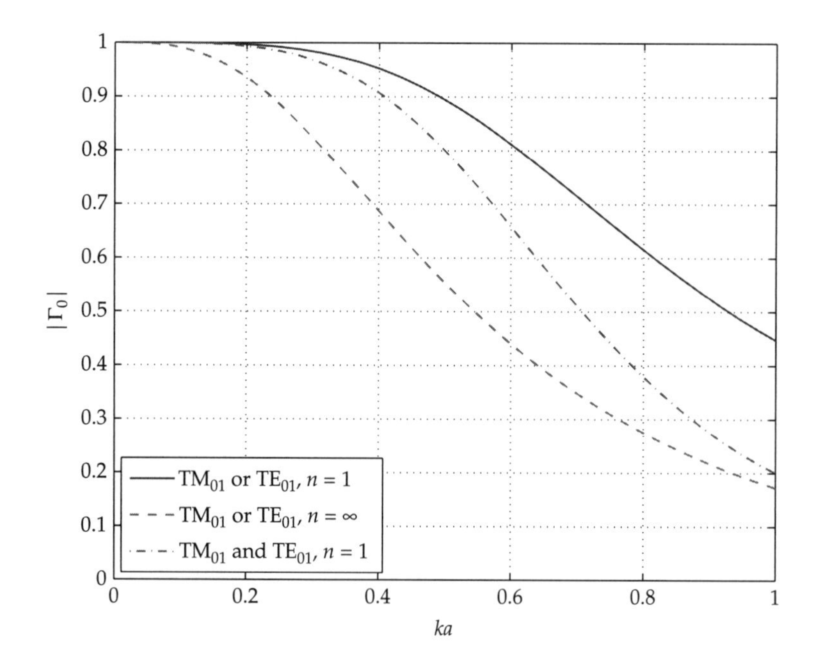

FIGURE 2.5 In-band reflection coefficient Γ_0 vs. the cutoff size ($ka = \frac{\omega_c a}{c}$) or cutoff frequency. The cutoff size is the smallest antenna size for which $|\Gamma| \leq |\Gamma_0|$.

are as follows

$$\int_0^{\infty} \omega^{-2} \ln\left(\frac{1}{|\Gamma_1|}\right) d\omega = \frac{\pi}{2}\left[A_1^0 - 2\sum_i s_{ri}^{-1}\right] \tag{2.10}$$

and

$$\int_0^{\infty} \omega^{-4} \ln\left(\frac{1}{|\Gamma_1|}\right) d\omega = \frac{-\pi}{2}\left[A_3^0 - \frac{2}{3}\sum_i s_{ri}^{-3}\right] \tag{2.11}$$

Since the matching network has infinite complexity, the form of $|\Gamma_1|$ is that of the ideal rectangular high-pass response (see Fig. 2.4). Therefore, the left-hand side of Eqs. (2.10) and (2.11) can be evaluated analytically. For the right-hand side, we must determine the A coefficients which depend upon the load. To begin, the equivalent circuit for the TE_{01} or TM_{01} mode is first normalized to the free space impedance η_0. The impedance, in terms of the complex frequency $s = j\omega$, seen looking into the equivalent circuit is then given by

$$Z_{TM_{01}}(s) = \frac{1}{Z_{TE_{01}}(s)} = \frac{s^2a^2 + sac + c^2}{s^2a^2 + sac} \tag{2.12}$$

The reflection coefficient normalized with respect to η is then given by

$$\Gamma(s) = \frac{Z-1}{Z+1} = \pm\frac{c^2}{s^2a^2 + sac + c^2} \tag{2.13}$$

where the plus sign refers to the TM_{01} mode and the minus sign to the TE_{01} mode. Regardless of the mode, Eq. (2.13) has two poles $s = \frac{-c \pm jc}{2a}$ and no zeros. Using Eq. (2.4), we have $A_1^0 = 2a/c$ and $A_3^0 = -(A_1^0)^3/6$. As Fano noted, to simultaneously maximize the matching area defined by each integral, the zeros of the matching network must be selected so that $\sum s_{ri}^{-3}$ is as large as possible while $\sum s_{ri}^{-1}$ is kept as small as possible. This is accomplished by using a single real zero, $s_r = \sigma_r$ [3]. Solving Eq. (2.10) for σ_r and substituting the result into Eq. (2.11) eliminates σ_r. The resulting equation is the cubic polynomial

$$\omega_c^3 - \omega_c^2\frac{K}{A_1^0} + \omega_c\frac{K^2}{\left(A_1^0\right)^2} - \frac{K^3 + 4K}{3\left(A_1^0\right)^3} = 0 \tag{2.14}$$

where $K = \frac{2}{\pi}\ln\left(\frac{1}{|\Gamma_0|}\right)$. The cubic equation was solved using Matlab to find the roots of ω_c for a given K. However, for any given K, only one of the roots of ω_c is real. The solution is shown graphically in Fig. 2.5 (case of $n = \infty$). We remark that in Fig. 2.5, ω_c was multiplied by a/c to convert it to the more useful parameter ka. The curve in Fig. 2.5 for the $n = \infty$ case should be interpreted as follows. For a given $|\Gamma_0|$, there is a corresponding k_ca defining the smallest electrical size for which the pass-band reflection coefficient can equal $|\Gamma_0|$. To achieve this cutoff size, the frequency response of Γ must be rectangular. That is, below $k_c a$ the magnitude of the reflection coefficient must be one and, for all frequencies above $k_c a$, it is exactly equal to $|\Gamma_0|$. Therefore, the curve in Fig. 2.5 corresponding to $n = \infty$ cannot be realized in its entirety.

The results in Fig. 2.5 can be used in conjunction with the directivity limits to define realized gain limit. For the cases where only first one of the lowest order spherical modes was excited, the directivity is limited to 1.5 ($\approx$1.76 dB) for all ka [13]. For the case where both TM_{01} and TE_{01} modes were excited equally, the maximum directivity that can be achieved for all ka is 3 [14]. Assuming antenna is lossless, the realized gain is calculated using the results from Fig. 2.5 and the corresponding directivity. The resulting realized gain curves are shown in Fig. 2.6.

In closing, some final comments about the impedance matching limits are in order. First, it is important to emphasize that the limits are for a *minimum Q antenna*. Therefore, for an antenna to reach these limits, first the antenna Q must approach the lower bound on Q. Second, the impedance matching limits do not apply if active circuit elements are employed or if there is loss in the antenna structure

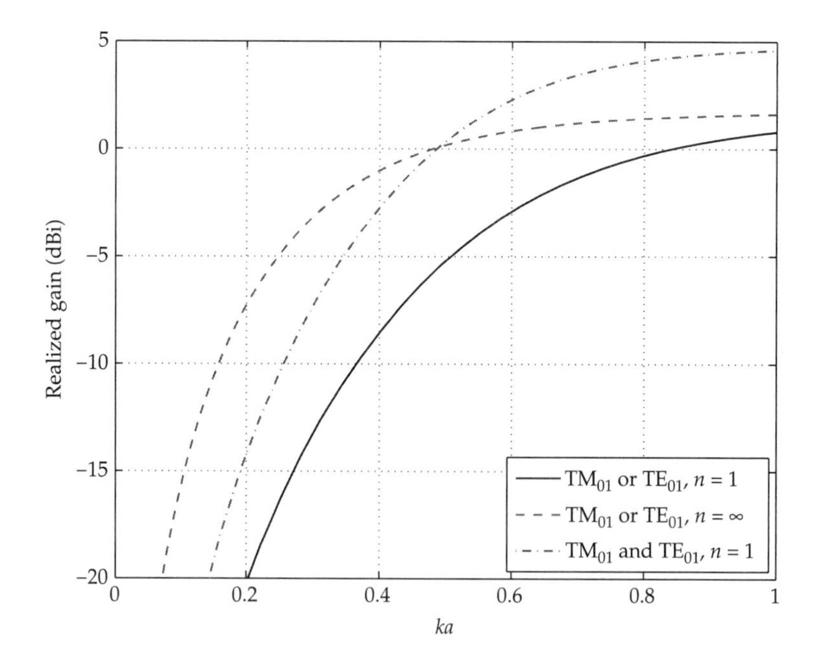

FIGURE 2.6 The lowest achievable cutoff frequency for a minimum Q antenna in terms of the realized gain.

(i.e., resistive loading or absorber). That is, Fano's method is only applicable to lossless passive matching networks. Furthermore, the inclusion of loss within the antenna structure permits the antenna Q to be lower than the minimum Q because Chu only considered the lossless case in his analysis. In fact, one can achieve an arbitrarily low antenna Q using resistive loading at the expense of efficiency.

2.5 Antenna Miniaturization

Having already discussed the fundamental limits on radiation Q, directivity and impedance matching for antennas, the remainder of the chapter focuses on how to approach the impedance matching limit with regards to electrically small UWB antennas. Since the impedance matching limit is based on a minimum Q antenna, it can only be approached using a low Q antenna design. The methods that can be used to approach this limit can be divided into two categories: (1) minimum Q antenna designs; (2) miniaturization of preexisting antenna designs to lower their Q. The first approach is about designing an antenna that utilizes the entire volume of the Chu sphere such that there is no stored energy inside the sphere and only the lowest spherical modes exist outside it (TM_{01} and TE_{01}) [13] (see Chap. 1 Fig. 1.1 for Chu sphere definition). Wheeler proposed two such antenna designs, the spherical inductor and the spherical cap dipole, that could come close to achieving this goal [15, 16] (see Chap. 1 Sec. 3.1). A characteristic of these designs is that they occupy a significant portion of the Chu sphere. Therefore, for applications that require conformal or low-profile antennas, the first approach is not attractive. The second approach involves the miniaturization of a preexisting antenna design to improve its Q at frequencies where it is electrically small. In the remainder of this chapter, we will discuss this approach. We begin by discussing the concept of antenna miniaturization. Subsequently, we demonstrate how the miniaturization of a physically small dipole can improve its Q. Using such concepts, we then develop basic guidelines for miniaturizing UWB antennas.

2.5.1 Concept

The concept of miniaturization involves reducing the phase velocity of the wave guided by the antenna structure to establish resonance or coherent radiation when the antenna is electrically small. To illustrate how this can be accomplished, we use the analogy between an antenna and transmission line. Consider a center-fed infinite biconical antenna, a type of a spherically radial waveguide, guiding a spherical wave [17]. This is analogous to an infinitely long uniform transmission line guiding a plane wave [18, 19].

It is well known that the phase velocity v_p and characteristic impedance Z_0 seen by the guided wave are given by

$$v_p = \frac{1}{\sqrt{LC}} = \frac{1}{\sqrt{\mu\epsilon}} \qquad Z_0 = G\sqrt{\frac{L}{C}} = G\sqrt{\frac{\mu}{\epsilon}} \tag{2.15}$$

where L is the series inductance per unit length, C is the shunt capacitance per unit length and G is a geometrical factor. Therefore, the phase velocity can be controlled using the series inductance and shunt capacitance per unit length of the antenna. For the biconical antenna, this means controlling the self-inductance of the cone and the capacitance between the two cone halves. By doing so, we can achieve the proper electrical delay to attain resonance or form a radiation band for a spiral antenna regardless of physical size. As we will show, this will improve the antenna Q. We remark that in general, the characteristic impedance of an antenna is frequency dependent unless the geometry naturally scales with frequency (constant geometrical factor).

Techniques that can be used to modify the inductance and capacitance of an antenna structure involve either material or reactive loading. Material loading refers to the application of materials which have $\epsilon_r > 1$ and/or $\mu_r > 1$. This is the most generic approach making it applicable to any antenna design. However, the material density can make this approach prohibitive for applications that require lightweight antenna designs. Additionally, material losses can play a significant role in their applicability. For example, the frequency dependent loss of currently available magnetic or magneto-dielectric materials prevents their use above VHF. On the other hand, reactive loading refers to any method which enhances the self-inductance and/or shunt-capacitance within the antenna structure. A classic example of inductive loading is the meandering of the conductor forming the antenna such as a meanderline dipole. Not only is this approach lightweight but it is also applicable for any frequency range. However, for some antennas, it can be difficult, if not impossible, to implement capacitive and/or inductive loading.

2.5.2 Dipole Antenna Example

To illustrate how miniaturization improves antenna Q, we consider a wire dipole which fits inside a sphere of radius a. Thus, the length of the dipole is equal to $2a$. Under these constraints, the dipole will have its first resonance at $ka \approx \pi/2$, which can be reduced by increasing the self-inductance of the wire. We note that the first resonance of a dipole occurs at $ka = \pi/2$ only if the current distribution is perfectly sinusoidal and the wire radius is infinitesimally small. Without physically modifying the dipole, this can be readily accomplished by using a method of moments code such as NEC [20]. In NEC, a distributed or lumped impedance (parallel or series circuit) can be assigned to each wire segment that forms the dipole. In this case, each segment was assigned the same distributed inductance L (uniform loading) thereby increasing the inductance per unit length of the dipole. In doing so, the resonant frequency is shifted to a lower frequency as the inductance is increased. The ratio of the resonant frequency for the unloaded case ($L = 0$) to the loaded case ($L > 0$) is defined here in as the miniaturization factor m $\left(m = f_u^{res}/f_l^{res}\right)$. Figure 2.7 shows the miniaturization factor as a function of the inductance per unit length. Note that we use this approach to miniaturize the dipole because it does not alter the physical structure of the antenna. Therefore, it is a very generic approach that can be used to demonstrate the impact of reducing the phase velocity.

To observe the impact of miniaturization on Q, we calculate the dipole Q at several values of ka below resonance. To do so, we used

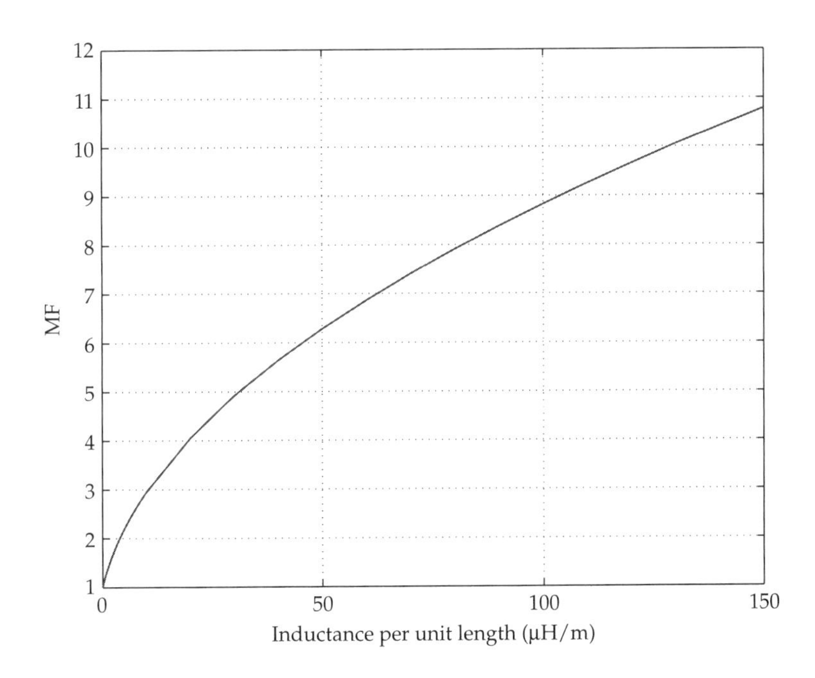

FIGURE 2.7 Miniaturization factor as a function of the wire inductance per unit length for a dipole antenna.

the following expression [21]

$$Q(\omega_0) \approx \frac{\omega_0}{2R(\omega_0)} \sqrt{[R'(\omega_0)]^2 + \left[X'(\omega_0) + \frac{|X(\omega_0)|}{\omega_0}\right]^2} \tag{2.16}$$

In Fig. 2.8, the Q is plotted as a function of the miniaturization factor for the various values of ka. For each ka, the dipole Q was normalized using the Q of the unloaded dipole ($m = 1$ case) to make all curves viewable on the same plot. From Fig. 2.8 it is evident that the Q decreases as m increases until a minimum is reached. The value of m for which the minimum Q occurs corresponds to the miniaturization factor required to make the dipole resonate at a given ka. This can be better depicted by replotting the data using the effective electrical size $k_m a = mka$ as shown in Fig. 2.9. Recalling that the unloaded ($m = 1$) dipole resonates when ka is slightly less than $\pi/2$, it is evident that the minimum Q occurs when $k_m a = 1.5$. This, of course, verifies our earlier statement that the Q is minimized when the dipole achieves resonance. Therefore, we can improve the Q of a physically small dipole by increasing the electrical size (reducing phase velocity).

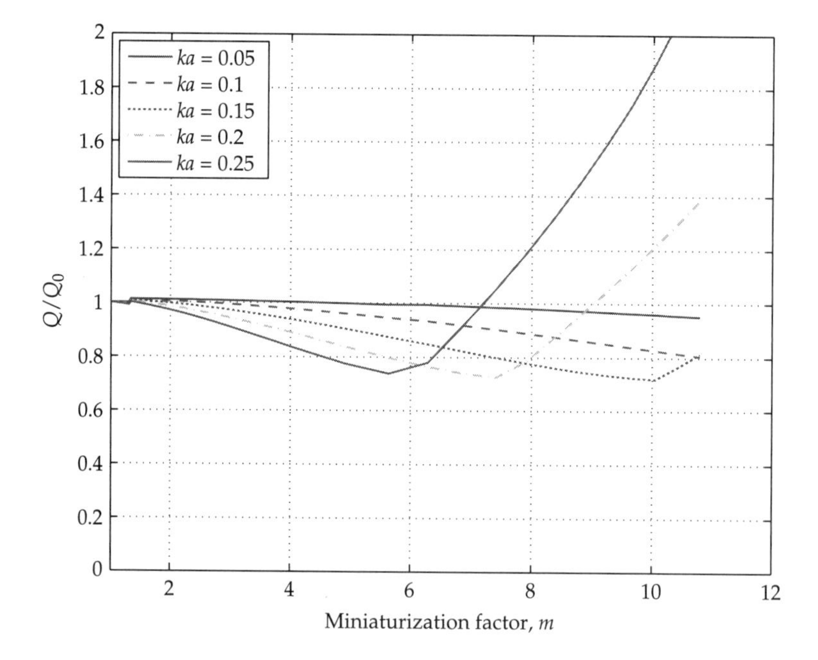

FIGURE 2.8 The normalized dipole Q (at a fixed ka) as a function of the miniaturization factor m.

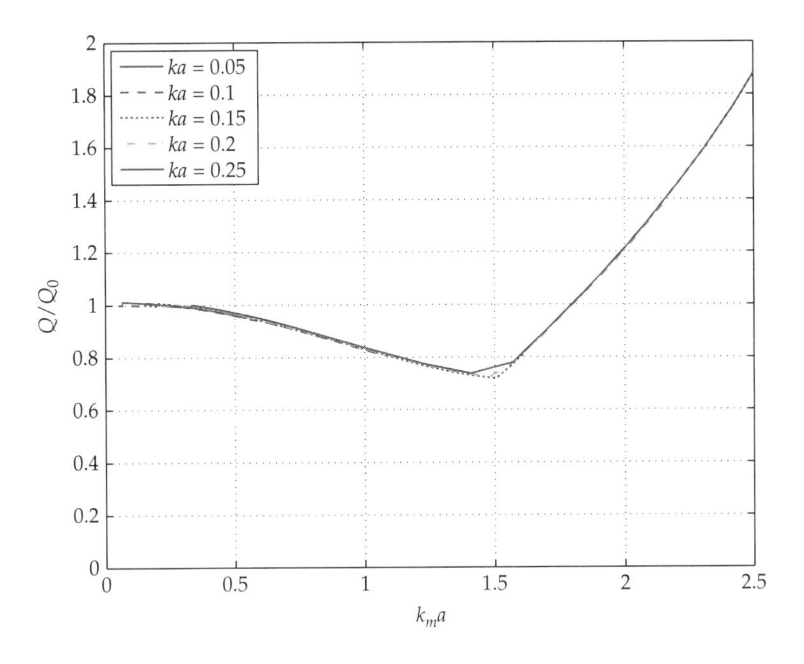

FIGURE 2.9 Dipole Q as a function of $k_m a = mka$.

2.6 Broadband Antenna Miniaturization

We now proceed to discuss broadband antenna miniaturization in general. Our goal is to illustrate some simple guidelines that antenna engineers can use to miniaturize a broadband antenna regardless of which miniaturization technique they choose to employ. To demonstrate these guidelines, we use a wire log-spiral antenna (see Fig. 2.11) and the inductive loading technique previously used for the dipole antenna. Examples of implementing inductive loading can be found in [22] where lumped surface mount inductors were utilized and in [23] where the spiral was inductively loaded by coiling its arm such that it resembles a helical waveguide. In this section, our focus is on demonstrating the concepts used in designing these antennas rather than a particular design. As a first step, we discuss how much the antenna should be miniaturized to minimize the Q at a given ka. We then examine how the loading profile impacts antenna performance and we discuss the benefit of equal inductive and capacitive loading.

2.6.1 Optimal Miniaturization Factor

For any type of antenna, it is important to know how much the phase velocity must be reduced to minimize the antenna Q at a given ka. For a band-pass or narrowband antenna, the phase velocity needs to be reduced such that the antenna is self-resonant at the desired ka. However, for a broadband antenna such as a spiral, determining the optimal miniaturization factor m_{opt} is not as straightforward. That is, consider a spiral antenna (see upcoming Fig. 2.11) that fits inside a sphere of radius a. From radiation band theory [24], we know that the spiral radiates primarily from an annular band whose circumference is one wavelength λ. For a circular spiral, the diameter of this radiation band is then $D = \lambda/\pi$, implying that the smallest electrical size a spiral can have and still support a radiation band is $ka = (2\pi/\lambda)(D/2) = 1$. Now, consider a frequency ω_1 for which a spiral antenna cannot establish a radiation band within its aperture because it is physically too small ($ka < 1$). To establish the radiation band within the spiral aperture, the phase velocity needs to be reduced at the very least by a factor of $m_1 = 1/k_1 a$. However, we could also establish the radiation band closer to the center of the spiral by reducing the phase velocity by any factor, m_2, which is larger than m_1. Therefore, the question is "Which miniaturization factor will minimize the Q at a given ka?" To answer this question, we miniaturize a spiral antenna (see Fig. 2.11) using the same inductive loading technique employed previously for the dipole antenna.

To determine the optimal miniaturization factor m_{opt} for a given ka, we follow the same procedure used for the dipole antenna. That is, we observed how the Q varies as a function of the effective size $k_m a$ for several values of ka. The results are shown in Fig. 2.10 and, for each case, the spiral Q has been normalized to the theoretical Q limit so that all of the curves can be displayed on the same plot. It is apparent that in each case the Q is minimized when $k_m a \approx 0.5\pi$, implying an optimal miniaturization factor of $m_{\text{opt}} \approx 0.5\pi/ka$. It is important to note that m_{opt} is larger than the required m to just fit the radiation band inside the spiral aperture. Such a result is not an unexpected result because a spiral does not radiate as effectively from the radiation bands located near the aperture edges. That is, if the radiation band is located too close to the edge of the aperture the current does not have sufficient space to decay before it reaches the end of the spiral arm. Consequently, a strong reflection occurs resulting in a standing wave type current distribution instead of the typical decaying current distribution. Just how close the radiation band can be to the aperture edge depends upon the growth rate. Thus, for a spiral, m_{opt} will most likely depend to some extent on the growth rate. Nevertheless, estimating m_{opt} in this manner still provides a useful result since it can be used as a starting point in the design process

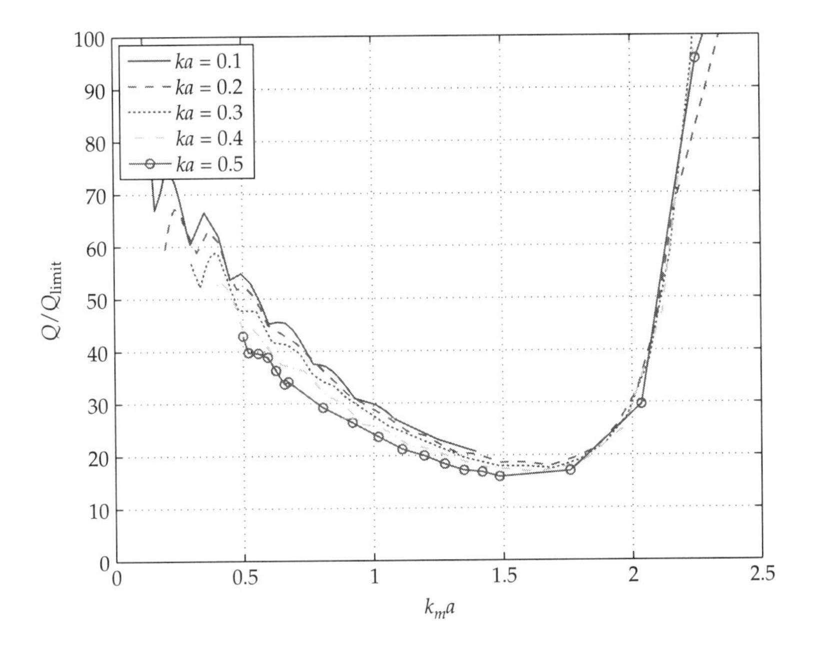

FIGURE 2.10 Spiral antenna Q as a function of the effective electrical size $k_m a$ for fixed values of ka.

for defining the maximum miniaturization factor that needs to be achieved.

2.6.2 Loading Profile

In designing a miniaturized broadband antenna it is important to consider the manner in which the antenna is miniaturized. As an example, consider the miniaturization of a spiral antenna by a factor m. This miniaturization factor can be achieved by uniformly loading the spiral such that the phase velocity is reduced by a factor m everywhere along the spiral. On the other hand, the same miniaturization factor can also be achieved by gradually decreasing the phase velocity along the spiral using a tapered loading profile. An obvious question is whether one approach is better than the other. To answer this question, let us consider the uniform loading of the 6″-diameter spiral shown in Fig. 2.11. For a frequency f_0, there is a radiation band whose location is shown in Fig. 2.11 prior to miniaturization. It is obvious that the spiral is electrically large enough to naturally establish a radiation band at f_0. Therefore, its performance at f_0 is sufficient and needs no further improvement. When the spiral is miniaturized by a

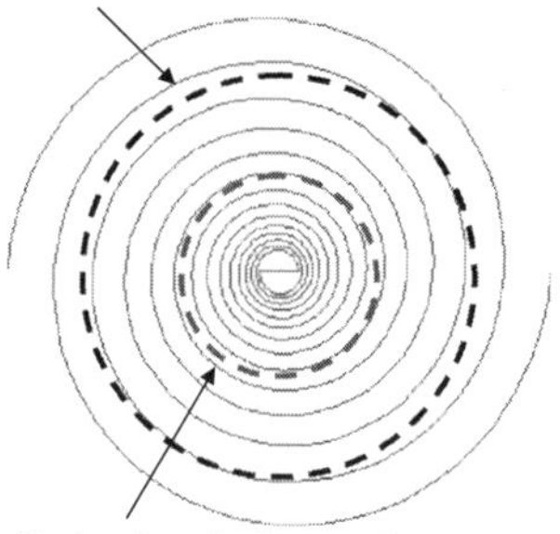

f_0 radiation band before miniaturization

f_0 radiation band after uniform loading

FIGURE 2.11 Illustration of the effect of miniaturization on the location and size of a radiation band. The wire log-spiral is 6″ in diameter with an expansion ratio $\tau = 0.525$ (growth rate $a \approx 0.1$) and angle $\delta = \pi/2$.

factor m using a uniform loading profile, the location of the radiation band associated with f_0 will then shift inward as shown in Fig. 2.11. However, in doing so, the size of the radiation band is also reduced by a factor of m. Therefore, after miniaturization, the spiral now radiates from an electrically smaller aperture at the frequency f_0, implying a higher antenna Q. Thus, while uniform loading can improve performance at frequencies where the spiral is electrically small, it can also lead to worse performance at those frequencies ($ka > 1$) where the spiral could already form a radiation band.

As a specific example of the above statement, let us uniformly load the spiral in Fig. 2.11 with an inductance per unit length of 5 μH ($m = 2.21$). The resulting Q is shown in Fig. 2.12. As seen, above 600 MHz the miniaturized spiral Q is significantly higher because the radiation bands are electrically smaller. The effect of the electrically smaller aperture on the directivity is shown in Fig. 2.13 for the uniformly loaded and unloaded spiral. It is apparent that the directivity for the uniformly loaded spiral is several dB lower than the unloaded spiral, and it exhibits an oscillatory behavior. The oscillatory behavior is caused by the spiral not being able to radiate as effectively (i.e., higher Q) from the electrically smaller radiation bands. That is, the current is not attenuated sufficiently as it passes through the electrically small radiation bands. Therefore, a large amount of current reaches the end of the spiral arm where it is reflected and gives rise to the oscillations.

An obvious solution to this problem is to only load the low-frequency portion of the spiral aperture or to use a tapered loading profile. In doing so, there will be minimal size reduction of the high-frequency radiation bands. However, it is impossible to miniaturize

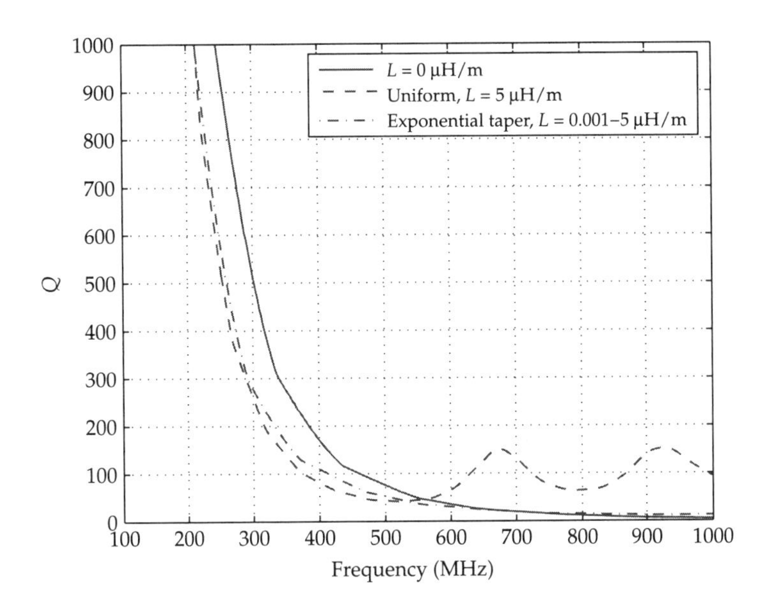

FIGURE 2.12 Comparison of the radiation Q for a 6″-diameter spiral antenna having uniform and exponential inductive loading profile.

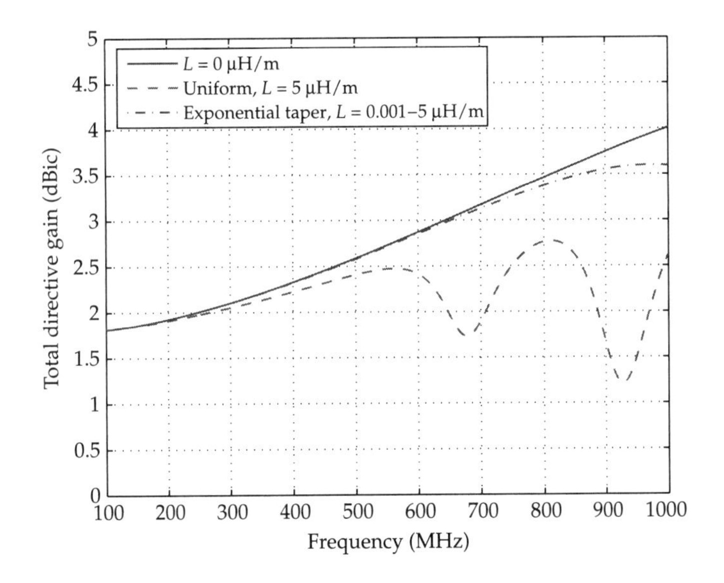

FIGURE 2.13 Comparison of the broadside directive gain for a 6″-diameter spiral antenna having a uniform and exponential inductive loading profile.

a broadband antenna without affecting some of the higher frequency components. To demonstrate the improvement obtained by tapering, the inductive loading was exponentially tapered along the length of the spiral arm starting with an inductance of 0.001 μH and increasing to 5 μH. The effect on the spiral Q and broadside directivity are shown in Figs. 2.12 and 2.13, respectively. It is clear that the Q for the tapered loading is almost identical to the unloaded spiral at high frequencies (above 600 MHz) as desired. Similarly, the directivity for the tapered loading is almost identical to the unloaded spiral. Therefore, a tapered loading profile is an attractive (if not necessary) choice for broadband antenna miniaturization because it simultaneously lowers the Q at low frequencies (below 600 MHz), while minimizing negative effects of miniaturization at higher frequencies.

It is important to note that the tapered loading profile provides the best overall performance. However, uniform loading results in the best low-frequency performance at the expense of high-frequency performance. This is illustrated in Fig. 2.14, which compares the broadside realized gain for a spiral antenna with a uniform and an exponentially tapered loading profile. In each case, the antennas are matched

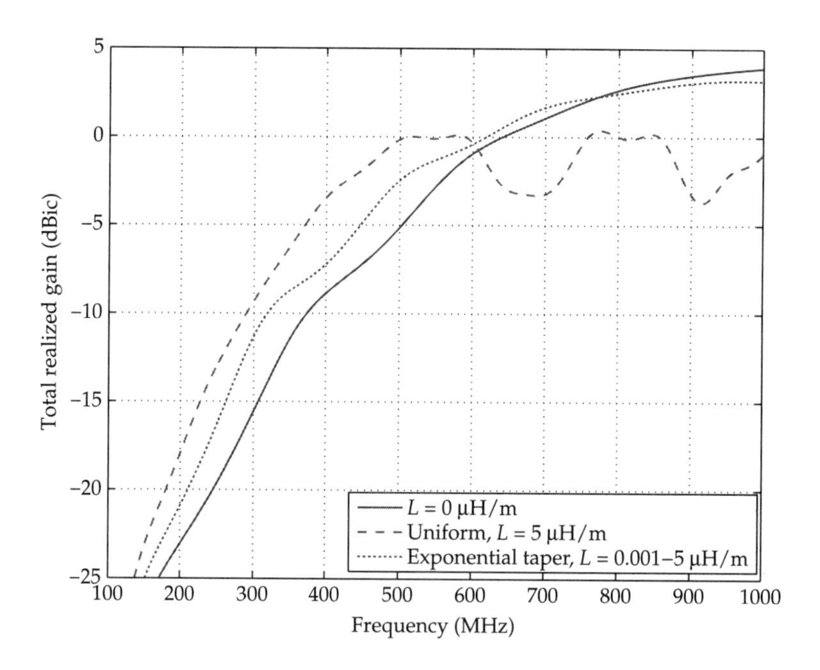

FIGURE 2.14 Comparison of the broadside realized gain for a 6″-diameter spiral antenna for three different inductive loadings: unloaded ($L = 0$ μH), uniformly ($L = 5$ μH) loaded, and exponentially ($L = 0.001$ to 5 μH) loaded profile.

to purely real load impedance that maximizes their overall performance. From Fig. 2.14, it is evident that below about 600 MHz the uniformly loaded spiral has the highest realized gain. As all three antennas are lossless with almost the same directivity (see Figure 2.13), this higher realized gain can be attributed to improved matching. That is, the uniformly loaded spiral has the lowest Q in this region as seen in Fig. 2.12. However, above 600 MHz, the realized gain of both the unloaded and tapered spiral is 5 to 8 dB better, which is a significant improvement. This improvement is a product of a slightly higher directivity (1–2 dB) and better matching in both cases due to their lower Q. Figure 2.14 illustrates an important aspect of miniaturizing a broadband antenna which was alluded to earlier in our discussion of broadband impedance matching. That is, the performance of a broadband antenna can be further improved at frequencies where it is electrically small as long as one is willing to sacrifice performance at frequencies where it is electrically large. Therefore, there exists an inherent trade-off in the miniaturization of broadband antennas and this is a consequence of the fundamental limits associated with the radiation Q.

2.6.3 Equal LC Loading

Even though a tapered loading profile alleviates many high-frequency issues, it can cause additional issues when the loading is purely inductive or capacitive. Similar to a tapered transmission line matching section, a tapered loading profile introduces impedance discontinuities along the length of the spiral arm that, in turn, cause reflections. The magnitude of these reflections depends upon how large the impedance discontinuity is from one section to the next. Of course, as long as the impedance differences are small, the reflections due to incremental impedance changes will also be small and will have minimal impact on the antenna input impedance. However, in theory, it is possible to eliminate reflections that may occur if both inductive and capacitive loadings are used simultaneously. That is, concurrent use of inductive and capacitive loading makes it possible to maintain the same impedance throughout the entire spiral structure as implied by Eq. (2.15). Therefore, ideally, an antenna should be miniaturized using equal inductive and capacitive loading.

2.7 Conclusion

In this chapter, the Fano-Bode impedance matching limitations were used to examine the fundamental limits of narrowband (band-pass) and wideband (high-pass) antennas. For the high-pass case, we showed how the Fano-Bode limits restrict the size of a UWB antenna subject to some desired pass-band reflection coefficient or realized gain. We then discussed how miniaturization techniques can be used with preexisting UWB antenna designs to approach this limit by minimizing the antenna Q. Finally, we outlined a methodology for miniaturizing a UWB antenna using a spiral antenna as an example.

References

1. H. W. Bode, *Network Analysis and Feedback Amplifier Design*. D. Van Nostrand Co., New York, NY, 1945.
2. R. M. Fano, "Limitations on the broadband matching of arbitrary impedances," *Journal of the Franklin Institute*, vol. 249, no. 1, January–February 1960, pp. 57–83, 139–154.
3. —, "Theoretical limitations on the broadband matching of arbitrary impedances," Ph.D. dissertation, Massachusetts Institute of Technology, 1947.
4. K. Fujimoto, A. Henderson, K. Hirasawa, and J. R. James, *Small Antennas*. Research Studies Press, John Wiley & Sons, Letchworth, England, 1987.
5. A. R. Lopez, "Review of narrowband impedance-matching limitations," *IEEE Antennas and Propagation Magazine*, vol. 46, no. 4, August 2004, pp. 88–90.
6. D. M. Pozar, *Microwave Engineering*, 2d ed. John Wiley & Sons, Inc., New York, NY, 1998.
7. R. C. Hansen, *Electrically Small, SuperDirective, and Superconducting Antennas*. John Wiley & Sons, Inc., Hoboken, New Jersey, 2006.
8. H. A. Wheeler, "The wide-band matching area for a small antenna," *IEEE Transactions on Antennas and Propagation*, vol. AP-31, no. 2, March 1983, pp. 364–367.
9. G. L. Matthaei, L. Young, and E. M. T. Jones, *Microwave Filters, Impedance-Matching Networks and Coupling Structures*. McGraw-Hill Book Company, Inc., New York, NY, 1964.
10. H. A. Wheeler, "Wideband impedance matching," Wheeler Labs, Technical Report 418, May 1950, report is available at http://www.arlassociates.net.
11. M. C. Villalobos, H. D. Foltz, J. S. McLean, and I. S. Gupta, "Broadband tuning limits on UWB antennas based on Fano's formulation," in *IEEE Antennas and Propagation Society International Symposium 2006*, July 2006, pp. 171–174.
12. H. L. Thal, "New radiation Q limits for spherical wire antennas," *IEEE Transactions on Antennas and Propagation*, vol. 54, no. 10, October 2006, pp. 2757–2763.
13. L. J. Chu, "Physical limitations of antenna Q," *Journal of Applied Physics*, vol. 19, December 1948, pp. 1163–1175.
14. R. F. Harrington, "Effect of antenna size on gain, bandwidth, and efficiency," *Journal of Research of the National Bureau of Standards*, vol. 64D, January 1960, pp. 1–12.
15. H. A. Wheeler, "Fundamental limitations of small antennas," *Proceedings of the IRE*, vol. 35, December 1947, pp. 1479–1484.
16. A. R. Lopez, "Fundamental limitations of small antennas: validation of Wheeler's formulas," *IEEE Antennas and Propagation Magazine*, vol. 48, no. 4, August 2006, pp. 28–36.
17. R. F. Harrington, *Time-Harmonic Electromagnetic Fields*. John Wiley & Sons, Inc., New York, NY, 2001.
18. S. A. Schelkunoff, "Theory of antennas of arbitrary size and shape," *Proceedings of the IRE*, vol. 29, no. 9, September 1941, pp. 493–521.
19. J. D. Kraus and R. J. Marhefka, *Antennas*, 3d ed. McGraw-Hill, Inc., New York, NY, 2001.

20. G. J. Burke, E. K. Miller, and A. J. Poggio, "The Numerical Electromagnetics Code (NEC)—a brief history," *Antennas and Propagation Society International Symposium, 2004 IEEE*, vol. 3, June 2004, pp. 2871–2874.
21. A. D. Yaghjian and S. R. Best, "Impedance, bandwidth, and Q of antennas," *IEEE Transactions on Antennas and Propagation*, vol. 53, no. 4, April 2005, pp. 1298–1324.
22. M. Lee, B. A. Kramer, C.-C. Chen, and J. L. Volakis, "Distributed lumped loads and lossy transmission line model for wideband spiral antenna miniaturization and characterization," *IEEE Transactions on Antennas and Propagation*, vol. 55, no. 10, October 2007, pp. 2671–2678.
23. B. A. Kramer, M. Lee, C.-C. Chen, and J. L. Volakis, "Size reduction of a low-profile spiral antenna using inductive and dielectric loading," *IEEE Antennas and Wireless Propagation Letters*, vol. 7, 2008, pp. 22–25.
24. R. Bawer and J. J. Wolfe, "The spiral antenna," *IRE International Convention Record*, vol. 8, March 1960, pp. 84–95.

CHAPTER 3

Overview of Small Antenna Designs

Yunqi Fu and John L. Volakis

3.1 Introduction

Wireless communication is now an important part of people's routine life. Besides cell phones, other wireless products abound. To mention a few, wireless computer and multimedia links, remote control units, satellite mobile phones, wireless internet, and radio frequencies identification devices (RFIDs).

As can be realized, mobile terminals must be light, small, and have low energy requirement. Indeed, developments in the chip industry over the 80s and 90s have allowed for dramatic size reduction in microelectronics and CPU computing. In contrast, miniaturization of the RF front-end has been a more recent focus (for the past 5 years) and has presented us with significant challenges due to small antenna limitation (see Chaps. 1 and 2). Wheeler [1] noted that antenna size limits radiation resistance values, efficiency and bandwidth. In other words, antenna design is a compromise among volume, bandwidth, gain, and efficiency. The best compromise is usually attained when most of the available volume is excited for radiation.

In this chapter, we give a general review of techniques already used in the literature to design small antennas. Such techniques can be separated in five categories [2–5]:

1. One of them is to load the antenna with high-contrast material, high permittivity and/or high permeability. Up to a point, the

size-reduction ratio is approximately related to $\sqrt{\mu_r \varepsilon_r}$ and the geometry of the loaded material.

2. Another technique is to modify and optimize the antenna geometry and shape. This can be typically illustrated in practice through the inverted-L antenna (ILA) with bending arms, helices and meander line antennas, and even antennas with volumetric curvature and fractal structure. Sometimes notch or slot loading is introduced as reactive loadings, shifting the resonance to lower frequencies.
3. A third technique is based on using lumped components. When the antenna size is reduced, the radiation impedance will have a large reactive part. This lumped component can be introduced to compensate for the reactive impedance. The latter two categories of miniaturization techniques are not independent to each other. That is, some of the structures mentioned in the second category may also be explained using lumped loading (such as the ILA and the slot loaded patches).
4. A recently emerging technique is based on artificially engineered electromagnetic metamaterials. Metamaterials include frequency bandgap structures, artificial magnetic conductors (AMC) and left-handed (LH) propagating components. Specifically, the AMC surface has an in-phase reflection for the incident plane wave (+1 reflection coefficient). Hence, a wire antenna can be placed very close and parallel to the AMC surface without experiencing the usual ground plane shorting (as in typical low-profile structure).
5. A more general technique is that based on formal optimization method. One can design an antenna subject to maximum achievable performance and minimum dimensions. For doing this, a formal optimization method must be adopted. Much work has been done in this area with genetic algorithm being a typical method. It should be noted that the antenna designed using this technique don't generally follow a conventional shape.

3.2 Miniaturization via Shaping

Shaping is the most extensively used technique in antenna miniaturization and there are many shaping approaches to consider. These include bending, folding, meandering, helical and spiral shaping, fractal folding, slot loading, etc. Figure 3.1 depicts some typical examples.

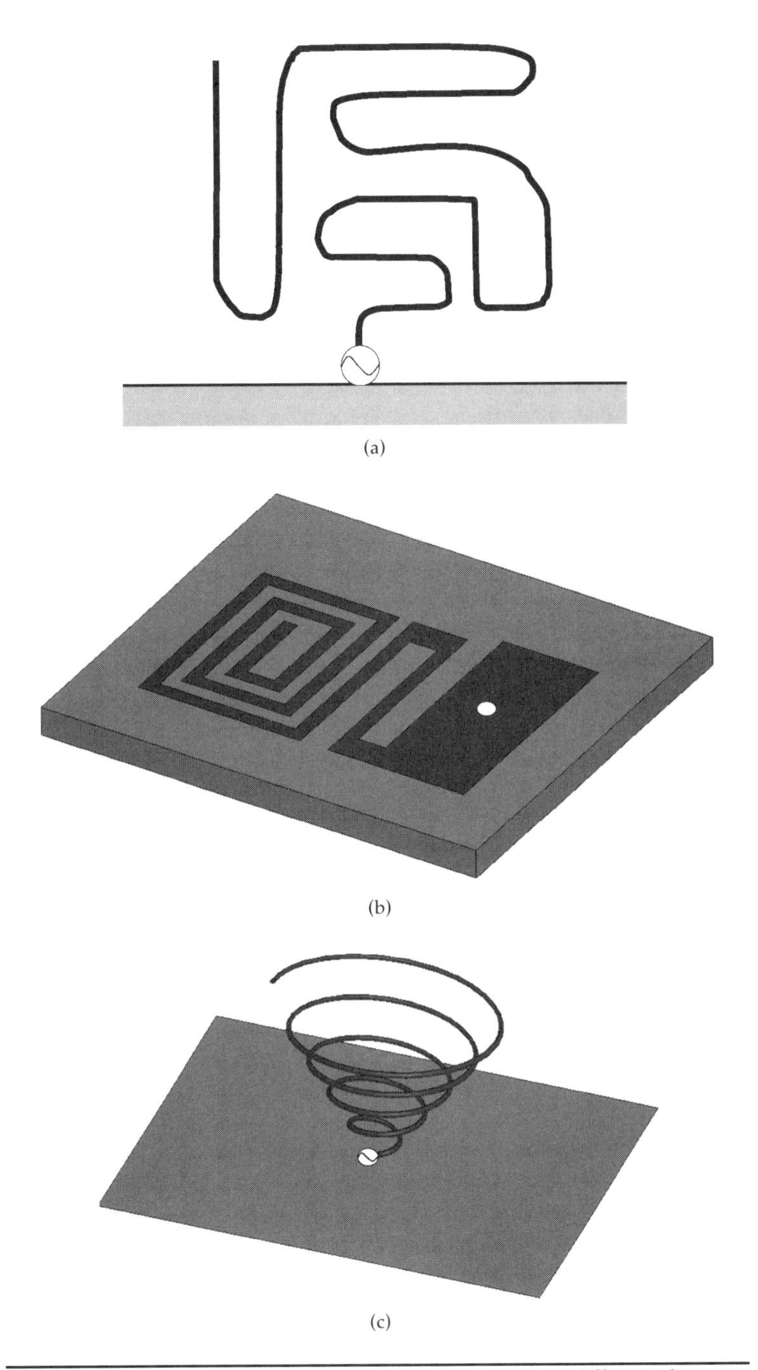

FIGURE 3.1 Illustration of antenna miniaturization using different shapes and geometries. (a) Monopole; (b) Planar antenna; (c) 3D helix.

In the following sections, we discuss representative examples of these approaches from recent literature. It should be noted that in every case, effective use of the ground plane mounting structure is critical. Notably, the λ/4 monopole takes advantage of the infinite ground plane to reduce the dipole size by one half.

3.2.1 Slot Loading

Slot loading is mainly used for planar antenna miniaturization (such as patch antennas) [6,7]. For a standard rectangular patch antenna, the current flows from one edge to the other. However, when slots are cut in the metal patch, the current is blocked, requiring a longer path (see Fig. 3.2) to reach the other edge. Thus, the resonant frequency of the patch is reduced. Nguyen et al. [8] investigated the relation between slot size and resonant frequency for a notch loaded patch antenna (see Fig. 3.2). In his study, he changed the slot length from 10 to 26.33 mm, while the notch width was kept at 5 mm. He observed that the resonant frequency decreased (see Fig. 3.2) as the notch length increased. Subsequently, he varied the notch width from 1 to 20 mm, while the length was kept at 26.33 mm. A conclusion was that narrower notches reduced the resonance frequency (i.e., improved miniaturization).

f_0 (GHz)	l (mm) (w = 5 mm)	f_0 (GHz)	w (mm) (l = 26.33 mm)
1.37	10	0.96	1
1.24	15	0.85	8
1.10	20	0.8	15
0.887	26.33	0.77	20

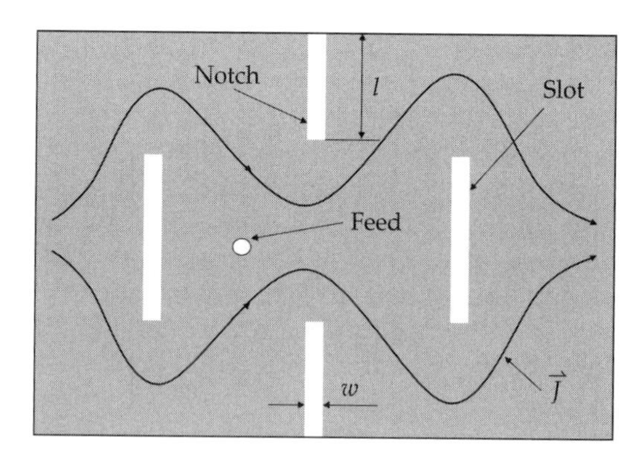

FIGURE 3.2 Patch antenna miniaturization using notch/slot loading. (*See Nguyen et al.* [8].)

There is a wide range of conceivable slot shapes. That is, slots can be modified to have round or smooth terminations, be triangular or circular in shape. Also slots may have cavities at the end, and multiple slots maybe included. As an example, Huang [9] loaded a circular patch antenna printed on a substrate with relative permittivity of $\varepsilon_r = 10$ with four slots. This led to a patch antenna that was 11 cm in diameter and operated at a ultra high frequency (UHF) frequency of 400 MHz, that is, its diameter was 0.15 λ.

Slot insertions to reduce printed global positioning system (GPS) antenna size abound. By cutting slits or slots on a patch printed on the ceramic substrates, the antenna size can be further reduced. Chen et al. [10] introduced symmetrical radial slits and L-shape slots (symmetrically at each of the four quarters) on the patch (see Fig. 3.3) printed on a 2.54 mm thick substrate with $\varepsilon_r = 9.8$. The patch size with slots loading was 22×22 mm^2. By comparison, mere use of the high-contrast substrate would have required $\varepsilon_r = 17.22$ to achieve a similar miniaturization. Zhou et al. [11] adopted four spiral slots to design a proximity-fed dual-band stacked patch (1575 MHz or L1 band, and 1250 MHz or L2 band). This design was for GPS reception and relied on a high-contrast substrate (K30 or $\varepsilon_r = 30$). To further reduce this GPS antenna size, slots were cut on the top patch, as shown in Fig. 3.3, to force meandering of the electric currents (miniaturization was needed for the top patch that radiated the L2 band with 25 MHz bandwidth). The eventual GPS antenna element was reduced down to 1″ aperture ($\lambda/10$ at the L2 band), that is, five-fold size reduction from the $\lambda/2$ size patch.

Slots can be also etched in the antenna ground plane to reduce size. An example [12] is given in Fig. 3.4. As depicted, for a patch antenna printed on a substrate having $\varepsilon_r = 4.4$, loss tangent $\tan\delta = 0.02$ and thickness $h = 1.6$ mm, three identical slots were introduced in the ground plane and aligned with equal spacing parallel to the patch's radiating edge. The embedded slots were narrow (1 mm in width) and had a length of $l_0 + l_i$, where l_0 and l_i refer to the slot lengths outside and inside the projection of the radiating patch on the ground plane, respectively. Four antenna designs with $l_0 = 10$ mm and $l_i = 8$, 10, 12, 14 mm were compared to a reference antenna without slots. The latter was resonant at 2387 MHz with a -10 dB return loss and a bandwidth of 2.0%. However, on setting $l_i = 14$ mm, the resonant frequency was reduced to 1587 MHz, implying a 56% size reduction. The bandwidth of the antenna with a slotted ground plane was also greater than that of the reference one. This behavior is largely due to meandering the ground plane slots, effectively lowering the quality factor. Of course, slots in the ground plane will often be undesirable to backplane leakage.

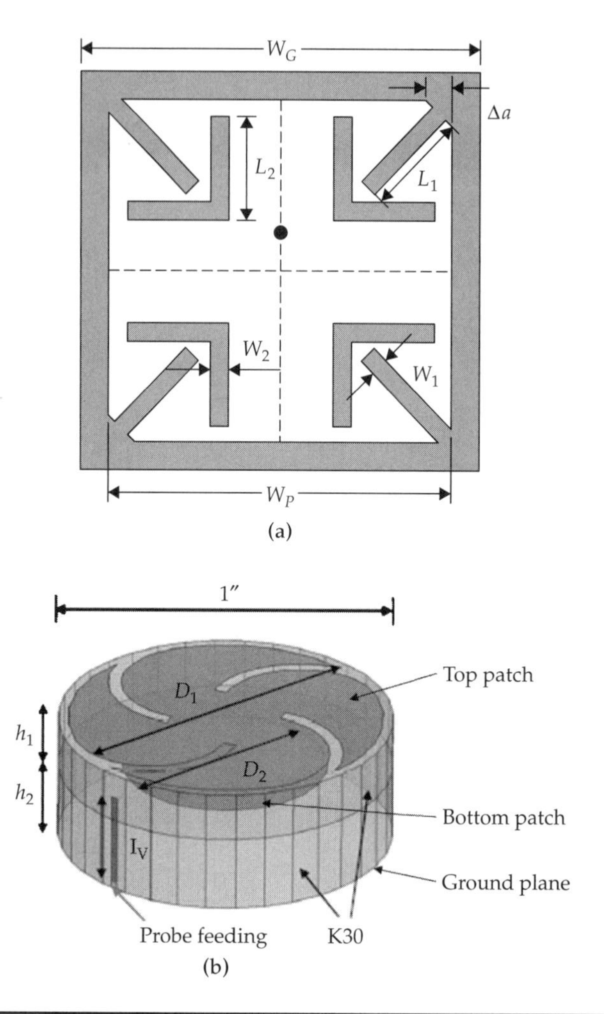

FIGURE 3.3 Slot loaded GPS antenna examples. (a) Radial slits and L-shape slots loaded rectangular patch. (*After Chen et al.* [10] © *IEEE 2007.*); (b) Spiral slots on the top of a dual-patch antenna. (*After Zhou et al.* [11] © *IEEE 2009.*)

3.2.2 Bending and Folding

Bending and folding (see Fig. 3.5) are popular techniques for antenna miniaturization. As noted, such modifications force the current to flow along a curved and longer path, resulting in lower resonant frequency. Usually, bending achieves miniaturization in height at the expense of larger planar dimension. A typical example is the inverted-L antenna (ILA), derived from the monopole. To reduce the original monopole height, the top section of the arm is bent down, making a portion of

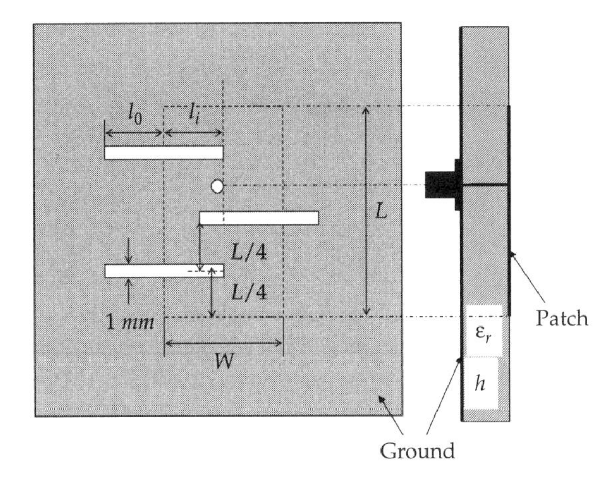

FIGURE 3.4 Slot loaded ground plane for antenna miniaturization. (*See Wong et al.* [12].)

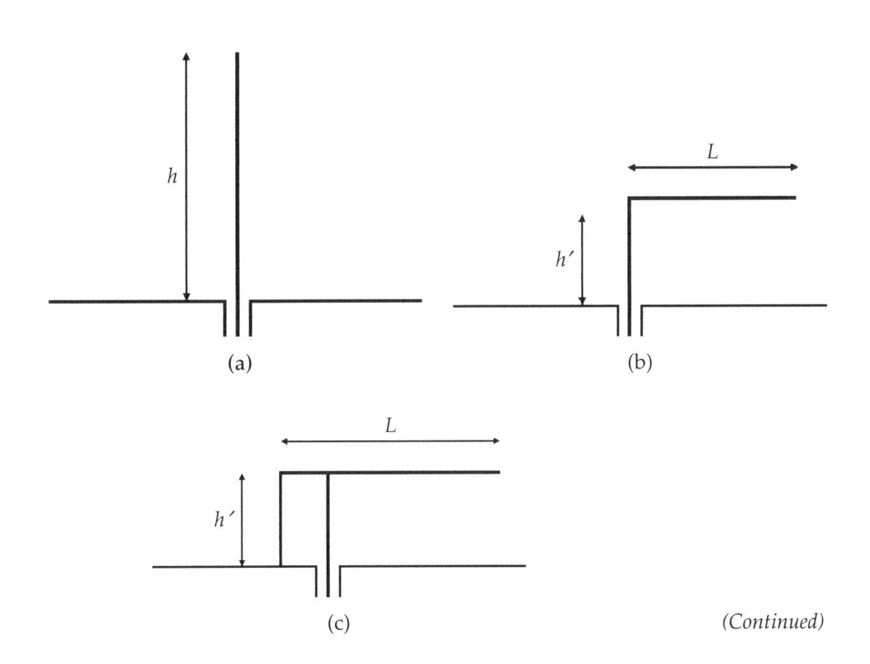

FIGURE 3.5 Some typical bending and folding antenna examples. (a) Original monopole; (b) Bent monopole (ILA); (c) Monopole Forming the IFA; (d) PIFA; (e) Folded PIFA; (f) Simplified Folded PIFA.

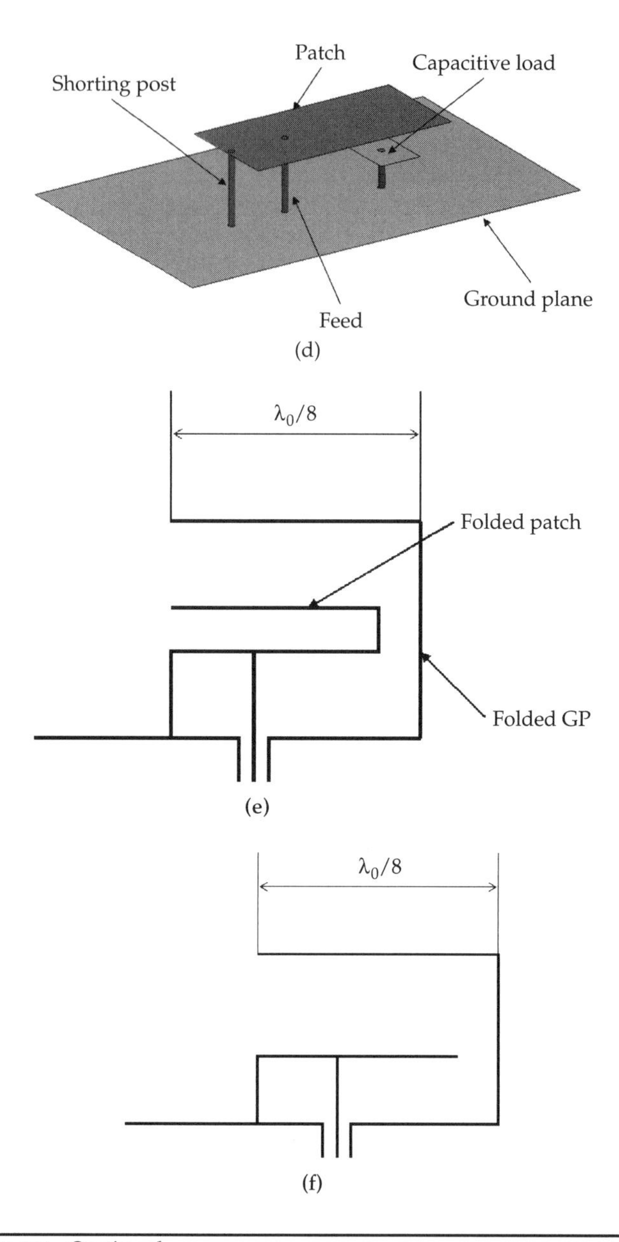

FIGURE 3.5 *Continued.*

it parallel to the ground plane (GP), as depicted in Fig. 3.5*a* and 3.5*b*. In this manner, the total height of the ILA is dramatically reduced. Consequently, the horizontal dimension is increased. Nevertheless, the lower profile makes it suitable for mounting on surfaces of the moving vehicles. It should be noted, though, that (due to ground plane effects) the radiation resistance of the ILA is very small, and its capacitance very large. To correct this issue, a second vertical post is added to the left (see Fig. 3.5*c*). This introduces inductive loading to compensate for the capacitive component of the ILA, leading to the well-known inverted-F antenna (IFA).

A similar structure is the planar inverted-F antenna (PIFA). In contrast to the IFA, the horizontal wire in Fig. 3.5*c* is replaced by a rectangular patch with a shorting pin (or plate), as depicted in Fig. 3.5*d*. PIFAs of many variations have been published and widely adopted in practical wireless devices [13]. As already noted, IFA and PIFA are typically quarter wavelength resonant antennas. However, their size can be further reduced as discussed next.

For size reduction, planar antennas (including PIFAs) can be folded into multilayer structures. As shown in Fig. 3.5*e*, by folding a quarter wavelength, wall-shorted rectangular patch or a PIFA, the antenna length can be reduced down to $\sim\lambda/8$ [14]. The combined ground plane and patch folding is important. Otherwise, the folded antenna would work much like a stacked patch giving dual-band operation [15]. The structure can be simplified as shown in Fig. 3.5*f* by adding an extra ground plane to the right, thus, eliminating patch folding. The resonant frequency of the folded stacked patch antenna can be further lowered by reducing the distance between the parallel surfaces of the folded patch (consisting of the lower and upper patches) or by reducing the width of the shorting walls. Both of these approaches result in lower profiles with likely less bandwidth and efficiency.

Holub and Polivka [16] generalized this folded PIFA to a multilayer meandered and folded patch structure depicted in Fig. 3.6*a*. The upper section of the structure now looks more like an interdigital capacitor. This technique enables a decrease of the original shorted (quarter-wavelength) patch dimension by a factor $1/N$, where N denotes the number of vertically placed patch plates above the ground plane. Two antennas operating at 1575 (L1 GPS band) and 869 MHz (RFID band) were designed and measured in [16]. The respective electrical lengths of these antennas were $\lambda/8.6$ and $\lambda/11.7$, with corresponding bandwidths of 2.98% and 1.15%.

The PIFA size can be also reduced via capacitive loading of the top plate. The goal with capacitive loading is to compensate for the inductive portion of the impedance (to reduce the reactive part of the impedance). This can be realized by folding the open end of the PIFA [17]. Other ways to introduce capacitance is to add a plate between the ground plane and the PIFA [18] as depicted in Fig. 3.5*d*. The capacitive load can be adjusted by changing the size and height of the middle patch. Of course, this will not increase the outer dimensions of the original structure.

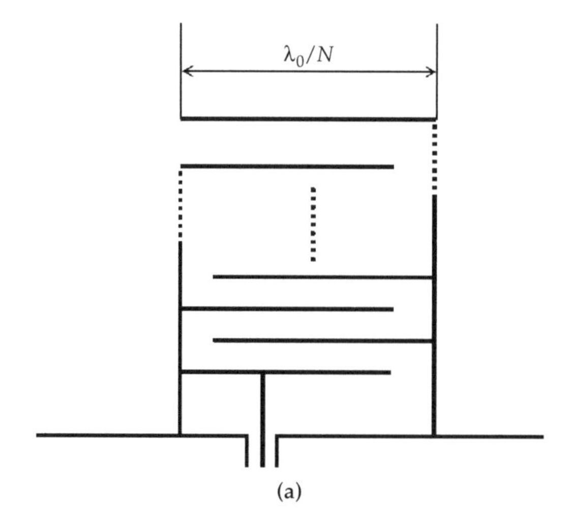

(a)

(b)

FIGURE 3.6 Meandered folded PIFA resulting in a horizontal size of λ_0/N. (a) Structure geometry; (b) Antenna sample (869 MHz). (*After Holub and Polivka* [16] © *IEEE 2008*.)

Control of the capacitive loading can be done by modifying the size and height of the capacitive plate as in Fig. 3.5*d*. Tuning of the PIFA can then be done by simply replacing the via with a screw and adjusting the height of the tuning screw. A tuning range of 2.5 to 3.3 GHz was shown in [19] using this approach.

Other methods to reduce the PIFA's size are based on dielectric loading between the ground plane and the plate [20], by optimization of the shorting pin locations [21, 22], or by introducing slots to increase the antenna's electrical length [23, 24]. As already noted, the ground plane plays a significant aspect in the overall antenna size and performance. In practice, it is comparable in size to the radiating element (as is the case with cell phone antennas). However, shape can be taken into account to reduce antenna size [25].

3.2.3 Space-Filling Curves

The concept of designing small antennas in the form of space-filling curves has been popular in recent years. Uses of space-filling curves date back to the late nineteenth century when several pioneering mathematicians, such as Hilbert and Peano, considered these curves [26]. For small antennas, space-filling curves are used to form a long curve within a small "surface" area. Thus, the structure can be resonant at wavelengths even if their structure is a fraction of a wavelength. This property has motivated antenna designers to explore planar resonant radiators in the form of space-filling curves within a small footprint.

As already noted in Chap. 1, small antennas are constrained in their behavior by fundamental size limits [27–29]. Several authors have already stated that one must effectively use the entire radial volume to obtain best performance. More specifically, Hansen writes [30] "*. . . it is clear that improving bandwidth for an electrically small antenna is only possible by fully utilizing the volume in establishing a TM or TE mode, or by reducing efficiency.*" Also, Balanis writes [31] "*. . . the bandwidth of an antenna (which can be closed within a sphere of radius r) can be improved only if the antenna utilizes efficiently, with its geometrical configuration, the available volume within the sphere.*"

Indeed, there are many ways to fill a defined area with curved wires (meander lines, helical antennas, and spirals). Chapter 5 is devoted to miniature spirals. Therefore, in the following sections we discuss other antenna types.

3.2.4 Meander Line Antennas

Meander line antennas (MLAs) have been widely used in many applications, including mobile handsets and wireless data links for laptops and PC cards. Fenwick [32] and Rashed and Tai [33,34] proposed early versions of the MLA as winding wire structures. A popular MLA structure was investigated by Nakano et al. [35].

A general configuration of the meander line dipole is shown in Fig. 3.7. It is a continuous periodically folded structure. Each unit section is composed of vertical and horizontal wire branches. The meander line has a width w, and the current path length for each unit section is $2(l + w)$. It is clear that if a straight monopole and a meander monopole have the same height, the latter will resonate at a lower frequency.

Figure 3.7 illustrates how the resonant frequency can be shifted using meandered structures. Four meander line dipoles M1 to M4

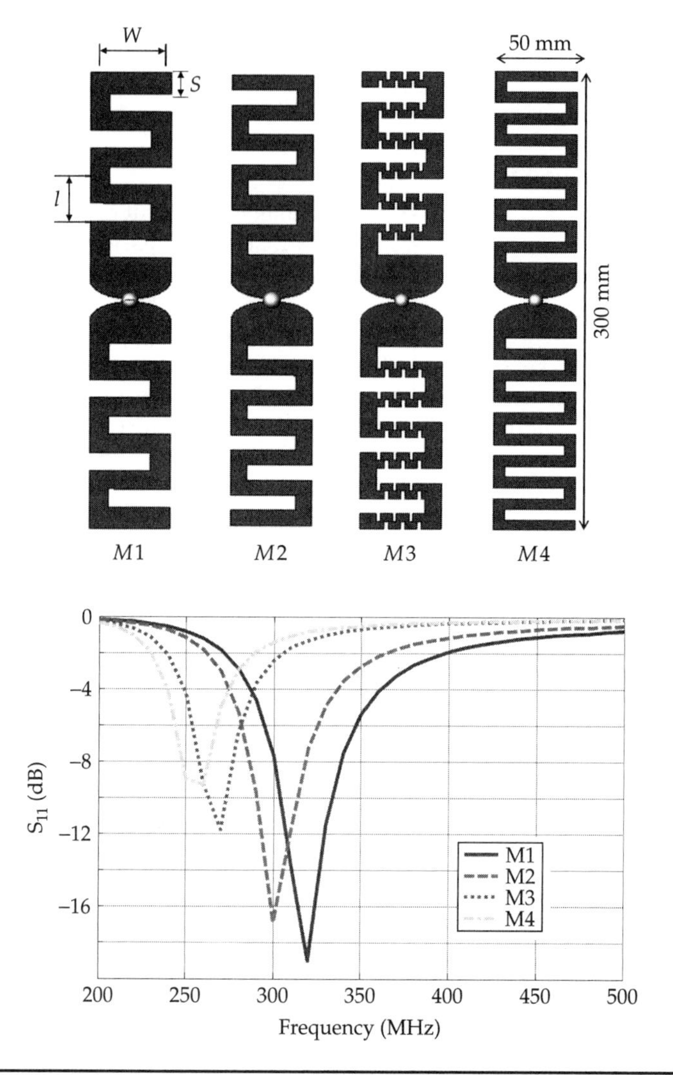

FIGURE 3.7 Meandered dipole examples and their performance. (*Courtesy of The Ohio State University, Electroscience Laboratory.*)

were considered (300 mm tall). A conventional dipole of the same height is resonant at 500 MHz. Correspondingly, the M1, M2, and M4 dipoles are resonant at 320, 300, and 260 MHz, respectively. Meandering can be done more than once, as in the M3 dipole (see Fig. 3.7). Tsutomu et al. [36] investigated the relationship between resonant frequency and geometrical parameters of a meander dipole and presented the formula

$$\eta = \frac{1}{1 + \frac{1}{4\pi} \frac{R_s}{R_d} \frac{\lambda}{s} \frac{Nw}{l}}$$

for the radiation efficiency. Here R_s is the conductor skin resistance, R_d is the radiation resistance of the half-wavelength dipole, s is the width of the meander line (wire diameter as in [36]), $2N$ represents the number of turns, l is the pitch of each meander and w is the antenna width.

The geometry of the meander line antenna does, of course, influence antenna performance. Figure 3.8 depicts several meander line dipoles [37] placed on a dielectric substrate of $\varepsilon_r = 3.38$ and $\tan\delta = 0.0027$. The rectangular meandered dipole in Fig. 3.8*a* could be enclosed within a $2\pi \times 6$ cm sphere (implying $ka = 0.499$) and resonates at 543.33 MHz. By comparison, a standard dipole within the same sphere size would resonate at 1.73 GHz. Thus, the rectangular MLA in Fig. 3.8*a* leads to a 70% size reduction when compared to a simple dipole. Figures 3.8*a* and 3.8*b* provide for alternate meandering approaches that lead to larger input resistances. That is, the input resistance (R_{in}) of the triangular profile dipole in Fig. 3.8*b* is $R_{\text{in}} = 120\,\Omega$ versus $R_{\text{in}} = 26.5\,\Omega$ of that in Fig. 3.8*a*. However, cross polarization radiation increased as noted in [37].

The sinusoidal meandering in Fig. 3.8*c* provides for additional frequency reduction (from 543 MHz down to 325 MHz), implying an 81% size reduction. This corresponds to a $ka = 0.296$ (largest dimension being $\sim 0.1\,\lambda$), approaching the Chu limits discussed in Chap. 1. However, the radiation resistance dropped to $R_{\text{in}} = 5.364\,\Omega$, even though the quality factor increased from $Q = 5.98$ (for Fig. 3.8*b*) to $Q = 74.6$ (for Fig. 3.8*c*).

From Table 3.1, it is clear that there is interplay among Q, radiation resistance, cross polarization, and geometry within a given volume. To observe this interplay, among others, Rashed-Mohassel et al. [37] considered six meandered (M1–M6) dipole shapes displayed in Fig. 3.9. Table 3.1 gives the performance for these meandered wire dipoles. We observe that the dipole in Fig. 3.9*f* has the lowest radiation resistance and highest miniaturization (resonates at 300.44 MHz, implying $ka = 0.274$). Concurrently, its bandwidth is the smallest as its Q is the largest.

3.2.5 Fractal Antennas

This is a form of space-filling antennas that follow fractal shapes (fractal shapes were originally proposed by Mandelbrot [38]). Fractal shapes were considered for electromagnetics in the 1990s [39] and more recently for antennas [40–42].

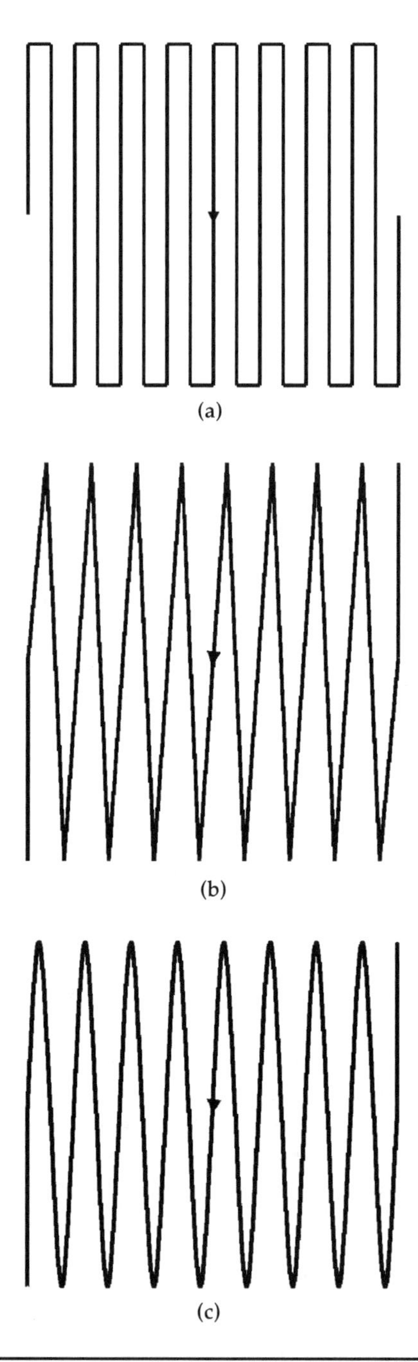

FIGURE 3.8 Dipole-like antennas with (a) rectangular, (b) triangular, and (c) sinusoidal meandering profiles. (*After Rashed-Mohassel et al.* [37] © *IEEE 2009.*)

Antennas	Radiation Resistance (Ω)	Resonant Frequency (MHz)	ka	Q	Q_{Chu}	Cross-Pol. (dB)
M1	11.68	512.92	0.645	114.2	5.277	−115.9
M2	18.32	474.5	0.67	66.25	4.871	−118.6
M3	15.7	423.9	0.6	102.5	6.296	−117.1
M4	5.89	414.95	0.588	54.39	6.62	−122.7
M5	4.44	380.45	0.535	67.8	8.4	−124
M6	2.87	300.44	0.425	280.8	15.38	−98

TABLE 3.1 Parameters of Different Rectangular Meandered Designed Antennas

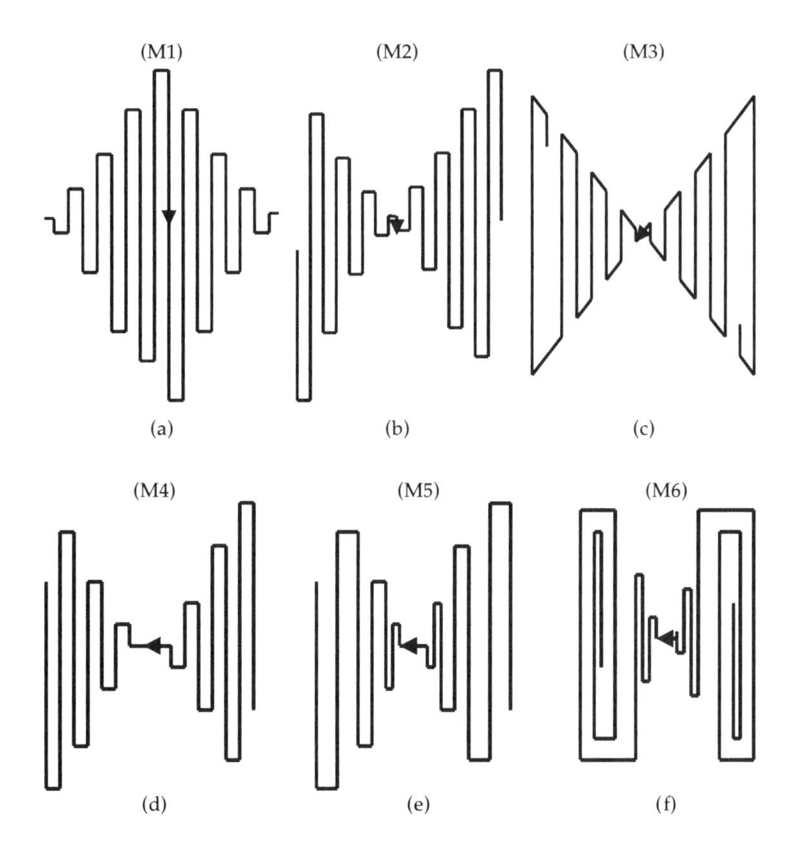

FIGURE 3.9 Meander line antennas with various space-filling geometries. (*After Rashed-Mohassel et al.* [37] © *IEEE 2009.*)

Fractal geometries are generated in an iterative fashion. The Hilbert curve shown in Fig. 3.10 is one of the most popular fractal elements. As depicted, the essence of fractal curves is the repeatability and scaling of unique shapes using predefined process to add additional sections of the same shape (after scaling) within a given boundary. As can be realized from Fig. 3.10, fractal shapes are a class of space-filling curves that enable the capability to define fine details and complex geometrical features by rotating and scaling the same shape.

Several fractal shapes (Koch curve, Hilbert curve, and Penao curve to mention a few) have been considered for multiband antenna designs [43–47]. Among them, the Koch curve is generated by replacing the middle third of each segment with an equilateral triangle without connecting its base. As depicted in Fig. 3.11*a*, the resulting curve after one iteration is comprised of four segments of equal length. Being ideally a nonrectifiable curve, its length grows as $(4/3)^n$ at each iteration. Baliarda et al. [43] measured six Koch monopoles (K0–K5) after 0 to 5 iterations. Figure 3.11*b* depicts the Koch monopole after five iterations (i.e., K5). This K5 monopole has an overall height $h = 6$ cm, but its stretched length is $l = h \cdot (4/3)^5 = 25.3$ cm. These Koch monopoles were measured when mounted on an 80 cm × 80 cm ground plane (see Fig. 3.11*b*). It was found that the resonance consistently shifted toward a lower frequency as its length increased (from 1.61 GHz for K0–0.9 GHz for K5).

For antenna miniaturization, it was already noted that size and performance interplay with each other. González-Arbesú et al. [48] considered several fractal shapes and sizes. These are depicted in Fig. 3.12 and include the Hilbert curves, variants of Peano curves and meandering antennas. Table 3.2 lists their geometrical parameters and performance (quality factor, input impedance vs. size, and resonance). For a more fair comparison, the resonance of all antennas was kept near 800 MHz (the GSM frequency for mobile phones). Clearly, the smallest

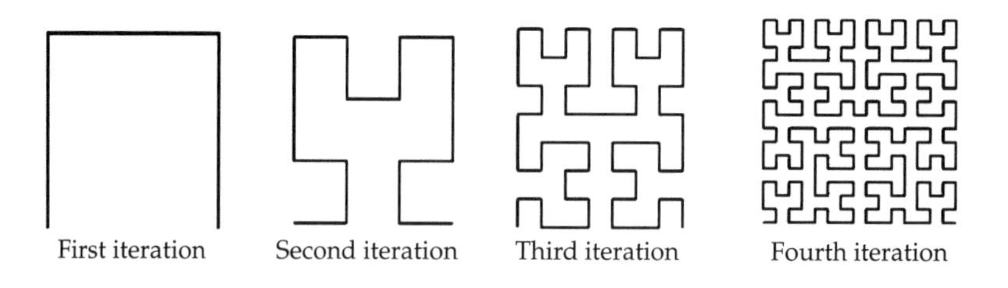

FIGURE 3.10 Hilbert curves with different iterations. (*After Werner et al.* [4] *@ McGraw-Hill 2007.*)

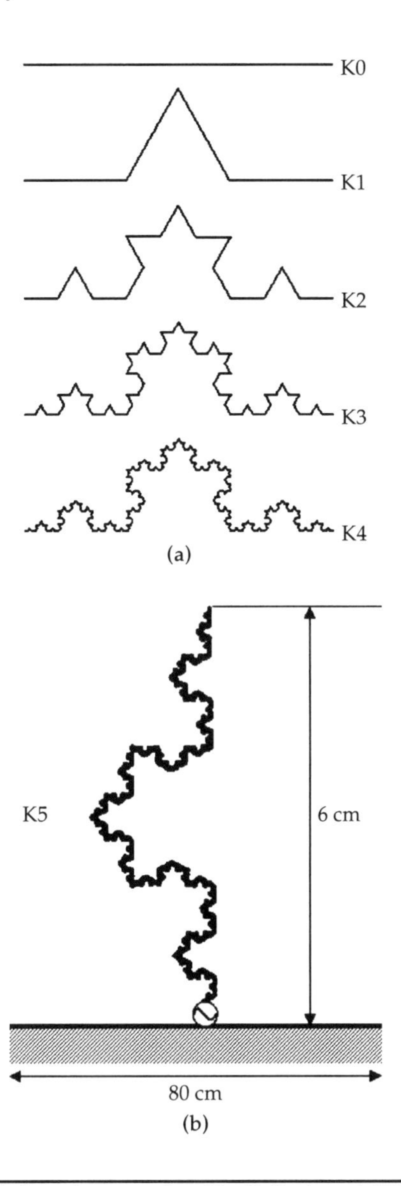

FIGURE 3.11 Koch fractal monopoles. (*See Baliarda et al.* [43].)

of all antennas is the meander line shape called MLM-8 in Table 3.2 and Fig. 3.12 (its length is 14% of the standard $\lambda/4$ monopole). This corresponds to $ka = 0.2$, implying an electrically small antenna based on our definition in Chap. 1. However, the Q of MLM-8 is rather large and its input resistance extremely small (only 1.1 Ω). This is true for all

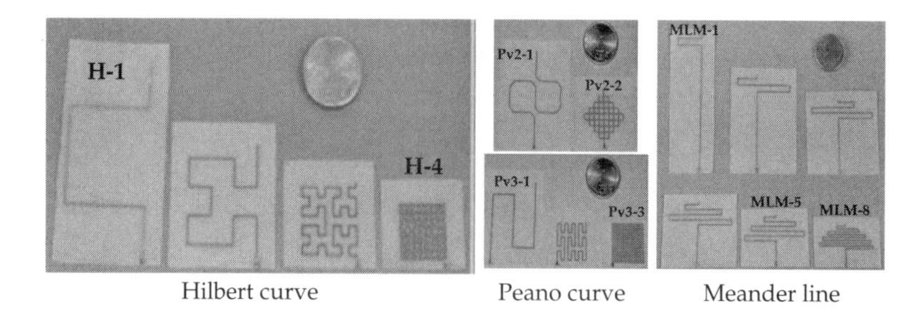

FIGURE 3.12 Fractal and meander line antenna samples. (*See González-Arbesú et al.* [48].)

highly meandered antennas in Table 3.2 (Hilbert-4, Peano 3-2). We also observe that fractal shapes do not necessarily have an advantage over other meandered shapes. A more detailed study on the effectiveness and performance of fractal shapes and other meandered antennas was given by Best [49–51].

As can be expected, an issue with the highly meandered shape is their higher conductor loss and lower radiation resistance. To better ascertain the reason for the lower performances of some meandered antennas, we refer to Fig. 3.13 showing three shapes: fourth order Hilbert curve, meandered dipole D1 and meandered dipole D2, all

Antenna	Height (mm)	*ka*	Wire Length (m)	Resonant Frequency (MHz)	Radiation Resistance (Ω)	*Q*
λ/4 monopole	90.6	1.506	0.091	793.1	35.8	7.8
Hilbert-1	53.5	0.8838	0.105	788.2	12.5	21.2
Hilbert-2	32.77	0.5503	0.125	786.4	4.7	54.6
Hilbert-3	23.3	0.4037	0.172	784.9	2.5	112.8
Hilbert-4	18.5	0.3277	0.266	788.6	1.7	197.2
Peano v2-1	44.0	0.7274	0.117	788.8	7.7	36.9
Peano v2-2	27.5	0.4555	0.209	790.3	3.3	105.3
Peano v3-1	37.7	0.6262	0.109	789.6	4.7	54.6
Peano v3-2	21.14	0.3714	0.174	791.8	1.4	209.7
MLM-1	66.8	1.1149	0.097	796.4	40.4	9.7
MLM-2	44.0	0.7325	0.105	794.3	17.0	17.5
MLM-3	32.8	0.5439	0.119	791.2	8.9	31.3
MLM-4	28.6	0.4754	0.146	793.1	6.2	45.0
MLM-5	20.2	0.3337	0.139	788.2	2.7	89.3
MLM-8	12.5	0.2033	0.141	775.9	1.1	160.5

TABLE 3.2 Size and Performance of the Fractal and Meandered Antennas

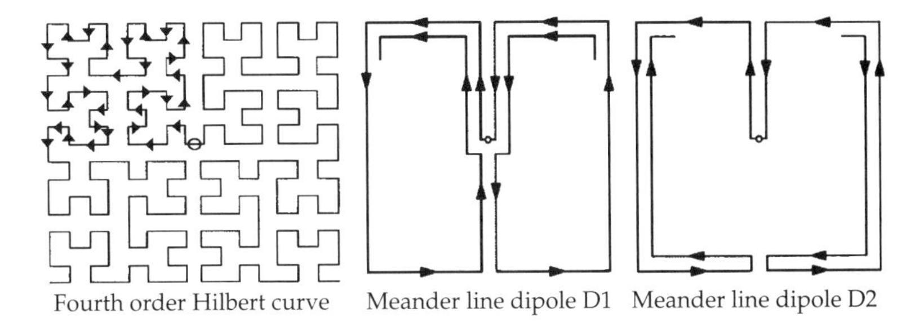

FIGURE 3.13 Hilbert curve and meander line dipoles. (*See Best and Morrow* [50].)

residing within an area of 7 × 7 cm. A key observation among these antennas is that the Hilbert curve resonates at a higher frequency than the dipole D1 (267.2 MHz vs. 154.9 MHz). That is, more meandering does not necessarily lead to lower frequency of operation. For the case in Fig. 3.13, the Hilbert curve includes many sections of cancelling currents from adjacent conductors. These sections do not radiate but do contribute losses and possible impedance mismatches. Such current cancellations occur in spiral antennas as well, but their inherent frequency scaling allows for sections of congruency that occurs at different radii (as the frequency changes), leading to broadband behavior. Spirals and their miniaturization are considered later in Chap. 5.

3.2.6 Volumetric Antennas

To improve planar (or purely printed) antenna performance, volumetric versions of planar designs have been considered. Several volumetric fractal antennas have already been studied (Koch, Hilbert curve antennas fabricated from wires and strips) [52–54].

As an example, multiband behavior has been delivered by tree-type fractal antennas as in Fig. 3.14. For these tree-type fractal shapes, the branches serve as reactive *LC* traps and may also incorporate switches for reconfigurations. For the shown shapes in Fig. 3.14, tunability up to 70% was demonstrated [55]. Similar (but more random) 3D fractal antennas were considered by Rmili et al. [56] displaying multiband performance across 1 to 20 GHz band.

Best [57] considered a spherical helical antenna with a goal to reach the optimal Q limit. His four-arm spherical spiral is depicted in Fig. 3.15 and is intended to more efficiently utilize the spherical Wheeler-Chu volume. As displayed in the table accompanying Fig. 3.15, a $Q = 32$ was achieved with an input resistance of $R = 43\,\Omega$. Correspondingly, the antenna height (off a ground plane) was only 0.0578 (ka = 0.36), that is, 1.5 times the Chu limit. The table shows

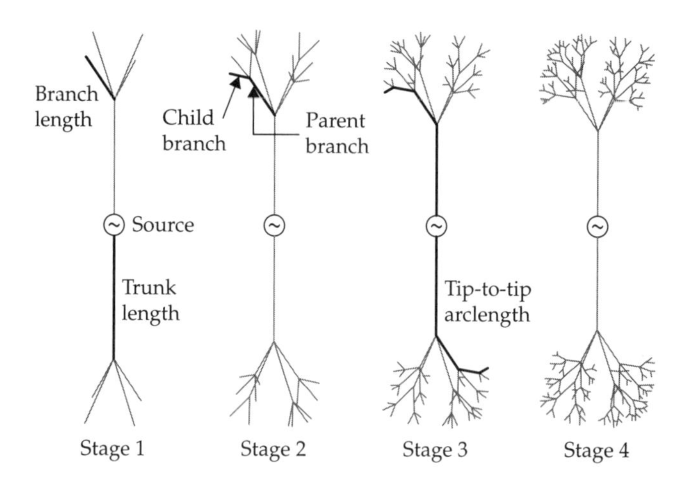

FIGURE 3.14 A fractal tree dipole antenna. (*After Petko and Werner* [55] © *IEEE 2004.*)

that more turns lead to higher Q (as expected) and less turns to lower Q. However, the input resistance also changes becoming smaller or larger. The one-turn example is attractive as its input resistance is 43.1 Ω (close to 50 Ω).

No. of Turns	Arm Length (cm)	f_R (MHz)	R_A (Ω)	Efficiency (%)	Q
0.5	17	515.8	87.6	99.6	5.6
1	30.9	300.3	43.1	98.6	32
1.5	45.07	210	23.62	97.6	88

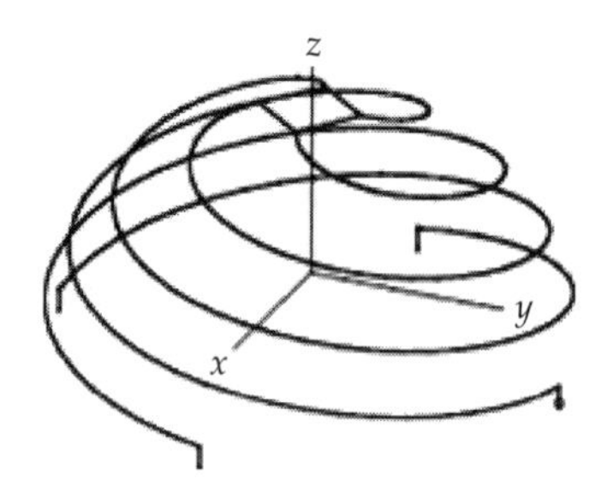

FIGURE 3.15 Four-arm folded spherical helix antenna with resonant performances. (*After Best* [57] © *IEEE 2004.*)

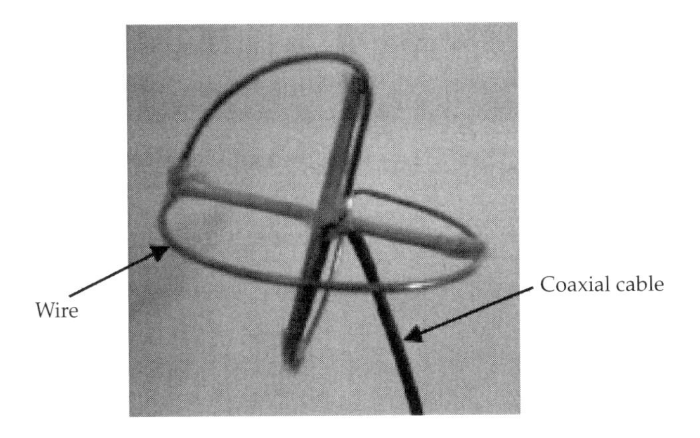

FIGURE 3.16 Photo of the fabricated spherical wire antenna. (*After Mehdipour et al.* [58] © *IEEE 2008.*)

A different volumetric structure was recently considered by Mehdipour et al. [58] As depicted in Fig. 3.16, this antenna was composed of four arms which were arranged in a perpendicular geometry to reduce the cancellation between each other. It demonstrated a resonance at 372.45 MHz with $ka = 0.62$ and $Q = 12.14$, that is, 2.09 times larger than the Wheeler-Chu limit ($Q_{\text{Chu}} = 5.81$). More importantly, its input resistance was nearly 50 Ω and had a high efficiency of 98.5% (i.e., comparable to the spherical helix by Best in Fig. 3.15).

3.2.7 Radio Frequency Identification Device Antennas

Meandering has been extensively used in designing antennas for RFIDs and keyless applications. Typical RFIDs operate at 16, 315, 433, and 911 MHz. At these frequencies, $\lambda/2$ or $\lambda/4$ size tags are very large to be useful in practice. Thus, miniaturization is necessary to make RFIDs practical.

Figure 3.17 shows a meandered rectangular loop structure. It was designed [59] to operate at 911.25 MHz and used as an RFID tag. It was printed on a foam 3.5 mm thick, having $\varepsilon_r = 1.06$. The overall tag size was only 31 × 31 mm (0.09 × 0.09 λ). By comparison, a nonmeandered loop operating at the same frequency would need to be $\lambda/4 \times \lambda/4$ in size. It is also important to note that the triangular shaping avoids current cancellation that would lead to lower resistance values. The meandered loop antenna was formed by strips 0.3 mm thick and separated by the same 0.3 mm distance (feed line is 6 mm long). This tag was found to show a −2.5 dB gain and had a −10 dB return loss bandwidth of 1.5% at 911.25 MHz.

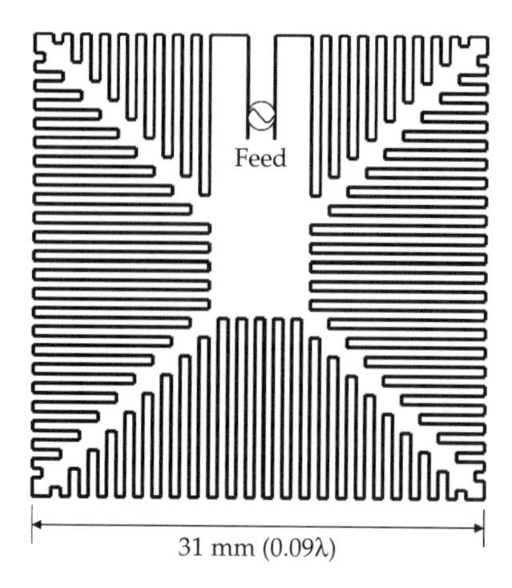

FIGURE 3.17 Geometrical structure of a meandered rectangular loop antenna. (*See Ryu and Woo* [59].)

Figure 3.18 depicts an electrically small monopole antenna for an active 433.92 MHz RFID system in a metallic container application [60]. It was a folded monopole antenna using a C-shaped meander and supported by Styrofoam ($\varepsilon_r = 1.06$) substrate. This antenna was shown [60] to deliver more than −1 dBd (dB above a dipole) gain relative to the folded monopole. However, the height and diameter of the antenna were 20 mm and 34 mm (0.0291 and 0.0491 λ at 433.91 MHz), respectively.

To reduce RFID antenna size, high substrate permittivity has been considered. Figure 3.19*a* shows [61] a slot-type antenna operating at 415 and 640 MHz. It was printed on a substrate having $\varepsilon_r = 197$,

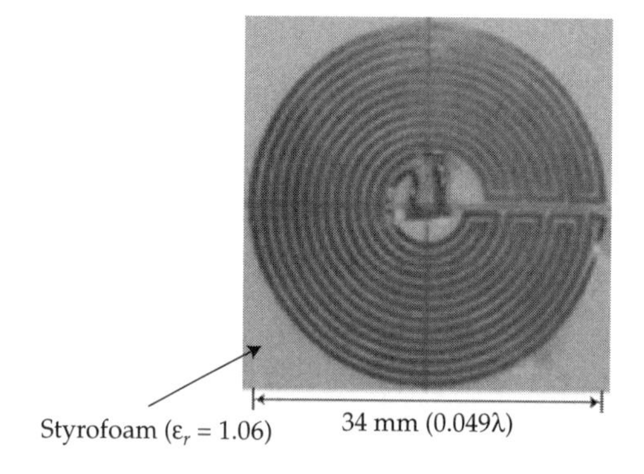

FIGURE 3.18 Fabricated folded monopole antenna using C-shaped meander. (*See Ryu et al.* [60].)

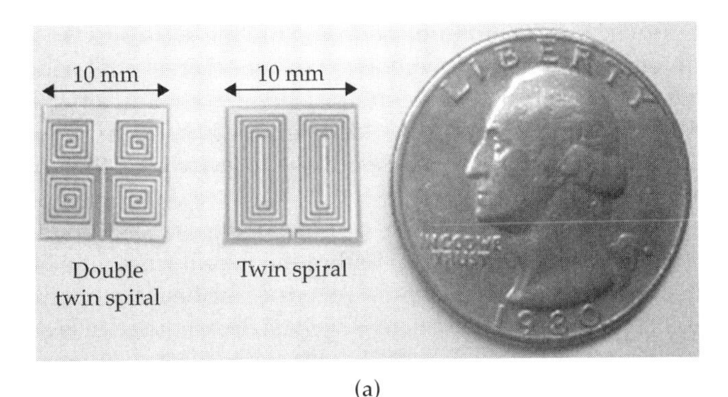

(a)

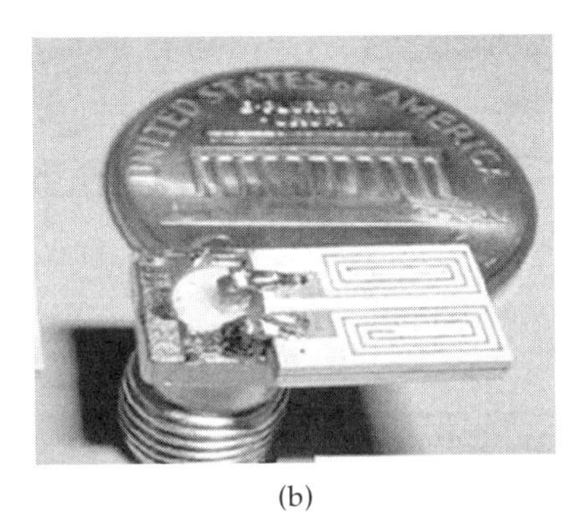

(b)

FIGURE 3.19 (a) Printed slot antennas etched on a high permittivity substrate. (*After Takigawa* [61] © *IEEE 2007.*) (b) Edge-fed planar meander line. (*See Gosalia* [62].)

10×10 mm^2 in size and thickness 2 mm (the slot width and spacing between them was 0.2 mm). For the shown double twin slot, the operating frequency was 640 MHz, implying a 0.022 λ size. The twin spiral operated at even lower frequency of 415 MHz, implying a 0.014 λ size. A similar meandered dipole (see Fig. 3.19*b*) was presented by Gosalia et al. [62] for use as an intraocular element in retinal prosthesis. It was designed to operate at 1.41 GHz with dimensions $5.25 \times 5.25 \times 1.5$ mm^3. A key aspect of the design is a small offset of the feed by shortening one of the dipole arms to induce current phase reversal. As such, it provided good impedance match while retained a symmetrical location of the feed point.

3.2.8 Small Ultra-Wideband Antennas

The above volumetric antennas are narrow band, but often of great interest are small UWB antennas for very high frequency (VHF) and high frequency (HF) communications on mobile platforms. To design such antennas, it is best to begin with the helical and spiral wideband antennas. Peng et al. [63] started with a standard helical antenna and introduced additional inductive loading along the arms. The latter was intended to further slowdown wave velocity and, thus, reduce helical antenna height.

To reduce the size of the helical antenna, Peng et al. [63] considered additional meandering (coiling) within the wire forming the helix as depicted in Fig. 3.20*a*. The potential to reduce the helical antenna's frequency of operation without increasing its size is depicted in Fig. 3.20*b*. As seen, each of the coiled (inductively loaded) helical antennas has

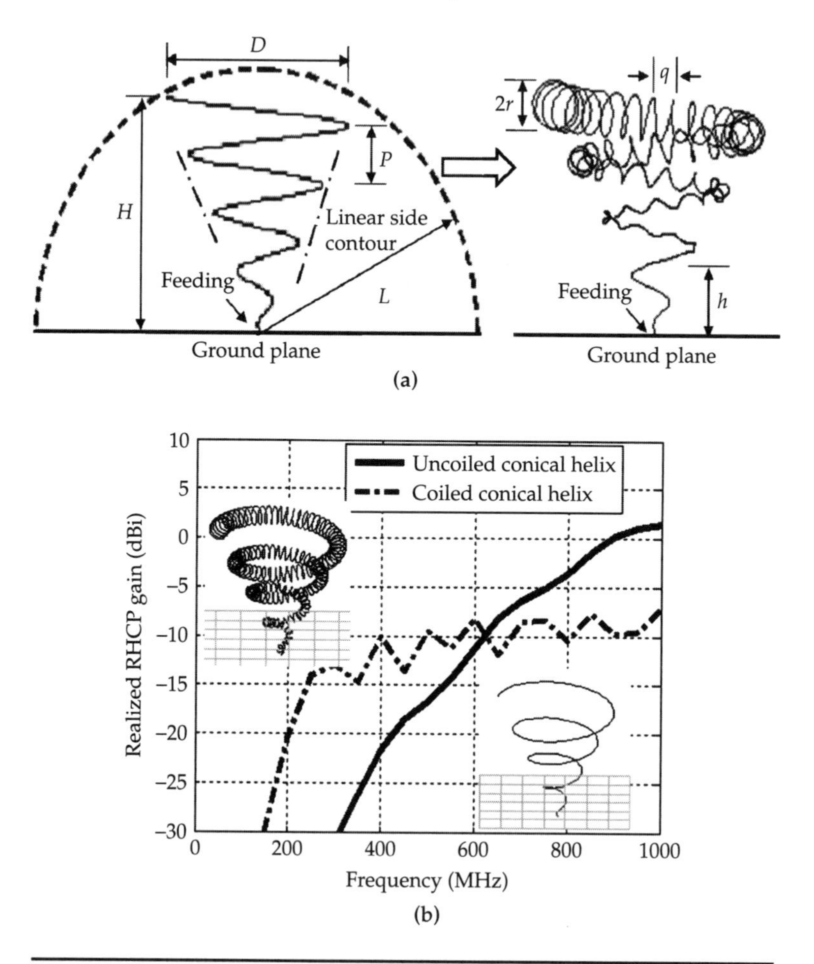

FIGURE 3.20 (a) Geometrical parameters of a conical helix antenna and coiling details. (b) Realized gain of the coiled and uncoiled conical helices. (*Courtesy The Ohio State University, Electroscience laboratory.*)

its −15 dBi point at a lower operational frequency. Specifically, the uncoiled helix (height $H = 5''$ and max. width $D = 5''$) has its −15 dBi gain point at ~500 MHz. By comparison, the coiled helices have the same performance at ~200 MHz. However, the gain at the higher frequencies is reduced after coiling, implying optimization of the coil pitch and radius is needed.

A reduced size UWB helical antenna was also considered by Yang et al. (see Fig. 3.21) [64]. As shown, the spherical helix encloses a truncated cone serving as a feed to the entire geometry, providing flexibility in input impedance control and bandwidth. The designed configuration delivered 10:1 bandwidth with its size being $\lambda/8$ at the lowest operational frequency. Thus, its size was $10/8\ \lambda$ at its highest operational frequency.

In contrast to helices, spirals provide for low-profile antennas. Kramer et al. [65] considered volumetric coiling of the traditional spiral to further reduce its size without much compromise in conformality. To enable conformality and still retain the bandwidth of the free-standing spiral, a ferrite ground plane was introduced. In one example, the ferrite was 0.25″ thick (from Trans-Tech) and was placed on the base of the metallic cavity to modify the reflection coefficient closer to +1 (instead of −1), and therefore avoid shorting at low frequencies. For testing, the spiral was weaved within (see Fig. 3.22*b*) on a 0.25″ thick Rogers' TMM4 laminate ($\varepsilon_r = 4.5$, $\tan\delta = 0.002$) and placed at the top of a cavity and 1″ above the ferrite-coated ground plane. The overall cavity height was 5″ (all included) and 6″ in diameter. As depicted in Fig. 3.22, the coiled spiral on a ferrite-coated ground

FIGURE 3.21 A UWB spherical helix with an enclosed tapered feed. (*Courtesy of Yang and Davis at Virginia Polytechnic Institute and State University.*)

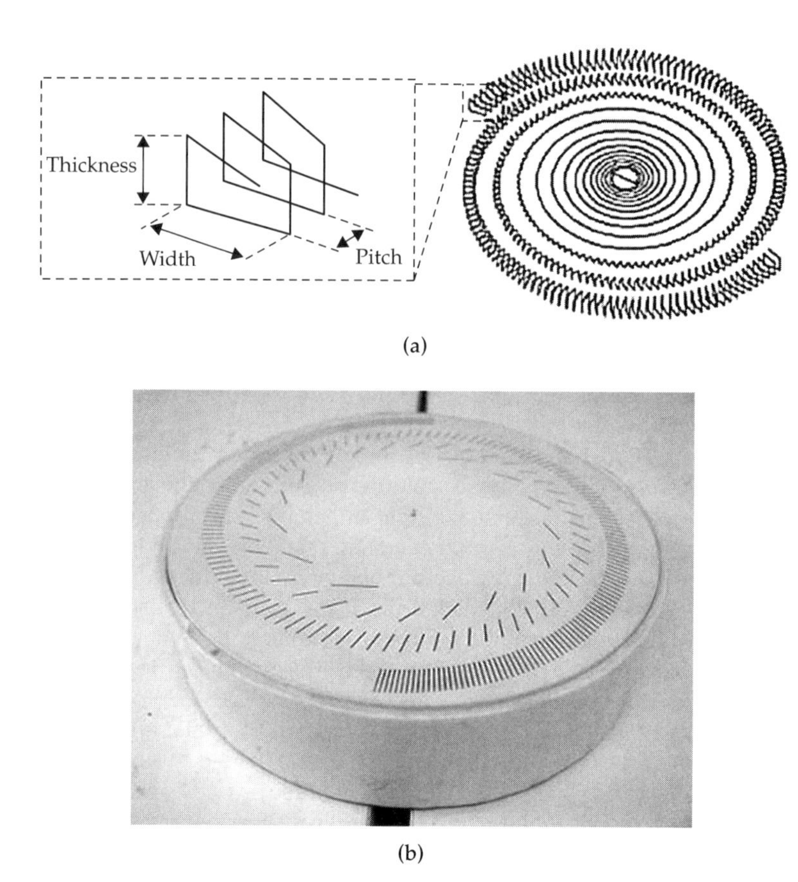

(a)

(b)

FIGURE 3.22 Photo of the fabricated 6″ inductively loaded spiral antenna and performance with/without the ferro-metallic ground plane; comparison to the standard Archimedean unminiaturized spiral (backed by a copper ground plane) is also provided for reference. (*After Kramer et al.* [65] © *IEEE 2008.*) This antenna is commercially available. (*www.appliedem.com*)

plane delivered −15 dB gain at 170 MHz, whereas the unminiaturized spiral had the same performance at ~400 MHz. Of importance is that the ferrite ground plane served well to recover the free-space gain of the spiral. We remark that the spiral looses its circular polarized (CP) performance at the low frequencies. As the operational frequency increases, its CP performance is recovered and its gain reaches around 5 dB. More details of the design of small spirals are given in Chap. 5.

3.2.9 Lumped Loading

Many of the miniaturization techniques by means of shaping, including meandering, can be explained using lumped loading. Referring to

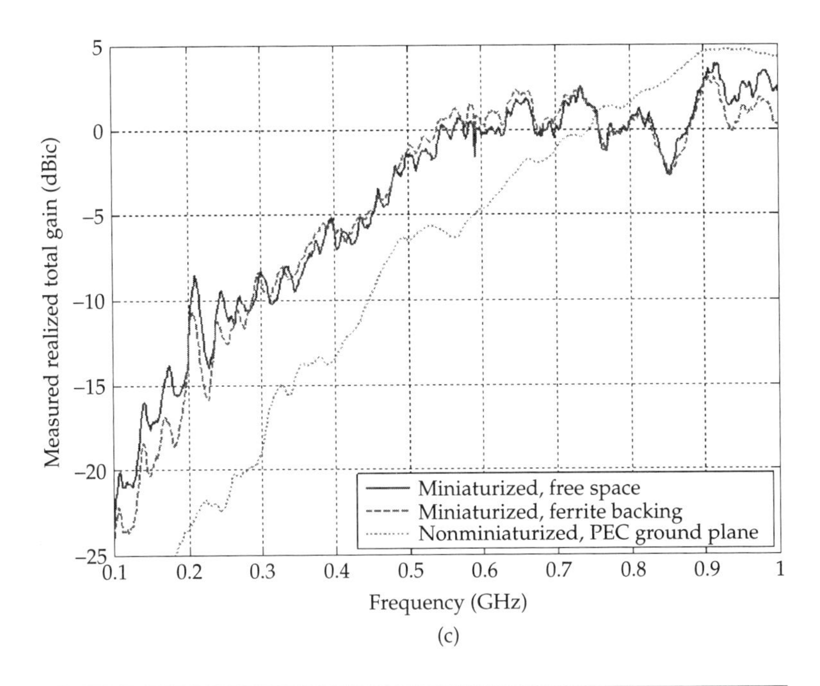

FIGURE 3.22 *Continued.*

Fig. 3.23*a*, the meandered monopole is equivalent to an inductively loaded one (right). Also, a top loaded monopole can be represented by a capacitively loaded monopole. Further, slot loaded patches [66] can be represented by equivalent LC circuits as depicted in Fig. 3.23*c*.

Figure 3.24 shows a low-profile GPS antenna within a cellular phone, that employs capacitive loading for size reduction. The actual antenna [67] was a folded plate forming a loop with overlapping plates to increase capacitance. The bottom of this folded loop antenna was connected to ground plane when the antenna was mounted on the PCB board. Feeding was done at the top forming a "tongue" on the top layer of the folded structure. The overall antenna was $25 \times 10 \times 2.5\ \text{mm}^3$ and operated at the L1 GPS band of 1575 MHz with a return loss of less than −10 dB over a 12 MHz bandwidth. Its gain was 2.5 dBi, implying a 75% overall efficiency. Of importance is that its frequency shifted only to 1579 MHz when human hand covered the handhold. That is, the human hand did not impact the operational frequency.

Lumped loads can be, of course, incorporated directly into antennas for size reduction. However, as noted in [68], use of lumped loads implies power handling limitations and additional losses. Their similarity is a main reason for using lumped loads to reduce antenna size. Several antennas have adopted use of lumped elements for miniaturization. A CPW-fed folded slot example [69] is shown in Fig. 3.25 and

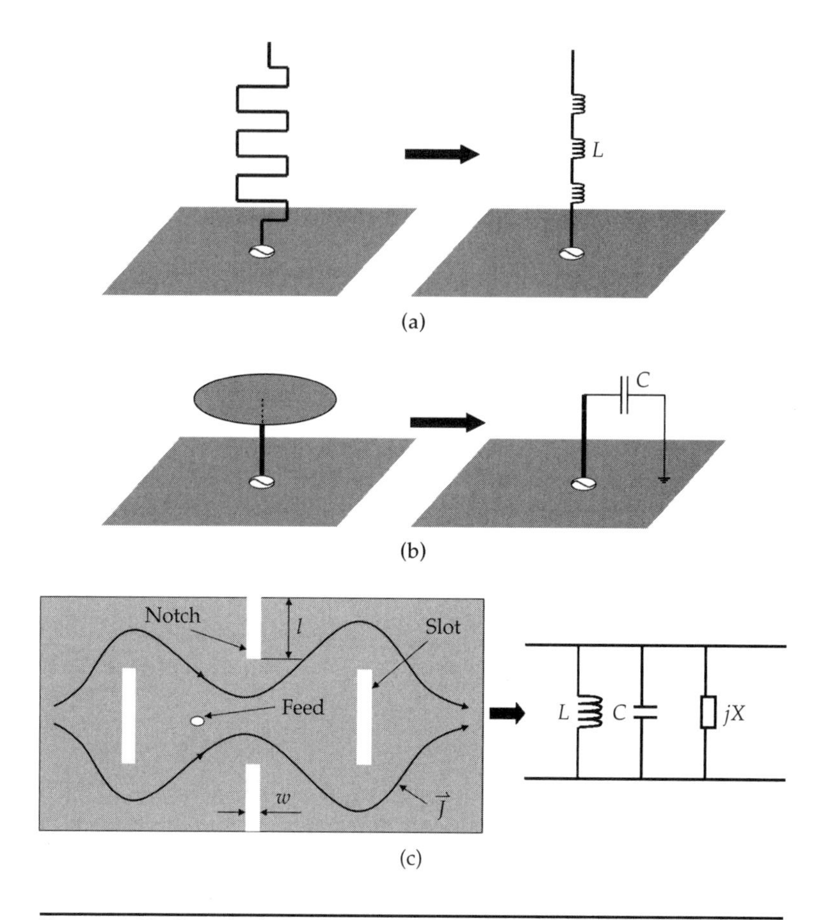

FIGURE 3.23 Reactively loaded monopoles and patches; their corresponding equivalent circuits are given to their left. (a) Meandered monopole; (b) Monopole with a top plate; (c) Slot loaded patch.

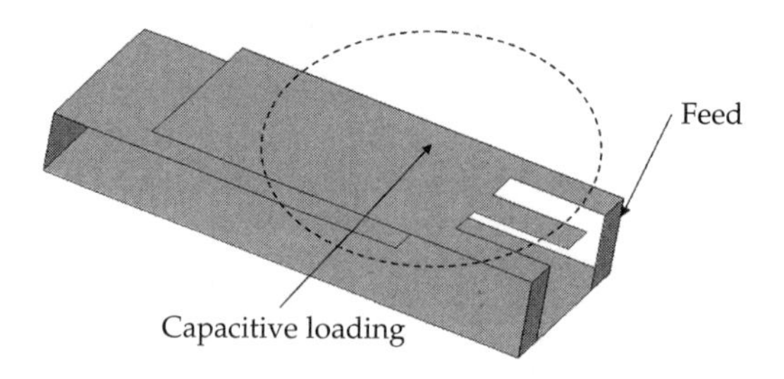

FIGURE 3.24 Geometrical configuration of a GPS antenna incorporated into the cell phone. (*See Rowson et al.* [67].)

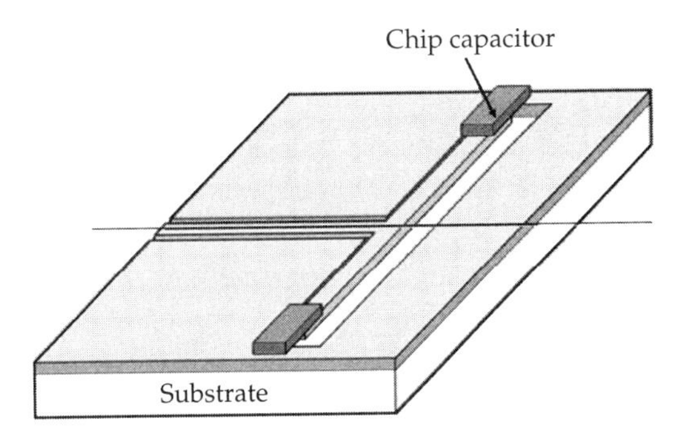

FIGURE 3.25 Slot antenna loaded with chip capacitor. (*See Scardelletti et al.* [69].)

incorporates two chip capacitors placed at both ends of the folded slots. These capacitors reduced the operational frequency by 22% and improved return loss as well.

Slot loaded with capacitors in a periodic fashion (see Fig. 3.26) is a popular loading approach in the context of metamaterial [70] structure for wave slowdown. Figure 3.26 shows a CPW-fed slot loop [71] printed on a Duroid 5880 substrate having $\varepsilon_r = 2.2$ and thickness $h = 1.5$ mm. For miniaturization, the loop was loaded with 14 capacitors, each 5 pF in capacitance. This capacitive loading resulted in a frequency reduction by a factor of 6 (from 3.57 GHz down to 0.56 GHz). The gain was a respectable −2 dBi.

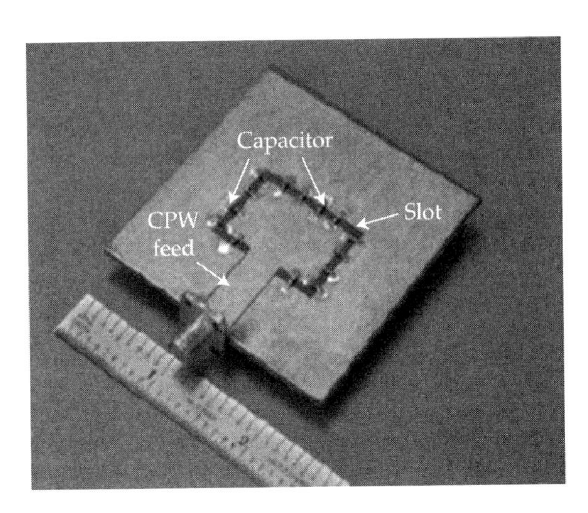

FIGURE 3.26 Miniaturization of a CPW-fed slot antenna loaded with 14 capacitors. (*See Chi et al.* [71].)

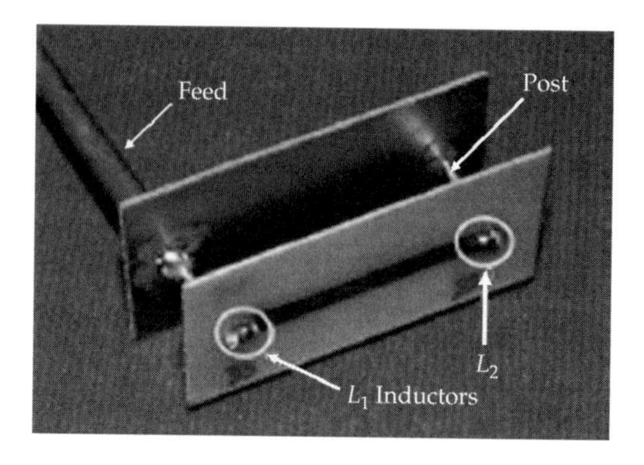

FIGURE 3.27 Vertical loop antenna incorporating chip inductors for miniaturization. (*After Lee et al.* [72] © *IEEE 2008.*)

Figure 3.27 shows a vertical loop antenna (much like the towel bag [4] antenna often used for HF radiation). It incorporated two chip inductors placed on the top strip forming the loop as shown in Fig. 3.27. Lee et al. [72] considered various values (denoted as $L1$ and $L2$) of the chip inductors to observe their impact on the resonant frequency. It was noted that as $L1$ increases, the resonant frequency decreases and so does the input impedance. However, as $L2$ increases the opposite occurs with the resonant frequency and input resistance. A fabricated version of the vertical loop antenna using $L1 = 33$ nH and $L2 = 33$ nH had an electrical size of 0.118 λ (length) $\times$ 0.013 λ $\times$ 0.047 λ (height) at the operational frequency of 1.2 GHz.

3.3 Miniaturization via Material Loading

A popular method for antenna miniaturization is to use high-contrast materials at strategic locations within the antenna volume. As is well known, material loading provides for wave slowdown in a manner proportional to $1/\sqrt{\varepsilon_r \mu_r}$. Material loading can also allow for impedance control as the medium impedance is proportional to $\sqrt{\mu_r/\varepsilon_r}$. For wideband impedance matching, it is important to retain $\varepsilon_r \approx \mu_r$ in as much as possible. However, commercially available magnetic materials are generally quite lossy above 300 MHz. This is due to domain wall resonances [73] as depicted in Fig. 3.28 which occurs at lower frequencies than gyromagnetic resonances. Careful fabrication and alignment of the magnetic moments is necessary to push the domain wall resonances at higher frequencies. However, as depicted in

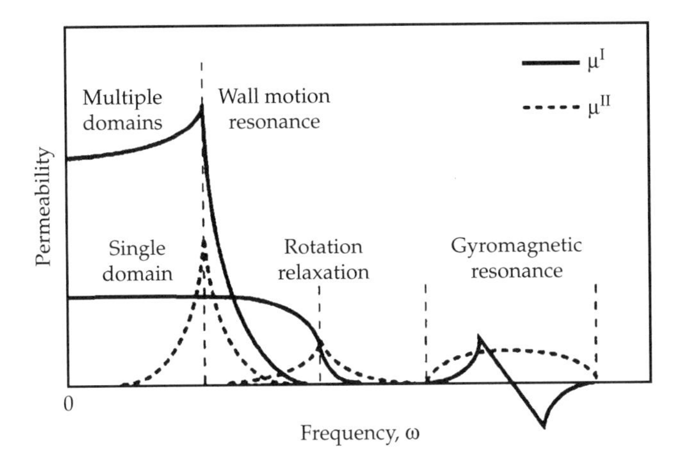

FIGURE 3.28 Schematic diagram of permeability variations (as a function of frequency) caused by various magnetic damping actions. The solid line is Mu-prime (real part) and the dash line is Mu-double-prime (image part). (*After Dionne* [73] @ *IEEE 2003*.)

Fig 3.29, there are low loss region between the μ_r'' peaks, and these can be exploited for printing antennas on magno-dielectric substrates [74]. One must also be aware of Snoek's law [75] which forces μ_r' to lower values when pursuing low loss substrates. Indeed, magnetic materials hold promise for improving antenna bandwidth and miniaturization,

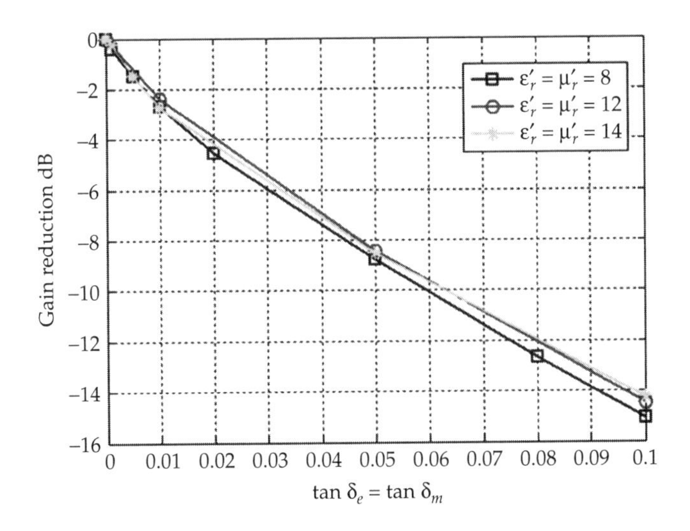

FIGURE 3.29 Gain reduction due to material losses. (*See Tzanidis et al.* [74].)

and are a topic of intense research. It is not therefore surprising that most efforts in using materials have focused on high-contrast (high ε_r) dielectrics. The latter are often ceramic structures and reach ε_r values well above $\varepsilon_r = 100$. Clearly, such values lead to mismatches and narrow bandwidth unless magnetic or inductive loading is introduced. In the following sections we discuss various literature efforts on using dielectric and magnetic materials. First, we discuss use of dielectric (ε_r only) materials for miniaturization and then consider magnetic substrates for conformal installations.

3.3.1 Dielectric Materials

As already noted, dielectric substrates or superstrates lead to miniaturization. However, there are diminishing returns as the dielectric's thickness increases. Kramer et al. [76] considered the miniaturization of 2 × 2″ spiral printed on two different dielectric substrates having $\varepsilon_r = 9$ and $\varepsilon_r = 16$. Specifically, the miniaturization factor defined as MF ($f_{\text{unloaded},-15\text{dBi}}/f_{\text{loaded},-15\text{dBi}}$) is a function of substrate thickness. As depicted in Fig. 3.30*a*, there are diminishing miniaturization returns after the substrate thickness exceeds 0.1 λ, with little to no impact in miniaturization after 0.2 λ. There are also issues with impedance matches. Therefore, depending on the antenna, one must be selective in choosing the substrate's dielectric constant [76]. Figure 3.30*b* provides an example study for choosing substrate dielectric constant. As seen, after $\varepsilon_r = 30$ to 40, there are again diminishing returns in miniaturization. That is, even half-space dielectric loading can theoretically provide fivefold miniaturization. Such a level of miniaturization cannot be achieved with a practical thickness dielectric superstrate.

Several small GPS antennas have been recently designed by Zhou et al. [77] using high-contrast dielectric substrates and superstrate. As noted, for the U.S. satellites, such antennas operate at 1575 MHz (L1 band), 1227 MHz (L2 band), and possibly the newer 1176 MHz (L5 band). In addition, the European Galileo satellites operate in the frequency ranges 1164 to 1215 MHz (E5a and E5b), 1215 to 1300 MHz (E6), and 1559 to 1592 MHz (E2-L1-E1). Further, the Russian GLONASS satellites transmit at 1602.56 ~ 1615.50 MHz (L1 band) and 1246.44 ~ 1256.50 MHz (L2 band). In all cases, the required bandwidth at each band does not exceed 25 MHz and feeding must ensure right hand circularly polarization (RHCP) radiation with a broadside gain typically around 2 dBi.

As can be realized from the preceding paragraph, GPS antennas must be small for integration into handhelds and small vehicle platforms, but must also allow for multiband operations. Zhou et al. [77,78], considered several small size GPS antennas. One of them is depicted in Fig. 3.31*a* and incorporated a dual patch design embedded

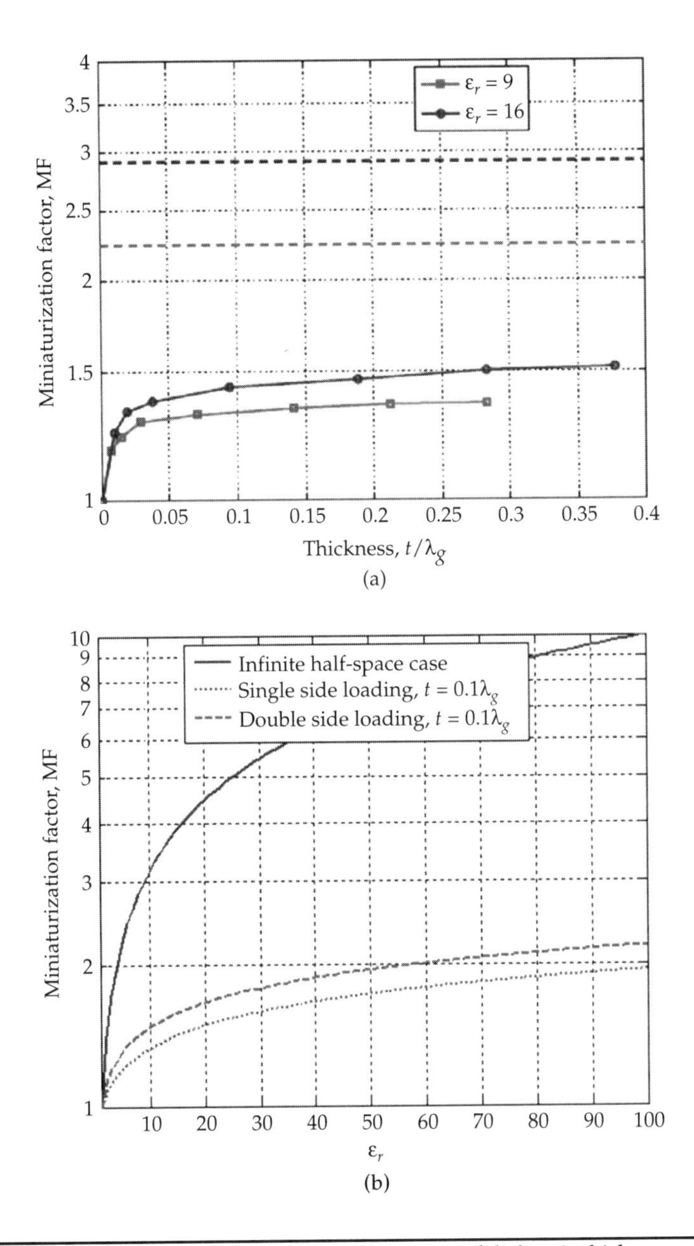

FIGURE 3.30 Miniaturization factor as functions of dielectric thickness and width. (a) MF as a function of dielectric slab thickness; (b) MF as a function of dielectric constant. (*After Kramer et al.* [76] @ *IEEE 2005.*)

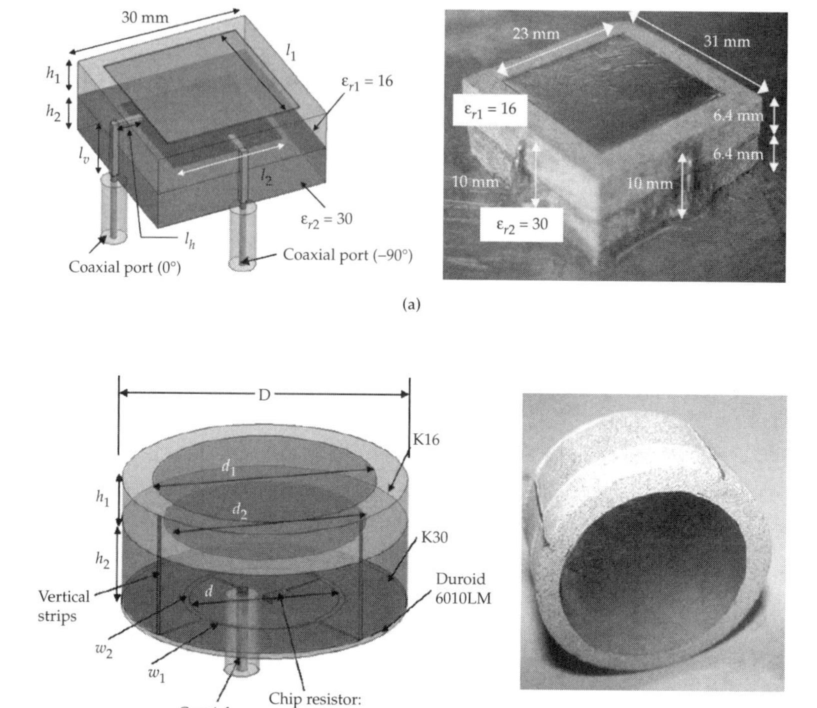

FIGURE 3.31 Geometry and fabricated proximity-fed stacked patch antenna: (a) Rectangular patch (*After Zhou et al.* [77] *@ IEEE 2007.*); (b) Circular patch (*After Zhou et al.* [78] *@ IEEE 2008.*)

in high-contrast dielectric. Figure 3.31*a* (right) shows the configuration indicating a bottom patch on an $\varepsilon_{r2} = 30$ substrate of thickness 6.4 mm and a top patch on a slab of $\varepsilon_{r1} = 16$, also of thickness 6.4 mm. This dual patch antenna resonated at 1575 and 1227 MHz, covering the GPS L1, L2, and L5 bands. It employed bent probes to excite the top and bottom patches. Of importance is that the aperture physical size was 1.2 × 1.2″, implying only $\lambda/8 \times \lambda/8$ at L5 band. Figure 3.31*b* depicts a circular patch version of the antenna in Fig. 3.31*a* and its performance is given in Fig. 3.32. The circular patch allowed for CP excitation and used a Wilkinson divider to permit a single-feed configuration but still enable CP radiation. The circular patch design was still 1.2″ in diameter, but a smaller version was developed recently, using inductive loading (in the form of slots for the top patch). This latter configuration (see Fig. 3.3) incorporated spiral-like slots for additional

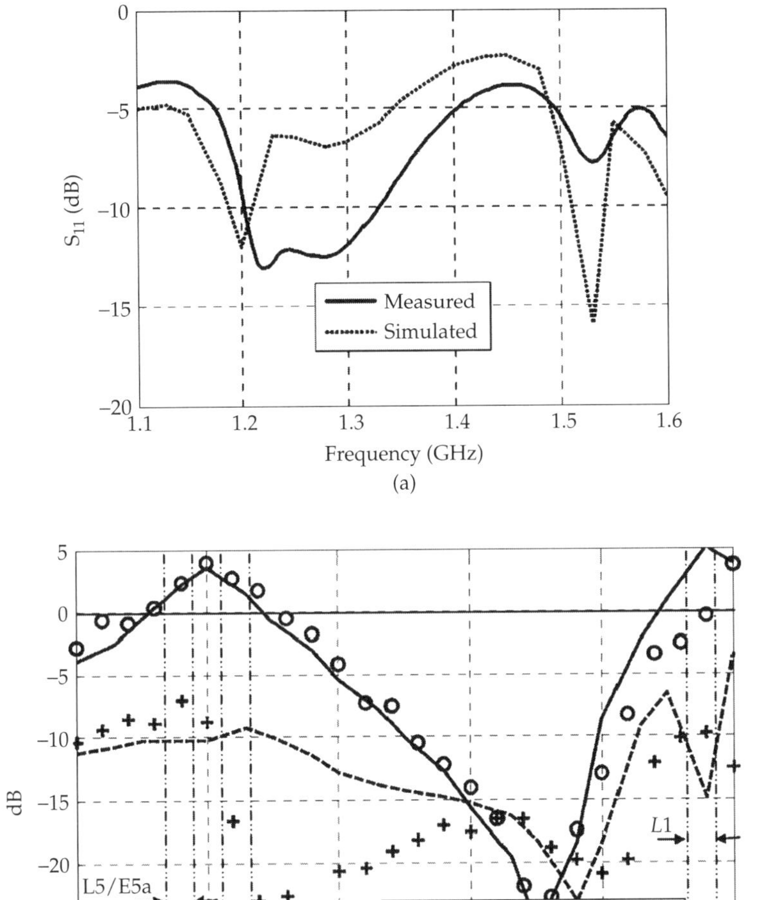

FIGURE 3.32 Performance of the circular proximity-fed stacked patch antenna in Fig. 3.31*b*. (a) Return loss; (b) Gain. (*After Zhou et al.* [78] @ *IEEE 2008.*)

miniaturization. The resulting aperture with the spiral-like slots was only 1″ (i.e., only $\lambda/10$ at 1227 MHz).

There are many antenna products in the literature and market (see Fig. 3.33) incorporating high-permittivity ceramic substrates. For example, Fig. 3.33*a* shows a helix wound on a high-contrast ceramic having $\varepsilon_r = 10$, intended for GPS reception [79–81]. Besides GPS, high-contrast materials are widely used for surface mounted on-chip antennas serving for USB, RFIDs, WLAN, WiFi, Bluetooth, etc., applications [82–85]. Figure 3.33*b* to *e* list some examples designed for these applications.

As already noted, miniaturization typically leads to lower gain antennas. Some of the gain loss can be recovered using thicker substrates and by incorporating superstrates. Alexopoulos and Jackson [86] concluded that a superstrate having ε_r higher than that of the substrate may increase radiation efficiency provided surface waves are not excited. A possible superstrate configuration is shown in Fig. 3.34. Huang and Wu [87] considered a specific superstrate configuration such as that in Fig. 3.34. The patch was printed on a substrate having $\varepsilon_r = 3.0$ and thickness $h_1 = 1.524$ mm (Duroid R03003). To improve

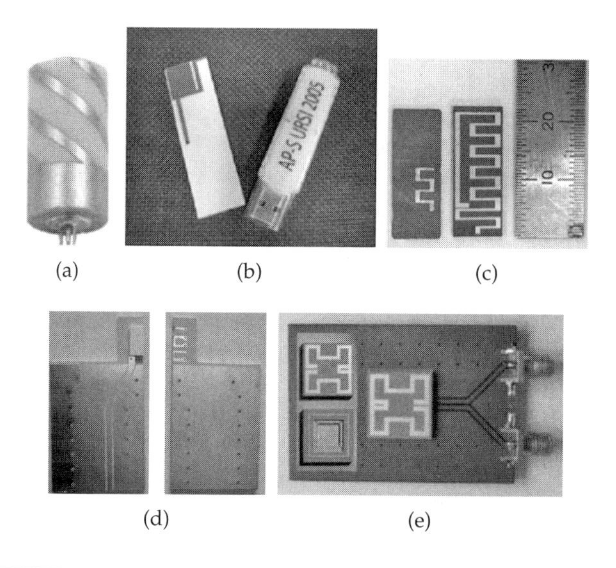

FIGURE 3.33 Antenna examples printed on high-contrast materials. (a) Quadrifilar GPS antenna on ceramic (*After Volakis @ McGraw-Hill 2007.*); (b) PCB antenna for a wireless USB dongle (*After Volakis @ McGraw-Hill 2007.*); (c) RFID/PCS/WiBro triple-band on-chip antenna (*After Lee et al. [83] © IEEE 2006.*); (d) 2.4/5.8-GHz dual ISM-band on-chip antenna (*After Moon and Park [84] © IEEE 2003*); (e) Single-chip RF transceiver on LTCC substrate. (*After Zhang et al. [85] © IEEE 2008.*)

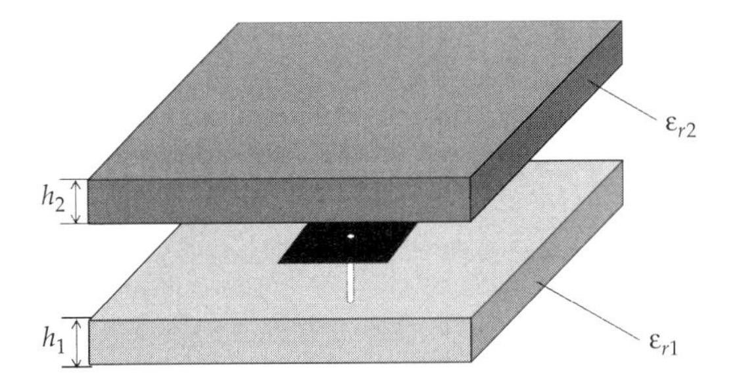

FIGURE 3.34 Structure of microstrip antenna covered with a high-permittivity material layer.

its gain, it was covered by a thicker superstrate ($h_2 = 3.05$ mm) having $\varepsilon_r = 79$. The resulting antenna resonated at 1.84 GHz and delivered a gain of 10.4 dB. This increase in gain is a result of modal field enhancement within the dielectric and pattern narrowing. More details on these modes are given in Chaps. 6 and 7. A 1 Ω resistor placed at one of the patch corners reduces the gain by 1.3 dB but can be used to increase bandwidth.

3.3.2 Magnetic Materials

As already noted, the magnetodielectric materials offer much advantage for antenna miniaturization, such as improved bandwidth and matching. Several antennas incorporating magnetodielectric substrates have been presented [88–96]. To avoid the issue of losses, researchers at The Ohio State University [95–98] utilized magnetodielectric layers as coatings on ground planes to enable conformal installation. That is, by maintaining a distance between the printed antenna and the ferrite coating, loss effects due to magnetic ferrites are minimized. More importantly, the ferrite coating modified the ground plane reflection coefficient between 0 and +1 to enhance conformal antenna radiation even at frequencies when the distance between the antenna and ground plane becomes less than $\lambda/20$.

Figure 3.35 shows possible antenna configuration on a ferrite-coated ground plane. The goal is to select materials which lead to a reflection coefficient near unity. Indeed it is shown in Fig. 3.36 that a magnetic layer on the ground plane with $\mu_r = 25$, $\varepsilon_r = 1$, and $\tan\delta = 0.01$ leads to significant gain improvement at low frequencies. The ferrite layer was only 0.5″ thick, but because $\Gamma = 2/3$ the destructive interference of the PEC ground plane was suppressed. In contrast, the treated ground plane does not cause cancelling interference, allowing for higher antenna gain even larger than the free standing antenna.

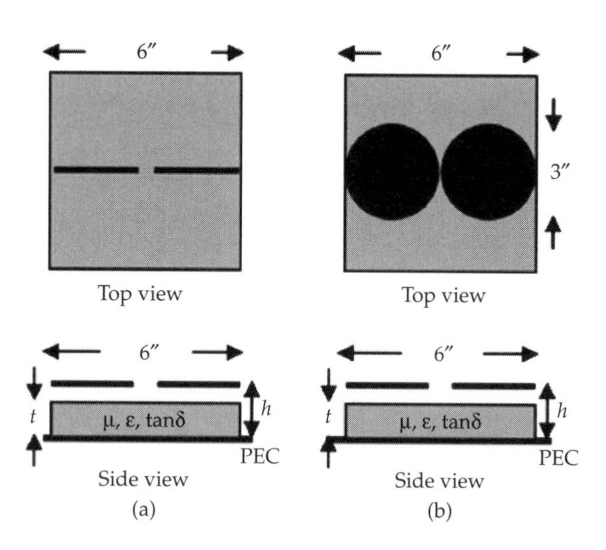

FIGURE 3.35 (a) Six-inch wire dipole antenna over a finite matched impedance layer (MIL); (b) Six-inch circular dipole over a finite MIL on a PEC surface. (*After Erkmen et al.* [96] © *IEEE 2009.*)

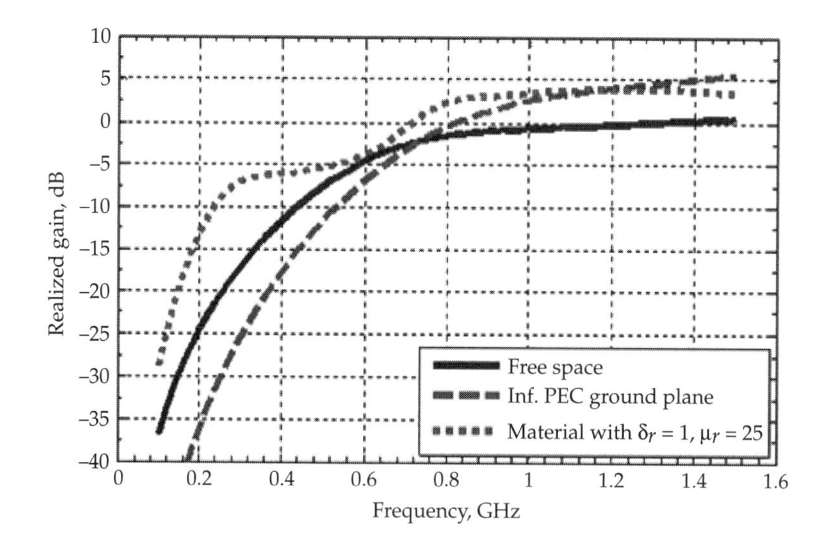

FIGURE 3.36 Realized boresight gain for a 6″ circular-disk dipole placed on a ferrite-coated ground plane. Comparisons are given for a freestanding disk dipole and when placed on ferrite-coated ground plane. (*After Erkmen et al.* [95] © *IEEE 2008.*)

More specifically, the -10 dB gain point was shifted from ~ 750 MHz for the PEC ground plane down to 250 MHz when the ferrite coating is inserted.

Clearly, the high μ_r in Fig. 3.36 is not realistic. In practice, the selection of magnetic material have a variety of (ε_r, μ_r) values and maybe lossy. It is therefore, more practical to focus on (ε_r, μ_r) which have an impedance $\eta = \eta_0\sqrt{\mu_r/\varepsilon_r}$ closer to the free-space impedance η_0. Such a material will therefore have as small reflection coefficient at its interface and, thus, referred to as matched impedance layer (MIL). The following three parameters must be considered in the design of the MIL:

- ε_r and μ_r of the layer
- Loss tangents $(\tan\delta_e, \tan\delta_m)$
- Angle of incidence

Figure 3.37 plots the calculated magnitude of the MIL reflection coefficient, $|\Gamma|$, as a function of ε_r and μ_r for a plane wave illuminating a homogeneous-material half space. The reflection coefficient was calculated for $8 \leq (\varepsilon_r, \mu_r) \leq 12$ and $\tan\delta_e = \tan\delta_m = 0.2$. Using $|\Gamma| < 0.1$ as a desirable goal, the material properties need to be within the region bounded by dashed lines. Indeed, some commercial ferrite materials fall within this acceptable range of (ε_r, μ_r) values. Three such examples are: (1) Co2Z by Trans-Tech (0.03–1.5 GHz), (2) TT2-101 by Trans-Tech (0.03–0.7 GHz), and (3) Ferrite-50 by National Magnetics (0.3–0.7 GHz).

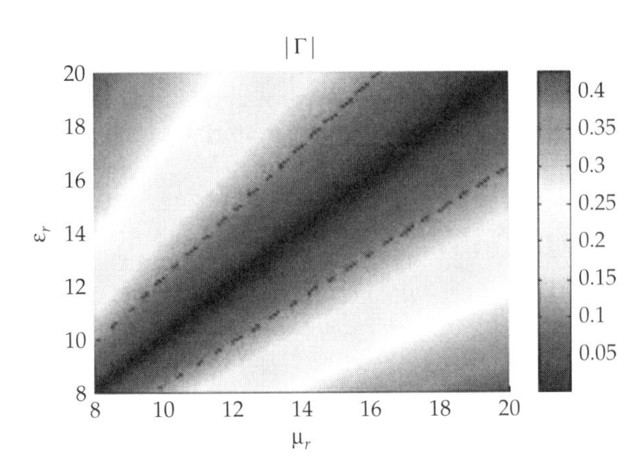

FIGURE 3.37 Reflection coefficient as a function of ε_r and μ_r for plane wave incidence on a homogeneous-material half space. (*After Erkmen* [95] © *IEEE 2008.*)

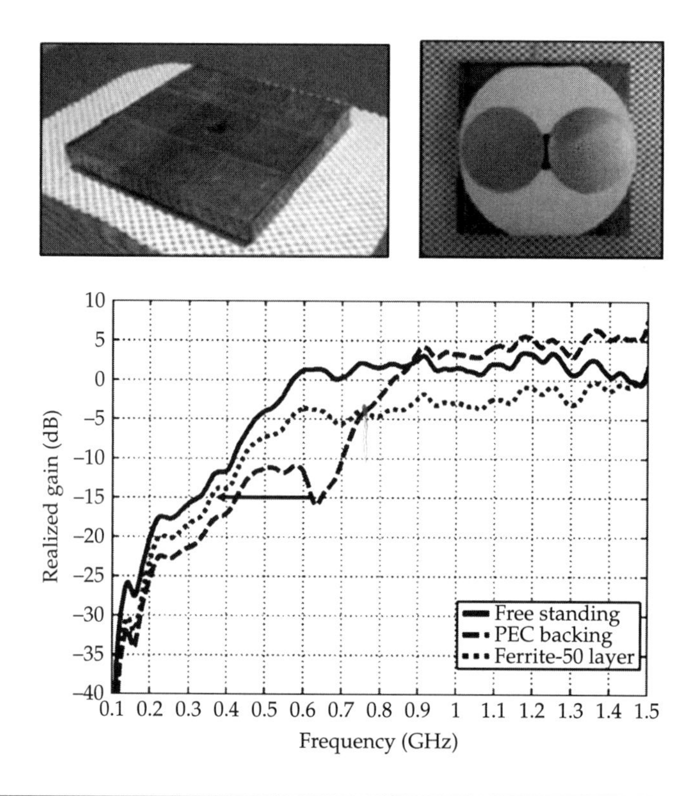

FIGURE 3.38 Photograph of a circular dipole placed on a Ferrite-50 slab of thickness 0.6″; its associated gain performance is shown from 100 to 1500 MHz. (*After Erkmen et al.* [96] © *IEEE 2009.*)

As the material properties can vary significantly with frequency, one must carefully select appropriate material for the performance of a circular dipole (same as shown in Fig. 3.38) placed on a Ferrite-50 layer of thickness 0.6″. The shown boresight gain displays significant improvement. Specifically, we observed that the −15 dB gain point shifted from 650 to 350 MHz, implying a miniaturization of 46.2%. However, the gain level above 750 MHz was lowered due to the varying ferrite properties at these frequencies. A way to address this issue is to taper the ferrite slab to reduce its thickness when it reaches the center region of the ground plane.

3.3.3 Polymer Ceramic Material

Polymer-ceramic composite is a good choice for antenna engineering [97,98] as they provide means for controlling the dielectric properties of the substrate from $\varepsilon_r = 2$ up to $\varepsilon_r = 20$ or even 30. Polymer-ceramic composites are light weight and can be mixed in a low-cost room temperature process for embedding into three-dimensional wire structures, allowing for miniaturization. Their dielectric loss tangent is typically $\tan\delta < 0.02$ for frequency up to several GHz. More recently, polymer substrates have been used to grow embed carbon nanotube (CNT) sheets [99,100] and fibers for load bearing/rugged applications.

To achieve higher dielectric constant using polymer mixtures, ceramic powder is dispersed into the polymer while wet. One such polymer is polydimethylsiloxane (PDMS), a silicone-based organic polymers (liquid crystal polymer or LCPs are other polymers but have a low dielectric constant). PDMS is hydrophobic and stable at high temperature (up to 200°C), but also of low cost. Also, PDMS has very low dielectric losses for frequencies up to several GHz, making it desirable for microwave applications. Various ceramic powders, namely barium titanate ($BaTiO_3$), Mg-Ca-Ti (MCT), strontium titanate ($SrTiO_3$ or D270) from Trans-Tech Inc., and Bi-Ba-Nd-Titanate (BBNT) from Ferro Corp., could be used for mixing with the polymer matrices. Specifically, $BaTiO_3$ exhibit a wide range of attainable dielectric permittivity (from $\varepsilon_r = 10$ to ε_r in the thousands), depending on its chemical form, grain size, temperature, and added dopants. The BBNT, MCT, D270 powders have dielectric constants of $\varepsilon_r = 95$, 140, and 270, respectively. The mixture process is described in the following paragraphs.

The process (see Fig. 3.39) starts with the preparation of PDMS (T2 Silastic from Dow Corning is one example) by adding one part of cross-link agent to ten parts of silicone gel (mass ratio). The resulting silicone

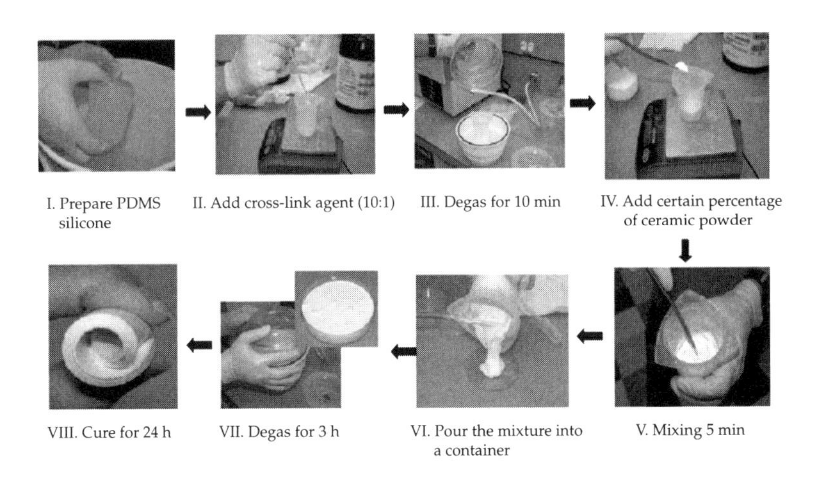

FIGURE 3.39 Fabrication process of the PDMS polymer composite mixtures. *(Courtesy of The Ohio State University, Electroscience Laboratory.)*

gel is then mixed thoroughly and placed into a vacuum chamber for around 10 min to remove excess gas by venting the surface bubbles within the prepared gel. The ceramic powder is slowly added next, up to the desired mixture and mixed. The resulting PDMS-ceramic slurry is subsequently poured into a plastic container (of some desired shape) and allowed to cool. However, degassing of the resulting mixture is necessary, first by placing the containers into a vented vacuum chamber as done for the silicone gel. This process, although tedious, plays a critical role in achieving homogeneous ceramic-reinforced PDMS substrates. An average degassing time for a dish (6 mm thick and 30 mm in diameter with 20% ceramic powder) is approximately 3 h. Typically, 24 h are needed for solidification or curing of the PDMS mixture.

Figure 3.40 provides example values of the polymer ceramic composite PDMS-$BaTiO_3$ after mixing PDMS with 5% up to 25% of barium titanate ($BaTiO_3$). The capacitance method [101] was used to measure the relative permittivity (ε_r) and associated loss tangent ($\tan\delta$). The stability of ε_r (real value) is quite good and reaches a value of $\varepsilon_r = 20$. As expected, the loss tangent increases with frequency. Specifically, the loss tangent with 25% $BaTiO_3$ mixture is $\tan\delta = 0.04$ at 1 GHz and about $\tan\delta = 0.018$ with 10% mixing.

Figure 3.41 considers comparisons of different ceramic mixtures with PDMS. Specifically, $BaTiO_3$, MCT, BBNT, and D270 are individually mixed (see Fig. 3.39) and the final dielectric constants are compared for the same percentage mixture. Of significance is that the MCT, BBNT, and D270 powders lead to $\tan\delta < 0.01$. Thus, they are more desirable as compared to BaTiO3. The PDMS-D270 composite is even more effective as it provides linear variation of ε_r versus powder volume with the PDMS.

Antennas on polymer composites (as those in Fig. 3.41) have been fabricated and tested (Fig. 3.42 shows some examples) [102, 103]. The rectangular patch to the left was printed on a 6.5% PDMS-D270 substrate (12 mm thick, with $\varepsilon_r = 4$ and $\tan\delta = 0.008$ in the 1~2 GHz band). The actual metallization of the polymer (to form the patch) was done by electroplating for 2 h at a current density of 20 mA/cm2. The inner pin of the feeding coaxial probe was then soldered onto the patch without melting the polymer substrate as that substrate could withstand 300°C temperature. The resulting S_{11} for the fabricated patch was -18 dB at resonance and had a realized gain of about 7 dB. To the right of Fig. 3.42 we depict a dielectric rod antenna, where concentric dielectric layers are adopted to readily form a tapered rod tip for ultra wideband radiation. The main dielectric core had $\varepsilon_{r1} = 9$ and the second layer (of $\varepsilon_{r2} = 6$) was poured around the core after forming a mold. Upon curing the two inner layers, third dielectric layer (of $\varepsilon_{r3} = 4$) was placed as the most outer shell. This was again done by creating a circular molding around the inner cores and pouring the

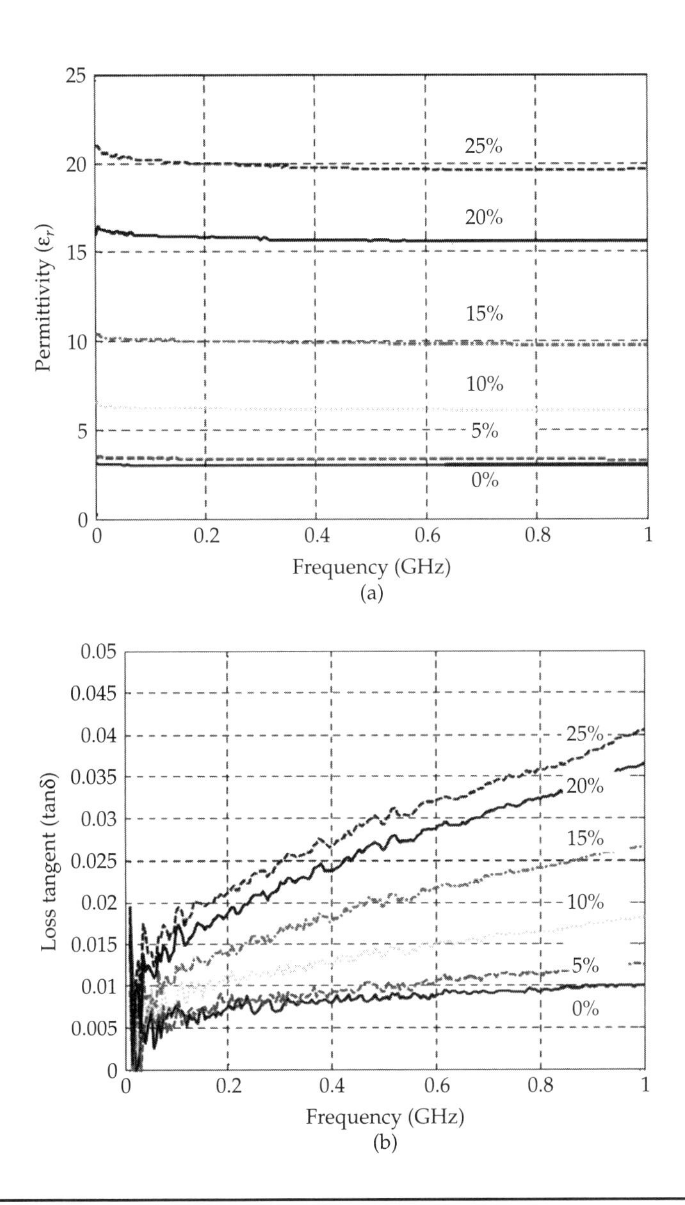

FIGURE 3.40 Dielectric properties of PDMS-$BaTiO_3$ composites. (a) Dielectric constant of PDMS-$BaTiO_3$ composites; (b) Loss tangent of PDMS-$BaTiO_3$ composites. (*Courtesy of The Ohio State University, Electroscience Laboratory.*)

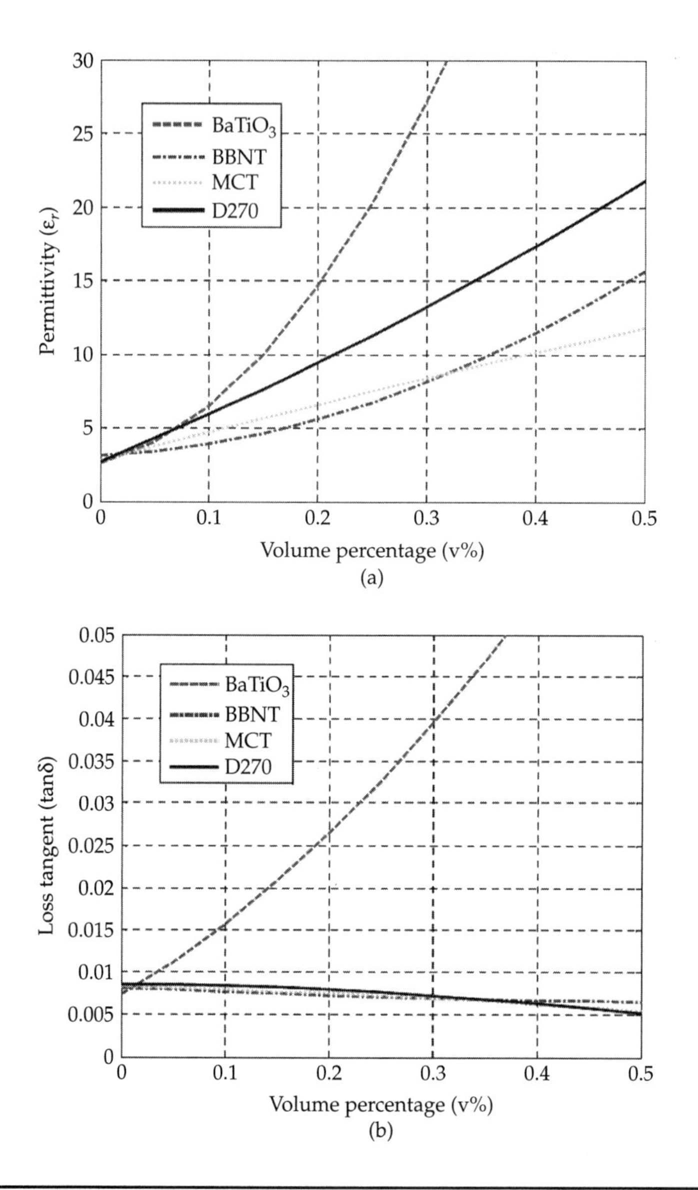

FIGURE 3.41 Dielectric properties versus volume percentage of ceramic powders (measured at 500 MHz). (a) Dielectric constant; (b) Loss tangent. (*Courtesy of The Ohio State University, Electroscience Laboratory.*)

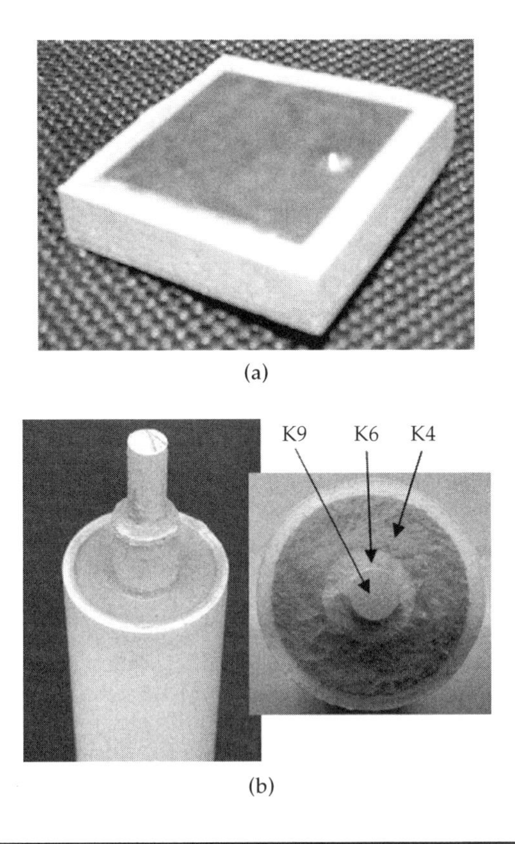

FIGURE 3.42 (a) Rectangular microstrip antenna printed on a polymer-ceramic composite substrate; (b) A three-layer dielectric rod antenna using polymer-ceramic composite. (*Courtesy of The Ohio State University, Electroscience Laboratory.*)

dielectric in its wet state. As is known [104], dielectric rods support traveling waves over a large bandwidth. Here, tapering the tip of the rod, a matching section is generated that aid wideband radiation.

There are still significant challenges when printing polymers. Printing on LCPs is now routine as this polymer has low dielectric constant and different composition. However, printing on PDMS (which allows for variable and large ε_r) is more challenging. Specifically, lithographic methods (using metal evaporation) do not work well due to poor metal-polymer adhesion. Further, interface incompatibilities can cause detachment of the printed layers under bending or tensile stress.

More recently, a printing technology utilizing carbon nanotube sheets was proposed [99,100]. Ensembles of carbon nanotubes (CNTs) can become conductive enough to satisfy antenna requirements [105,106]. As can be expected, the length of the nanotube, its diameter, composition, and entanglement can have significant impact on the conductivity and radio frequency (RF) properties. Their strength and suitability for applications in extreme environment has attracted significant interest in the RF and optical community [107–110]. However, so far, reported antennas using CNTs have low radiation efficiencies due to their higher resistance (wirelike CNT antenna was reported to have resistances as much as 6.45 kΩ/μm). To reduce their resistance, CNT ensembles were proposed. For example, nonaligned CNT ensembles were reported to reduce sheet resistance down to around 20 Ω/μm [111]. However, even this lower resistance is still too high to realize efficient CNT antennas. The so-called E-textile CNT surface [99] and vertically aligned CNT sheet [101] were recently proposed to decrease resistance and increase conductivity.

The E-texile CNT sheet (see Fig. 3.43) is much like any woven cloth except that CNTs are used to replace the cotton or nylon strings. Single-walled CNTs (SWCNTs) or multiwalled CNTs (MWCNTs) can be used in conjunction with metal particles to increase the E-textile sheet conductivity. For the latter, the CNT thread is just prepared by dispersing SWNTs in the diluted nafionethanol [112]. A cotton textile thread is then dipped into the CNT-SWNT dye dispersion for 10 s and then drying for 1 h at 60°C. After repeating this dying process 10 times (using a SWCNT solution), the sheet resistance of the E-textile dropped down to around 10 Ω/μm. To further increase the E-textile conductivity, silver particles (Ag) were sputtered for 200 s. The CNT coated fiber was then treated in a hot press overnight for 24 h at 100°C to achieve a strong adhesion of SWNTs and Ag particles on the cotton textile. However, after Ag sputtering, the CNT fiber becomes stiffer and this leads to be accounted for in its end applications. The resulting E-textile had a thickness of 150 μm and a resistance of around to only 1 Ω/μm.

Figure 3.44 displays the process of printing an E-textile CNT sheet on a polymer-ceramic substrate [113]. The E-textile is cut into a desired

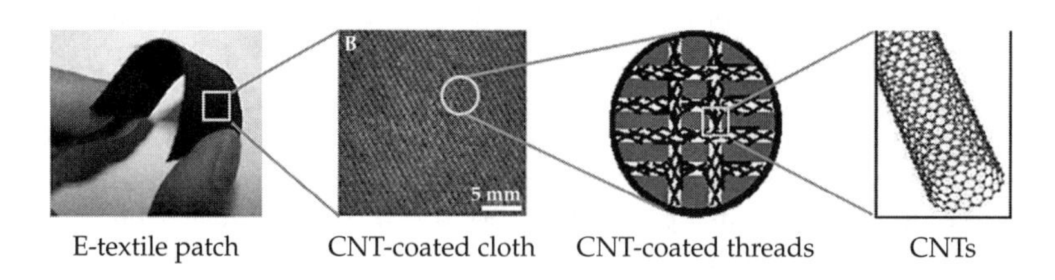

FIGURE 3.43 Photograph of the single carbon nanotube (CNT) and its weaving into a dense textile section. (*Courtesy of The Ohio State University, Electroscience Laboratory.*)

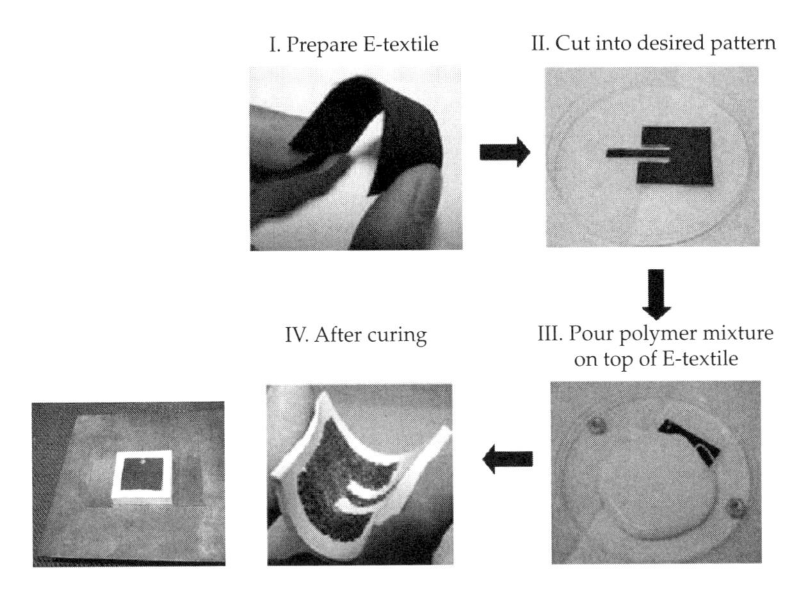

FIGURE 3.44 E-textile printing on a polymer-ceramic composite. (*See Zhou et al.* [113].)

pattern at the final step to form the antenna geometry's surface. To place the textile on a substrate, the polymer-ceramic composite mixture is poured over the E-textile fabric. After curing, the E-textile adheres strongly to the polymer-ceramic leading to strong mechanical compliance. As an example, an E-textile CNT patch (35 × 35 mm) was printed on a polymer substrate of thickness 300 mils and permittivity $\varepsilon_r = 4.0$. This textile patch had a sheet resistance of $2\,\Omega/\mu\text{m}$ and delivered a gain of 6 dBi at 2 GHz (i.e., ~1–2 dB less than the standard metal patch). Due to its higher loss, the E-textile patch had a bit wider bandwidth as well.

An alternative to the E-textile process is to grow short nanotubes on the polymer [114,115]. This improves adhesion and flexibility and has also led to lower conductivities. The growth of CNTs over the polymer results in a surface of extremely dense vertically aligned nanotubes (in much the same way body hairs grow, but much denser). Figure 3.45 displays a scanning electron microscope (SEM) image of the subject CNT sheet formed by "printing" 3×10^9 nanotubes per cm^2 (100 nm tall).

Figure 3.46 shows the process used to synthesize and "print" the CNT sheets on polymers. The process is as follows: First an array of ferrous particles is sputtered on a silicon wafer to serve as catalysts for CNT growth. Next the silicon substrate is placed inside a tube furnace (Thermolyne 79400) and methane gas (CH_4) is blown into the furnace

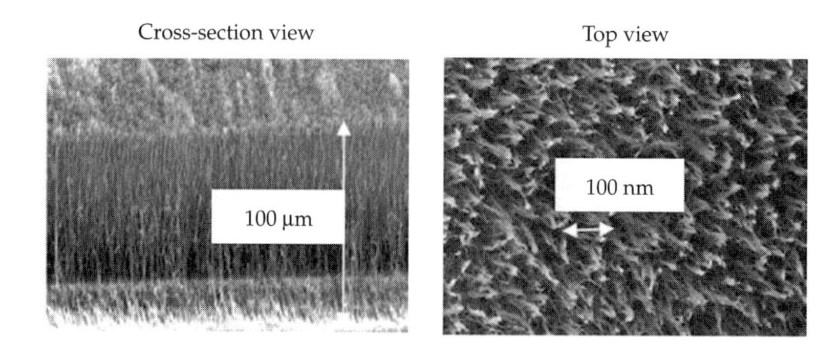

FIGURE 3.45 SEM photo of vertically aligned nanotubes (100 nm tall) having density of 3×10^9 nanotubes per cm^2. (*See Zhou et al.* [114].)

via a carrier argon flow. At high temperatures (1000°F), methane gases are decomposed into carbon atoms, aligned along the catalyst Fe particles in the cylinder. By controlling the furnace temperature (1000°F) and deposition time (2 h), a vertically aligned CNT array is grown on the silicon wafer. The length of the CNT array depends on the

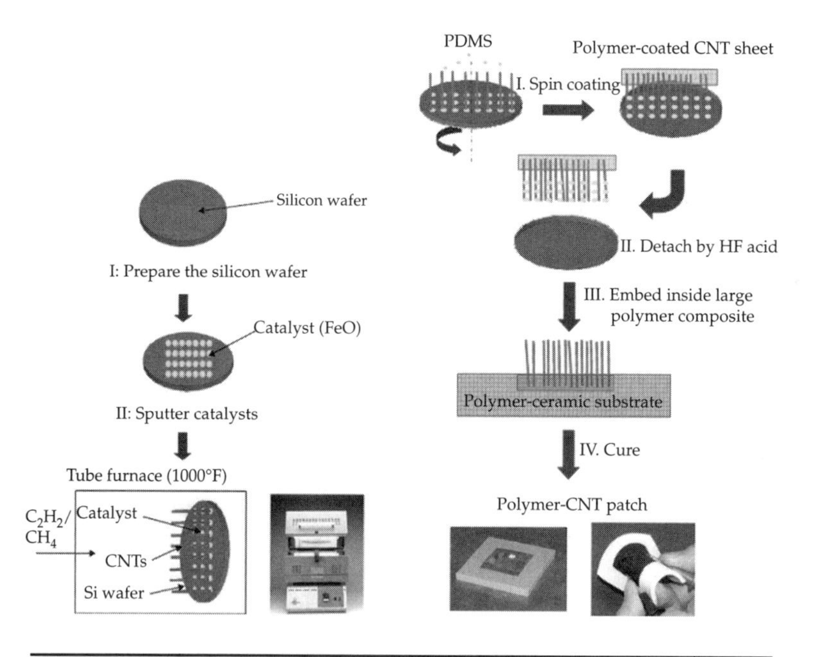

FIGURE 3.46 Process for growing vertically aligned CNT sheet and transferring CNT sheet onto polymer-ceramic composites. (*Courtesy of The Ohio State University, Electroscience Laboratory.*)

deposition time. For the case [115] shown in Fig. 3.45, it was about $3 \times 10^9/\text{cm}^2$.

The vertically aligned CNT arrays residing on the Si substrate are transferred onto the polymer surface using a two-stage curing process. First a thin PDMS composite layer is spin-coated onto the CNT sheet as displayed in Fig. 3.46. After curing, the CNTs are implanted inside a thin polymer layer to form the CNT sheet. The polymer-coated CNT sheet is subsequently detached from the silicon wafer by dissolving SiO_2 on the Si surface using hydrofluoric acid. The final step is to embed the polymer-coated CNT sheet onto a larger customized polymer-ceramic substrate for antenna loading. During this curing stage, the polymer-ceramic substrate is cross-linked with the coated polymer, leading to strongly bonded CNT sheets on polymer ceramic substrates.

A 31 × 31 mm patch antenna was fabricated [115] using the CNT sheet printing. The PDMS-MCT substrate was 56 × 56 mm and had $\varepsilon_r = 3.8$, $\tan\delta = 0.015$ at the resonant frequency of 2.25 GHz. This antenna was measured on a 150 × 150 mm ground plane and delivered a gain of ~6 dB, verifying the low surface resistivity of 0.9 Ω/μm for the CNT sheet (lower resistivities are pursued at this point).

Conformal mounting of the flexible polymer patch in Fig. 3.46 was also considered [116] (see Fig. 3.47). The goal was to evaluate the impact of bending on resonance and conductivity of the CNT sheet. It was observed that bending resulted in 13% stretching and a resonance

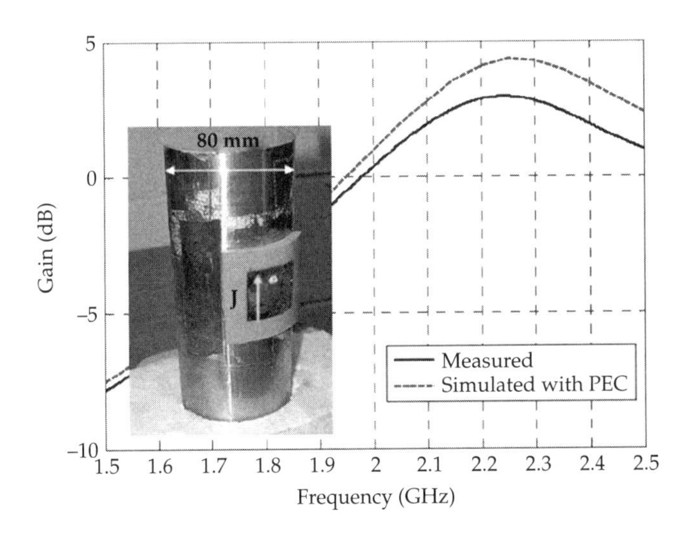

FIGURE 3.47 Illustration of cylindrically mounted polymer-CNT patch antennas. (*After Zhou et al.* [116] @ *IEEE 2009.*)

shift from 2.25 to 1.95 GHz. The E-plane gain was also reduced from 4.7 dB down to 1.7 dB due to bending. This gain reduction was due to the increased resistance as the CNTs were pulled away from each other. The corresponding H-plane CNT patch had a measured gain of 2.9 dB at 2.25 GHz, viz. 1.5 dB lower than the simulated PEC patch on the same cylinder. The H-plane bent patch had higher gain than the E-plane one since currents flowed along the unbent direction of the CNT patch. Thus, they were not subject to the stretching effects. Nevertheless, for most practical applications, the patches will not be subjected to such large bending or stretching (about 2% of bending/stretching is typical).

3.4 Optimization Methods

3.4.1 Introduction

From the above, antenna miniaturization is often done at the expense of efficiency, bandwidth, and gain. As noted in Chap. 1, small antenna design is a compromise between performance, dimension as well as materials, operational practice, and manufacturing ease. Experience and intuition are essential in the antenna design process. However, when a certain level of design complexity is reached, design optimization tools become valuable, if not necessary. In this section, we discuss optimization methods adopted for antenna design. Specifically, genetic algorithm (GA) and particle swarm optimization (PSO) schemes are considered for antenna design improvements.

3.4.2 Genetic Algorithm

Genetic algorithms (GAs) are search methods based on the concepts of natural selection and evolution. These optimization methods consider a set of trial solutions (in parallel) based on a parametric variation of a set of coded geometric and material features. GA employs known concepts, such as chromosomes, genes, mating, and mutation, to code the antenna geometry pixels for best design selection (see Fig. 3.48). More details can be found in papers and books covering the application of GA in electromagnetics [117–121]. The most studied application of GAs within the electromagnetic areas refers to antenna design. Specifically, GAs have been used to reduce sidelobes [122, 123], aperture amplitude and phase tapering [124, 125], and for adaptive array optimization [126, 127]. Several uses of GA in single element designs have also been reported [128–130] in finding the best locations of circuit elements for antenna loading. GAs have also been used to optimize the performance of standard antennas such as reflectors [131]

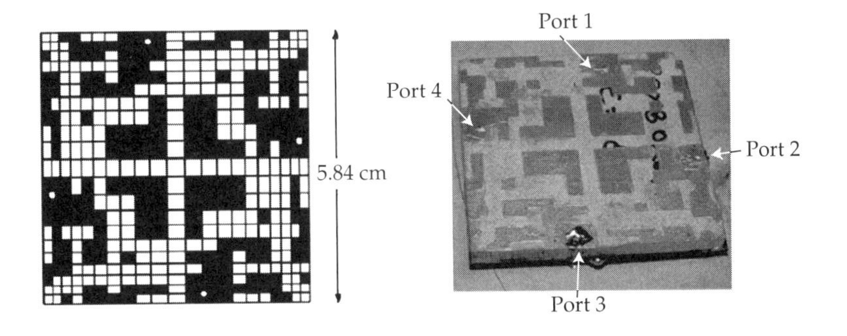

FIGURE 3.48 Metallization map of the optimized textured patch (feeds are denoted as white dots) and the fabricated prototype. (*After Koulouridis and Volakis* [133] @ *IEEE 2008.*)

and Yagis [132]. As is the case with all optimization algorithms, a search for an optimal solution is carried out subject to prespecified performance criteria (such as bandwidth, gain, and return loss, etc).

Several applications of antenna optimization via GA have been demonstrated. In the following paragraphs we consider some examples.

First a GA application to reduce microstrip patch size is demonstrated [133]. As depicted in Fig. 3.48, prior to optimization, the microstrip patch is divided into $N \times M$ equal rectangular cells. Through the GA-based optimization design process, the variables are the number of metal cells and their locations (all other parameters remain constant during the antenna optimization process). The coding of "1" or "0" is chosen to select if the pixel is metallized or not, respectively. Only the pixel connected to the coax probe was forced to be metallized from the start. The optimized antenna structures and their performance are shown in Fig. 3.48. The antenna was optimized within an aperture area of 5.84×5.84 cm^2 and printed on a 0.46 cm thickness substrate having a relative dielectric constant $\varepsilon_r = 18$ and loss tangent $\tan \delta = 0.0001$. The final optimized prototype antenna delivered good VSWR throughout the 1.5 to 1.7 GHz band except at 1.5 GHz where it briefly increases to VSWR 2.2. Also the realized gain was 2.5 to 3 dBi.

As mentioned in Sec. 3.2, the space-filling curves based on meandering, helical or spiral geometries and fractals have the potential to reduce operational frequency. But, we also noted that radiation performance depends strongly on the shape of the filling curves. Optimization can be used to adjust/modify the filling curve for concurrent miniaturization, best gain or input impedance control. In the following paragraphs we consider such optimized form of the meandered line antenna (MLA).

As already mentioned in Sec. 3.2, typical MLAs do not exhibit optimum gain [134], especially when conductor losses are considered. Thus, optimization can be employed to improve gain for a prespecified MLA size [135]. A typical MLA can be defined by its height, width, and number of turns. However, we can consider more parameters in a GA design as depicted in Fig. 3.49. They include the number of turns N, the length of the horizontal (w_{n1} and w_{n2}) and vertical (h_{n1} and h_{n2}) segments of the nth turn and the length of the central segments w_{00} and h_{00}. To minimize loss and maximized miniaturization all vertical and horizontal segments will be independently designed using a GA-based optimization. With this goal in mind, for each pth antenna ("individual") of the GA population at the kth generation ("iteration"), the following fitness function was evaluated:

$$f_p^{(k)}(G_p, H_p, X_p) = r_1 \frac{G_p}{G_0} + r_2 \frac{H_p}{H_{\max}} + r_3 \frac{X_0}{|X_p| + X_0}$$

Here, H_p, G_p, X_p are the corresponding pth antenna height, maximum gain, and input reactance. Also, $G_0 = 1.63$ (maximum gain of the $\lambda/2$ dipole), $X_0 = 1\,\Omega$ with the constraint $r_1 + r_2 + r_3 = 10$. We note that the fitness function reduces to $f_p = 10$ when $G = G_0$, $H_p = H_{\max}$ and $X_p = 0$ (i.e., antenna is at resonance).

Several MLA designs were considered in the GA optimization by Marroco [135], subject to maximum size constraint. Specifically, the product $H_{\max} \times W_{\max}$ was set to different values from $3 \times 3\ \text{cm}^2$ to $6 \times 6\ \text{cm}^2$ (the antenna wires are made of copper). As depicted in Fig. 3.50, GA-optimization is more effective in optimizing the MLA as the area

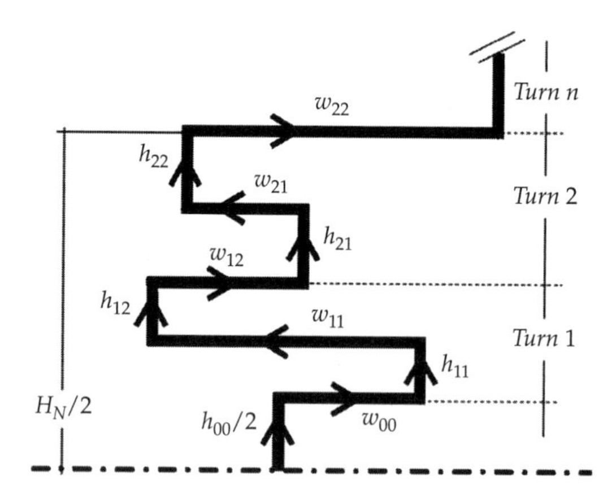

FIGURE 3.49 Parameters for optimizing the MLA dipole (only the upper half of the antenna is shown). (*After Marrocco et al.* [135] @ *IEEE 2003.*)

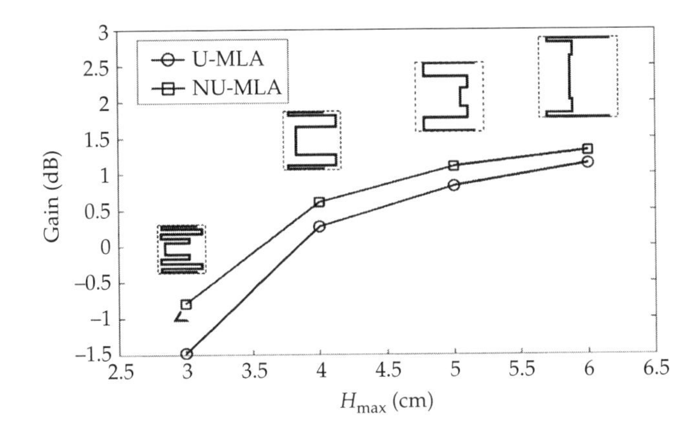

FIGURE 3.50 GA-optimized MLA dipole in antennas shown in Fig. 3.49 subject to given area $H_{max} \times W_{max}$ constraint (NU-MLA: optimized, U-MLA: not optimized). (*After Marrocco et al.* [135] © *IEEE 2003.*)

$H_{max} \times W_{max}$ decreases. We note that as the MLA area increases, shorter horizontal segments are required for turning and meandering is mostly localized at the antenna extremities. The latter serves to minimize losses since the current goes to zero at the MLA ends.

GA techniques have also been applied to optimize fractal antenna elements. For example, a second order Koch-like antenna was optimized using GAs by Werner et al [136]. Pantoja et al. [137] extended this work using multiobjective optimization to find optimal solutions subject to resonance frequency, bandwidth, and efficiency. In general, though, optimization tools can be best used to define shapes that do not obey specific geometrical restrictions. For a given wire length and antenna size, wire geometry optimization can lead to Eucledian geometries that can perform better than fractal shapes. As an example, Altshuler [138] considered the optimization of a seven-segment wire antenna. He employed GAs to find the lengths and locations of the segments to form a wire antenna radiating right-hand-circular polarization at elevation angles above 10° [139]. Later Altshuler [140] considered a general multisegment wire optimization (see Fig. 3.51). The optimization goal was to design a wire antenna having minimum Q and maximum bandwidth. At the resonance frequency (where the input impedance is real), this was done subject to filling the wire geometry within a prespecified cube as depicted in Fig. 3.51. Altshuler found that as he reduced the cube size, more wire segments were needed to reach resonance. Figure 3.51 displays the optimized geometries with and without inserted impedance loads. For the latter case, the delivered polarization was elliptical but the pattern provided hemispheric

(a)

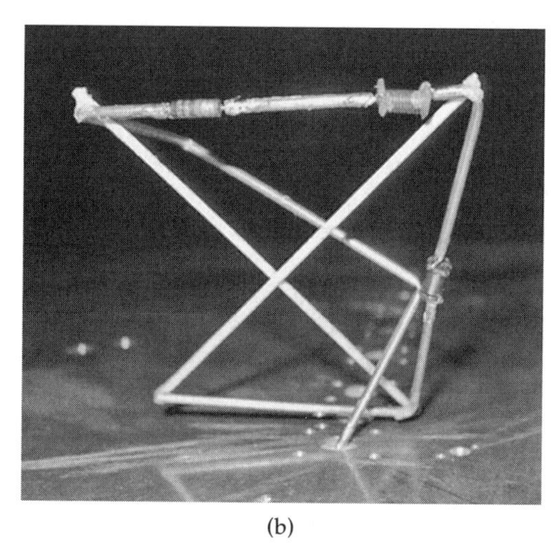

(b)

FIGURE 3.51 Optimized multisegment wire antenna on a ground plane filling within a predefined cubic volume. (a) Optimized wire geometry with no load operating at 384 MHz filling within a cube of size of 0.057 λ (after Altshuler [140] © IEEE 2002.) and (b) Optimized wire geometry with inserted impedance loads to increase bandwidth. (*After Altshuler and Linden* [141] © *IEEE 2004.*)

coverage [141]. Also, the optimized wire antenna with lumped loads has an impressive bandwidth 50:1 (from 300 MHz to 15 GHz) with VSWR < 4.5.

With the goal of exploiting more designing freedoms, Kiziltas et al. [142, 143] considered optimizations of the entire antenna volume,

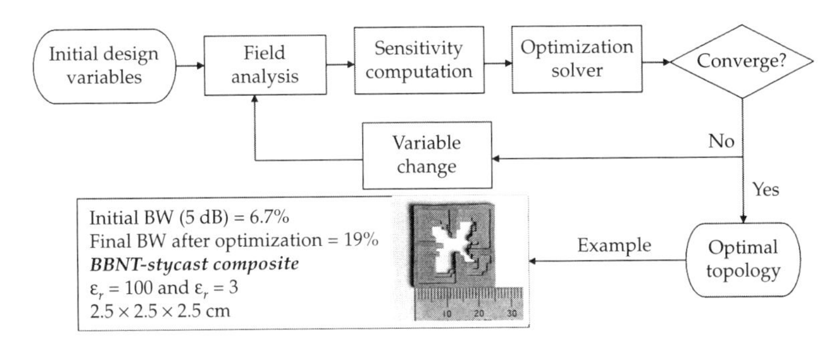

FIGURE 3.52 Design optimization flowchart and example optimal dielectric design. (*See Kiziltas et al.* [142].)

including the dielectric constants within the volume and metallization shapes, etc. It is reasonable to expect that designs based on optimal selection of materials, metallization shape, and matching circuits have a much better chance of achieving optimal narrowband and wideband performance. Concurrently, robust and fast computational tools are required to achieve such design [144].

The work in [142,143] focused on optimum topology and material optimization (see Fig. 3.52). That is, the material volume was allowed to have a combination of two or more dielectric materials with the overall volume synthesized as a set of small cubical material elements (see Fig. 3.53). Concurrently the metallization was modified till the optimization design was reached subject to bandwidth and size requirement. This approach led to significant improvements of the simple patch bandwidth. A design flowchart of the topology optimization process is given in Fig. 3.52. The design reported in [142,143] refers to a patch 1.25×1.25 cm residing on a $2.5 \times 2.5 \times 2.5$ cm substrate.

An example design based on material and metallization shape optimization is depicted in Fig. 3.53 [145]. A focus on this design was to achieve good bandwidth using an extremely thin substrate of 0.01 λ at mid-frequency. Indeed, the design reported in [145] achieved a 3.4% bandwidth with a 4.7 dBi gain using an aperture of only 0.11 λ in size.

3.4.3 Particle Swarm Optimization

Apart from GA optimization, particle swarm optimization (PSO) has been employed for antenna design [146–149]. Unlike GAs, PSO allows for additional "intelligence" in the design. PSO is based on the principle that each solution can be represented as a particle (agent) in a swarm. It has been used in designing array [150–152], and stand alone

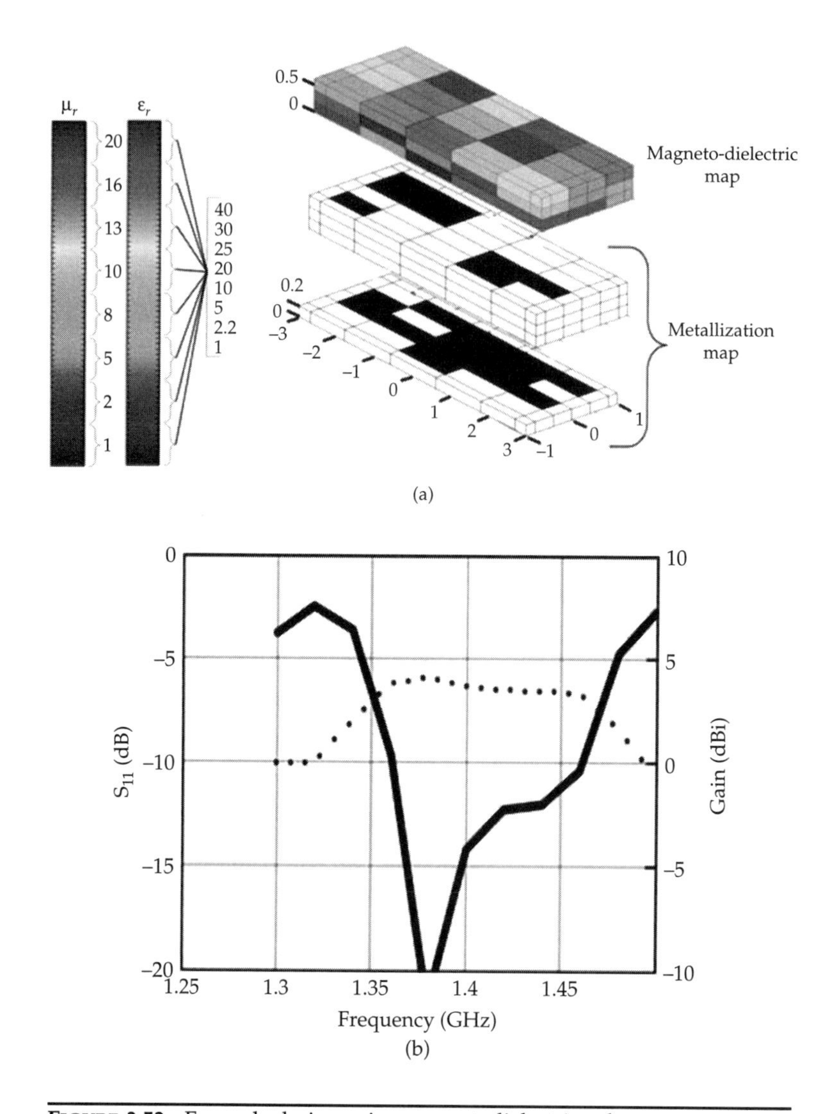

FIGURE 3.53 Example design using magnetodielectric substrates. (a) Material and metallization layout (color coding refers to different dielectric values); (b) Gain (dashed) and return loss from 1.3 to 1.5 GHz (*See Koulouridis et al.* [145].)

antennas [153–155]. PSO and other optimization algorithms have been employed to design small antennas. As an example, a mobile handset antenna design [156] via PSO is depicted in Fig. 3.54. This dual band handset antenna operated at 1.8 GHz and 2.4 GHz, with its maximum area dimensions being 0.23 × 0.13 λ at 1.8 GHz.

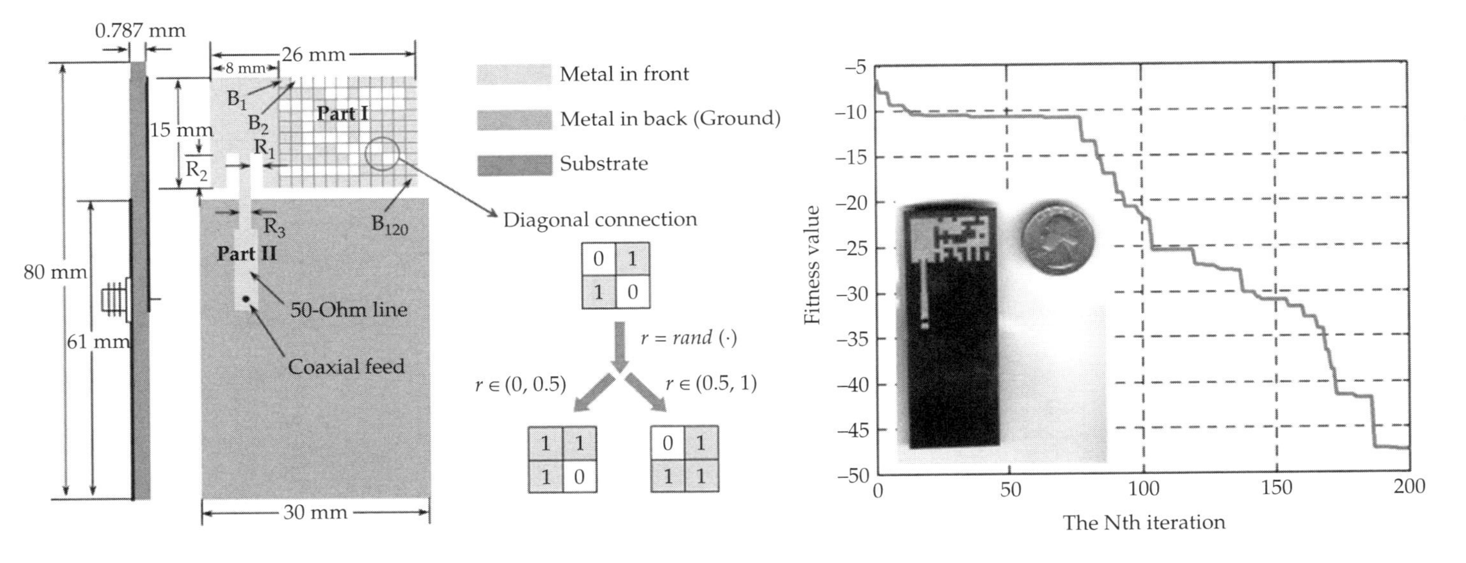

FIGURE 3.54 Illustration of a cell-phone antenna design using PSO. (*After Jin and Rahmat-Samii* [156] © *IEEE 2008.*)

3.5 Antennas on Electromagnetic Bandgap Ground Planes

In the recent years, strong interest has been on using periodic structures to lower antenna profile. These designs exploit resonance phenomena which are unique to the make up of periodic (planar or volumetric) structure and not necessarily the constituent materials. The latter are referred to as "metamaterials (from the Greek as "next" materials, i.e., next generation materials), engineered or simply artificial materials. Among them, electromagnetic bandgap (EBG) material and surfaces have been found successful as ground planes to reduce the profile of antennas mounted on ground planes. A specific example of EBG surface is the mushroom-like structure in Fig. 3.55 [157]. This structure consisted of a metallic ground plane at the bottom, a dielectric substrate, periodic metal patches and connecting vias.

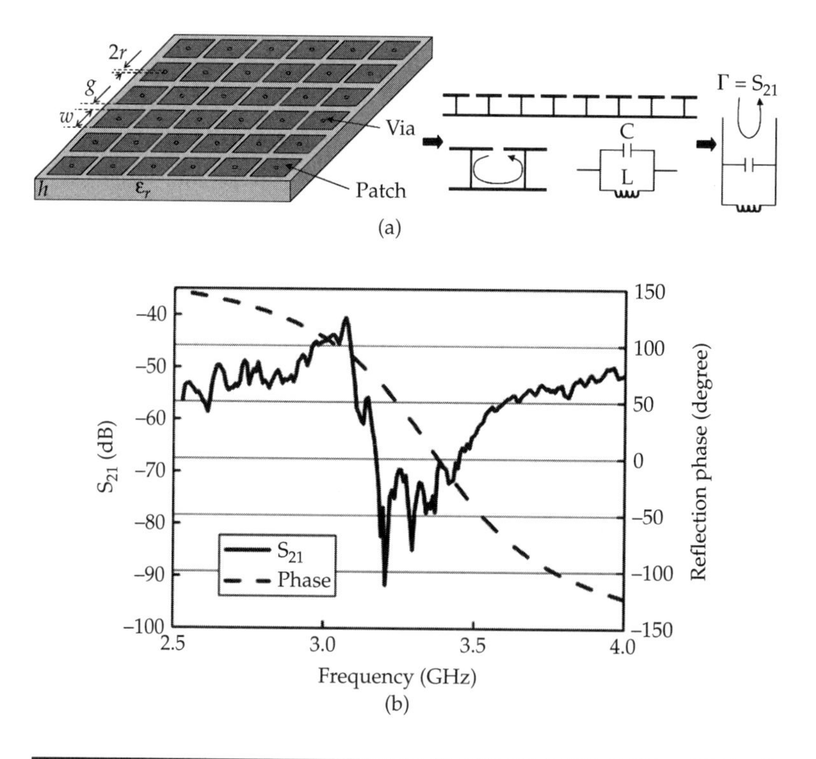

FIGURE 3.55 (a) Geometrical configuration of the mushroom EBG structure and corresponding equivalent circuit model; (b) Transmission across the EBG surface (S_{21}) and phase of the plane wave reflection coefficient.

The periodic nature of the patches and vias constitute a new effective medium. By controlling the geometry of the patches and vias, the surface can be viewed as an equivalent load that can alter the reflection coefficient Γ. If $\Gamma \approx +1$, antennas placed conformal to the EBG ground plane can exhibit enhanced performance as compared to the metallic ground plane with $\Gamma = -1$ (by exploiting the new design degrees of freedom). Also, they can be placed much closer to the ground plane.

Of importance is that the EBG in Fig. 3.55 is easy to fabricate using printed circuit board (PCB) technology and exhibits two characteristics. One is suppression of surface waves over a given frequency band (referred to as the bandgap) [158,159]. This property is also responsible for reducing coupling among antenna elements in arrays [160] to possibly eliminate scan blindness [161,162]. Key characteristic of the EBG in Fig. 3.55 is the nearly zero phase of the reflection coefficient over a portion of the bandgap. Specifically, achieving $\Gamma = S_{21} \approx 1e^{j\phi}$ (with $\phi \approx 0$) in a given band is critical to increasing the gain of low-profile antennas. This latter property has been extensively examined in the literature [4,163].

In the following sections we provide examples of small antennas in presence of EBG surfaces.

3.5.1 Performances Enhancement via Surface-Wave Suppression

As already noted, antennas on metallic ground planes (typically coated with dielectrics) suffer from surface-wave radiation. Specifically, at discontinuities, surface-waves scatter and create undesired pattern distortion. In addition, ground plane reflections cancel direct radiation, significantly reducing the bandwidth of low-profile antennas. Specifically, without ground plane treatments, the antenna must be placed about $\lambda/4$ away from the ground plane to ensure that reflected waves add in congruence (rather than cancelling each other). EBG ground planes can bring the phase of the reflection coefficient closer to $+1$, suppress surface waves and even suppress diffraction from edges of finite ground planes. This is demonstrated in Fig. 3.56 and is important in reducing interference from nearby structures.

An example application of EBGs (to treat ground planes) is shown in Fig. 3.57. In this case, it is used to suppress backward lobes in antennas (1575 MHz) for Galileo (also for GPS L1 band) [164]. As compared to the traditional choke ring, the EBG ground plane shows as good performance (in terms of axial ratio and multipath mitigation). However, the ground plane is only 1 cm ($\lambda/19$) thickness. The specific antenna is 30×30 cm ($1.58 \times 1.58\ \lambda$) in footprint and less than 0.5 kg in weight. As expected, the radiation pattern in presence of the EBG ground plane is smoother and exhibits a less cross polarization.

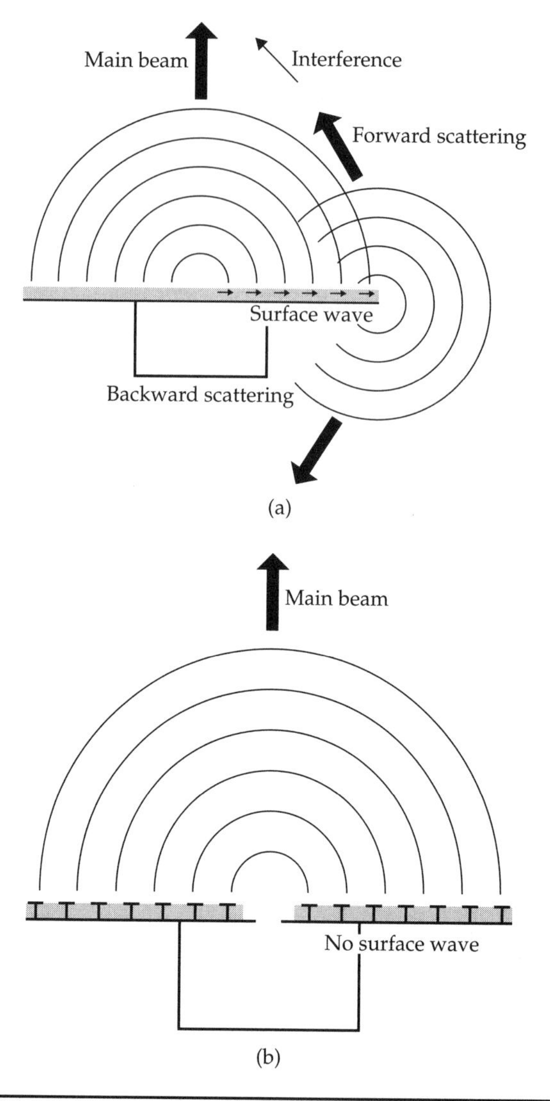

FIGURE 3.56 Illustration of surface-wave suppression using EBG ground plane. (a) Diffraction caused by untreated ground plane; (b) Treated or EBG ground plane suppressing diffraction and surface waves.

3.5.2 Low-Profile Antennas on EBG Ground Plane

As noted, the in-phase reflected field from the EBG ground plane makes them suitable for realizing low-profile antennas [165, 166]. The advantage of EBG ground planes is illustrated in Fig. 3.58.

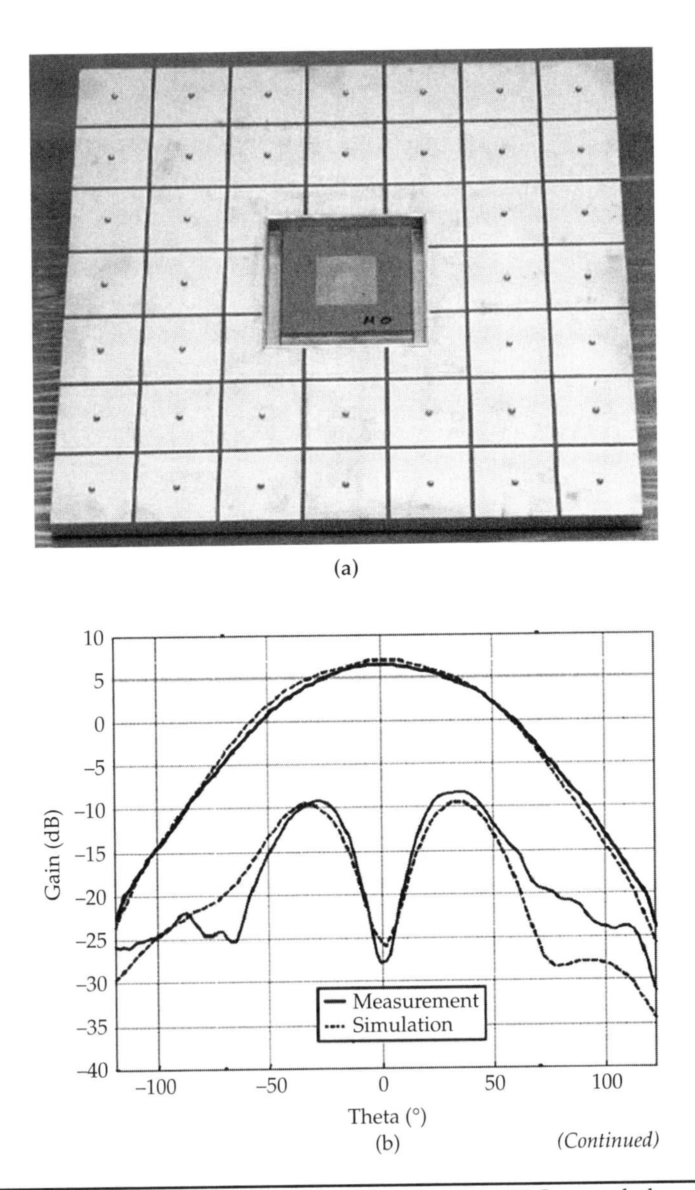

FIGURE 3.57 Performance of a Galileo antenna on an EBG ground plane: (a) Geometry of a low-profile L1 band (~1575 MHz) antenna on a mushroom EBG ground plane; (b) Gain pattern (co-pol and cross-pol) when mounted on the EBG ground plane, and (c) Gain and pattern performance when mounted on a PEC ground plane. (*After Baggen et al.* [164] @ *IEEE 2008.*)

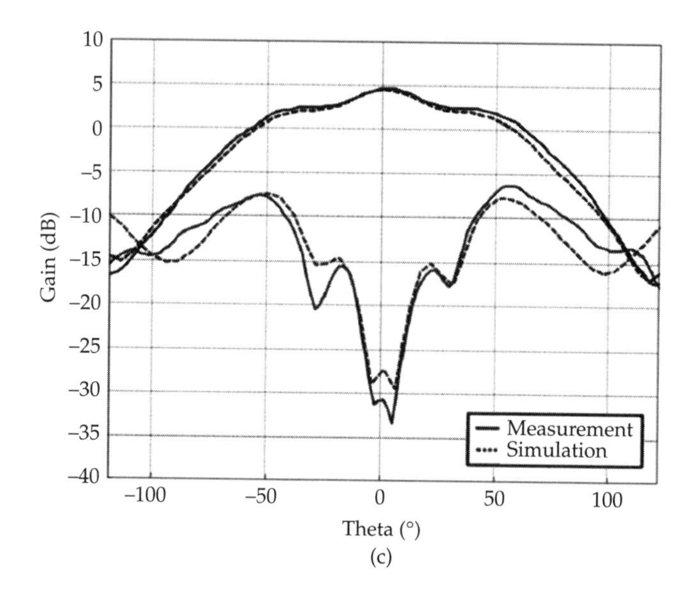

FIGURE 3.57 *Continued.*

As depicted and noted earlier, the PEC ground plane needs to be $\lambda/4$ away from the antenna to cause the reflected field to be in-phase with the direct (antenna radiated) field. As depicted in Fig. 3.58, the EBG has a reflection coefficient $\Gamma = 1e^{j\Delta\phi} \approx 1$ at its resonance frequency. Thus, the antenna can be placed very close to the EBG ground plane, implying a much lower profile realization.

A typical example of a conformal antenna on an EBG ground plane is depicted in Fig. 3.59 [167]. It refers to a dipole placed parallel to the EBG ground plane. As the EBG behaves similar to a perfectly magnetic conducting (PMC) surface at resonance (in phase reflection), the dipole radiated fields are reinforced by the presence of the EBG ground plane. The stronger resonance ($S_{21} < -10$ dB) caused by the presence of the EBG ground plane is clearly seen in Fig. 3.59*b*. By comparison, the PEC ground plane deteriorates the resonance and radiation characteristics of the dipole.

A discussion on the operational characteristics of EBG ground plane is given in [168]. Generally, it is important to operate the antenna at the same frequency when the phase of the EBG reflection coefficient is in the range of $90° \pm 45°$ (or within $-90° \pm 45°$). The radiation pattern also benefits from the bandgap of the EBG ground plane. As noted, this is due to surface-wave suppression.

A parametric study of the width (W), gap between patches (g), height of patches (h), and dielectric constant (ε_r) for the EBG in

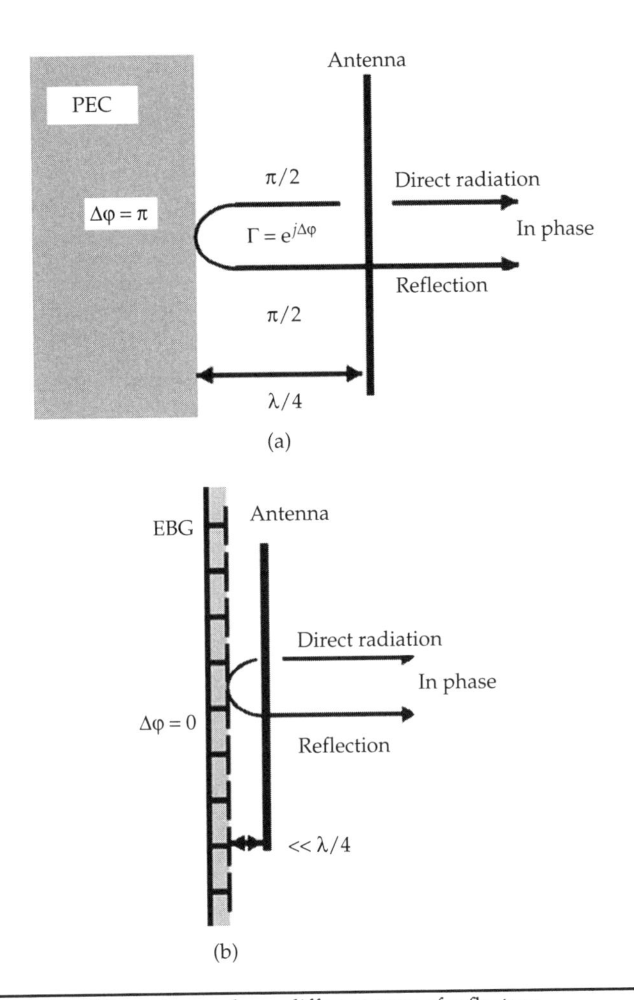

FIGURE 3.58 Wire antenna above different type of reflectors.

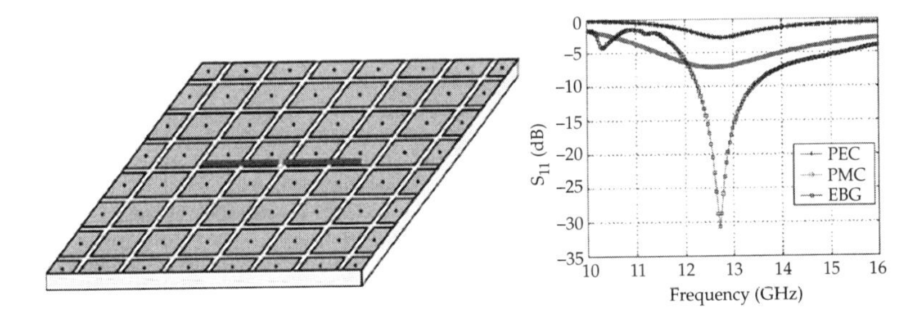

FIGURE 3.59 Dipole on an EBG ground plane and return loss as compared to a dipole on a PEC ground plane. (*After Yang et al.* [4] @ *McGraw-Hill 2007.*)

Fig. 3.55 has been performed [4] and some conclusions are:

- As the patch width W increases, the resonant frequency decreases and the slope of the reflection coefficient phase curve becomes steeper (vs. frequency).
- As the gap between patches increases, the resonant frequency also increases, and the slope of the reflection coefficient phase curve becomes flatter.
- As the substrate thickness h increases, the resonant frequency decreases but the slope of the reflection coefficient phase curve becomes flatter. This implies greater bandwidth (as would be expected with larger h values).
- As the dielectric constant, ε_r, increases, the resonant frequency decreases, but the slope of the reflection coefficient phase curve becomes steeper (smaller bandwidth). Indeed, as is well known, higher ε_r leads to lower bandwidths.

Several low-profile antennas have been published using EBG surfaces (see Fig. 3.60). Examples include polarization diversity antennas [169], spirals [170,171], resonant cavity antennas [172], low-profile circular polarized antennas [173], tunable low-profile antennas [174], and so on. Also, a polarization dependent EBG structure was proposed using rectangular metal patches (instead of square ones). The latter was employed in designing circularly polarized antenna using a single linear (dipole) antenna parallel to the EBG as in Fig. 3.60*c* [175].

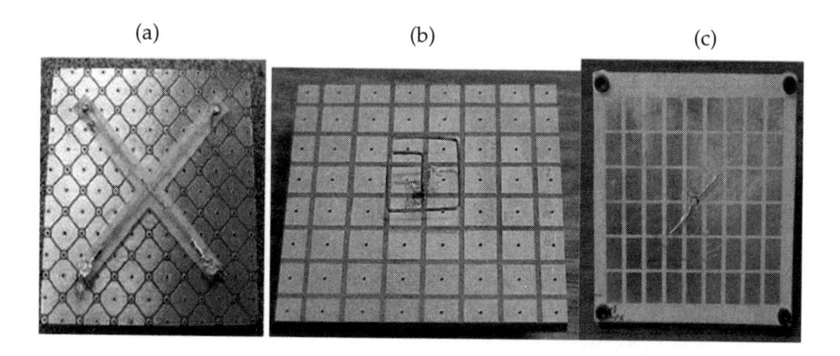

FIGURE 3.60 Example antennas placed conformal to an EBG ground plane: (a) Polarization diversity antenna (*See McKinzie and Fahr* [169] @ *IEEE 2002.*); (b) CP spiral-like antenna (*After Yang et al.* [4] @ *McGraw-Hill 2007.*); (c) CP antenna using a single straight wire on an EBG ground plane. (*After Yang et al.* [4] @ *McGraw-Hill 2007.*)

3.5.3 Wideband EBG Design

As the EBG properties are typically observed only over a narrow bandwidth, they are not suitable for multiband and wideband antennas.

A way to increase the bandwidth and/or multiband behavior of EBG ground planes is to employ additional tunable parameters in constructing it. As depicted in Fig. 3.61*a*, lumped circuit elements (L and C) can be placed between the patches and tuned to alter the EBG ground plane electrical behavior (phase of the reflection coefficient). Specifically, tunable capacitors can be introduced using varactor diodes [176,177]. In this case, the capacitance value is determined by the bias voltage. To supply the required voltage to the varactors, half of the vias are grounded and the others are connected to a voltage control network through a hole in the ground plane. It is important to note that the varactors are oriented in opposite directions at alternate rows. Thus, when a positive voltage is applied to the control lines, the diodes are reversely biased. Further, the EBG ground plane reflection phase can be controlled at each varactor location. Thus, the reflection coefficient phase can be varied over the surface of the ground plane.

Bray and Werner [178] presented an approach to independently tune a dual-band EBG surface. Their EBG geometry consisted of two reactively loaded concentric square loops. As expected, the outer loop was responsible for the lower frequency and the inner one controlled the upper frequency. The frequency of each loop was then tuned

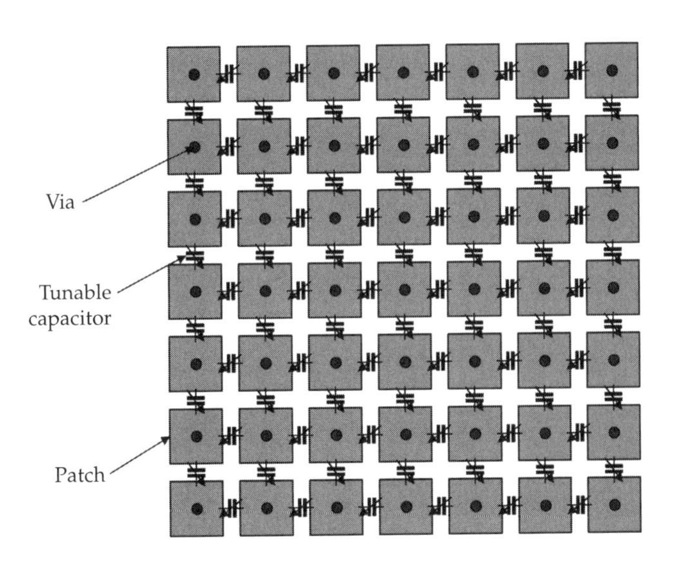

FIGURE 3.61 Examples of varactor diodes to design frequency tunable EBG ground plane. (*See Sievenpiper et al.* [179].)

independently using inserted capacitors and inductors. Such EBG surfaces can be used for electronically scanned antennas [179], or as a ground plane of multiband antennas [180].

Considering the fabrication cost and reliability of varactor diodes, exploring EBG element shapes that are broadband or multiband is a more attractive approach. Among them, patch shapes having multiple resonance [181] and spirals [182] are possible elements. Using optimization routines, we can concurrently optimize the geometry and size of such periodic elements (as well as the substrate thickness and dielectric constants). As an example, Fig. 3.62 shows the phase of the reflection coefficient of a GA-optimized EBG element. This EBG displayed zero phase at 1.575 GHz (GPS L1 band) and 1.96 GHz (cellular PCS band). The unit cell size for the EBG was 2.96 × 2.96 cm printed on a 2.93 mm thick substrate having a dielectric constant of $\varepsilon_r = 13$. We note that the reflection coefficient bandwidth (frequency band over which the phase is within ±45°) is 4.43% at 1.575 GHz and 2.2% at 1.96 GHz.

Multiband behavior can also be achieved by simply employing multilayer EBG surfaces [183, 184], each layer resonating at a different frequency.

Use of variable size (multiperiodic) unit cells is another method to design multiband EBG surfaces. Such structures are alternates to

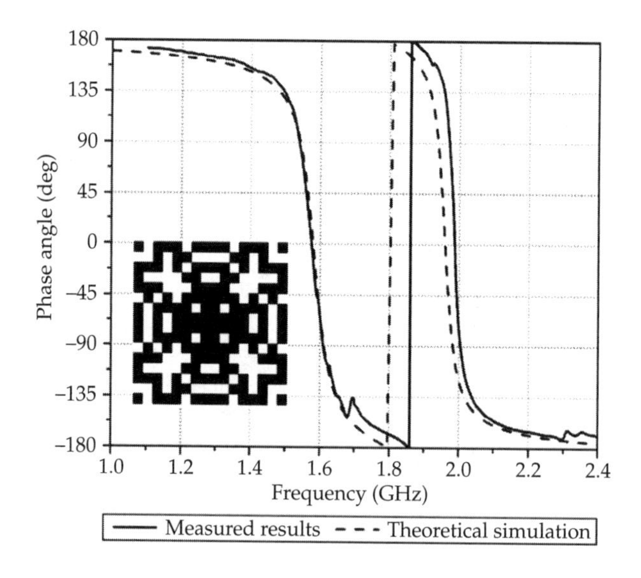

FIGURE 3.62 Reflection coefficient phase response of an engineered EBG (see inset of the unit cell) displaying zero phase at two frequencies. (*After Kern et al.* [181] © *IEEE 2005.*)

the multilayer EBG geometries. However, they would typically require larger surfaces and be likely applicable to lower frequencies. To realize wideband EBGs, impedance loading of the unit cell has been considered. An example of circuit loading is depicted in Fig. 3.63. Kern et al. [185] implemented the load using an amplifier with a positive feedback. Thus, a negative impedance was obtained modeled by a capacitor and an inductor in parallel (see Fig. 3.63). By varying the values of the negative capacitor C_{neg} and inductor L_{neg}, the reflection coefficient bandwidth can be increased considerably. Specifically, while the unloaded EBG had a bandwidth of only 4.4%, the circuit loaded EBG displayed a bandwidth of 175% using $L_{neg} = -0.204$ nH and $C_{neg} = -0.196$ pF.

The EBG bandwidth can be further increased using magnetic materials of high permeability (as part of the substrate) [186,187]. As is well known (see Sec. 3.3), the phase bandwidth of the reflection coefficient for the EBG surface is proportional to $\sqrt{L/C}$ (and, of course, the resonance frequency is proportional to $1/\sqrt{LC}$). Magnetic loading will increase the equivalent L, resulting in an increase of $\sqrt{L/C}$ and large in-phase bandwidth. Increased L has the additional advantage of reducing patch size as smaller capacitance is needed to achieve the same resonant frequency. Therefore, the in-phase reflection coefficient bandwidth increases further due to having smaller equivalent C.

An EBG structure in a magnetic substrate is depicted in Fig. 3.64. The specific patch geometry has the following parameters:

$$w = 2.44 \text{ mm}, g = 0.15 \text{ mm}, h = 1.57 \text{ mm}, r = 0.25 \text{ mm},$$

$$\varepsilon_r = 2.51, \mu_r = \begin{cases} 1 \\ 6 \end{cases}$$

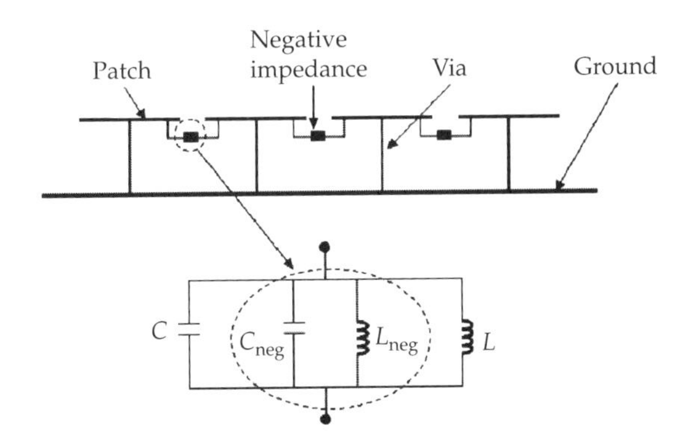

FIGURE 3.63 Active negative impedance loading to realize a wideband EBG structure and equivalent circuit. (*See Kern et al.* [185].)

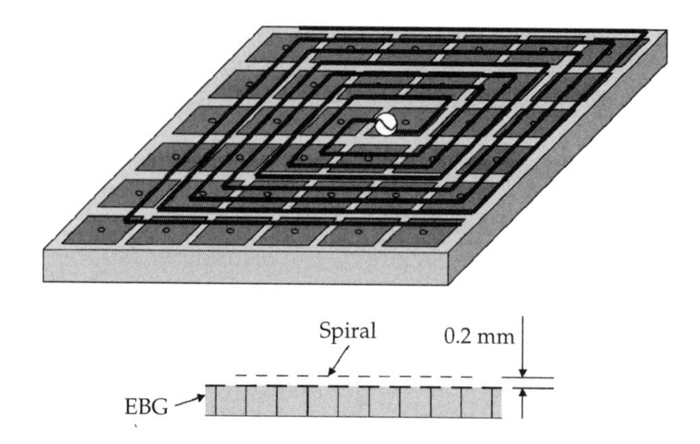

FIGURE 3.64 Square spiral antenna on an EBG surface loaded with magnetic material. (*See Yousefi et al.* [187].)

The results show that the EBG [187] with $\mu_r = 6$ substrate displayed an in-phase bandwidth of 70%. However, the bandwidth of the conventional EBG with $\mu_r = 1$ was only 18%. This EBG (on a $\mu_r = 6$ magnetic substrate) was used as a ground plane for a spiral antenna [187]. The spiral was placed only 0.2 mm ($\lambda/105$ at the center frequency of

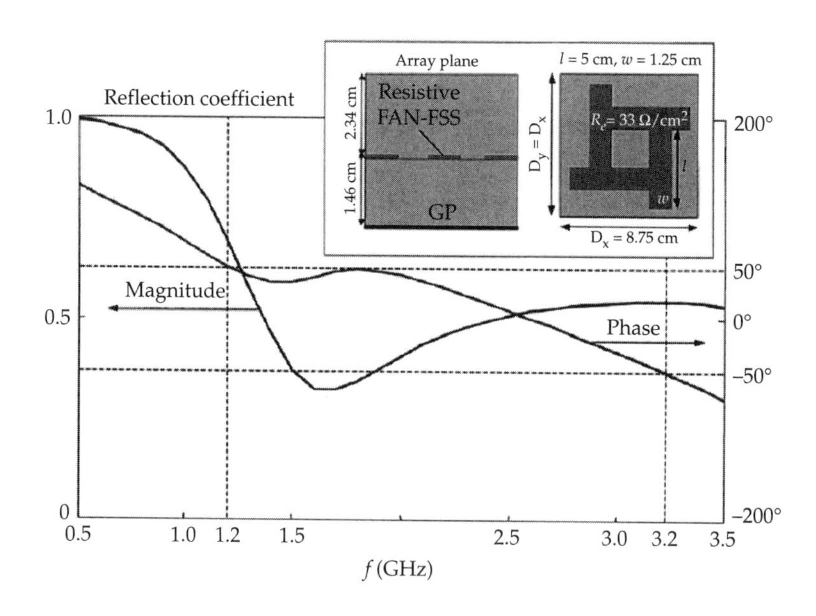

FIGURE 3.65 Fanlike EBG structure to improve phase flatness over a large bandwidth 1.2 to 3.2 GHz. (*After Erdemli et al.* [188] © *IEEE 2002.*)

14 GHz) above the EBG ground plane. Even though of low profile, the spiral exhibited nearly 100% bandwidth with VSWR<2 (9 – 19 GHz). By comparison, the EBG without a magnetic substrate had the same performance from 13 to 16 GHz.

More broadband elements can be used to construct surfaces aimed at increasing the in-phase reflection coefficient bandwidth. This was demonstrated by Erdemli et al. [188,189] (see Fig. 3.65) where a resistively loaded fanlike element (a combination of loop and dipole) was used to form the periodic EBG. This surface provided a relatively flat phase response for the reflection coefficient over a large band. A 3:1 in-phase bandwidth was achieved at the expense of some efficiency loss due to the resistive loading. As shown in Fig. 3.65, the resulting phase response is quite flat from 1.2 to 3.2 GHz.

References

1. H. A. Wheeler, "Fundamental limitations of small antennas," *Proceedings of the IRE*, vol. 35, December 1947, pp. 1479–1484.
2. K. Fujimoto, A. Henderson, K. Hirasawa, and J. R. James, *Small Antennas*, John Wiley and Sons, Research Studies Press, New York, 1987.
3. K. Fujimoto and J. R. James, *Mobile Antenna Systems Handbook*, Artech House, Norwood, Mass., 1994.
4. J. L. Volakis (editor), *Antenna Engineering Handbook*, 4th ed., McGraw-Hill, New York, 2007.
5. K. Skrivervik, J. F. Zurcher, O. Staub, and J. R. Mosig, "PCS antenna design: the challenge of miniaturization," *IEEE Antennas Propagat. Mag.*, vol. 43 (August 2001): 12–26.
6. C. A. Balanis, *Antenna Theory: Analysis and Design*, 3d ed., John Wiley & Sons Inc., New York, 2005.
7. K. Carver and J. Mink, "Microstrip antenna technology," *IEEE Transactions on Antennas and Propagation*, vol. 29, January 1981, pp. 2–24.
8. H. T. Nguyen, S. Noghanian, and L. Shafai, "Microstrip patch miniaturization by slots loading," *IEEE Antennas and Propagation Society International Symposium*, 2005, pp. 215–218.
9. J. Huang, "Miniaturized UHF microstrip antenna for a Mars mission," *IEEE Antennas and Propagation Society International Symposium*, 2001, pp. 486–489.
10. S. Y. Chen, H. T. Chou, and Y. L. Chiu, "A size-reduced microstrip antenna for the applications of GPS signal reception," *IEEE Antennas and Propagation Society International Symposium*, 2007, pp. 5443–5446.
11. Y. Zhou, C. C. Chen, and J. L. Volakis, "A compact 4-element dual-band GPS array," *IEEE Antennas and Propagation Society International Symposium*, vol. 1–5, 2009, pp. 1–4.
12. K. L. Wong, J. S. Kuo, and T. W. Chiou, "Compact microstrip antennas with slots loaded in the ground plane," *IEE Antennas and Propagation Society International Symposium*, 2001, pp. 623–626.
13. K. L. Wong, *Planar Antenna for Wireless Communications*, Wiley-InterScience, Hoboken, New Jersey, 2003.
14. R. L. Li, G. Jean, M. M. Tentzeris, and J. Laskar, "Development and analysis of a folded shorted-patch antenna with reduced size," *IEEE Transactions on Antennas and Propagation*, vol. 52, February 2004, pp. 555–562.
15. L. Zaid, G. Kossiavas, J. Y. Dauvignac, J. Cazajous, and A. Papiernik, "Dual-frequency and broad-band antennas with stacked quarter wavelength

elements," *IEEE Transactions on Antennas and Propagation*, vol. 47, April 1999, pp. 654–660.
16. A. Holub and M. Polivka, "A novel microstrip patch antenna miniaturization technique: a meanderly folded shorted-patch antenna," *COMITE*, vol. 23–24, 2008, pp. 1–4.
17. C. R. Rowell and R. D. Murch, "A capacitively loaded PIFA for compact mobile telephone handsets," *IEEE Transactions on Antennas and Propagation*, vol. 45, May 1997, pp. 837–842.
18. C. Y. Chiu and C. H. Chan, "A miniaturized PIFA utilizing via-patch loading," *Asia-Pacific Microwave Conference*, vol. 1, 2005, pp. 4–7.
19. C. Y. Chiu, K. M. Shum, and C. H. Chan, "A tunable via-patch loaded PIFA with size reduction," *IEEE Transactions on Antennas and Propagation*, vol. 55, January 2007, pp. 65–71.
20. Y. Hwang, Y. P. Zhang, G. X. Zheng, and T. K. C. Lo, "Planar inverted F antenna loaded with high permittivity material," *Electronics Letters*, vol. 31, September 1995, pp. 1710–1711.
21. R. B. Waterhouse, S. D. Targonski, and D. M. Kokotoff, "Design and performance of small printed antennas," *IEEE Transactions on Antennas and Propagation*, vol. 46, November 1998, pp. 1629–1633.
22. Y. J. Wang, C. K. Lee, W. J. Koh, and Y. B. Gan, "Design of small and broadband internal antennas for IMT-2000 mobile handsets," *IEEE Transactions on Microwave Theory and Techniques*, vol. 49, August 2001, pp. 1398–1403.
23. T. K Lo, J. Hoon, and H. Choi, "Small wideband PIFA for mobile phones at 1800 MHz," *Vehicular Technology Conference*, vol. 1, 2004, pp. 27–29.
24. K. L. Wong and K. P. Yang, "Modified planar inverted F antenna," *Electronics Letters*, vol. 34, January 1998, pp. 7–8.
25. M. C. Huynh and W. Stutzman, "Ground plane effects on the planar inverted-F antenna (PIFA) performance," *IEE Proceedings—Microwave, Antennas and Propagation*, vol. 150, August 2003, pp. 209–213.
26. H. Sagan, *Space-Filling Curves*, Springer-Verlag, New York, 1994.
27. L. J. Chu, "Physical limitations of omni-directional antennas," *Journal of Applied Physics*, vol. 19, December 1948, pp. 1163–1175.
28. H. A. Wheeler, "The radian sphere around a small antenna," *Proceedings of the IRE*, August 1959, pp. 1325–1331.
29. J. S. McLean, "A re-examination of the fundamental limits on the radiation Q of electrically small antennas," *IEEE Transactions on Antennas and Propagation*, vol. 44, May 1996, pp. 672–675.
30. R. C. Hansen, "Fundamental limitations in antennas," *Proceedings of the IEEE*, vol. 69, February 1981, pp. 170–182.
31. C. A. Balanis, *Antenna Theory: Analysis and Design*, 2d ed., John Wiley & Sons, Inc., New York, 1997.
32. R. C. Fenwick, "A new class of electrically small antennas," *IEEE Transactions on Antennas and Propagation*, vol. 13, May 1965, pp. 379–383.
33. J. Rashed and C. T. Tai, "A new class of wire antennas," *IEEE Antennas and Propagation Society International Symposium*, vol. 20, 1982, pp. 564–567.
34. J. Rashed and C. T. Tai, "A new class of resonant antennas," *IEEE Transactions on Antennas and Propagation*, vol. 39, September 1991, pp. 1428–1430.
35. H. Nakano, H. Tagami, A. Yoshizawa, and J. Yanauchi, "Shortening ratios of modified dipoles," *IEEE Transactions on Antennas and Propagation*, vol. 32, April 1984, pp. 385–386.
36. E. Tsutomu, Yonehiko, S. Shinichi, and K. Takashi, "Resonant frequency and radiation efficiency of meander line antennas," *Electronics and Communications in Japan*, vol. 83, 2000, pp. 52–58.
37. J. Rashed-Mohassel, A. Mehdipour, and H. Aliakbarian, "New schemes of size reduction in space filling resonant dipole antennas," *3rd European Conference on Antennas and Propagation*, vol. 23–27, 2009, pp. 2430–2432.
38. B. B. Mandelbrot, *The Fractal Geometry of Nature*, W. H. Freeman, New York, 1983.

39. D. L. Jaggard, "On Fractal Electrodynamics," in H. N. Kritikos and D. L. Jaggard, eds. *Recent Advances in Electromagnetic Theory*, Springer-Verlag, New York, 1990, pp. 183–224.
40. D. H. Wener and R. Mittra, *Frontiers in Electromagnetics*, IEEE Press, Piscataway, New Jersey, 2000.
41. D. H. Werner and S. Gangul, "An overview of fractal antenna engineering research," *IEEE Antennas and Propagation Magazine*, vol. 45, February 2003, pp. 38–57.
42. J. P. Gianvittorio and Y. Rahmat-Samii, "Fractal antennas: a novel antenna miniaturization technique and applications," *IEEE Antennas and Propagation Magazine*, vol. 44, February 2002, pp. 20–36.
43. C. P. Baliarda, J. Romeu, and A. Cardama, "The Koch monopole: a small fractal antenna," *IEEE Transactions on Antennas and Propagation*, vol. 48, November 2000, pp. 1773–1781.
44. J. Zhu, A. Hoorfar, and N. Engheta, "Bandwidth, cross-polarization, and feed-point characteristics of matched Hilbert antennas," *IEEE Antennas and Wireless Propagation Letters*, vol. 2, 2003, pp. 2–5.
45. X. Chen, S. S. Naeini, and Y. Liu, "A down-sized printed Hilbert antenna for UHF band," *IEEE Antennas and Propagation Society International Symposium*, 2003, pp. 581–584.
46. J. Zhu, A. Hoorfar, and N. Engheta, "Peano antennas," *IEEE Antennas and Wireless Propagation Letters*, vol. 3, 2004, pp. 71–74.
47. S. R. Best, "A comparison of the resonant properties of small space-filling fractal antennas," *IEEE Antennas and Wireless Propagation Letters*, vol. 2, 2003, pp. 197–200.
48. J. M. González-Arbesú, S. Blanch, and J. Romeu, "Are space-filling curves efficient small antennas?" *IEEE Antennas and Wireless Propagation Letters*, vol. 2, 2003, pp. 147–150.
49. S. R. Best, "Comments on hilbert curve fractal antenna: a small resonant antenna for VHF/UHF applications," *Microwave and Optical Technology Letters*, vol. 35, December 2002, pp. 420–421.
50. S. R. Best and J. D. Morrow, "The effectiveness of space-filling fractal geometry in lowering resonant frequency," *IEEE Antennas and Wireless Propagation Letters*, vol. 1, 2002, pp. 112–115.
51. S. R. Best, "A comparison of the performance properties of the hilbert curve fractal and meander line monopole antennas," *Microwave and Optical Technology Letters*, vol. 35, November 2002, pp. 258–262.
52. H. Elkamchouchi and M. A. Nasr, "3D-fractal rectangular koch dipole and Hilbert dipole antennas," *International Conference on Microwave and Millimeter Wave Technology*, vol. 18–21, 2007, pp. 1–4.
53. J. Romeu and S. Blanch, "A three dimensional Hilbert antenna," *IEEE Antennas and Propagation Society International Symposium*, 2002, pp. 550–553.
54. H. Elkamchouchi and M. A. Nasr, "3D-fractal hilbert antennas made of conducting plates," *International Conference on Microwave and Millimeter Wave Technology*, vol. 18–21, 2007, pp. 1–4.
55. J. S. Petko and D. H. Werner, "Miniature reconfigurable three-dimensional fractal tree antennas," *IEEE Transactions on Antennas and Propagation*, vol. 52, August 2004, pp. 1945–1956.
56. H. Rmili, O. Mrabet, J. M. Floc'h, and J. L. Miane, "Study of an electrochemically-deposited 3-D random fractal tree-monopole antenna," *IEEE Transactions on Antennas and Propagation*, vol. 55, April 2007, pp. 1045–1050.
57. S. R. Best, "The radiation properties of electrically small folded spherical helix antennas," *IEEE Transactions on Antennas and Propagation*, vol. 52, April 2004, pp. 953–960.
58. A. Mehdipour, H. Aliakbarian, and J. Rashed-Mohassel, "A novel electrically small spherical wire antenna with almost isotropic radiation pattern," *IEEE Antennas and Wireless Propagation Letters*, vol. 7, 2008, pp. 396–399.

59. H. K. Ryu and J. M. Woo, "Miniaturization of rectangular loop antenna using meander line for RFID tags," *Electronics Letters*, vol. 43, March 2007, pp. 372–374.
60. H. K. Ryu, S. Lim, and J. M. Woo, "Design of electrically small, folded monopole antenna using C-shaped meander for active 433.92 MHz RFID tag in metallic container application," *Electronics Letters*, vol. 44, December 2008, pp. 1445–1447.
61. Y. S. Takigawa, S. Kashihara, and F. Kuroki, "Integrated slot spiral antenna etched on heavily-high permittivity piece," *Asia-Pacific Microwave Conference*, vol. 11–14, 2007, pp. 1–4.
62. K. Gosalia, M. S. Humayun, and G. Lazzi, "Impedance matching and implementation of planar space-filling dipoles as intraocular implanted antennas in a retinal prosthesis," *IEEE Transactions on Antennas and Propagation*, vol. 53, August 2005, pp. 2365–2373.
63. T. Peng, S. Koulouridis, and J. L. Volakis, "Miniaturization of conical helical antenna via coiling," submitted to *Microwave and Optical Technology Letters.*
64. T. Yang, W. A. Davis, and W. L. Stutzman, "The design of ultra-wideband antennas with performance close to the fundamental limit," *URSI*, France, 2008.
65. B. A. Kramer, C. C. Chen, and J. L. Volakis, "Size reduction of a low-profile spiral antenna using inductive and dielectric loading," *IEEE Antennas and Wireless Propagation Letters*, vol. 7, 2008, pp. 22–25.
66. W. J. R. Hoefer, "Equivalent series inductivity of a narrow transverse slit in microstrip," *Transactions on Microwave Theory and Techniques*, vol. 25, October 1977, pp. 822–824.
67. S. Rowson, G. Poilasne, and L. Desclos, "Isolated magnetic dipole antenna: application to GPS," *Microwave and Optical Technology Letters*, vol. 41, June 2004, pp. 449–451.
68. M. Lee, B. A. Kramer, C. C. Chen, and J. L. Volakis, "Distributed lumped loads and lossy transmission line model for wideband spiral antenna miniaturization and characterization," *IEEE Transactions on Antennas and Propagation*, vol. 55, October 2007, pp. 2671–2678.
69. M. C. Scardelletti, G. E. Ponchak, S. Merritt, J. S. Minor, and C. A. Zorman, "Electrically small folded slot antenna utilizing capacitive loaded slot lines," *IEEE Radio and Wireless Symposium*, vol. 22–24, 2008, pp. 731–734.
70. G. V. Eleftheriades and K. G. Balmain, *Negative Refraction Metamaterials: Fundamental Principles and Applications*, John Wiley and Sons Inc., Hoboken, New Jersey, 2005.
71. P. L. Chi, K. M. Leong, R. Waterhouse, and T. Itoh, "A miniaturized CPW-fed capacitor-loaded slot-loop antenna," *International Symposium on Signals, Systems, and Electronics*, 2007, pp. 595–598.
72. D. H. Lee, A.Chauraya, Y. Vardaxoglou, and W. S. Park, "A compact and low-profile tunable loop antenna integrated with inductors," *IEEE Antennas and Wireless Propagation Letters*, vol. 7, 2008, pp. 621–624.
73. G. F. Dionne, "Magnetic relaxation and anisotropy effects of high-frequency permeability," *IEEE Transactions on Magnetics*, vol. 39, September 2003, pp. 3121–3126.
74. I. Tzanidis, C. C. Chen, and J. L. Volakis, "Smaller UWB conformal antennas for VHF/UHF applications with ferro-dielectric loadings," *IEEE Antennas and Propagation Society International Symposium*, vol. 5–11, 2008, pp. 1–4.
75. J. L. Snoek, "Dispersion and absorption in magnetic ferrites at frequencies above one Mc/s," *Physica*, vol. 14, 1948, pp. 207–217.
76. B. A. Kramer, M. Lee, C. C. Chen, and J. L. Volakis, "UWB miniature antenna limitations and design issues," *IEEE Antennas and Propagation Society International Symposium*, vol. 3–8, 2005, pp. 598–601.
77. Y. Zhou, C. C. Chen, and J. L. Volakis, "Dual band proximity-fed stacked patch antenna for tri-band GPS applications," *IEEE Transactions on Antennas and Propagation*, vol. 55, January 2007, pp. 220–223.

78. Y. Zhou, C. C. Chen, and J. L. Volakis, "Single-fed circularly polarized antenna element with reduced coupling for GPS arrays," *IEEE Transactions on Antennas and Propagation*, vol. 56, May 2008, pp. 1469–1472.
79. S. Y. Chen, H. T. Chou, and Y. L. Chiu, "A size-reduced microstrip antenna for the applications of GPS signal reception," *IEEE Antennas and Propagation Society International Symposium*, vol. 9–15, 2007, pp. 5443–5446.
80. O. Leisten, J. C. Vardaxoglou, P. McEvoy, R. Seager, and A. Wingfield, "Miniaturized dielectrically-loaded quadrifilar antenna for Global Positioning system (GPS)," *Electronics Letters*, vol. 37, October 2001, pp. 1321–1322.
81. Y. S. Wang and S. J. Chung, "Design of a dielectric-loaded quadrifilar helix antenna," *IEEE International Workshop on Antenna Technology: Small Antennas and Novel Metamaterials*, 2006, pp. 229–232.
82. W. G. Yeoh, Y. B. Choi, L. H. Guo, A. P. Popov, K. Y. Tham, B. Zhao, and X. Chen, "A 2.45-GHz RFID tag with on-chip antenna," *IEEE Radio Frequency Integrated Circuits (RFIC) Symposium*, vol. 11–13, 2006, pp. 1–4.
83. Y. Lee, H. Lim, and H. Lee, "Triple-band compact chip antenna using coupled meander line structure for mobile RFID/PCS/WiBro," *IEEE Antennas and Propagation Society International Symposium*, vol. 9–14, 2006, pp. 2649–2652.
84. J. I. Moon and S. O. Park, "Small chip antenna for 2.4/5.8-GHz dual ISM-band applications," *IEEE Antennas and Wireless Propagation Letters*, vol. 2, 2003, pp. 313–315.
85. Y. P. Zhang, M. Sun, and W. Lin, "Novel antenna-in-package design in LTCC for single-chip RF transceivers," *IEEE Transactions on Antennas and Propagation*, vol. 56, July 2008, pp. 2079–2088.
86. N. G. Alexopoulos and D. R. Jackson, "Fundamental superstrate (cover) effects on printed circuit antennas," *IEEE Transactions on Antennas and Propagation*, vol. 32, 1984, pp. 807–816.
87. C. Y. Huang and J. Y. Wu, "Compact microstrip antenna loaded with very high permittivity superstrate," *IEEE Antennas and Propagation Society International Symposium*, vol. 2, 1998, pp. 680–683.
88. Z. Lin, X. Ni, and J. Zhao, "New type organic magnetic materials and their application in the field of microwaves," *Journal of Microwaves and Optoelectronics*, vol. 15, December 1999, pp. 329–333.
89. S. Zhong, J. Cui, R. Xue, and J. Niu, "Compact circularity polarized microstrip antenna on organic magnetic substrate," *Microwave and Optical Technology Letters*, vol. 40, March 2004, pp. 479–500.
90. S. Bae and Y. Mano, "A small meander VHF & UHF antenna by magneto-dielectric materials," *Asia-Pacific Microwave Conference*, vol. 4, no. 4–7, 2005, p. 3.
91. T. Tanaka, S. Hayashida, K. Imamura, H. Morishita, and Y. Koyanagi, "A Study on Miniaturization of a Handset Antenna Utilizing Magnetic Materials," *Asia-Pacific Conference on Communications and the 5th International Symposium on Multi-Dimensional Mobile Communications*, vol. 2, 2004, pp. 665–669.
92. F. He and Z. Wu, "Modelling of a Slot Loop Antenna on Magnetic Material Substrate," *International Workshop on Antenna Technology: Small and Smart Antennas Metamaterials and Applications*, vol. 21–23, 2007, pp. 412–415.
93. J. L. Volakis, C. C. Chen, J. Halloran, and S. Koulouridis, "Miniature VHF/UHF conformal spirals with inductive and ferrite loading," *IEEE Antennas and Propagation Society International Symposium*, June 2007, pp. 5–8.
94. B. A. Kramer, C. C. Chen, and J. L. Volakis, "Size reduction of a low-profile spiral antenna using inductive and dielectric loading," *IEEE Antennas and Wireless Propagation Letters*, vol. 7, 2008, pp. 22–25.
95. F. Erkmen, C. C. Chen, and J. L. Volakis, "UWB magneto-dielectric ground plane for low-profile antenna applications," *IEEE Antennas and Propagation Magazine*, vol. 50, August 2008, pp. 211–216.
96. F. Erkmen, C. C. Chen, and J. L. Volakis, "Impedance matched ferrite layers as ground plane treatments to improve antenna wide-band performance," *IEEE Transactions on Antennas and Propagation*, vol. 57, January 2009, pp. 263–266.

97. Y. Zhou, "Polymer-ceramic composites for conformal multilayer antenna and RF systems," Dissertation of the Ohio State University, 2009.
98. S. Koulouridis, G. Kiziltas, Y. Zhou, D. Hansford, and J. L. Volakis, "Polymer ceramic composites for microwave applications: Fabrication and performance assessment," *IEEE Transactions on Microwave Theory and Techniques*, vol. 54, December 2006, pp. 4202–4208.
99. Y. Bayram, Y. Zhou, J. L. Volakis, B. S. Shim, and N. A. Kotov, "Conductors textile and polymer-ceramic composites for load bearing antennas," *IEEE Antennas and Propagation Society International Symposium*, vol. 5–11, 2008, pp. 1–4.
100. Y. Zhou, Y. Bayram, L. Dai, and J. L. Volakis, "Conductive polymer-carbon nanotube sheets for conformal load bearing antennas," *URSI Radio Science Meeting*, Boulder, CO, January 2009.
101. S. B. Cohn and K. C. Kelly, "Microwave measurement of high-dielectric constant materials," *IEEE Transactions on Microwave Theory and Techniques*, vol. 14, 1966, pp. 406–410.
102. E. Apaydin, D. Hansford, S. Koulourids, and J. L. Volakis, "Integrated RF circuits design and packaging in high contrast ceramic-polymer composites," *IEEE Antennas and Propagation Society International Symposium*, vol. 9–15, 2007, pp. 1733–1736.
103. E. Apaydin, Y. Zhou, D. Hansford, S. Koulouridis, and J. L. Volakis, "Patterned metal printing on pliable composites for RF design," *IEEE Antennas and Propagation Society International Symposium*, vol. 5–11, 2008, pp. 1–4.
104. J. Y. Chung and C. C. Chen, "Two-layer dielectric rod antenna," *IEEE Transactions on Antennas and Propagation*, vol. 56, June 2008, pp. 1541–1547.
105. M. S. Dresselhaus, G. Dresselhaus, and P. Avouris, *Carbon Nanotubes: Synthesis, Structure, Properties and Applications*, Springer-Verlag, New York, 2001.
106. L. Dai, ed., *Carbon Nanotechnology: Recent Developments in Chemistry, Physics, Materials Science and Device Applications*, Elsevier, Amsterdam Netherlands, 2006.
107. G. W. Hanson, "Fundamental transmitting properties of carbon nanotube antennas," *IEEE Transactions on Antennas and Propagation*, vol. 53, November 2005, pp. 3426–3435.
108. P. J. Burke, S. Li, and Z. Yu, "Quantitative theory of nanowire and nanotube antenna performance," *IEEE Transactions on Nanotechnology*, vol. 5, July 2006, pp. 314–334.
109. Y. Huang, W. Y. Yin, and Q. H. Liu, "Performance prediction of carbon nanotube bundle dipole antennas," *IEEE Transactions on Nanotechnology*, vol. 7, November 2008, pp. 331–337.
110. J. Hao and G. W. Hanson, "Infrared and optical properties of carbon nanotube dipole antennas," *IEEE Transactions on Nanotechnology*, vol. 5, November 2006, pp. 766–775.
111. L. Wang, R. Zhou, and H. Xin, "Microwave (8-50 GHz) characterization of multiwalled carbon nanotube papers using rectangular waveguides," *IEEE Transactions on Microwave Theory and Techniques*, vol. 56, February 2008, pp. 499–506.
112. J. Wang, M. Musameh, and Y. Lin, "Solubilization of carbon nanotubes by nafion toward the preparation of amperometric biosensors," *Journal of the American Chemical Society*, vol. 125, February 2003, pp. 2408–2409.
113. Y. Zhou, E. Apaydin, S. Koulouridis, Y. Bayram, D. Hansford, and J. L. Volakis, "High conductivity printing on polymer-ceramic composites," *IEEE Antennas and Propagation Society International Symposium*, vol. 5–11, 2008, pp. 1–4.
114. Y. Zhou, Y. Bayram, L. Dai, and J. L. Volakis, "Conformal load-bearing polymer-carbon nanotube antennas and RF front-ends," *IEEE Antennas and Propagation Society International Symposium*, vol. 1–5, 2009, pp. 1–4.
115. Y. Zhou, Y. Bayram, F. Du, L. Dai, and J. L. Volakis, "Polymer-carbon nanotube sheets for conformal load bearing antennas," accepted to be published by *IEEE Transactions on Antennas and Propagation*.

116. Y. Zhou, Y. Bayram, L. Dai, and J. L. Volakis, "Conformal load-bearing polymer-carbon nanotube antennas and RF front-ends," *IEEE Antennas and Propagation Society International Symposium*, vol. 1–5, 2009, pp. 1–4.
117. R. L. Haupt, "An introduction to genetic algorithms for electromagnetics," *IEEE Antennas and Propagation Magazine*, vol. 37, April 1995, pp. 7–15.
118. D. S. Whiele and E. Michielssen, "Genetic algorithm optimization applied to electromagnetics: a review," *IEEE Transactions on Antennas and Propagation*, vol. 45, March 1997, pp. 343–353.
119. J. M. Johnson and Y. Rahmat-Samii, "Genetic algorithms in engineering electromagnetics," *IEEE Antennas and Propagation Magazine*, vol. 39, August 1997, pp. 7–21.
120. Randy L. Haupt and Douglas H. Werner, *Genetic Algorithms in Electromagnetics*, John Wiley & Sons, Inc., Hoboken, New Jersey, 2007.
121. Yahya Rahmat-Samii and Eric Michielssen, *Electromagnetic Optimization by Genetic Algorithms*, John Wiley & Sons, Inc., New York, 1999.
122. R. L. Haupt, "Thinned arrays using genetic algorithms," *IEEE Transactions on Antennas and Propagation*, vol. 42, July 1994, pp. 993–999.
123. M. J. Buckley, "Linear array synthesis using a hybrid genetic algorithm," *IEEE Antennas and Propagation Society International Symposium*, 1996, pp. 584–587.
124. R. L. Haupt, "Optimum quantised low sidelobe phase tapers for arrays," *Electronics Letters*, vol. 31, 1995, pp. 1117–1118.
125. M. Shimizu, "Determining the excitation coefficients of an array using genetic algorithms," *IEEE Antennas and Propagation Society International Symposium*, 1994, pp. 530–533.
126. A. Tennant, M. M. Dawoud, and A. P. Anderson, "Array pattern nulling by element position perturbations using a genetic algorithm," *Electronics Letters*, vol. 30, 1994, pp. 174–176.
127. D. S. Weile and E. Michielssen, "The control of adaptive antenna arrays with genetic algorithms using dominance and diploidy," *IEEE Transactions on Antennas and Propagation*, vol. 49, October 2001, pp. 1424–1433.
128. Z. Altman, R. Mittra, J. Philo, and S. Dey, "New designs of ultrabroadband antennas using genetic algorithms," in *Proceedings of IEEE Antenna and Propagation Society International Symposium*, Baltimore, MD, July 1996, pp. 2054–2057.
129. M. Bahr, A. Boag, E. Michielssen, and R. Mittra, "Design of ultrabroadband loaded monopoles," in *Proceedings of IEEE Antenna and Propagation Society International Symposium*, Seattle, WA, June 1994, pp. 1290–1293.
130. A. Boag, A. Boag, E. Michielssen, and R. Mittra, "Design of electrically loaded wire antennas using genetic algorithms," *IEEE Transactions on Antennas and Propagation*, vol. 44, May 1996, pp. 687–695.
131. S. L. Avila, W. P. Carpes, and J. A. Vasconcelos, "Optimization of an offset reflector antenna using genetic algorithms," *IEEE Transactions on Magnetics*, vol. 40, March 2004, pp. 1256–1259.
132. D. S. Linden and E. E. Altschuler, "The design of Yagi antennas using a genetic algorithm," *USNC/URSI Radio Science Meeting*, 1996, p. 283.
133. S. Koulouridis and J. L. Volakis, "L-band circularly polarized small aperture thin textured patch antenna," *IEEE Antennas and Wireless Propagation Letters*, vol. 7, 2008, pp. 225–228.
134. G. Marrocco, A. Fonte, and F. Bardati, "Evolutionary design of miniaturized meander-line antennas for RFID applications," *IEEE Antennas and Propagation Society International Symposium*, vol. 2, 2002, pp. 362–365.
135. G. Marrocco, "Gain-optimized self-resonant meander line antennas for RFID applications," *IEEE Antennas and Wireless Propagation Letters*, vol. 2, 2003, pp. 302–305.
136. D. H. Werner, P. L. Werner, and K. H. Church, "Genetically engineered multiband fractal antennas," *Electronics Letters*, vol. 37, 2001, pp. 1150–1151.
137. M. Fernández Pantoja, F. García Ruiz, A. Rubio Bretones, R. Gómez Martín, J. M. González-Arbesú, J. Romeu, and J. M. Rius, "GA design of wire pre-fractal

antennas and comparison with other euclidean geometries," *IEEE Antennas and Wireless Propagation Letters*, vol. 2, 2003, pp. 238–241.
138. E. E. Altshuler and D. S. Linden, "Wire-antenna designs using genetic algorithms," *IEEE Antennas and Propagation Magazine*, vol. 39, April 1997, pp. 33–43.
139. E. E. Altshuler, "Design of a vehicular antenna for GPS/Iridium using a genetic algorithm," *IEEE Transactions on Antennas and Propagation*, vol. 48, June 2000, pp. 968–972.
140. E. E. Altshuler, "Electrically small self-resonant wire antennas optimized using a genetic algorithm," *IEEE Transactions on Antennas and Propagation*, vol. 50, March 2002, pp. 297–300.
141. E. E. Altshuler and D. S. Linden, An ultra-wideband impedance-loaded genetic antenna, *IEEE Transactions on Antennas and Propagation*, vol. 52, November 2004, pp. 3147–3151.
142. G. Kiziltas, D. Psychoudakis, J. L. Volakis, and N. Kikuchi, "Topology design optimization of dielectric substrates for bandwidth improvement of a patch antenna," *IEEE Transactions on Antennas and Propagation*, vol. 51, October 2003, pp. 2732–2743.
143. G. Kiziltas and J. L. Volakis, "Shape and material optimization for bandwidth improvement of printed antennas on high contrast substrates," *IEEE International Symposium on Electromagnetic Compatibility*, vol. 2, no. 11–16, 2003, pp. 1081–1084.
144. J. Volakis, A. Chatterjee, and L. Kempel, *Finite element method for electromagnetics: antenna, microwave circuits and scattering applications*, Wiley-Interscience, New York, 1998.
145. S. Koulouridis, D. Psychoudakis, and J. L. Volakis, "Multiobjective optimal antenna design based on volumetric material optimization," *IEEE Transactions on Antennas and Propagation*, vol. 55, March 2007, pp. 594–603.
146. J. Kennedy and R. Eberhart, "Particle swarm optimization," *IEEE International Conference on Neural Networks*, vol. 4, 1995, pp. 1942–1948.
147. R. Poli, J. Kennedy, and T. Blackwell, "Paticle swarmoptimization: an overview," *Swarm Intelligence*, vol. 1, 2007, pp. 33–57.
148. J. Robinson and Y. Rahmat-Samii, "Particle swarm optimization in electromagnetics," *IEEE Transactions on Antennas and Propagation*, vol. 52, 2004, pp. 397–407.
149. N. Jin and Y. Rahmat-Samii, "Advances in particle swarm optimization for antenna designs: real-number, binary, single-objective and multiobjective implementations," *IEEE Transactions on Antennas and Propagation*, vol. 55, March 2007, pp. 556–567.
150. N. Jin and Y. Rahmat-Samii, "Analysis and particle swarm optimization of correlator antenna arrays for radio astronomy applications," *IEEE Transactions on Antennas and Propagation*, vol. 56, May 2008, pp. 1269–1279.
151. D. W. Boeringer and D. H. Werner, "Particle swarm optimization versus genetic algorithms for phased array synthesis," *IEEE Transactions on Antennas and Propagation*, vol. 52, March 2004, pp. 771–779.
152. P. J. Bevelacqua, C. A. Balanis, "Minimum sidelobe levels for linear arrays," *IEEE Transactions on Antennas and Propagation*, vol. 55, December 2007, pp. 3442–3449.
153. F. Afshinmanesh, A. Marandi, and M. Shahabadi, "Design of a single-feed dual-band dual-polarized printed microstrip antenna using a Boolean particle swarm optimization," *IEEE Transactions on Antennas and Propagation*, vol. 56, July 2008, pp. 1845–1852.
154. N. Jin and Y. Rahmat-Samii, "Parallel particle swarm optimization and finite-difference time-domain (PSO/FDTD) algorithm for multiband and wide-band patch antenna designs," *IEEE Transactions on Antennas and Propagation*, vol. 53, November 2005, pp. 3459–3468.
155. L. Lizzi, F. Viani, R. Azaro, and A. Massa, "A PSO-driven spline-based shaping approach for ultrawideband (UWB) antenna synthesis," *IEEE Transactions on Antennas and Propagation*, vol. 56, August 2008, pp. 2613–2621.

156. N. Jin and Y. Rahmat-Samii, "Particle swarm optimization for multi-band handset antenna designs: A hybrid real-binary implementation," *IEEE Antennas and Propagation Society International Symposium*, vol. 5–11, 2008, pp. 1–4.
157. D. Sievenpiper, L. Zhang, R. F. J. Broas, N. G. Alexopoulos, and E. Yablonovitch, "High-impedance electromagnetic surfaces with a forbidden frequency band," *IEEE Transactions on Microwave Theory and Techniques*, vol. 47, November 1999, pp. 2059–2074.
158. R. F. J. Broas, D. F. Sievenpiper, and E. Yablonovitch, "A high-impedance ground plane applied to a cellphone handset geometry," *IEEE Transactions on Microwave Theory and Technology*, vol. 49, July 2001, pp. 1261–1265.
159. P. Maagt, R. Gonzalo, Y. C. Vardaxoglou, and J. M. Baracco, "Electromagnetic bandgap antennas and components for microwave and (sub)millimeter wave applications," *IEEE Transactions on Antennas and Propagation*, vol. 51, October 2003, pp. 2667–2677.
160. F. Yang and Y. Rahmat-Samii, "Micro-strip antennas integrated with electromagnetic band-gap structures: a low mutual coupling design for array applications," *IEEE Transactions on Antennas and Propagation*, vol. 51, October 2003, pp. 2936–2946.
161. Y. Fu and N. Yuan, "Elimination of scan blindness of microstriop phased array using electromagnetic bandgap structures," *IEEE Antennas and Wireless Propagation Letters*, vol. 3, 2004, pp. 63–65.
162. I. Zeev, S. Reuven, and B. Reuven B, "Micro-strip antenna phased array with electromagnetic band-gap substrate," *IEEE Transactions on Antennas and Propagation*, vol. 52, 2004, pp. 1446–1453.
163. Fan Yang and Y. Rahmat-Samii, *Electromagnetic Bandgap Structures in Antenna Engineering*, Cambridge University Press, Cambridge, 2009.
164. R. Baggen, M. Martinez-Vazquez, J. Leiss, S. Holzwarth, L. S. Drioli, and P. de Maagt, "Low profile GALILEO antenna using EBG technology," *IEEE Transactions on Antennas and Propagation*, vol. 56, March 2008, pp. 667–674.
165. S. Clavijo, R. E. Diaz, and W. E. McKinzie, "Design methodology for Sievenpiper high-impedance surfaces: an artificial magnetic conductor for positive gain electrically small antennas," *IEEE Transactions on Antennas and Propagation*, vol. 51, October 2003, pp. 2678–2690.
166. M. F. Abedin and M. Ali, "Effects of EBG reflection phase profiles on the input impedance and bandwidth of ultrathin directional dipoles," *IEEE Transactions on Antennas and Propagation*, vol. 53, November 2005, pp. 3664–3672.
167. Z. Li and Y. Rahmat-Samii, "PBG, PMC and PEC ground planes: a case study for dipole antenna," *IEEE Antennas and Propagation Society International Symposium*, vol. 4, 2000, pp. 2258–2261.
168. F. Yang and Y. Rahmat-Samii, "Reflection phase characterizations of the EBG ground plane for low profile wire antenna applications," *IEEE Transactions on Antennas and Propagation*, vol. 51, October 2003, pp. 2691–2703.
169. W. E. McKinzie and R. R. Fahr, "A low profile polarization diversity antenna built on an artificial magnetic conductor," *IEEE Antennas and Propagation Society International Symposium*, vol. 1, no. 16–21, 2002, pp. 762–765.
170. J. M. Bell and M. F. Iskander, "A low-profile Archimedean spiral antenna using an EBG ground plane," *IEEE Antennas and Wireless Propagation Letters*, vol. 3, 2004, pp. 223–226.
171. A. M. Mehrabani and L. Shafai, "A dual-arm Archimedean spiral antenna over a low-profile artificial magnetic conductor ground plane," *ANTEM/URSI*, vol. 15–18, 2009, pp. 1–4.
172. S. Wang, A. P. Feresidis, G. Goussetis, and J. C. Vardaxoglou, "Artificial magnetic conductors for low-profile resonant cavity," *IEEE Antennas and Propagation Society International Symposium*, vol. 2, 2004, pp. 1423–1426.
173. F. Yang and Y. Rahmat-Samii, "Curl antennas over electromagnetic band-gap surface: a low profileddesign for CP applications," *IEEE Antennas and Propagation Society International Symposium*, vol. 3, 2001, pp. 372–375.

174. F. Costa, A. Monorchio, S. Talarico, and F. M. Valeri, "An active high-impedance surface for low-profile tunable and steerable antennas," *IEEE Antennas and Wireless Propagation Letters*, vol. 7, 2008, pp. 676–680.
175. F. Yang and Y. Rahmat-Samii, "A low-profile single dipole antenna radiating circularly polarized waves," *IEEE Transactions on Antennas and Propagation*, vol. 53, 2005, pp. 3083–3086.
176. D. Sievenpiper, J. Schaffner, B. Loo, G. Tangonan, R. Harold, J. Pikulski, and R. Garcia, "Electronic beam steering using a varactor-tuned impedance surface," *IEEE Antennas and Propagation Society International Symposium*, vol. 1, 2001, pp. 174–177.
177. D. Sievenpiper, "Chapter 11: review of theory, fabrication, and applications of high impedance ground planes," in N. Engheta and R. Ziolkowski, eds., *Metamaterials: Physics and Engineering Explorations*, John Wiley & Sons Inc., 2006.
178. M. G. Bray and D. H. Werner, "A novel design approach for an independently tunable dual-band EBG AMC surface," *IEEE Antennas and Propagation Society International Symposium*, vol. 1, 2004, pp. 289–292.
179. D. F. Sievenpiper, J. H. Schaffner, H. J. Song, R. Y. Loo, and G. Tangonan, "Two-dimensional beam steering using an electrically tunable impedance surface," *IEEE Transactions on Antennas and Propagation*, vol. 51, 2003, pp. 2713–2722.
180. M.G. Bray and D. H. Werner, "A broadband open-sleeve dipole antenna mounted above a tunable EBG AMC ground plane," *IEEE Antennas and Propagation Society International Symposium*, vol. 2, no. 20–25, 2004, pp. 1147–1150.
181. D. J. Kern, D. H. Werner, A. Monorchio, L. Lanuzza, and M. J. Wilhelm, "The design synthesis of multiband artificial magnetic conductors using high impedance frequency selective surfaces," *IEEE Transactions on Antennas and Propagation*, vol. 53, January 2005, pp. 8–17.
182. Y. Yao, X. Wang, and Z. Feng, "A novel dual-band compact electromagnetic bandgap (EBG) structure and its application in multi-antennas," *IEEE Antennas and Propagation Society International Symposium*, vol. 9–14, 2006, pp. 1943–1946.
183. W. McKinzie and S. Rogers, "A multi-band artificial magnetic conductor comprised of multiple FSS layers," *IEEE Antennas and Propagation Society International Symposium*, vol. 2, no. 22–27, 2003, pp. 423–426.
184. K. Inafune and E. Sano, "Multiband artificial magnetic conductors using stacked microstrip patch layers," *IEEE International Symposium on MAPE*, vol. 1, no. 8–12, 2005, pp. 590–593.
185. D. J. Kern, D. H. Werner, and M. J. Wilhelm, "Active negative impedance loaded EBG structures for the realization of ultra-wideband Artificial Magnetic Conductors," *IEEE Antennas and Propagation Society International Symposium*, vol. 2, 2003, pp. 427–430.
186. R. Diaz, V. Sanchez, E. Caswell, and A. Miller, "Magnetic loading of artificial magnetic conductors for bandwidth enhancement," *IEEE Antennas and Propagation Society International Symposium*, vol. 2, 2003, pp. 427–430.
187. L. Yousefi, B. Mohajer-Iravani, and O.M. Ramahi, "Enhanced bandwidth artificial magnetic ground plane for low-profile antennas," *IEEE Antennas and Wireless Propagation Letters*, vol. 6, 2007, pp. 289–292.
188. Y. E. Erdemli, K. Sertel, R. Gilbert, D. Wright, and J. L. Volakis, "Frequency selective surfaces to enhance performance of broadband reconfigurable arrays," *IEEE Transactions on Antennas and Propagation*, vol. 50, 2002, pp. 1716–1724.
189. Y. Erdemli, R. A. Gilbert, and J. L. Volakis, "A reconfigurable slot aperture design over a broad-band substrate/feed structure," *IEEE Transactions on Antennas and Propagation*, vol. 52, November 2004, pp. 2860–2870.

CHAPTER 4

Antenna Miniaturization via Slow Waves

Chi-Chih Chen

4.1 Introduction

Antenna miniaturization will continue to be a key issue in wireless communications, navigation, sensors, and radio frequency identification (RFIDs). For instance, each cellular tower is often populated with many antennas to cover different angular sectors and different frequency bands. Each modern notebook computer is likely embedded with multiple antennas to provide service in WWAN (824–2170 MHz) and WLAN (2.4 and 5.5 GHz), bluetooth, etc. Also automobiles, vessels, and aircrafts will require more antennas to compete for very limited real estate. This dire situation is changing antenna designer worldwide with a goal to develop a new generation of physically small antennas that multibands or wideband.

Numerous new small-antenna designs have been developed by researchers and engineers worldwide. Chapter 3 provides a survey of these designs. This chapter is dedicated to provide fans insights on generic small-antenna design approaches. This will be done using an equivalent artificial transmission line (ATL) theory that governs the impedance and wavelengths within an antenna. A few examples will be included to show such theory at work in miniaturizing antennas.

4.2 Miniaturization Factor

The most generic definition of antenna miniaturization is to reduce overall antenna dimensions while maintaining its key characteristics such as impedance and radiation patterns. For narrowband antennas, this means achieving resonance at physical dimensions much less than the half free-space wavelength (λ_0) at resonance. This is illustrated in Fig. 4.1*a* which compares antenna gain curves as a function of frequency before and after miniaturization of a narrowband antenna. Of course, any miniaturized passive antenna is still bounded by the theoretical performance limits discussed in Chap. 1. For wideband antennas, it is convenient to choose the "onset" frequency above which the realized gain level satisfies a minimal gain requirement (see Chap. 2) that often varies from application to application. For the sake of discussion, the definition in Eq. (4.1) for "miniaturization factor" (*MF*) will be adopted throughout this chapter to indicate the degree of miniaturization. This definition allows users to choose the reference frequency in accordance with the application at hand. The larger the miniaturization factor, the greater the degree of miniaturization.

$$\mathrm{MF} = \frac{f_{\mathrm{ref}}^{\mathrm{original}}}{f_{\mathrm{ref}}^{\mathrm{min\,iaturized}}} \tag{4.1}$$

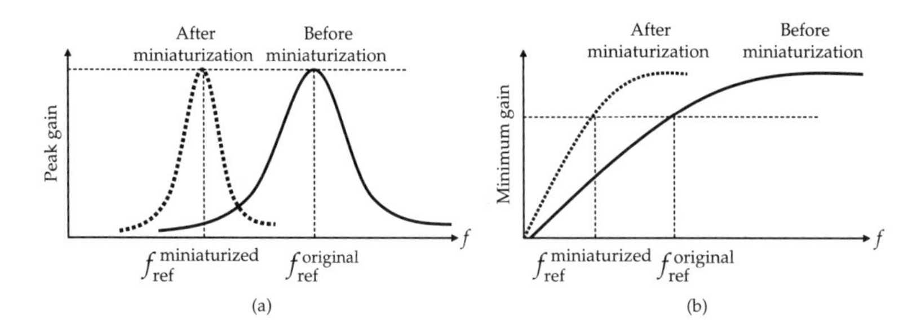

FIGURE 4.1 Ideal results of antenna miniaturization. (a) Miniaturization of narrowband antenna (b) Miniaturization of wideband antenna.

4.3 Basic Antenna Miniaturization Concept

4.3.1 Phase Coherent Radiation Conditions

For any antenna the total radiated field in a given direction is the sum of direct radiation from the source and scattered fields from all electromagnetic discontinuities and the ground plane in the vicinity of the antenna. The polarization, intensity, and phase of scattered fields depend on antenna geometry, and location of source. For example, a center-fed thin-wire dipole made of perfectly conducting wire shown in Fig. 4.2 contains two diffraction locations at both ends of the wires. In order for this dipole to radiate effectively, the first, second, and higher order of diffraction terms from both ends should have the same polarity and coherent phase. Figure 4.2 illustrates field polarity and time delays of the first three diffraction terms along the broadside direction. Notice that the odd number terms and even number terms have opposite polarity. Therefore, in order to have all diffracted fields to be added constructively, the following phase relationship in Eq. (4.2) needs to be satisfied. This condition naturally leads to a half-wavelength dipole. When the length L is much shorter than half

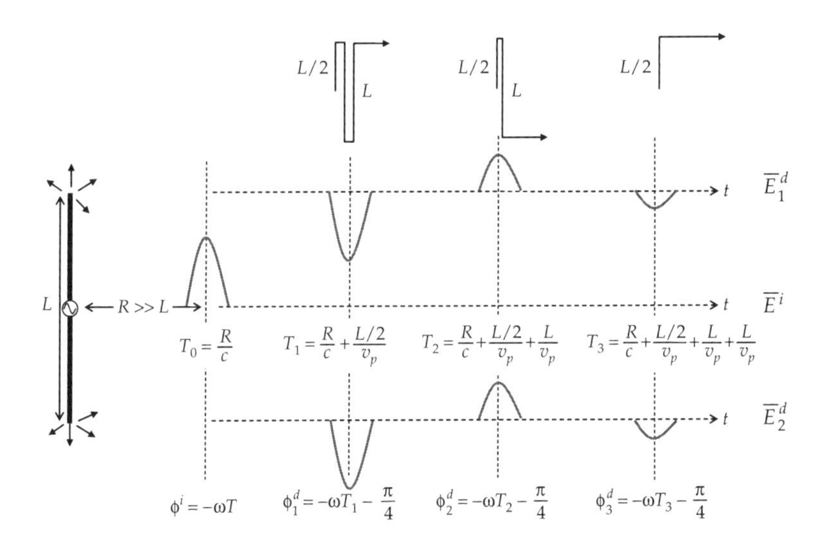

FIGURE 4.2 Radiated fields from a wire dipole in normal direction.

wavelength, all diffracted fields add destructively and this results in weak net radiation. This situation can be improved by artificially reducing the phase velocity of electromagnetic fields propagating along antenna structure from one end to the other such that the in-phase condition in Eq. (4.2) can be achieved even when the physical length is much short than half wavelength. This simple example demonstrates the basic concept of antenna miniaturization to be discussed in this chapter.

$$\left|\phi_n^d - \phi_{n+1}^d\right| = \frac{2\pi L}{\lambda_p} = \pi \quad \text{or} \quad L = \frac{\lambda_p}{2} \tag{4.2}$$

4.3.2 Equivalent Transmission Line (TL) Model of an Antenna

The radiated fields in the far-field region can be obtained from integrating equivalent currents over the source region containing the antenna of interest. That is

$$\bar{E}(\vec{r}) = -\frac{jk^2}{\omega\varepsilon}\frac{e^{-jkr}}{4\pi r}(\bar{\bar{I}} - \hat{r}\hat{r})\int_V \bar{J}(\bar{r}')e^{jk\hat{r}\bar{r}'}dv' \tag{4.3}$$

For antenna made of PEC structures and an excitation source like most antennas, we have

$$\bar{J}(\bar{r}') = 2\hat{n} \times \bar{H}^i \tag{4.4}$$

where $\bar{H}^i$ is incident field upon antenna structure from excitation source in region $\bar{r}'$ which contains the source and antenna. Therefore, an antenna is practically a collection of scattering structures near the source. Note that an antenna does not have to be made of electrical conductors. From Eq. (4.3) it is apparent that the radiated fields in the far zone depend on net radiations from all radiation points in the source region. Therefore, the relative amplitude and phase relationships among all radiation points in the source region determine the final outcome of net radiation. The dipole antenna discussed in the previous section serves as a good example. Let us consider another PEC structure (see Fig. 4.3) made of two thin parallel wires with a voltage source at one end and the other end extends to infinite. Is this structure an antenna or a transmission line? Well, if the separation d is much smaller than the wavelength, that is, $d \ll \lambda_0$, the radiated fields from two opposing wire segments ($\bar{r}_1'$and $\bar{r}_2'$) will then have similar magnitude and propagation phase but opposite polarization at any observation point in the far field. This leads to weak net radiation due to cancellation effect from opposite current directions. In this case, this structure behaves mostly like a transmission line. On the other

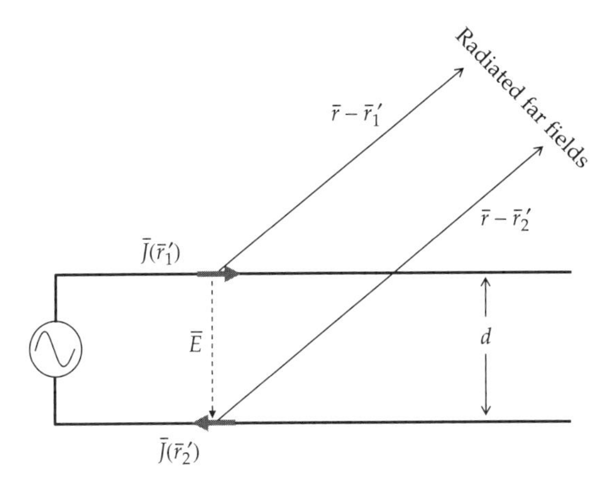

FIGURE 4.3 A two-wire "transmission line" or a two-wire "antenna"?

hand, if $d \approx \lambda_0/2$, the radiated fields from opposing wire segments become constructively added in upward and downward directions in the plane containing the structures. In this case, this structure becomes a good antenna.

4.3.3 Artificial Transmission Line (ATL) of Antennas

The artificial transmission line (ATL) concept was first introduced by Hiraoka [1] in various microwave applications [2, 3]. More recently, it was exhibited to reduce the size of wideband antennas utilizing a simple transmission-line (TL) model to describe the propagation behavior of electromagnetic waves [4]. ATL models have also been used to construct the so called "effective negative materials" when special transmission-line parameters are chosen [5–7]. ATL can also be applied to describe the impedance behavior of an antenna. Note that an ATL model of an antenna differs from an "equivalent-circuit model" which was used to describe impedance behavior model of dipole and loop antennas [8, 9] in that ATL segments (see Fig. 4.4) are usually less than $\lambda_{effective}/20$ in length and it accounts for proper phase variation between segments. Therefore, ATL model is capable for modeling ultra-wideband antennas and is especially useful in predicting frequency-independent antennas such as spiral antennas, conical spiral antennas, log-periodic antennas, bow-tie dipoles, bi-conical antennas, etc. The importance of analyzing antenna impedance and resonant behavior using ATL is that it naturally leads to antenna miniaturization via properly modifying transmission-line parameters. This section will review basic

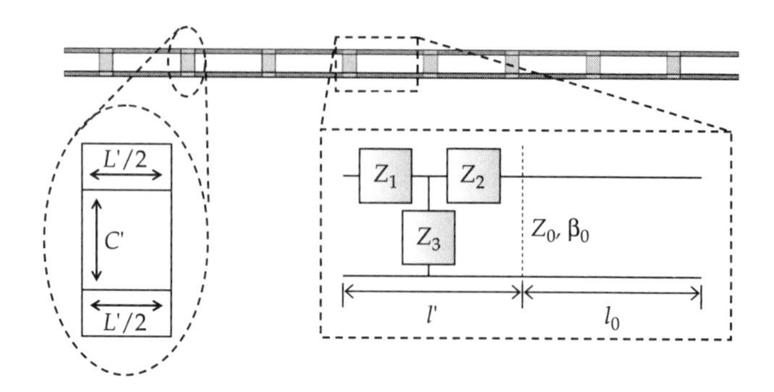

FIGURE 4.4 ATL example using a small-unit segment of series inductors and shunt capacitors periodically inserted into a two-wire TL. (*After Lee et al. [4] ©IEEE, 2007.*)

transmission-line parameters and formulas using a two-wire transmission line as an example. We will then show an approach to artificially alter these parameters and change the overall characteristic impedance and propagation constant. In the later sections, we will discuss specific antenna examples miniaturized by applying ATL concept. The main purpose of these discussions is to provide a physical understanding of the miniaturization principles beyond a trial-and-error using computation tools.

First, let us examine some basic characteristics of a transmission line shown in Fig. 4.4. This TL is composed of many unit cells in series. Each contains an ATL segment and a conventional TL segment. In this case, the ATL segment has a length l' and is characterized by a series inductance and a shunt capacitance (L' and C'). The unit cell is also composed of a regular transmission line (TL) segment of length l_0 having characteristic impedance Z_0 and propagation constant β_0. The ATL section can be represented by an equivalent T-network [10] as shown in Fig. 4.4.

The resultant T-network parameters are

$$\begin{cases} Z_1 = Z_2 \equiv \dfrac{j\omega(L'l')}{2} \\ Z_3 \equiv \dfrac{1}{j\omega(C'l')} \\ Z' \equiv \sqrt{\dfrac{L'}{C'}} \qquad Z_0 \equiv \sqrt{\dfrac{L_0}{C_0}} \end{cases} \tag{4.5}$$

$$\text{phase factors} \quad \theta' \equiv \beta' l' \quad \text{and} \quad \theta_0 \equiv \beta_0 l_0$$

where Z_1, Z_2, and Z_3 are depicted in Fig. (4.4) and are computed from the inductance (L') and capacitance (C') loading, and the loading length (l) within each unit cell. Combining Eq. (4.5) with the TL segment, the entire ($l_0 + l'$) section of the ATL can be characterized via ABCD matrix [11]. The resultant ABCD values are

$$\begin{cases} A = \left(1 - \dfrac{\theta'^2}{2}\right)\cos(\theta_0) - \dfrac{Z'}{Z_0}\theta'\left(1 - \dfrac{\theta'^2}{4}\right)\sin(\theta_0) \\ B = jZ_0\left(1 - \dfrac{\theta'^2}{2}\right)\sin(\theta_0) + j\theta' Z'\left(1 - \dfrac{\theta'^2}{4}\right)\cos(\theta_0) \\ C = j\dfrac{\theta'}{Z'}\cos(\theta_0) + \dfrac{j}{Z_0}\left(1 - \dfrac{\theta'^2}{2}\right)\sin(\theta_0) \\ D = \left(1 - \dfrac{\theta'^2}{2}\right)\cos(\theta_0) - \dfrac{Z_0}{Z'}\theta'\sin(\theta_0) \end{cases} \tag{4.6}$$

In a special case when $Z' \approx Z_0$ in Eq. (4.5) (matched impedance between the ATL and TL segments), and θ' is small, then Eq. (4.6) reduces to (i.e., electrically short ATL segment),

$$\begin{bmatrix} A & B \\ C & D \end{bmatrix} \approx \begin{bmatrix} \cos(\theta_0 + \theta') & jZ_0\sin(\theta_0 + \theta') \\ \dfrac{j}{Z_0}\sin(\theta_0 + \theta') & \cos(\theta_0 + \theta') \end{bmatrix} \tag{4.7}$$
$$\text{for } Z_0 \approx Z' \quad \text{and} \quad \text{small } \theta'$$

The corresponding propagation constant, k_{eff}, can then be found by applying Bloch's theorem [12] (see Chap. 6). This gives

$$\begin{aligned} &e^{-jk_{\text{eff}}(l_0+l')} \\ &= \cos(\theta_0) - \frac{\theta'}{2}\sin\theta_0\left(\frac{Z_0}{Z'} + \frac{Z'}{Z_0}\right) - \frac{\theta'^2}{2}\cos\theta_0 + \frac{\theta'^3}{8}\frac{Z'}{Z_0}\sin\theta_0 \\ &\quad + \sqrt{\left[\frac{\theta'}{2}\sin\theta_0\left(\frac{Z_0}{Z'} + \frac{Z'}{Z_0}\left(1 - \frac{\theta'^2}{4}\right)\right) - \cos\theta_0\left(1 - \frac{\theta'^2}{2}\right)\right]^2 - 1} \\ &e^{-jk_{\text{eff}}(l_0+l')} \approx e^{\pm j(\beta' l' + \beta_0 l_0)} \quad \text{when } Z_0 = Z' \text{ and small } \theta' \end{aligned} \tag{4.8}$$

with the resultant effective propagation constant being

$$k_{\text{eff}} = \frac{\beta' l' + \beta_0 l_0}{l' + l} \tag{4.9}$$

It is interesting to observe that Eq. (4.9) can represent the average propagation constants between the ATL and TL segments. Similarly, the effective inductance and capacitance per unit length of the whole unit cell is found from

$$L_{\text{eff}} \approx \frac{l_0 \times L_0 + l' \times L'}{l_0 + l'} \quad C_{\text{eff}} \approx \frac{l_0 \times C_0 + l' \times C'}{l_0 + l'} \tag{4.10}$$

The corresponding characteristic impedance and phase velocity of the unit cell ($l_0 + l'$ in length) is then

$$Z_{\text{eff}} = \sqrt{\frac{L_{\text{eff}}}{C_{\text{eff}}}} \quad v_{\text{eff}} = \frac{1}{\sqrt{L_{\text{eff}} C_{\text{eff}}}} \tag{4.11}$$

We note that the inductive and capacitive loading provide two degrees of freedom for controlling impedance and phase velocity simultaneously. This is important for antenna miniaturization where one must reduce phase velocity but also ensure good impedance matching. Without impedance matching, miniaturization will result in nonoptimal performance. As can be understood, the weighted average formula Eqs. (4.9) and (4.10) are valid only for electrically small unit cells and for TEM mode. However, this is almost always true for unit cells and necessary to ensure wideband operation and avoid high-order modes.

As an example, let us consider a transmission line shown at the top of Fig. 4.5, created using the finite element software package (HFSS). Each wire of the transmission line is modeled as a narrow PEC strip of zero thickness, and total length of 15 cm. It is divided into 8 unit cells with each containing a series inductor and a shunt capacitor in the ATL segment that is 1.5 cm in length. Each unit cell has a length of 1.5 cm, or $\lambda_0/20$ at 1 GHz. The input and output ports are located at the left and right ends of the entire TL, designated as port "A" and port "B", respectively. From full-wave simulations, the total phase delay from port A to port B and effective TL characteristic impedance were extracted from S parameters. The phase velocity values versus phase delay is given in Fig. 4.5 and the corresponding TL impedance versus frequency is plotted in Fig. 4.6. The effective phase velocity can be derived from the slope of the phase delay, and is used to determine MF by comparing with free-space phase velocity. It was found that the effective miniaturization factor (MF) ranges from 1.7 up to as much as 8.6. For MF = 6.3 and MF = 8.6, the unit cell is no longer electrically short at frequencies greater than 0.3 GHz, and results in nonlinear behavior as a function of frequency. This nonlinear phenomenon allows one to determine whether the unit cell is sufficiently small at the desired operation frequencies.

The extracted effective characteristic impedance normalized with Z_0 of TL prior to miniaturization is plotted in Fig. 4.6 which gives several curves of Z_{eff}/Z_0 as a function of MF. The goal is to show that unless L_{eff} and C_{eff} in Eq. (4.10) are chosen properly, Z_{eff} can become large or very small, causing mismatches. As expected, when L_{eff}

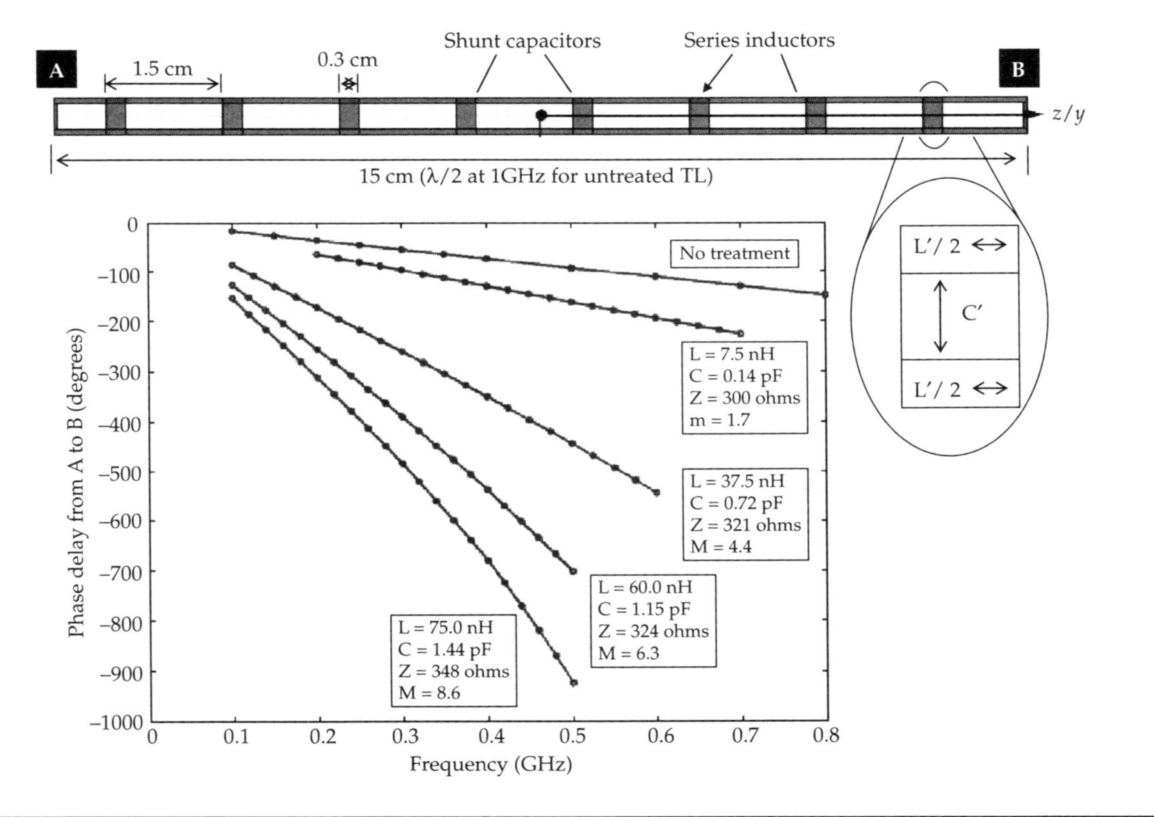

FIGURE 4.5 Phase delay of an ATL with different degree of miniaturization using LC loading. (*After Lee et al. [4] ©IEEE, 2007.*)

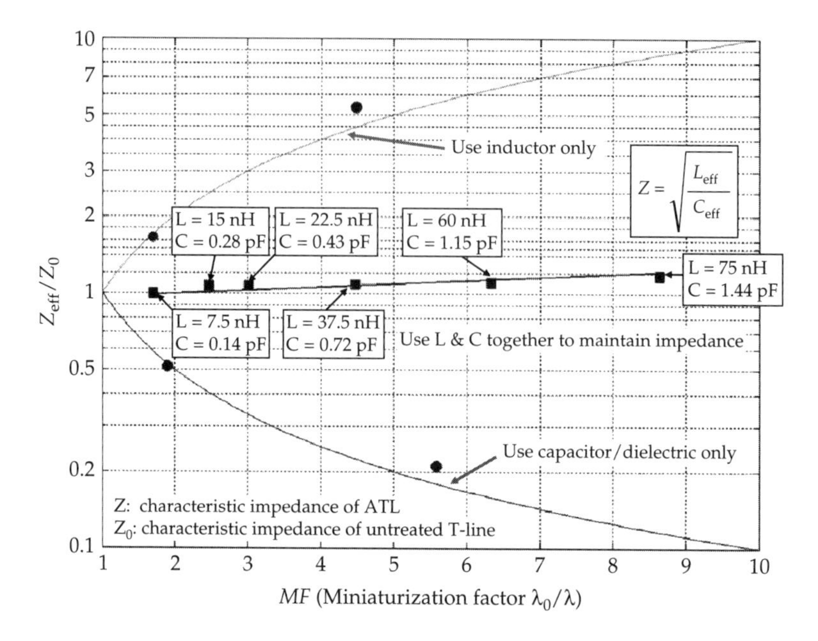

FIGURE 4.6 Effective characteristic impedance of the TL shown in Fig. 4.5 for different (L′, C′) loading to produce different miniaturization factors. (*After Lee et al. [4] ©IEEE, 2007.*)

is large, Z_{eff} is also large. Correspondingly, when C_{eff} is large, then Z_{eff} is small value. The middle curve in Fig. 4.6 is most attractive and shows the L_{eff} and C_{eff} need to be changing with frequency to achieve broadband miniaturization. This will be exploited later to design wideband conformal antenna. Above all, Fig. 4.6 demonstrates that it is possible to simultaneously maintain impedance matching miniaturization by employing series inductance and shunt capacitance loading.

The above findings (from simulations) were validated experimentally. A two-wire transmission line on a printed circuit board (Rogers 43, thickness 0.5 mm) was fabricated without any reactive loading to serve as a reference. This measurement showed that the unloaded line's characteristic impedance was around 140 Ω. Another similar transmission line with lumped LC components added with a goal to achieve a miniaturization factor of three and a characteristic impedance of 100 Ω (for matching to balanced feed line) was also fabricated as shown in Fig. 4.7 which shows two series inductors and one shunt capacitor. Standard surface-mount chip inductors and capacitors (size code 0402) were used to realize the LC loading. The conducted measurements of the phase delay with and without LC loading is shown in Fig. 4.8. As seen, the calculated values are in good

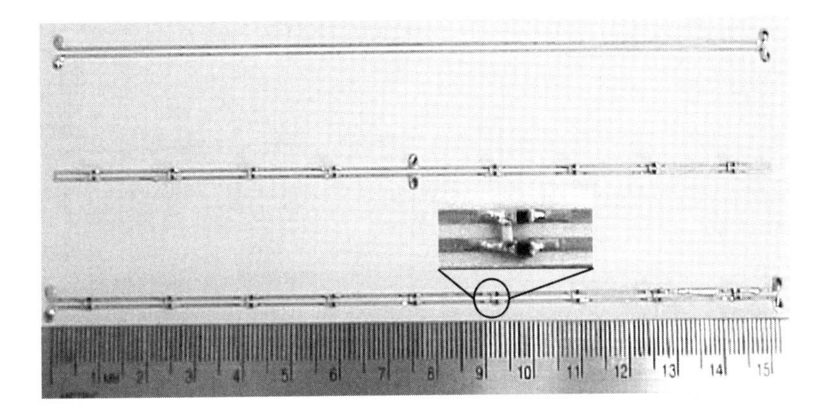

FIGURE 4.7 Fabricated transmission-line shown in Fig. 4.5. The enlarged section shows the placement of the shunt capacitor and serial inductor chips. (*After Lee et al. [4] ©IEEE, 2007.*)

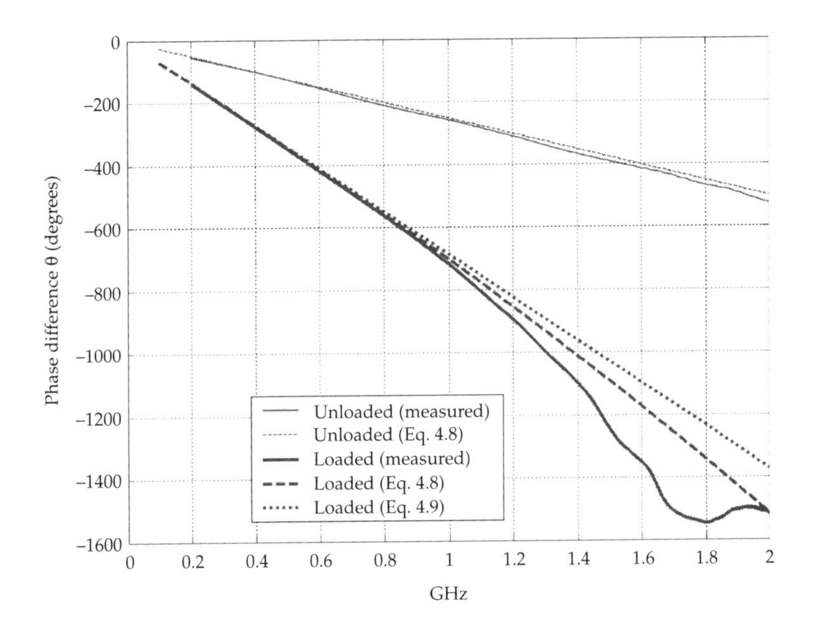

FIGURE 4.8 Measured phase delay across the ATL test board shown in Fig. 4.7 ($C' = 1.6$ pF and $L' = 5.6$ nH). (*After Lee et al. [4] ©IEEE, 2007.*)

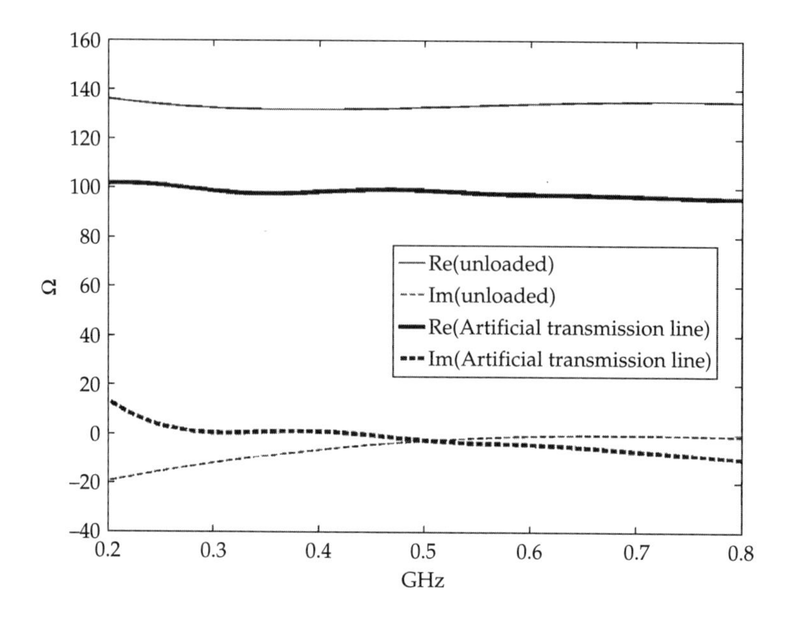

FIGURE 4.9 Measured characteristic impedance of the ATL test board shown in Fig. 4.7 with and without reactive loading. (*After Lee et al. [4] ©IEEE, 2007.*)

agreement with measurements. It is also observed that the complete dispersion relation in Eq. (4.8) predicts the nonlinearities and fits better to the measured data as compared to the simplified expression in Eq. (4.9). Discrepancies between measured and predicted data are likely due to parasitic effects of the lumped elements. The characteristic impedance of the tested ATL lines with and without reactive loading was also determined from reflection measurement and is shown in Fig. 4.9. We observe that the final effective impedance agrees well with the designed 100 Ω line (the imaginary part was also negligible).

4.4 Antenna Miniaturization Examples

4.4.1 Two-Wire Loop Antenna

To demonstrate the ATL concept we consider the miniaturization of a loop antenna as shown in Fig. 4.10. This simple antenna is formed by a square loop of a pair of wires forming a transmission line. The pair of wires forms an easy way to load series inductors and shunt

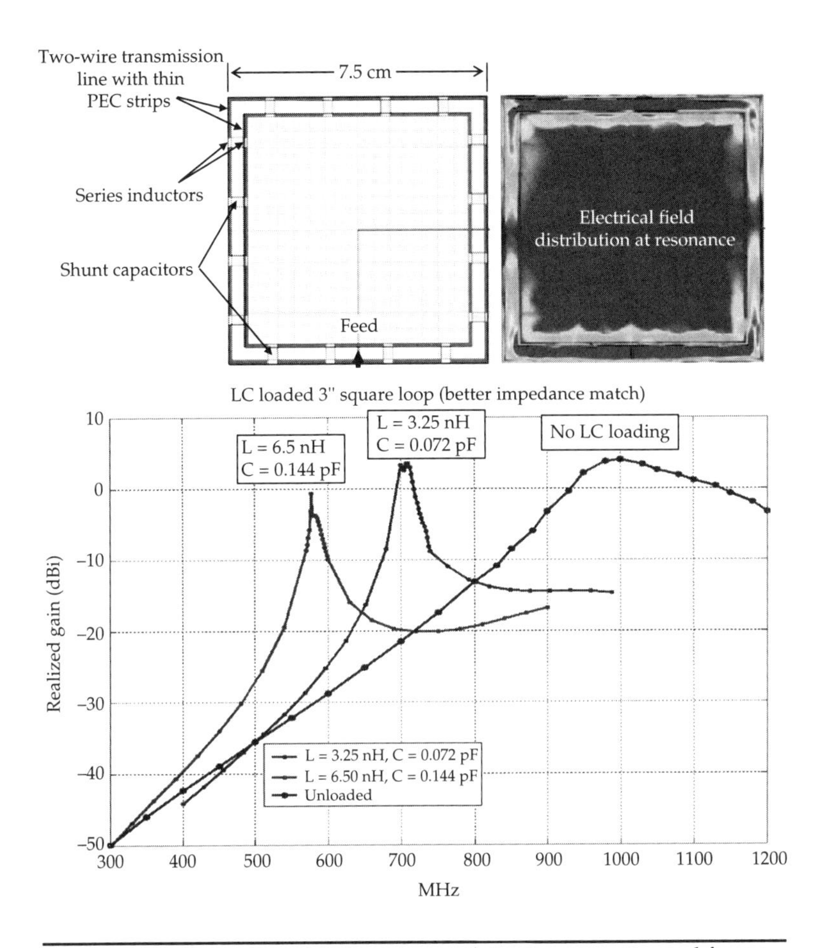

FIGURE 4.10 Geometry setup for a square slot antenna constructed from a two-wire transmission line loaded with 16 ATL segments using $L'C'$ loading as noted. (*After Lee et al. [4] ©IEEE, 2007.*)

capacitors to form ATL. A voltage-gap source is also applied at arrow location across two wires for excitation. As before, each ATL segment contains a series inductor and a shunt capacitor. The broadside realized gain as a function of frequency was calculated using full-wave simulations (HFSS) for three different loading conditions: unloaded, ($L' = 3.25$ nH, $C' = 0.0725$ pF), and ($L' = 6.5$ nH, $C' = 0.144$ pF). The gain results are plotted at the bottom of Fig. 4.10 [4]. As indicated when the loop is unloaded, the gain peaks appear at approximately 990 GHz when the loop circumference is λ_0. For the loaded cases, the resonant frequency decreases as reactive values increase. The resultant miniaturization factors are approximately MF $= 1.39$ for ($L' = 3.25$ nH, $C' = 0.0725$ pF) and MF $= 1.71$ for ($L' = 6.5$ nH, $C' = 0.144$ pF). Nevertheless, although ideal lump L/C components can achieve miniaturization, practical lump components suffer from parasitic resistance, frequency-dependent, and low-power handling. The accumulated resistance from all components could reduce antenna efficiency significantly. Therefore, in practice, it is better to increase series inductance and shunt capacitance via antenna design without the lump components.

4.4.2 Antenna Miniaturization by Increasing Shunt Capacitance

4.4.2.1 Dielectric Loaded Patch Antenna

Patch antennas have been used widely due to their simple and compact geometry. Using a dielectric substrate is a well-known method to reduce the patch size. Such dielectric loading increases the shunt capacitance of the patch antenna, causing slower wave velocity within the patch. However, increasing shunt capacitance results in lower antenna impedance as predicted by Eq. (4.11), and degrading impedance matching. As a result, the realized gain decreases with miniaturization. This effect is demonstrated in a simple design example shown in Fig. 4.11. The patch antenna is made of a square PEC. Patch is 84×84 mm in size and placed 4 mm above an infinite PEC ground plane. A simple 50 Ω feeding port was used for feeding and placed 16.8 mm from center. The right plot shows the realized gain versus frequency in the broadside direction for three different cases: no substrate, substrate with dielectric constant of $\varepsilon_r = 2$ and $\varepsilon_r = 4$. As expected, when ε_r increases the gain peak moves toward lower frequencies. Concurrently, the peak gain level decreases. In fact, the resonant frequency for the high dielectric constant case ($\varepsilon_r = 4$) approaches $c/(2w\sqrt{\varepsilon_r})$ with "w" being the patch width.

To retain the same degree of miniaturization, but retain good gain performance, impedance matching must be ensured by maintaining $\mu_r/\varepsilon_r = 1$. Specifically, if the substrate is chosen such that $\varepsilon_r = 2$, $\mu_r = 2$ instead of $\varepsilon_r = 4$, $\mu_r = 1$, the gain and return loss performance is maintained with an improved bandwidth as demonstrated in Fig. 4.12. Concurrently, the miniaturization factor is maintained since $\sqrt{\mu_r/\varepsilon_r}$ is remained the same. In practice, the challenge is to find a material that has the desired permeability over the needed frequency range (with very low loss of course).

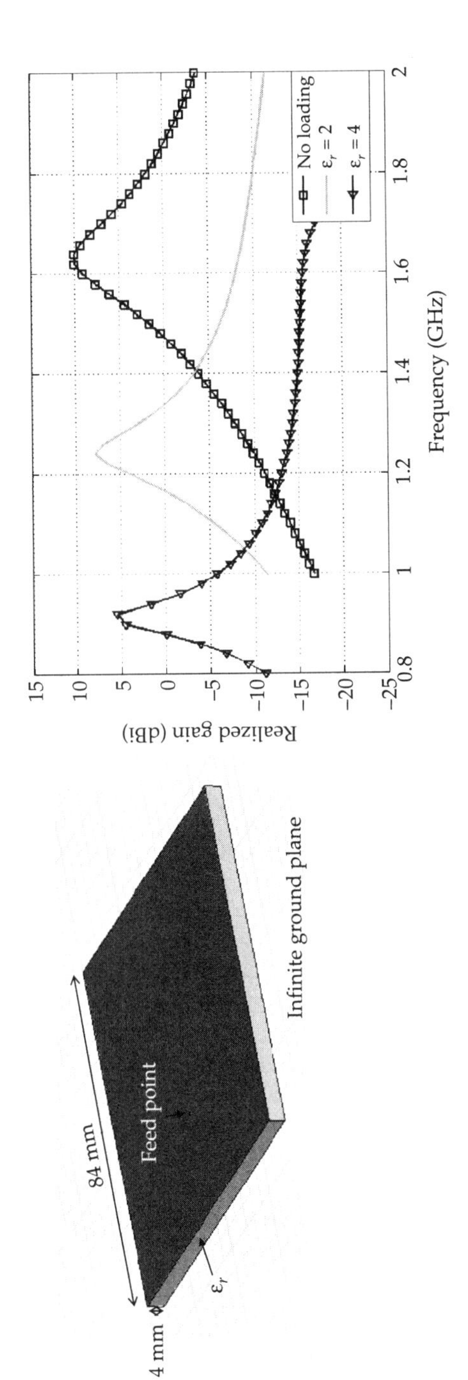

FIGURE 4.11 Patch antenna miniaturization via increasing shunt capacitance using dielectric substrate.

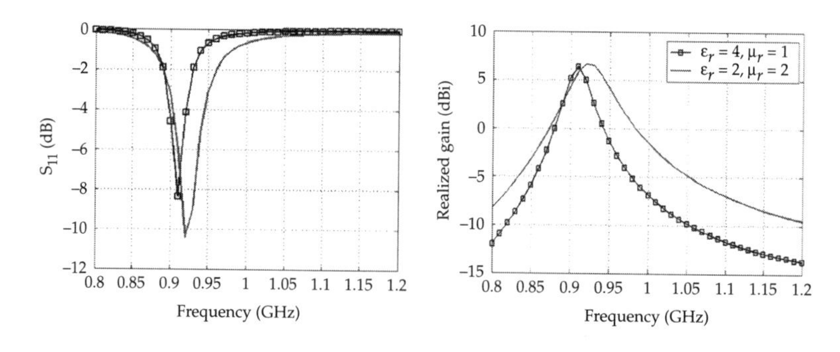

FIGURE 4.12 Comparison of gain performance versus miniaturization when substrate with $\mu_r = 1$ and $\mu_r \neq 1$ are loaded.

4.4.3 Antenna Miniaturization by Increasing ATL Series Inductance

4.4.3.1 Slotted Patches

In addition to increasing shunt capacitance to miniaturize a patch antenna, one can also achieve miniaturization by increasing series inductance. This can be done by adding inductive slots on patch as demonstrated in the next design example shown in Fig. 4.13 where uniformly distributed hexagonal holes are cut out of the original patch. The locations and sizes of these holes need to be chosen properly to effectively interact with fields propagating underneath the patch to cause current to detour around the slot and results in local magnetic flux (see inset in Fig. 4.13) to increase per unit length, thus increasing series inductance. If the size of holes is too small as shown in case "B", undesired series capacitance can occur due to coupling across the hole and offset the series inductance. In case of "C", the size of the hexagonal holes is increased from 2 to 5 mm to reduce the effect of series capacitance and further increase the miniaturization factor as indicated by more shifting of the resonant peak toward left. However, if these holes become too large, they may significantly obstruct the flow of resonant fields and currents and cause undesired reflections. This situation is similar to the case when the size of unit cells in an ATL is no longer much smaller than the guided wavelengths. Case "D" shows how one can achieve even more miniaturization by adding shunt capacitance using small conducting cylinders/pins hung from the patch as will be discussed in the next section.

It should be pointed out that the slot treatment discussed here differs from many slot-loaded patch antennas found in literatures [13–15] where slots are introduced to produce additional slot resonant mode for dual band operations. For miniaturization, we need to avoid

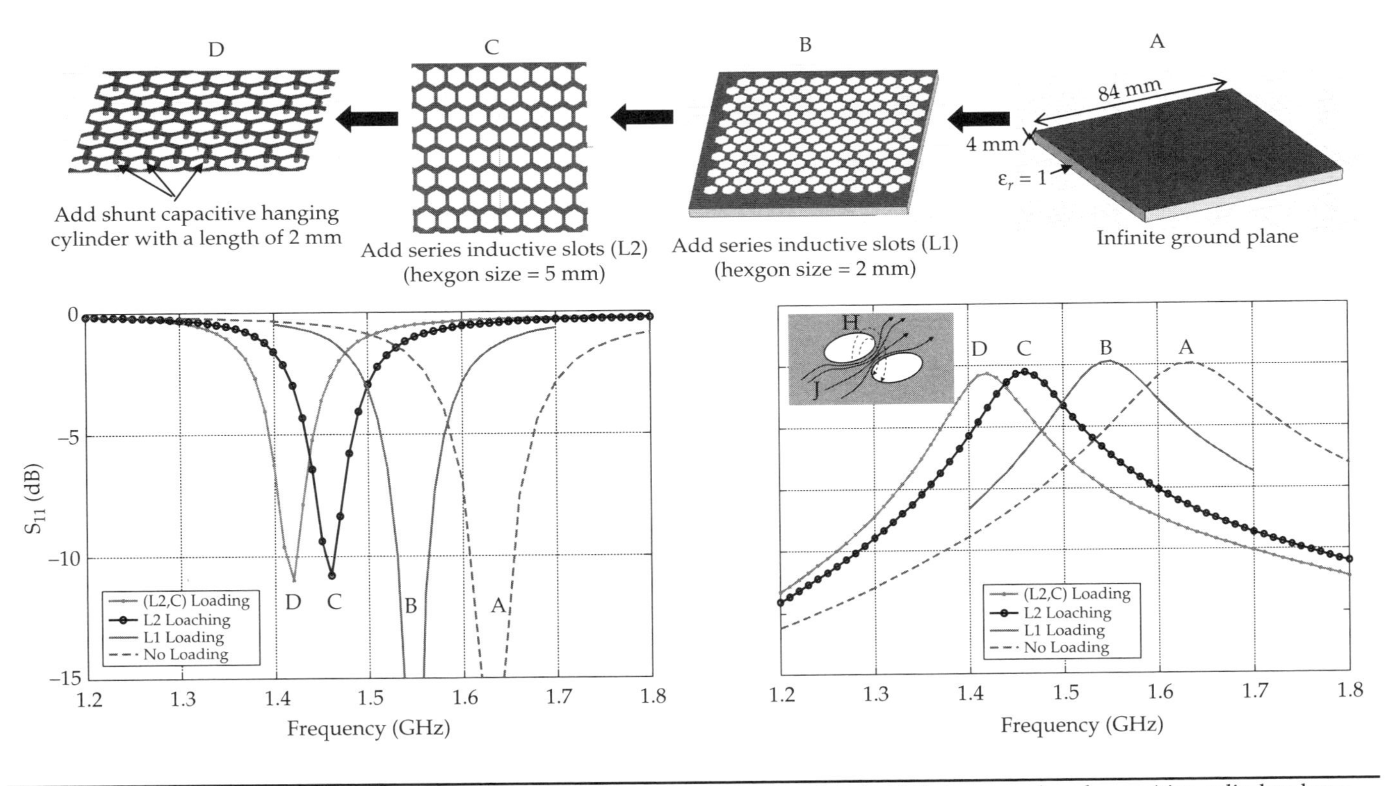

FIGURE 4.13 Patch antenna miniaturization via ATL treatment using inductive slots (middle two cases) and capacitive cylinders hung from slotted patch (left most case).

multiple and high-order modes. Therefore, each slot size in an ATL design should be kept much shorter than operational wavelengths.

4.4.3.2 Spiral with Coiled Arms

As is well known, the length of monopole or dipole antennas can be reduced by meandering or zigzagging antenna arms [16]. Effectively, these designs achieve miniaturization by increasing series inductance. In the following paragraphs, we show how broadband spiral antennas can be miniaturized (effectively shifting their operational frequency to lower values) by including inductive loading using coiling along their arms. The reader should be aware that, in some cases, shunt capacitance and series capacitance are unintentionally introduced. The former is caused by coupling between opposite arms (or antenna arm and ground in the monopole's case). The latter is caused by coupling between adjacent meandering or zigzagging sections in the same arm. This undesired series capacitance may counteract series inductance, rendering ineffective miniaturization (also limiting the maximum attainable miniaturization factor).

Figure 4.14 shows an example of adding series inductance in a wire spiral antenna by zigzagging, meandering, and coiling antenna

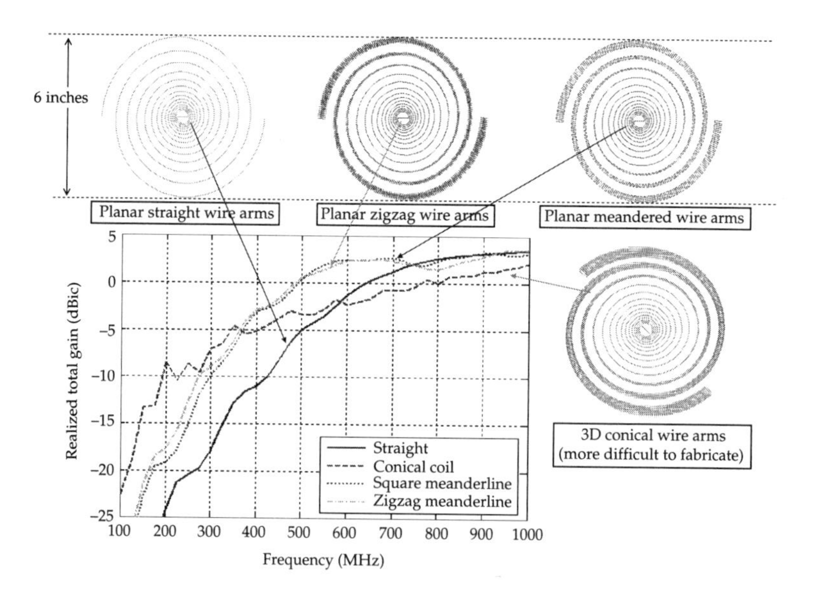

FIGURE 4.14 Spiral antenna miniaturization via increasing ATL series inductance using meandered or coiled arms. Broadside realized gain improved after miniaturization.

arms [17]. The realized gain for untreated-wire case and loaded cases in broadside directional were calculated using Numerical Electromagnetic Cod (NEC) code and are compared in Fig. 4.14. All four antennas have the same dimension of 6 in. The left most design is made of thin wires and has no intentional reactive loading. For the other three spirals shown in Fig. 4.14, the three outmost turns include zigzagging, meandering, and coiling treatments to increase series inductance. As seen from the gain curves, the gain improves with the higher inductive loading exhibited by the coiled version of the antenna. It is understandable that the coiled design produces most inductance and therefore achieves the best performance at frequency below 350 MHz. The coil design also shows lower gain compared to planar inductive treatment cases due to the specific choice inductive profile. Optimization of such coil design involves selecting starting point, coil size variation from beginning to end, pitch variation from beginning to end, etc. More details about spiral antenna miniaturization using this approach will be discussed in the next chapter.

4.4.4 Antenna Miniaturization by Increasing both ATL Series Inductance and Shunt Capacitance

Applying ATL technology with both series inductance and shunt capacitance treatments to miniaturize antenna has many advantages over using only inductance or capacitance treatment along. This is especially true in the applications that call for aggressive miniaturization factor. For instance, one would not need materials with extremely high dielectric constant (which is often heavy, expensive, and not flexible) if series inductance is also used to help slowing down the waves. Vice versa, using capacitance loading such as dielectric substrate can help achieve desired miniaturization factor without solely relying on inductive loading along which has its own effectiveness limitations. In addition, applying both serial inductance and capacitance treatment can help maintain impedance matching and bandwidth as previously demonstrated in Fig. 4.12 where hypothetical lossless magnetic material was used. Using series inductance loading approach offers a nice alternative without the need of wideband, low loss magnetic material as predicted in Eq. (4.11).

Fig. 4.15 shows an example of increasing both series inductance and shunt capacitance for lowering the operation frequency of a patch antenna by adding uniformly distributed small conducting cylinders hanging down from the top patch. As illustrated in the figure, the width of the cylinder perturbs current flow and raises series inductance. The height and width of the gap between the bottom of cylinder and ground plane raises shunt capacitance. In another example shown in Fig. 4.16, the lengths of cylinders vary spatially to achieve tapered

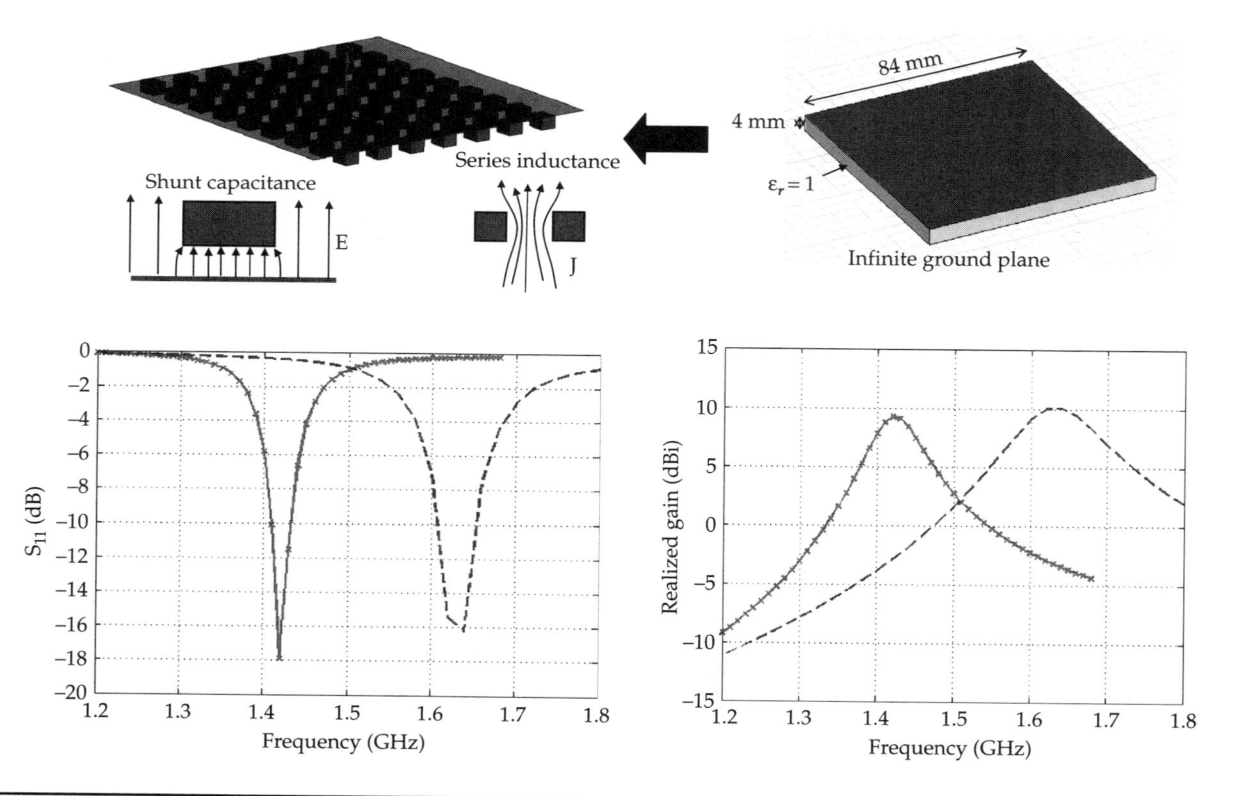

FIGURE 4.15 Patch antenna miniaturization via ATL treatment inductive/capacitive windows.

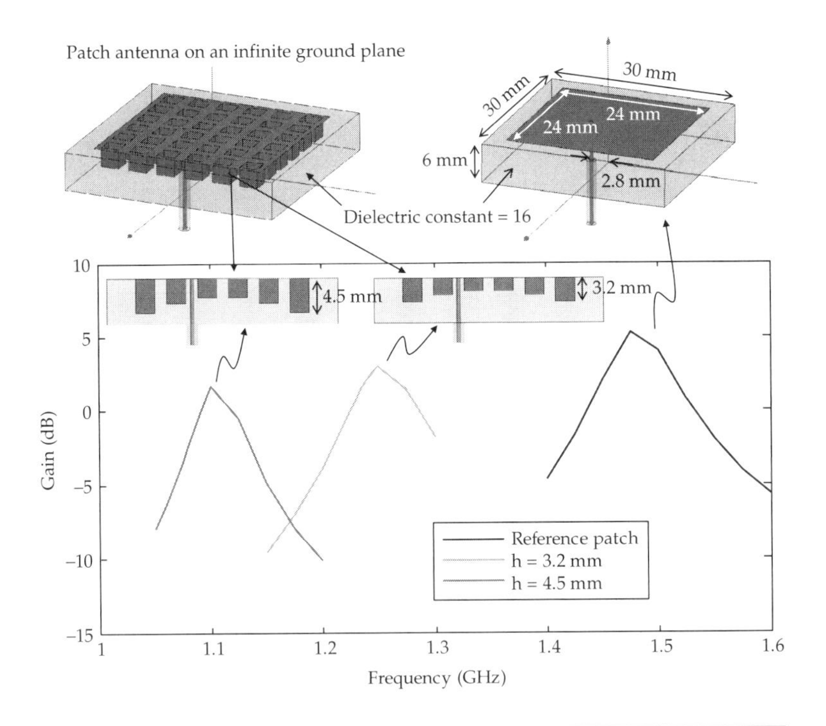

FIGURE 4.16 A patch antenna miniaturized via increasing ATL shunt capacitance.

loading for possible bandwidth optimization. Similar spatial varying capacitive loading can also be achieved using a more complex textured dielectric material [18] which showed improved bandwidth. These are only two of many ways to increase the series inductance and shunt capacitance.

Fig. 4.17 shows an example of aggressive miniaturization with final antenna size being only approximately 1/10 of wavelength. This design employs both inductive and capacitive treatments. Initial capacitive loading along using dielectric substrate with a dielectric constant of 16 reduce resonant frequency from 6.25 GHz down to 1.48 GHz. Additional inductive loading using slots further reduce resonant frequency down to 1.09 GHz. Most importantly, the impedance matching condition maintains fairly stable after the miniaturization process such that the feed position does not have to be moved. The return loss and realized gain curves in the figure shows that the resonant frequency decreases as the slot length increases. The current vector plot in Fig. 4.17 clearly shows the curl of the induced currents.

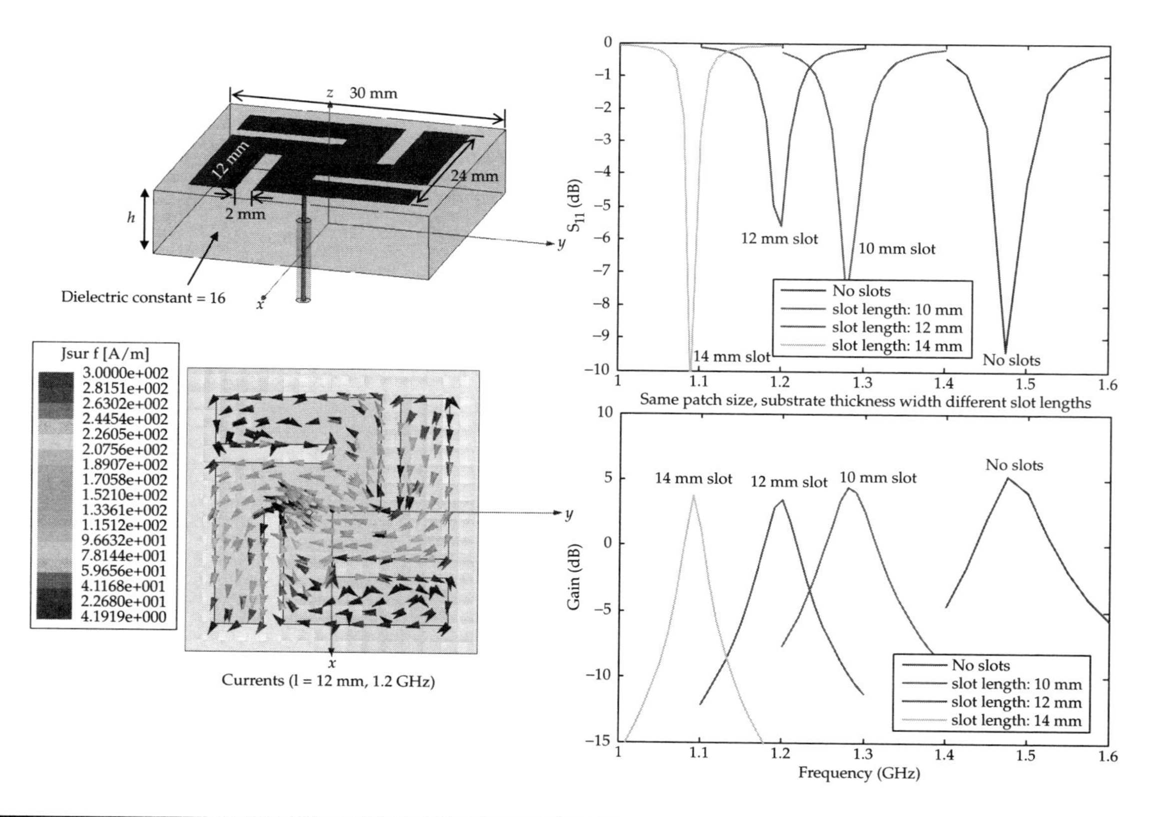

FIGURE 4.17 Patch antenna miniaturization via increasing series inductance using slots.

References

1. T. Hiraoka, T. Tokumitsu, and M. Aikawa, "Very small wide-band MMIC magic T's using microstrip lines on a thin dielectric film," *IEEE Transactions on Microwave Theory and Techniques*, vol. 37, no. 10, 1989, pp. 1569–1575.
2. Y. Xu and R. G. Bosisio, "Analysis and design of novel structures of artificial transmission lines for MMIC/MHMIC technology," *IEEE Transactions on Microwave Theory and Techniques*, vol. 47, no. 1, 1999, pp. 99–102.
3. P. Kangaslahti, P. Alinikula, and V. Porra, "Miniaturized artificial-transmission-line monolithic millimeter-wave frequency doubler," *IEEE Transactions on Microwave Theory and Techniques*, vol. 48, no. 4–1, 2000, pp. 510–518.
4. M. Lee, B. A. Kramer, C. C. Chen, and J. L. Volakis, "Distributed lumped loads and lossy transmission line model for wideband spiral antenna miniaturization and characterization," *IEEE Transactions on Antennas and Propagation*, vol. 55, no. 10, October 2007, pp. 2671–2678.
5. G. V. Eleftheriades and K. G. Balmain, *Negative–Refraction Metamaterial*, John Wiley & Sons, IEEE Press, New York, 2005.
6. M. A. Antoniades and G. V. Eleftheriades, "Compact linear lead/lag metamaterial phase shifters for broadband applications," *IEEE Antennas and Wireless Propagation Letters*, vol. 2, no. 7, 2003, pp. 103–106.
7. Qing Liu, P. S. Hall, and A. L. Borja, "Efficiency of electrically small dipole antennas loaded with left-handed transmission lines," *IEEE Transactions on Antennas and Propagation*, vol. 57, no. 10, October 2009, pp. 3009–3017.
8. T. G. Tang, Q. M. Tieng, and M. W. Gunn, "Equivalent circuit of a dipole antenna using frequency-independent lumped elements," *IEEE Transactions on Antennas and Propagation*, vol. 41, no. I, January 1993, pp. 100–103.
9. M. Hamid and R. Hamid, "Equivalent circuit of dipole antenna of arbitrary length," *IEEE Transactions on Antennas and Propagation*, vol. 45, no. 11, November 1997, pp. 1695–1696.
10. W. F. Egan, *Practical RF System Design*, Wiley-IEEE, Hoboken, New Jersey, 2003, pp. 12–15.
11. G. L. Matthaei and L. Young, *Microwave Filters, Impedance-Matching Networks, and Coupling Structures*, McGraw-Hill, New York, 1964, pp. 26.
12. C. Kittel, *Introduction to Solid State Physics*, John Wiley & Sons, Hoboken, New Jersey, 7th ed., 1996.
13. B. F. Wang and Y. T. Lo, "Microstrip antennas for dual frequency operation," *IEEE Transactions on Antennas and Propagation*, vol. AP-32, November 1984, pp. 938–943.
14. S. Maci, G. B. Gentili, P. Piazzesi, and C. Salvador, "Dual-band slot-loaded patch antenna," *Proceedings of the IEE—Microwaves, Antennas, and Propagation*, vol. 142, June 1995, pp. 225–232.
15. J. H. Lu, C. L. Tang, and K. L. Wong, "Novel dual-frequency and broad-band designs of slot-loaded equilateral triangular microstrip antennas," *IEEE Transactions on Antennas and Propagation*, vol. 48, July 2000, pp. 1048–1054.
16. T. J. Warnagiris and T. J. Minardo, "Performance of a meandered line as an electrically small transmitting antenna," *IEEE Transactions on Antennas and Propagation*, vol. 46, December 1998, pp. 1797–1801.
17. B. Kramer, C-C. Chen and J. L. Volakis, "Size Reduction of a low-profile spiral antenna miniaturization using inductive and dielectric loading," *IEEE Antennas and Wireless Propagation Letters*, vol. 7, pp. 22–25, 2008.
18. D. Psychoudakis, Y. H. Koh, J. L. Volakis, and J. H. Halloran, "Design method for aperture-coupled microstrip patch antennas on textured dielectric substrates," *IEEE Transactions on Antennas and Propagation*, vol. 52, no. 10, October 2004, pp. 2763–2766.

CHAPTER 5
Spiral Antenna Miniaturization

Chi-Chih Chen

5.1 Spiral Antenna Fundamentals

Spiral antennas were introduced in the 1950s by Edwin Turner who experimentally demonstrated that an Archimedean spiral delivered constant input impedance and circular polarization (CP) over a wide range of frequencies. His work ignited great interest in spiral and frequency independent antennas [1,2]. Some early works include those of Dyson for the planar and equiangular antennas [3–6], but most practical spirals are of the Archimedean type [7,8] due to their better CP properties. Spiral in array configuration have also been considered (see [2] and [8–13]). Much recent work has also focused on conformal spiral installations (slot and printed spirals) [14–21], including miniaturization [22–25]. Four-arm spirals have also been found particularly attractive for direction finding applications due to their polarization agility [26].

5.1.1 Basic Planar Spiral Antenna Geometry

Since its first documented introduction by Edwin Turner, spiral antennas were rigorously analyzed and widely used due to their desirable impedance, polarization, and pattern properties over a very wide frequency range. Many variations of spiral antennas [1, 2] have been considered, but the Archimedean and equiangular spirals, shown in Fig. 5.1, are most commonly used. With tight windings, both designs

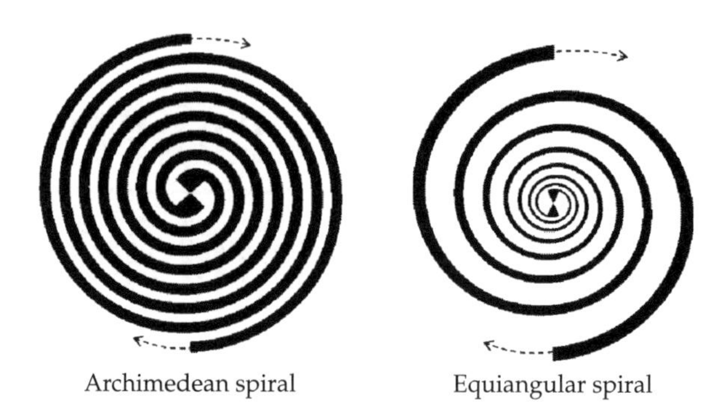

FIGURE 5.1 Two-arm Archimedean and equiangular spirals.

exhibit similar characteristics. However, unlike an infinite equiangular spiral, a loosely wound finite size Archimedean spiral antenna exhibits frequency-dependent behavior.

Each arm of an equiangular spiral and Archimedean spiral antenna is wound along a planar spiral path defined by

$$\rho = e^{a(\varphi+\varphi_{\text{start}})} \quad \text{Equiangular spiral} \tag{5.1}$$

and

$$\rho = a(\varphi + \varphi_{\text{start}}) \quad \text{Archimedean spiral} \tag{5.2}$$

where ρ is the radius from the center and φ is the sweeping angle defined in Fig. 5.2.

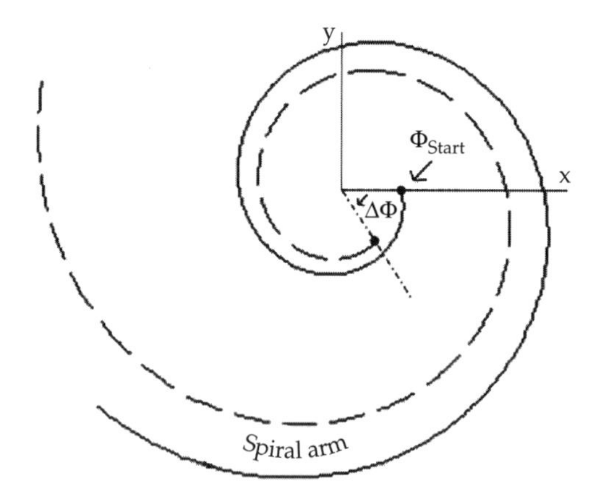

FIGURE 5.2 Generating an equiangular spiral arm by tracing the curve $\varphi = e^{a(\varphi+\varphi_{\text{start}})}$.

We note that the growth factor a in Eqs. 5.1 and 5.2 control the rate of spiral expansion, and, thus the tightness of the windings. Also, φ_{start} denotes the starting angle of the spiral growth. The width of each arm can be generated in the area between two similar spiral curves but with different start angles as shown in Fig. 5.2. Clearly the equiangular spiral exhibits quick expansion from the center. But the Archimedean is more tightly wounded as depicted in Fig. 5.1. The second spiral arm can then be generated by rotating the first spiral arm, that is, setting $\varphi = 180^\circ$ at the start.

5.1.2 Spiral Radiation

As early as 1960, Curtis [27] gave an explanation of the spiral's radiation using a series of semicircles, a concept also employed by Wheeler [28] to generate the pattern of the equiangular spiral. In a typical operation of a planar two-arm spiral antenna, a balanced excitation of equal amplitude and opposite phase is applied to the inner terminals of the two arms meeting at the center (see Fig. 5.3 top left).

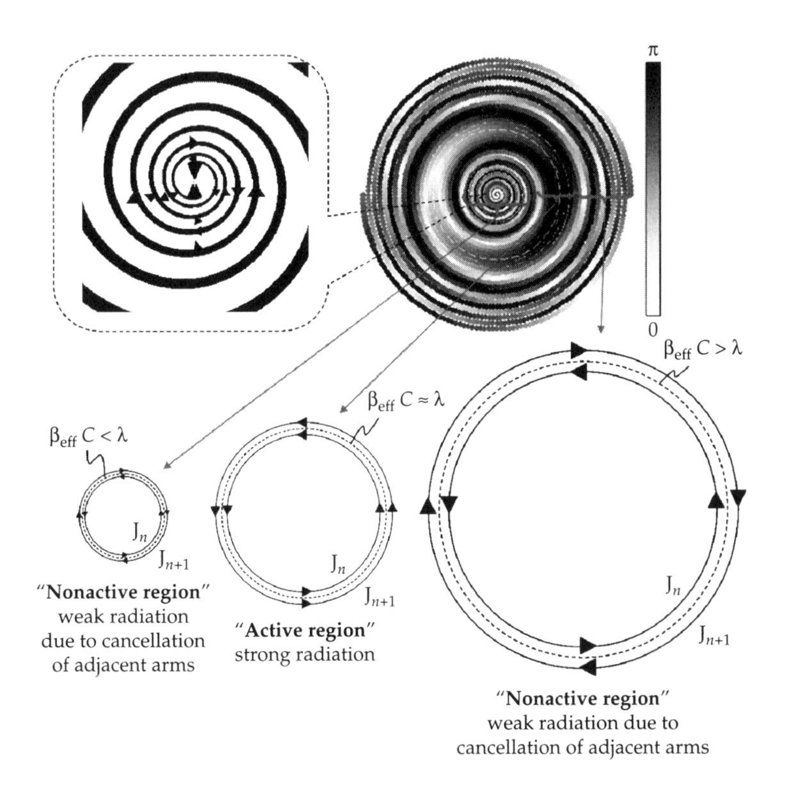

FIGURE 5.3 Radiation region of a two-arm spiral antenna.

The corresponding phase along the spiral arms (for this excitation) at a given frequency is plotted in the upper right of Fig. 5.3 where the dark grayscale indicates negative phase and the bright grayscale refers to positive phase. If the spiral is electrically large as compared to the wavelength, its operation can be separated into "active" and "nonactive" two regions as shown in Fig. 5.3. The former refers to the region where adjacent and opposite arms have similar phase, resulting in coherent radiation. This active region occurs when $\beta_e C = 2\pi$ where β_e is the effective phase velocity of the current along spiral arms and C is the circumference length in the active region loop (see Fig. 5.3). Since energy continues to travel along the spiral arms, the radiation of a spiral antenna resembles to a 1λ rotating loop antenna, producing circularly polarized (CP) radiated fields. The sense of polarization follows the outward spiraling direction of spiral arms during the transmitting mode. During the receiving mode, the sense of polarization follows the inward spiraling direction. The "nonactive" regions refer to regions outside the active region. We will discuss more about these different regions in Sec. 5.2.

Based on the above description of the spiral's radiation, we can summarize that tighter arms enhance effective radiation in the active region and effective cancellation of radiation from adjacent arms in the nonactive regions. The former condition minimizes "leftover" currents beyond the active region, thus minimizing truncation effect in a finite spiral antenna. These leftover currents in turn are reflected back at truncation and propagate toward the spiral center. As they pass through the active region again, they produce extraneous radiation that deteriorates the spiral's CP properties and bandwidth. Resistive loading [22] may be needed to suppress such reflected currents. The latter condition minimizes leakage radiation from inner nonactive region and ensures frequency-independent gain and impedance. As the frequency varies, the size and location of the active region changes automatically according to $\rho_{\text{active}} = 1/\beta_{\text{eff}} = \lambda_{\text{eff}}/2\pi$. Therefore, a large, tightly wound spiral antenna should produce frequency-independent radiation characteristics such as impedance, gain, and patterns. The lower frequency limit is determined by the maximum electrical circumference of the spiral. As it would be expected, at the arms truncation, traveling wave, and current reflections occur. On the other hand, the upper frequency limit is determined by the size, tightness, and precision of spiral arms near the center region.

5.1.3 Input Impedance

Finite spiral antennas exhibit complex impedance behavior over a wide frequency range. Figure 5.4 shows an example of input impedance of a finite planar Archimedean spiral antenna [29] whose conductor and slot widths are the same (i.e., a complementary spiral).

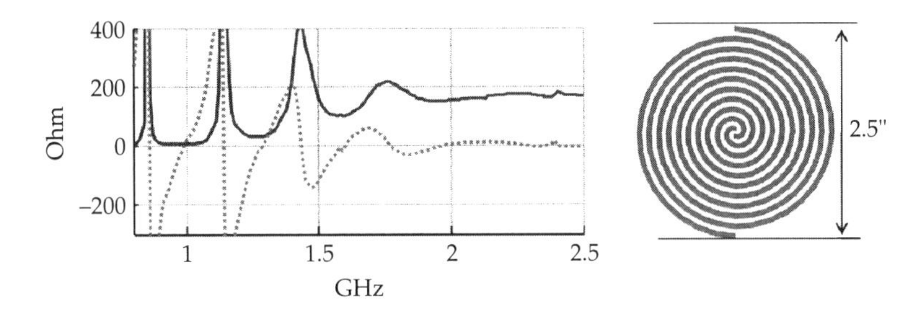

FIGURE 5.4 Example of the input impedance of an Archimedean spiral antenna.

The oscillatory behavior at low frequencies is caused by significant inward-propagating energy reflected from the arm ends. These reflections are similar to those in an open-ended transmission line with observation point being located at the feed point. Such strong reflections are resulted from poor radiation as energy propagates outward due to the lack of active region for insufficient circumference length at these wavelengths.

At high frequencies where the maximum antenna circumference is much greater than wavelengths, the impedance converges to a constant resistive value (a key characteristic of frequency-independent antennas). For example, the spiral antenna shown in Fig. 5.4 has a diameter of 2.5" and the impedance was found to converge to 189 Ω (as expected from a complementary antenna) above 1.9 GHz referred to as the spiral's "characteristic impedance." In this case, the impedance converges approximately when $C_{max} > 1.25\,\lambda$. Above this frequency, the spiral's characteristic impedance remains constant as a result of the frequency-scaled active region. Note that such convergence only occur in large equiangular spirals and tightly wound Archimedean spirals. It is also important to note that the characteristic impedance of a planar equiangular spiral antenna is determined by the ratio of conductor width to slot width.

5.1.4 Radiation Patterns

Cheo et al. [30] gave the expression of the far-zone field of an infinite two-arm equal-angular spiral antenna as

$$E_\phi \approx E_0 \beta^3 A(\theta)\, e^{j(n(\phi+\frac{\pi}{2})-\psi(\theta)]} \frac{e^{-j\beta r}}{r} \tag{5.3}$$

where E_0 is a constant related to the excitation strength and "n" is an integer indicating the order of the Bessel function of the first kind used to expand the near-fields as a function of distance in the spiral plane.

As the radiated fields are perfectly circularly polarized, we can also note that $E_\theta = \pm j E_\phi$ if the Eq. 5.1 winding is adopted. The $n = 1$ mode corresponds to the most common spiral antenna mode that produces a single lobe along the broadside. The θ-dependent magnitude and phase functions, $A(\theta)$ and $\psi(\theta)$, respectively, were also given as

$$A(\theta) = \frac{\cos\theta \left(\tan\frac{\theta}{2}\right)^n e^{\left(\frac{n}{a}\right)\tan^{-1}(a\cos\theta)}}{\sin\theta\sqrt{1 + a^2\cos^2\theta}} \tag{5.4}$$

and

$$\psi(\theta) = \frac{n}{2a}\ln|1 + a^2\cos^2\theta| + \tan^{-1}(a\,\cos\theta) \tag{5.5}$$

where θ is measured from the z-axis (see Fig. 5.2).

The directivity gain pattern computed from Eq. 5.4 for various growth rates are plotted in Figs. 5.5 and 5.6. As seen, when the growth rate is less than 0.1, that is, tighter winding, the pattern stabilizes and eventually becomes independent of growth rate as predicted from Eq. 5.5 which also predicts a 3-dB beam width of approximately 70°. It should be pointed out that radiation may not vanish in the plane of

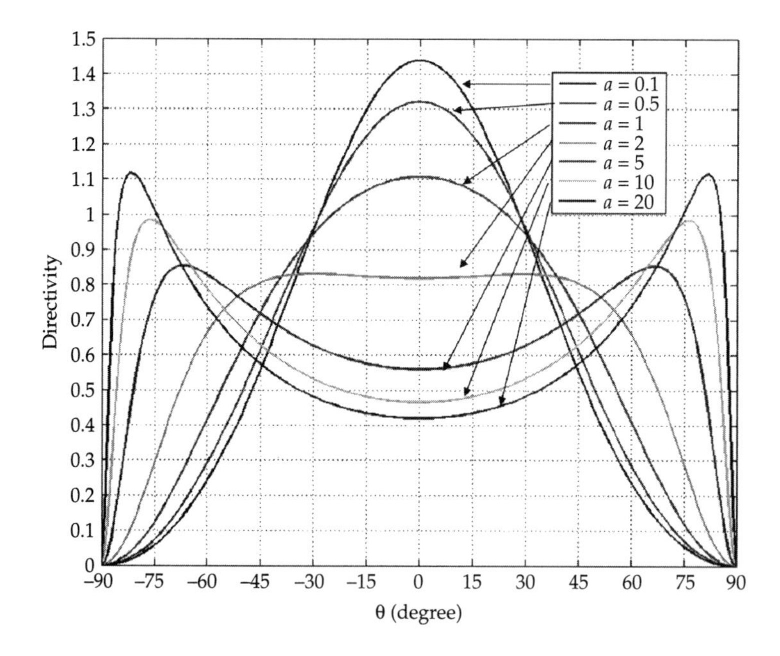

FIGURE 5.5 Directivity patterns of an equiangular spiral antenna for various growth rates; the plots correspond to the normal mode ($n = 1$) and θ refers to the angle from the z-axis, the later being normal to the spiral's plane.

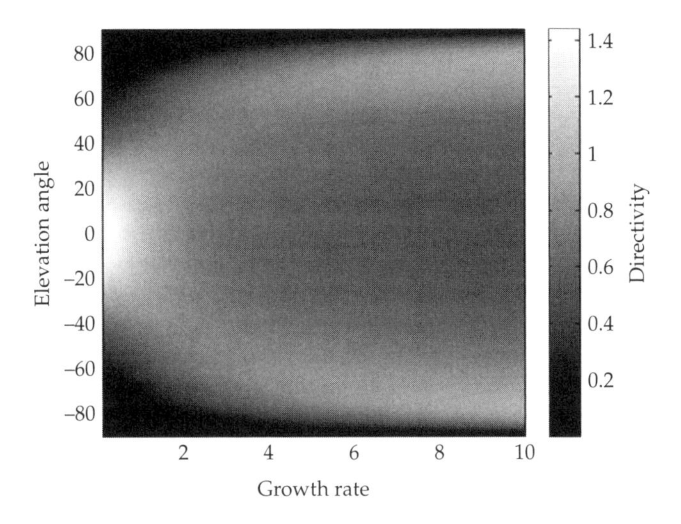

FIGURE 5.6 Grayscale directivity (radiation pattern) of the equiangular spiral as a function of the growth rate a; refer to Eq. 5.3 and Figs. 5.2 and 5.3 for the spiral geometry.

antenna structure, that is, at $\theta = \pm 90^\circ$, in a finite spiral antenna due to diffractions from truncations. It is also interesting to observe that for larger growth rates, the pattern spreads out more away from z-axis, similar to end-fire traveling-wave antennas.

5.1.5 Radiation Phase Center

In some applications (phase arrays, near-field probing, and reflector feeds), it is important to know the effective phase center of radiation patterns. For spiral antennas, the phase center varies with pattern angle and growth rate as predicted from Eq. 5.5. For the normal mode ($n = 1$), the rotational symmetry of the spiral antenna implies that the phase center is along the vertical axis as illustrated in Fig. 5.7. If we assume that the effective phase center for $z > 0$ patterns is located at a distance d below the spiral, then the phase center is associated with a phase delay

$$\psi(\theta) = k_0 d \cos\theta \tag{5.6}$$

The phase center distance, d, from the spiral plane can then be determined from Eqs. (5.5) and (5.6) and is given by

$$\frac{d(\theta)}{\lambda} = \frac{\frac{1}{2a}\ln|1 + a^2\cos^2\theta| + \tan^{-1}(a\cos\theta)}{2\pi\cos\theta} \tag{5.7}$$

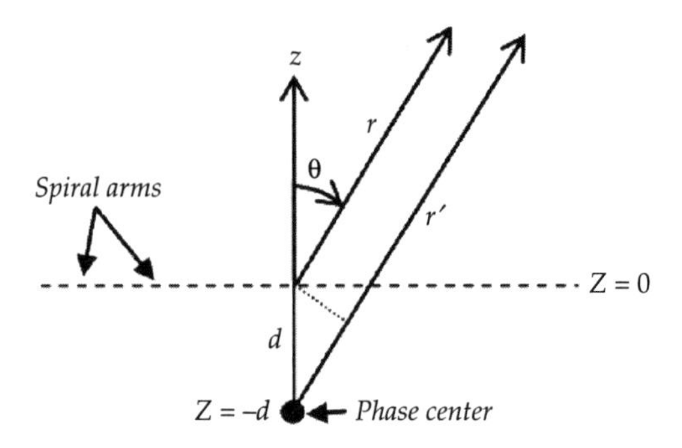

FIGURE 5.7 Effective phase center location in a spiral antenna.

The expression in Eq. 5.7 indicates that the phase center location is a function of pattern angle, θ, and growth factor, a (see Eq. 5.1).

Figure 5.8 plots the normalized phase center distance as a function of θ for four different growth factors ($a = 0.01, 0.1, 1, 2$). It shows that the phase center moves closer to the spiral plane as the growth factor gets smaller. It is also observed that the phase center, d, becomes nearly independent of radiation angle (a desired effect) when the growth rate

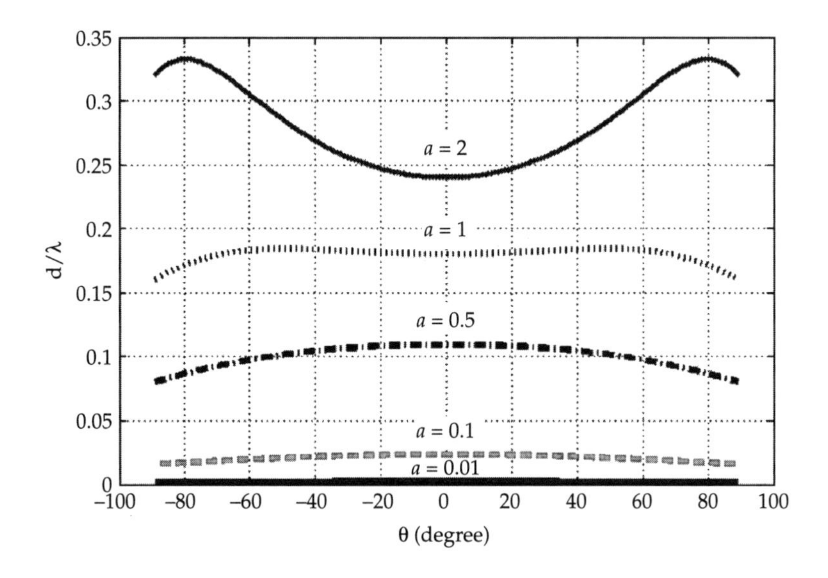

FIGURE 5.8 Phase center (d/λ) location for an equiangular spiral as a function of θ and growth rate, a.

is small ($a < 0.1$). More specifically,

$$\frac{d}{\lambda} \approx \frac{a}{2\pi} \quad \text{for} \quad a \ll 1 \tag{5.8}$$

The relation in Eq. 5.8 is particularly useful for antenna calibration.

5.2 Truncation Effect in Finite Spiral Antennas

Practical spiral antennas have finite dimensions. As already noted, such truncation imposes a lower bound on the spiral's operating frequency. Here, we will briefly discuss the effects of spiral truncation on axial ratio, gain, and impedance for tightly wound equiangular spirals.

First, let us consider a finite two-arm equiangular spiral antenna made of thin perfectly conducting wires (see Fig. 5.9). In this case, the minimum and maximum radii of the spiral arms are 0.5 and 76.2 mm, respectively. The growth factor a in this case is 0.0385. A voltage source is then applied to the feed point (across the tips of two spiral wires) at the center. Figure 5.10 plots the resultant current amplitude profile as a function of normalized radius, $\beta\rho$, for three different growth rates (a = 0.02, 0.04, and 0.06). This amplitude behavior exhibits four distinct regions discussed next.

Source region This is the region within the first turn of spiral arms where nonradiation and rapidly decayed fields dominate.

Transmission-line region In this region, fields propagate outward along the spiral arms with no net radiation due to cancellations from adjacent arms. The current amplitude in this region remains fairly constant

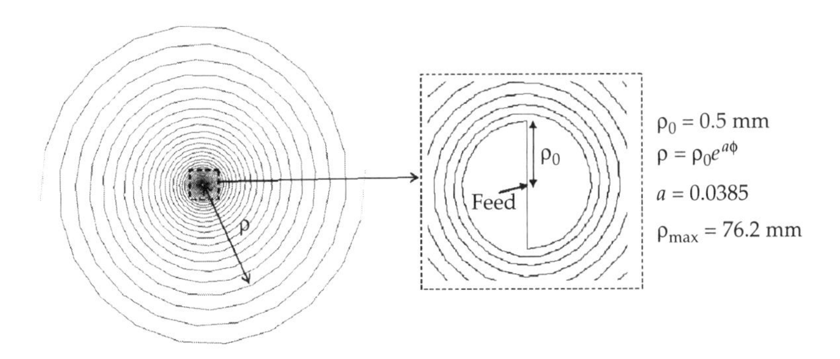

FIGURE 5.9 A two-arm finite-size equiangular spiral antenna formed of perfectly conducting wires.

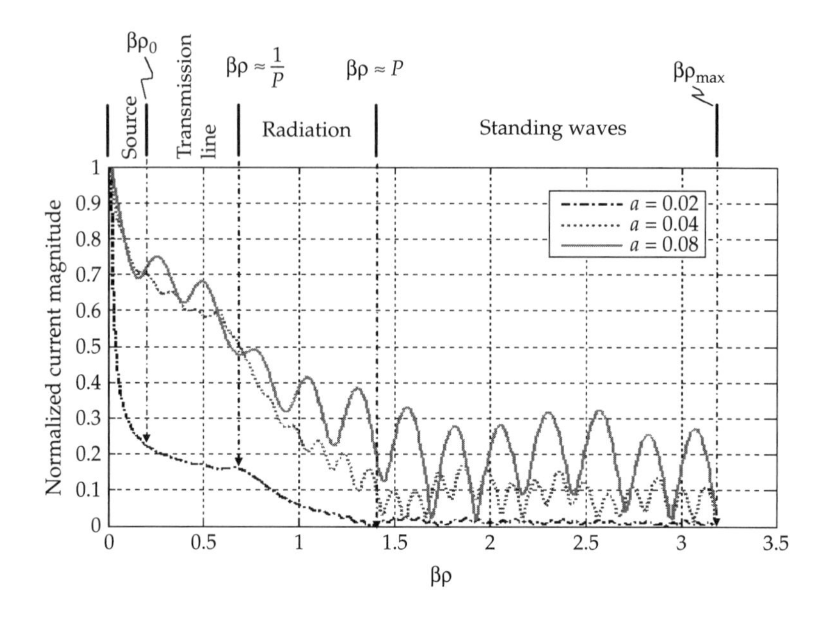

FIGURE 5.10 Normalized magnitude of a current measured on one of the spiral arms in Fig. 5.9.

especially in tightly wound case ($a = 0.02$) since there is no radiation loss. A small amount of radiation leakage may occur when the windings are not tight enough.

Radiation region ("active region") This region spans approximately from $\beta\rho = 1/P$ to $\beta\rho = P$ where $P \approx 1.4$ in tightly wound spirals (as in the $a = 0.02$ case). As discussed earlier in Sec. 5.1.2, tight windings produce effective radiations and result in rapid current amplitude decay due to radiation loss. That is, the current amplitude beyond this radiation region becomes small and is least impacted by the spiral truncation. As the growth rate increases, radiation becomes less effective and more "leftover" currents will continue propagating beyond the active region toward spiral truncation.

Standing-wave region When the "leftover" nonradiated currents reach the truncation, most of them are reflected back. These inward traveling waves interfere with outward traveling waves and cause standing waves type of current amplitude variations (see Fig. 5.10). As the same current propagates inward and reaches the radiation region again, secondary radiation is produced in the opposite sense of the circular polarization (left handed CP in this case) as seen in the $a = 0.08$ case in Fig. 5.12. Such undesired opposite sense radiation increases the axial ratio and gets worse as growth rate increases. It is concluded from these results that the desirable size of a tightly wound spiral antenna should have $\beta\rho_{max} > P$ or $2\pi\rho_{max} > 1.4\lambda$. Here, ρ_{max} is the maximum radius of the spiral arms.

When significant current amount reaches both ends of a truncated spiral antenna, it is not unusual for the radiation pattern to be significantly different from those obtained from an infinite spiral antenna (i.e., Figs. 5.5 and 5.6). Among possible pattern changes, we note that horizontal patterns will no longer be uniform due to additional diffraction from the arm ends. For instance, Fig. 5.11 plots the azimuth (in the spiral plane) and elevation patterns of three finite spirals with the same maximum aperture size of $2\pi\rho_{max} = 1.4\lambda$ (but three different growth rates). As the growth rate increases ($a > 0.02$), the azimuth patterns ceases to be omni-directional due to significant diffraction from the arm ends. Increasing the number of spiral arms could improve pattern

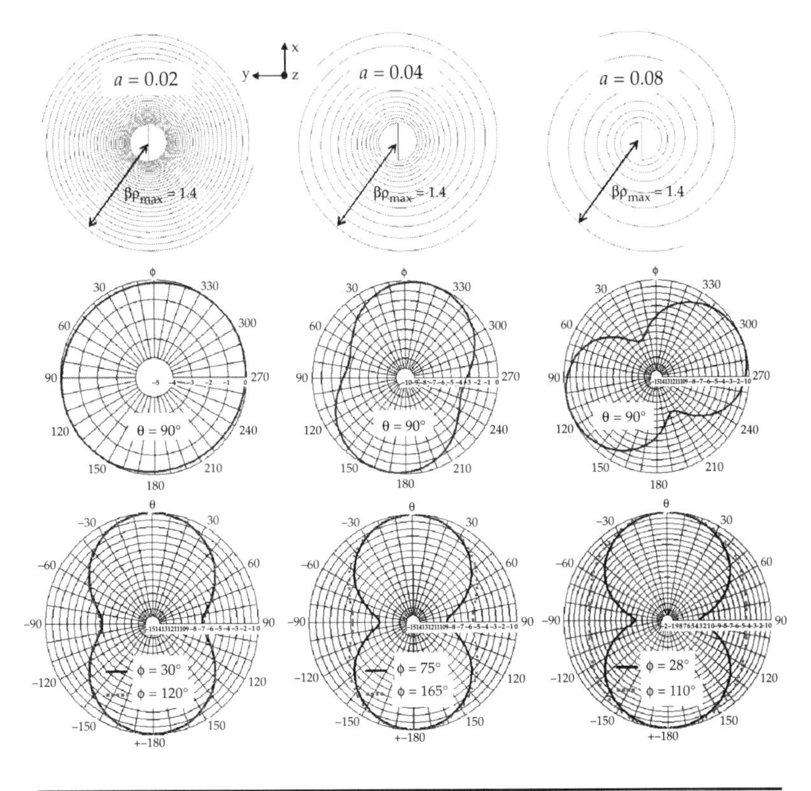

FIGURE 5.11 Spiral radiation patterns for different growth rates. Top row: spiral geometry for different growth rates. Middle row: azimuth radiation patterns. Bottom row: elevation patterns.

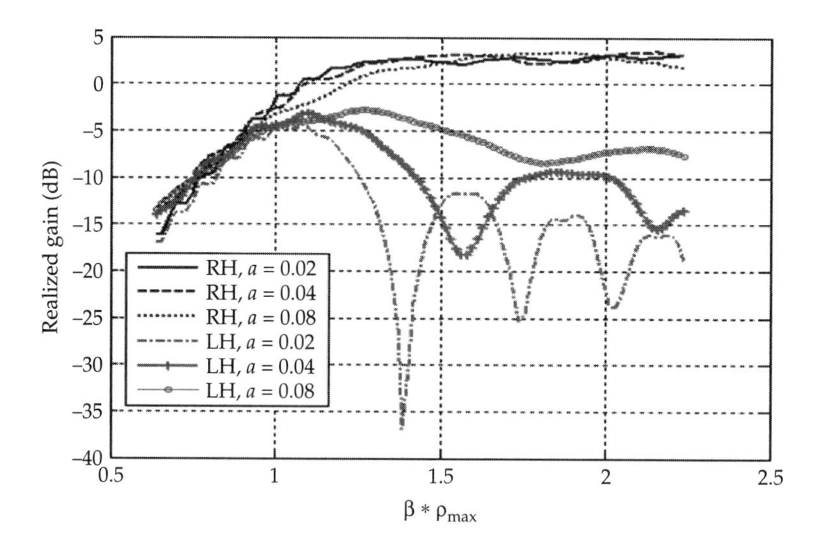

FIGURE 5.12 Broadside realized gain for different growth rates of the spirals shown in Fig. 5.11.

uniformity in the azimuth plane but adds feeding complexity. Figure 5.11 also shows that the elevation pattern does not vanish at the horizon as in the infinite spirals case. This is, again, due to additional diffractions at truncations. The realized gain curves along the broadside direction (z-axis) as a function of (normalized) frequency for the above three spiral cases are also shown in Fig. 5.12. It can be seen that the truncation diffractions in larger growth factor cases cause more fluctuations in right hand circularly polarization (RHCP) gain and raise the left hand circularly polarization (LHCP) gain. Some resistive loading near the arms can be used to reduce the undesired opposite CP fields. Such treatment should have little impact on performance at high frequencies [31].

5.3 Spiral Antenna Backed with a PEC Ground Plane

Many applications require spiral antennas to be mounted conformally on electrically conducting surface. In such cases, the separation between the antenna and conducting backing is usually very small (less than $\lambda/20$). However, the radiation pattern, impedance, and bandwidth degrade severely for low-profile antennas on a PEC ground plane due to strong high-order coupling modes. Figure 5.13 gives the

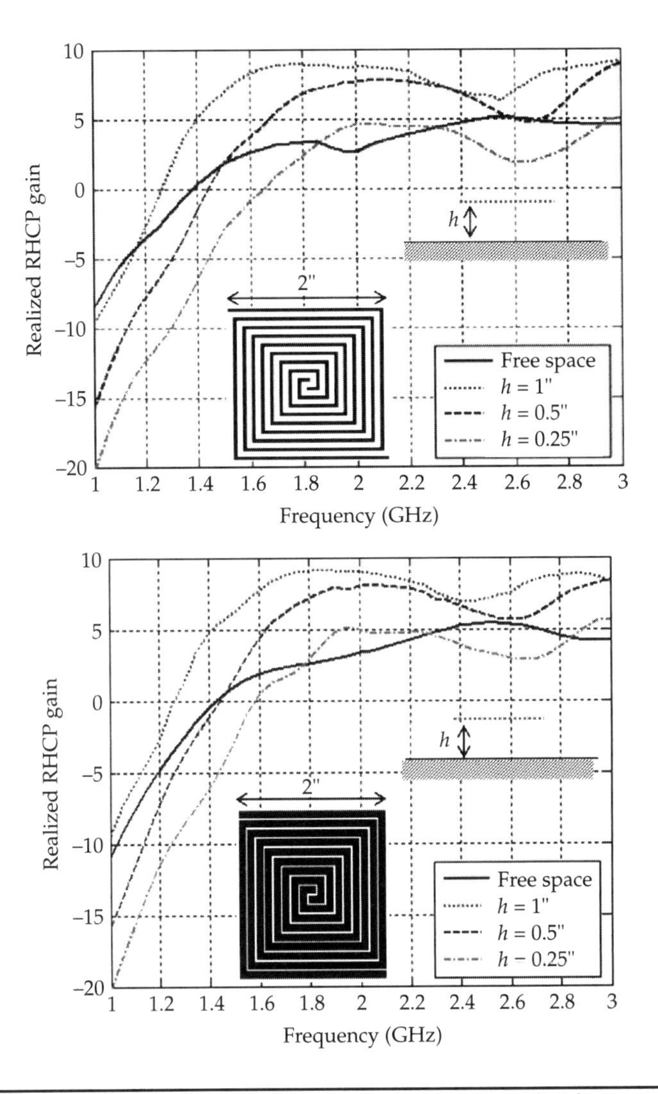

FIGURE 5.13 Influence of the PEC ground plane on the realized gain (including mismatch loss) of 2″ × 2″ Archimedean strip (top) and slot (bottom) spiral antennas in free space and above a ground plane [29].

gain performance of a 2″ × 2″ strip and slot [14] spirals on an infinite ground plane. As depicted, calculations were carried out for three different heights, $h = 1''$, 0.5″, and 0.25″. We also note that the given realized gains were referenced to 225 Ω for the strip spiral and to 110 Ω for the slot spiral. These impedances were chosen so that the spirals are matched when in free space.

When separation between the ground and the antenna's surface is reasonably large (say above $\lambda/8$) we can express the broadside gain as

$$G_{\text{on ground plane}} = G_{\text{free space}} \left| 1 - e^{-j2\pi\frac{2h}{\lambda}} \right|^2 \tag{5.9}$$

This expression can help explain the large variation in gain where h denotes the separation between the antenna and ground plane (see Fig. 5.13 above 1.5 GHz). Of course, as $h \to 0$, the coupling between antenna aperture and ground plane is dominant, and, thus, Eq. (5.9) becomes oversimplified. If we use 0 dBi gain level as a reference, it is observed that the gain remains higher than its free-standing value when the height is greater than $\lambda/15$ and becomes lower than its free-standing values below 2 GHz for $h = 0.25''$ ($\lambda/24$ at 2 GHz). Notice that similar degree of ground plane effect is observed in both strip and slot spirals.

The gain reduction in Fig. 5.13 can be understood by applying image theory as illustrated in Fig. 5.14. When the separation between a laterally structured antenna and its image becomes small (in terms of λ), the image antenna currents are in opposite direction, thus, cancelling the fields of the actual antenna above the ground plane. For slot spirals, radiation can be represented by magnetic currents along the slots. However, these magnetic currents are sustained by electrical currents near both edges of slots, and thus experiencing same effect as strip spirals.

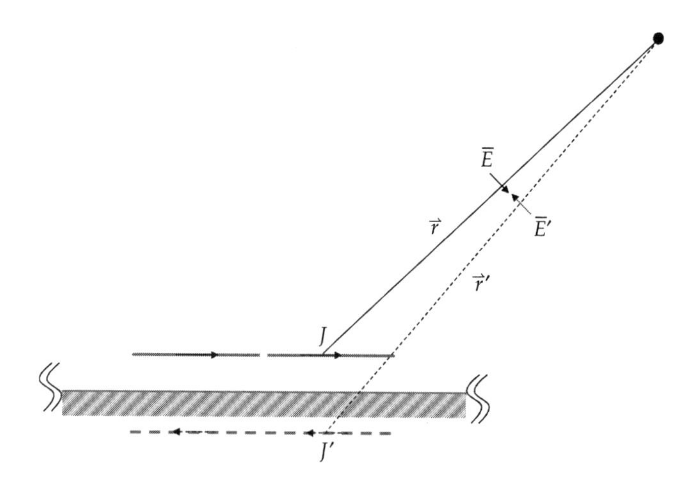

FIGURE 5.14 Spiral current image to explain low gains for conformal installations.

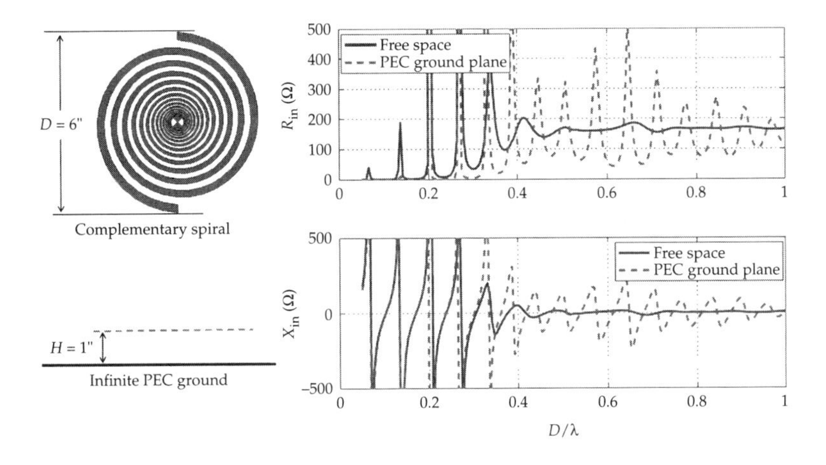

FIGURE 5.15 A comparison of the input impedance for a 6″ circular log-spiral in free-space and placed approximately 1″ above an infinite PEC ground plane.

Figures 5.15 and 5.16 show the calculated impedance, gain, and axial ratio of a 6″ log-spiral antenna placed 1″ above an infinite ground plane. As mentioned earlier, cancellation from image antenna reduces the gain at low frequency as shown in Fig. 5.15. The reduction in gain means more energy to travel to spiral ends and reflect back to the source, leading to oscillatory input impedance behavior (see Fig. 5.15) as commonly observed in an open-ended transmission line. The main radiation source in the extremely low profile case will be from diffractions at the spiral ends. These reflected currents that propagate inward toward the feed also produce some radiation with the opposite sense of polarization, resulting in poorer axial ratio compared to the values obtained in free-stand case (see Fig. 5.16).

5.4 Spiral Antenna Miniaturization Using Slow Wave Treatments

Antenna miniaturization techniques such as dielectric [22, 32–37] or reactive loading [19,31,38–40] are commonly used and were discussed in Chap. 3. However, each of these miniaturization techniques faces important performance trade-offs especially when significant miniaturization is pursued. In this section, we discuss the effectiveness, advantages and disadvantages of using inductive arms or/and dielectric material loading to miniaturize spiral antennas. The former increases

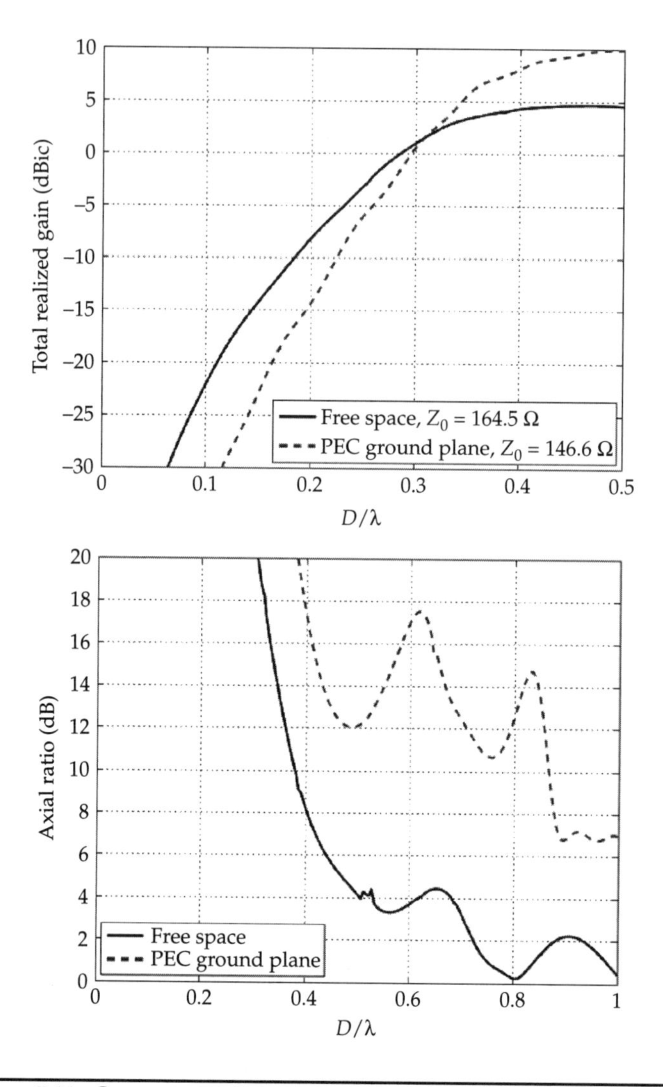

FIGURE 5.16 Comparison of broadside realized gain (top) and axial ratio (bottom) of a 6″ circular log-spiral in free-space and placed approximately 1″ above an infinite PEC ground plane.

the series inductance and the latter increases the shunt capacitance of the equivalent transmission line representing the electromagnetic energy propagating along the spiral arms. As it will be shown, a hybrid approach using both inductive and capacitive loading not only allows for reduction in wave velocity (along the spiral arms), but also retains the desirable impedance over a large bandwidth. This is done by maintaining the ratio of series inductance to shunt capacitance close to unity. In practice, the capacitive loading approach amounts to using low-loss dielectric materials on one or both sides of a spiral antenna [21] and inductive loading using meandering or coiling to the spiral arms [41].

For spiral antennas, the miniaturization treatments should be gradually introduced away from the feed in an effort to maintain good high-frequency performance as well. Such tapered loading approach is also important in avoiding excitation of undesired reflection, scattering, local resonances, and surface waves [21, 22].

As a convenient measure of miniaturization of a UWB antenna like a spiral, we introduce the following miniaturization factor (see Chap. 3) [17, 29]

$$MF = \frac{f_{G=-15\,\text{dBi}}^{\text{unminiaturized}}}{f_{G=-15\,\text{dBi}}^{\text{miniaturized}}} \tag{5.10}$$

Here, the subscript "−15 dBi" refers to the choice of a minimal gain level of −15 dBi. This choice was made for our operational needs. Other gain values could have been used to meet different systems' minimum gain requirement. However, the miniaturization factor defined in Eq. (5.10) remains approximately the same as long as it is less than −5 dBi. This miniaturization factor measures the shift of the frequency where the minimum gain requirement is met before and after miniaturization treatment.

5.5 Spiral Miniaturization Using Dielectric Material Loading (Shunt Capacitance)

Let us consider the miniaturization of two spirals configurations loaded with dielectric superstrate (single-side loading) and substrate (double-sided loading) as depicted in Fig. 5.17. In this figure, we plot the miniaturization factor Eq. (5.10) as a function of the dielectric constant, ε_r. For the calculations, we also used the characteristic impedance (i.e., resistance at high frequencies where the gain is greater than 0 dBi) as reference.

In Fig. 5.17, the left and right figures refer to a circular equiangular and a square Archimedean spiral, respectively. The left figure compares the miniaturization factor for a single-side (dotted line) and a double-side (dashed line) dielectric slab loading of thickness of $\lambda_g/10$ where

$$\lambda_g = \frac{\lambda_{G=-15\,\text{dBi}}^{\text{unminiaturized}}}{\sqrt{\varepsilon_r}} \tag{5.11}$$

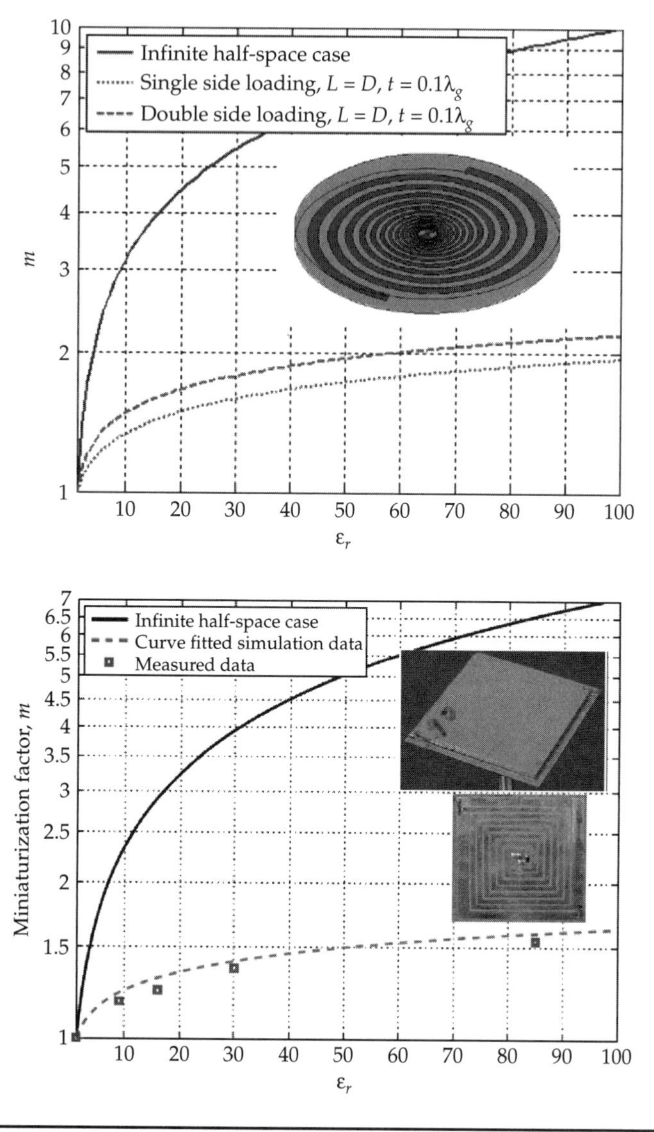

FIGURE 5.17 Spiral antenna miniaturization using dielectric loading.

For reference, the miniaturization factor obtained when the entire half space (on one side of the spiral) is loaded with a lossless dielectric is also plotted (solid line). As expected, the miniaturization factor is $MF = \sqrt{(1 + \varepsilon_r)/2}$ if the half side of antenna is completely loaded with a homogeneous dielectric material (solid line). Likewise, if the entire space (on both sides of the spiral) is loaded with dielectric material, we expect that $MF = \sqrt{\varepsilon_r}$. Of importance in these plots is that finite dielectric loading is not capable of reaching the miniaturization factors of the infinite space loading since the loading material does not capture all the electromagnetic fields. For the two cases shown here with a loading thickness of $0.1\lambda_g$, the maximum miniaturization factor of only 2 is reached even when the dielectric constant is as high as 50. For reasonable ε_r values of $\varepsilon_r \approx 20$, the MF factor is only 1.4 to 1.6. This simulation results is validated with measurements (see right of Fig. 5.17) and due to that a small portion of the near fields can be contained within the $0.1\lambda_g$ thick superstrate or substrate.

As suspected, dielectric loading lowers the antenna impedance. Figure 5.18 demonstrates this effect by plotting the characteristic impedance of a square Archimedean spiral calculated from full-wave models. This spiral is loaded with a dielectric slab of thickness t. The data for two different dielectric constants, $\varepsilon_r = 9$ and $\varepsilon_r = 16$, is shown. We observe that as the thickness increases from 0 to $\lambda_g/50$, the spiral's characteristic impedance decreases rapidly from its free space value. When the superstrate thickness reaches $\lambda_g/10$, the impedance converges to the value corresponding with half-space dielectric loading. The rate of such impedance reduction as a function of thickness should be related to capturing of near-zone fields by the superstrate. However, the dielectric thickness needed to reach impedance convergence depends on antenna types. For dipoles, the thickness may need $\lambda_g/5$ to converge to its half-space value.

A better dielectric loading approach for a spiral antenna is to gradually increase the loading effect as the distance from the feed increases.

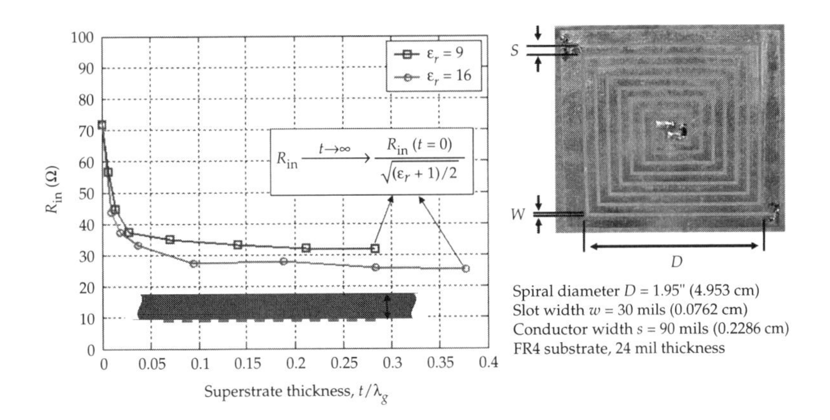

FIGURE 5.18 Lowering of spiral's characteristic impedance due to dielectric loading.

This can be achieved by either increasing the dielectric constant or the thickness of the material (as a function of distance from the feed). Such a tapered loading approach enables miniaturization at low frequencies, while minimizing loading at high frequencies (inner spiral region), where miniaturization is not needed. An example of tapered dielectric loading is shown in Fig. 5.19. In this figure, a composite dielectric superstrate of a constant thickness but linearly varying dielectric constant from $\varepsilon_r = 3$ to $\varepsilon_r = 9$ is applied on a top of 2″ Archimedean slot spiral antenna. The calculated realized gain for this loading is compared to that obtained using a uniform loading with $\varepsilon_r = 9$ (dashed line) and no-loading (solid line). As noted in the figure inset, the uniform dielectric slab reduces the characteristic impedance from 93 Ω down to 33 Ω. In contrast, the tapered dielectric loading reduces the characteristic impedance only down to 53 Ω due to less loading near the center. That is, tapered loading approach provides an additional degree of freedom to improve impedance matching at high frequencies for miniaturization. Of course, such tapered loading is only practical if it can achieve the same degree of miniaturization at low frequency as compared to uniform loading. Figure 5.19

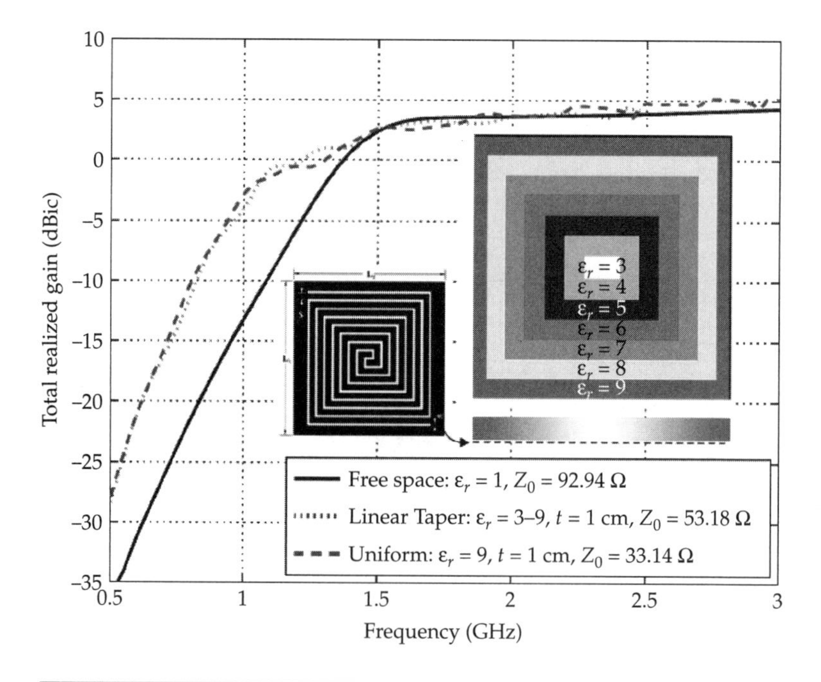

FIGURE 5.19 Effect of dielectric superstrate tapering on the spiral's input impedance and realized gain [41].

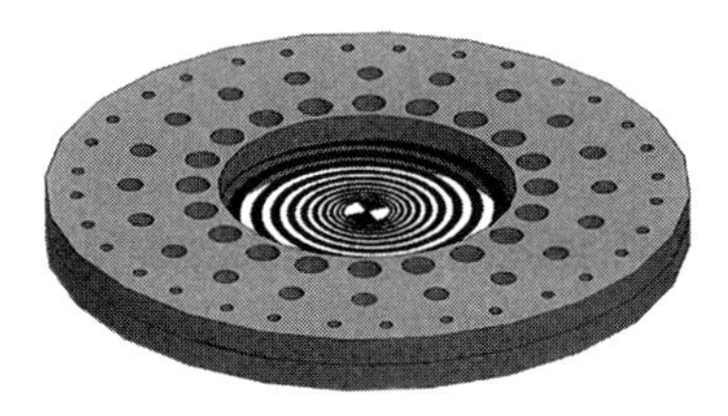

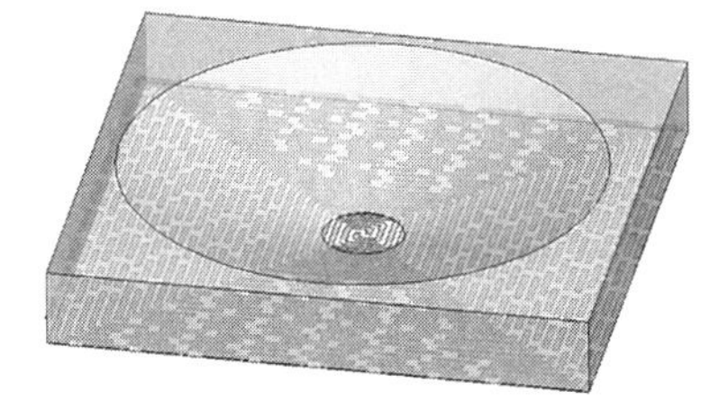

FIGURE 5.20 Practical tapering of the dielectric loading (superstrate) by varying the material density (left) or thickness (right).

demonstrates that indeed for this example both types of loading achieve similar miniaturization. This also assumes matching to their corresponding input impedance.

Tapered dielectric loading is also advantageous to avoid excitation of undesired resonances and surface waves. In practice, it is easier to implement a tapered dielectric loading by varying the thickness or the density of a constant thickness material as illustrated in Fig. 5.20. The measured gain data for a tapered-dielectric loading example is given in Fig. 5.21. This design refers to a 6″ Archimedean square spiral sandwiched between two dielectric composite layers. The dielectric layers have $\varepsilon_r = 9$ and a maximum thickness of 0.625″ and 0.75″ for the superstrate and substrate, respectively. The actual thickness of the superstrate and substrate layers increase linearly along the two orthogonal axes. We observed that the measured −15 dBi gain in the tapered dielectric loading is achieved at 142 MHz as compared to same

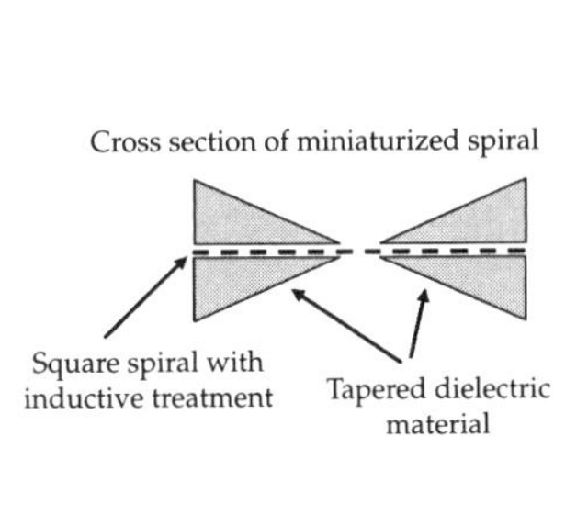

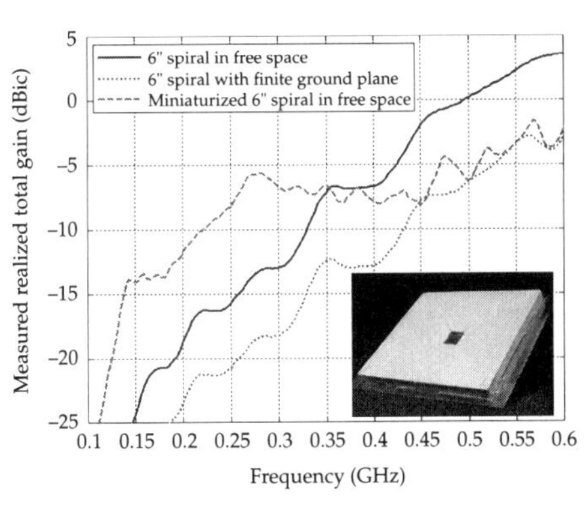

FIGURE 5.21 Measured gain of a 6″ Archimedean square spiral with and without a tapered dielectric loading ($\varepsilon_r = 9$) whose thickness increases linearly [41].

gain at 255 MHz without loading. This implies miniaturization factor of 1.8.

The examples given in this section demonstrate that proper dielectric loading leads to significant miniaturization away from the center. However, in practice, dielectric loading is associated with higher costs, weight, and surface wave losses. The lowering of antenna impedance also presents an issue to be overcome. Conventional ceramic or LTCCs of high dielectric constants are brittle and may crack under vibration. This particular issue could be overcome by employing utilization of flexible polymer-ceramic [42].

5.6 Spiral Antenna Miniaturization Using Inductive Loading (Series Inductance)

An alternative to dielectric loading (or possibly a concurrent miniaturization approach) is to increase the spiral's self-inductance via inductive loading (achieved in a variety of ways). One option is to use a series of small surface mount lump inductors along each spiral arm [31]. Although this technique works in theory, the ohmic loss in off-the-shelf inductors makes the approach less attractive. Another option is to modify the spiral arms via planar zigzagging or meandering. The following subsection discusses this commonly used planar meandering technique and its limitations in miniaturizing spiral antennas. Subsequently, we pursue a novel volumetric coiling of the spiral (see Fig. 5.22) to introduce more inductance and, thus, greater miniaturization.

5.6.1 Planar Inductive Loading

A common way to realize inductive loading is by introducing meandering. Figure 5.22 shows an example of this approach applied to an equiangular wire spiral antenna having 6″ diameter. Here, the wire arm is meandered in a zigzag manner with a pitch of 120 mils. However, the zigzag width is varied linearly from 22 to 275 mils to generate a smooth impedance transition from the untreated (nonmeandered) portion to the inductively loaded section of the spiral arm. The max width of the zigzagging is bounded by the two spiral curves which define the edges of the spiral arms as illustrated by the dashed lines in Fig. 5.22. The equations for the two spiral curves are as follows

$$\rho = \rho_0 e^{a\varphi} \quad \text{and} \quad \rho = \rho_0 e^{a(\varphi+\delta)} \tag{5.12}$$

where ρ_0 is the initial radius, a is the growth rate ($\tau = e^{-2\pi|a|}$) and the angle δ controls the width of the arm. We note that $\delta = \pi/2$ produces

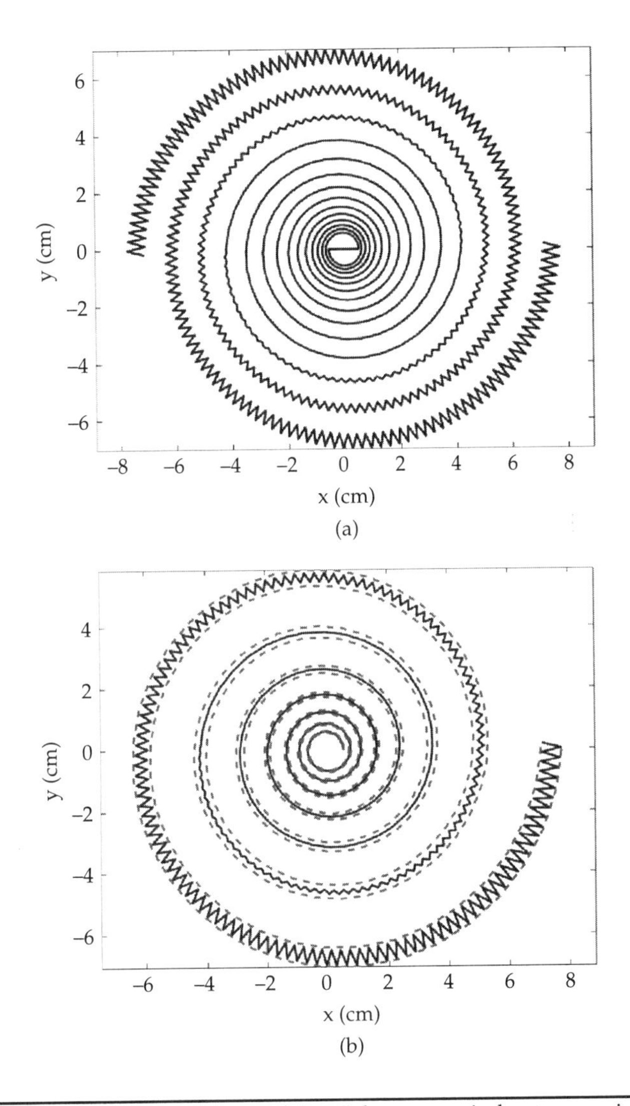

FIGURE 5.22 Inductive loading of the 6″ diameter spiral antenna using planar meandering.

self-complementary structure. All spirals discussed in the subsection have the same parameter values of $\rho_0 = 0.5$ cm, $a = 0.602$ ($\tau = 0.685$) and $\delta = \pi/2$.

Figure 5.23 compares the gain and attains miniaturization using planar meandering to achieve wave slowdown in the spiral arms. Three curves are shown in Fig. 5.23: (1) straight wire spiral, (2) meandered spiral arms with pitch = 30 mils, and (3) meandered spiral arms with

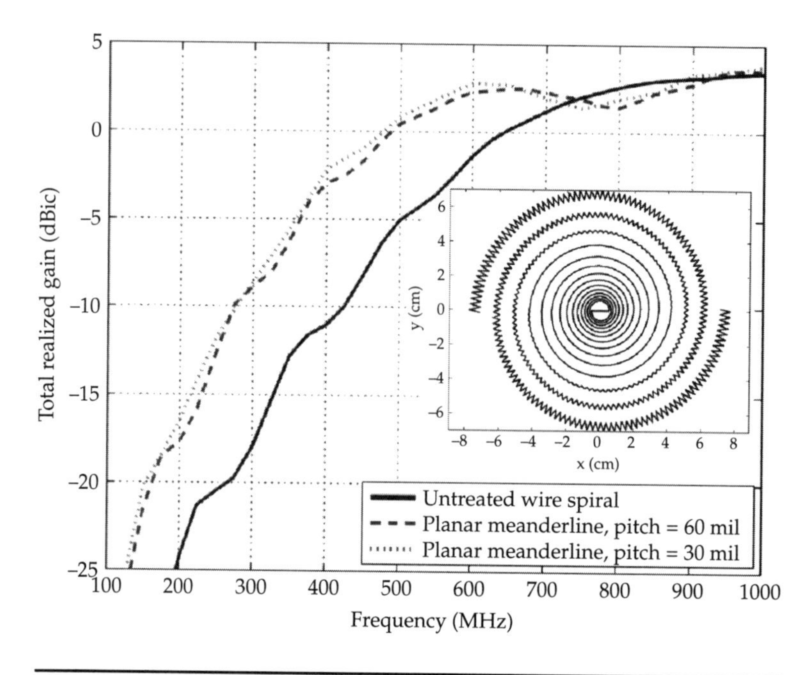

FIGURE 5.23 Spiral antenna miniaturization using planar meandered arms. (*see [46]*.)

pitch = 60 mils. The curves were obtained using numerical electromagnetic code (NEC-2) [43] based on the method of moments. It is observed that the chosen planar meandered spiral achieves a miniaturization factor of about 1.5 based on the definition in Eq. (5.10). However, further miniaturization using this approach is minimal. To demonstrate this, the pitch of the meandered spiral was reduced by half in an attempt to increase the self-inductance of the spiral arm. However, from Fig. 5.23 it is apparent that decreasing the pitch by a factor of 2 leads to negligible miniaturization (i.e., gain curves are similar for both pitch curves). Our studies showed the different meandering shapes leading to similar miniaturization performance.

The limitation of planar meandering can be illustrated by considering a 2D meandered dipole that occupies a fixed area. For a wire dipole, meandering reduces its resonant frequency by increasing the self-inductance of the wire [44,45]. Indeed, the self-inductance of the wire can be increased by increasing the number of meander sections. However, as more meandering is added, strong coupling between adjacent wire segments leads to higher self-capacitance that counteracts the increase in self-inductance [44]. As the self-capacitance increases, the reduction in resonant frequency begins to diminish. Also, when the self-capacitance becomes larger, the resonant frequency begins to increase. In essence, as more meandering is used, the meandered dipole (see Chap. 3) will become more like a strip dipole. Therefore, for a fixed area, the resonant frequency can only be reduced to a certain extent using meandering. Since planar meandering cannot even achieve as much miniaturization as dielectric loading, a more aggressive approach to realize more inductive loading (and therefore higher miniaturization) is needed. In the following subsection, we introduce three-dimensional (3D) volumetric coiling for greater miniaturization.

5.6.2 Volumetric Inductive Loading

The concept of volumetric inductive loading is to employ 3D coiling of spiral arms (see Fig. 5.24). The goal is to increase inductance per unit length of the spiral arm and therefore further slowdown the spiral currents. That is by exploiting all three dimensions, it is possible to achieve larger inductance than using planar meandering.

For the spiral, this approach involves coiling the arms such that they resemble a helix as shown in Fig. 5.24. The coil for the spiral arm can have rectangular cross-section allowing for control of the inductance using the pitch (separation between turns), width, and height of the coil. These parameters can be related to the self-inductance of the coil using the equivalent inductance per unit length of a lossless helical

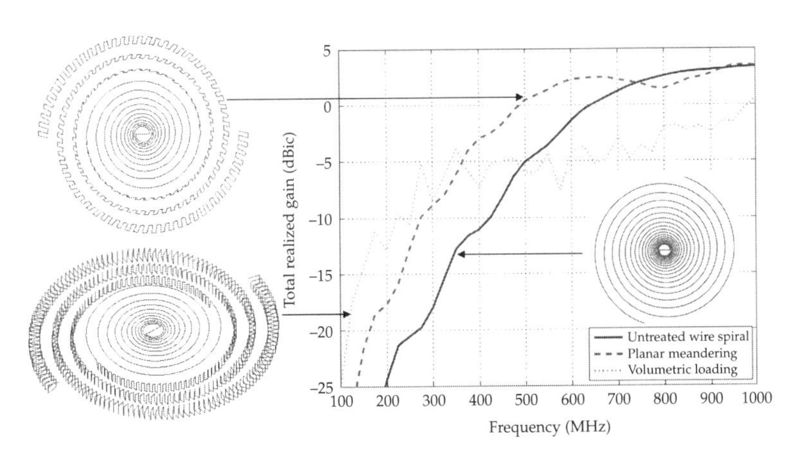

FIGURE 5.24 Gain comparison between planar meandering and 3D coiling for constant pitch and height spiral weaving but varying meandering width. (*see [46].*)

waveguide. The inductance per unit length of a coil wire is given by [47]

$$L_e = \frac{\mu_0}{2\pi}\left(\frac{2\pi a}{p}\right)^2 I_1(\gamma a)K_1(\gamma a) \qquad \text{(H/m)} \tag{5.13}$$

where a is the coil radius, p is the pitch, I is the modified Bessel function of the first kind and K is the modified Bessel function of the second kind. The variable γ is determined from the dispersion equation

$$\gamma^2 = k^2\left(\frac{2\pi a}{p}\right)^2 \frac{I_1(\gamma a)K_1(\gamma a)}{I_0(\gamma a)K_0(\gamma a)} \tag{5.14}$$

where k refers to the free space propagation constant. From Eq. 5.13, it is clear that the inductance per unit length is linearly proportional to the cross-sectional area and inversely proportional to the pitch squared. Therefore, it provides a physical insight into how the coil inductance can be controlled.

To demonstrate the advantage of volumetric meandering, in Fig. 5.24 we compare the realized gain of 6″ diameter spirals with planar and volumetric meandering. The meandered arms have the same pitch of 60 mils and same width from 22 to 275 mils. Also, for the volumetric coiling the coil height was chosen to be 250 mils for practical fabrication considerations. Further, for both meanderings, the inductive loading starts at the middle of the arm as high frequencies do not need miniaturization. As the realized gain is plotted, as before, for each case the high frequency impedance was used as reference for matching. Indeed, the gain curves in Fig. 5.24 show that the volumetrically coiled spiral has significantly higher gain below 300 MHz as compared to the planar meandered spiral. Specifically, the −15 dBi gain point was shifted from 320 MHz in the untreated spiral to 147 MHz. This corresponds to a miniaturization factor of 2.18 (as compared to 1.8 with dielectric loading).

An issue with the coil spiral design is identified in Fig. 5.24. That is, the gain is reduced at higher frequencies (> 500 MHz) as compared to the no-loading case. This is caused by reflection between the unloaded and loaded spiral regions, as well as the nonoptimized inductance profile. One can overcome this issue by adopting a better tapered loading profile with gradual height increase. This is illustrated in Fig. 5.25. Here, the constant height coil design is compared to an improved conical coil design. For this, the coil height (or thickness) increases linearly from 20 to 250 mils along the spiral arms of the inductive loading section. This new conical arrangement leads to a smoother inductive taper (throughout the whole loading section). Thus, as shown in Fig. 5.25 it recovers the realized gain at higher frequencies.

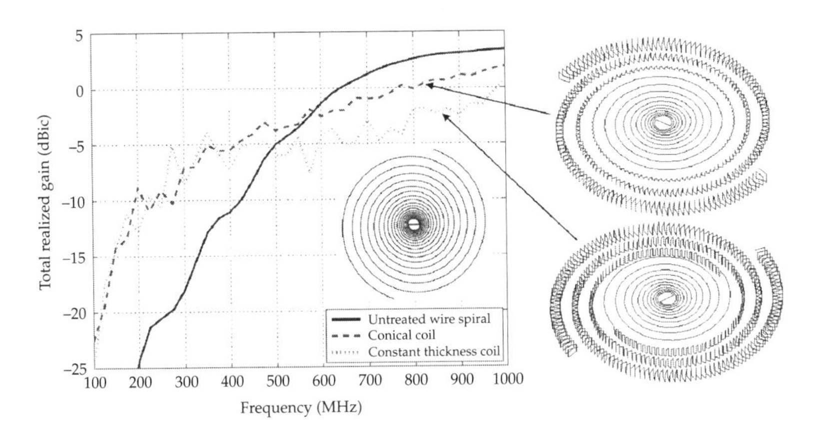

FIGURE 5.25 Gain comparison of constant height coils and conical coils to improve low-frequency performance without appreciably compromising high-frequency gain. (*see [46].*)

The main reason in tapering the inductive loading is to avoid detrimental effects at high frequencies (similar to tapering the dielectric loading). The other reason is to provide a smooth transition from the unloaded center to the more heavily loaded outer rim of the spiral. To achieve the best overall performance, an optimal tapering profile of inductance values along the arms is required. Although, such optimization is, in many aspects, similar to the problem of impedance tapering between two transmission line sections, it is far more complex since the inductance in this case is both spatial and frequency dependent. Therefore, formal optimization methods were used to find the optimal coiling. Kramer et al. [23] conducted such an optimization on the different coil parameters such as pitch, length, and inductance taper profile based on a fifth order polynomial function using genetic algorithms in conjunction with full-wave NEC simulations. It was found that an exponential taper with the total loading length of approximately $1.5\lambda_0$ ($\lambda_0 = 2\pi r_{max}$) provides sufficient miniaturization without compromising performance at high frequencies (see Fig. 5.25).

5.7 Fabricated Miniature Spiral Antennas

Several practical fabrication issues such as feeding, coil generation, supporting material, etc., all need to be considered to fabricate previous miniature spiral designs into a realizable article for measurements.

Based on studies by Kramer et al. [48], a good taper length could be $1.5\lambda_0$, implying use of the last 1.5 turns of a spiral. For a 6″ spiral the starting radius for applying inductive loading will be 1.5654″, implying that the inductive loading will not affect the gain above 1200 MHz. Also, since a standard printed circuit board (PCB) fabrication process is desired, a constant coil height with a thickness $t = 0.5''$ was chosen. This choice forces one to generate a smooth inductance taper by only varying the coil width and pitch. From Eq. (5.13), we know that the inductance is inversely proportional to the pitch squared and approximately proportional to the width. Thus, the inductance is predominately determined by the pitch. Therefore, for an exponential taper profile, the pitch could be decreased gradually in accordance with the following expression

$$p(s) = p_0 e^{\alpha s} = p_0 e^{\left[\frac{\ln(p_e/p_0)}{L_{\text{taper}}}\right] s} \tag{5.15}$$

Here, s = distance from the beginning of the taper ($0 < s < L_{\text{taper}}$) and p_0/p_e refer to the initial and end pitch of the spiral, respectively. For this design, p_e = 45 mils and $p_0 = 20p_e$. Further, to simplify the coil fabrication, the coil width varied in accordance with the geometry defined by Eq. (5.12). That is

$$w(s) = (as + r_{\text{start}})|1 - e^{a\delta}| \sin\psi \tag{5.16}$$

where $\psi = \tan^{-1}(1/a)$. Therefore, the coil width is a linear function of the distance, s. We note that for the initial design the coil inductance was calculated at a frequency of 150 MHz using Eqs. (5.13) and (5.14) and an effective radius $a = \sqrt{wt/\pi}$. The latter was based on a circle with the same area as that of a rectangular coil. Figure 5.26 shows the top view of the final 6″ coiled spiral fabricated on a thick PCB.

To feed the antenna from a coaxial cable, a broadband balun is also needed. For this, a commercially available broadband (30–3000 MHz) 0° to 180° hybrid was used to split the coaxial input to two equal-amplitude and out-of-phase output currents (see Fig. 5.27). This balun also transforms the 50 Ω input impedance to 100 Ω at the output. Since the characteristic impedance of the inductive-loaded spiral (shown in Fig. 5.26) was found to be approximately 138 Ω, it had to also be reduced to 100 Ω for matching to the balun. This could be achieved by changing the dielectric constant of the PCB substrate.

Figure 5.28 shows the characteristic impedance of the spiral antenna in Fig. 5.26 implemented using a $t = 0.25''$ board using different commercially available dielectric constants. In this case, the 3D coils are formed by connecting traces on both sides of the board via holes (see Fig. 5.29). It is observed that a substrate with a dielectric constant $\varepsilon_r \approx 4$ produces a desired characteristic impedance of 100 Ω. We therefore used a Rogers TMM-4 laminate board (with $\varepsilon_r = 4.5$ and $\tan\delta = 0.002$) for fabrication. Figure 5.29 shows a photo of the

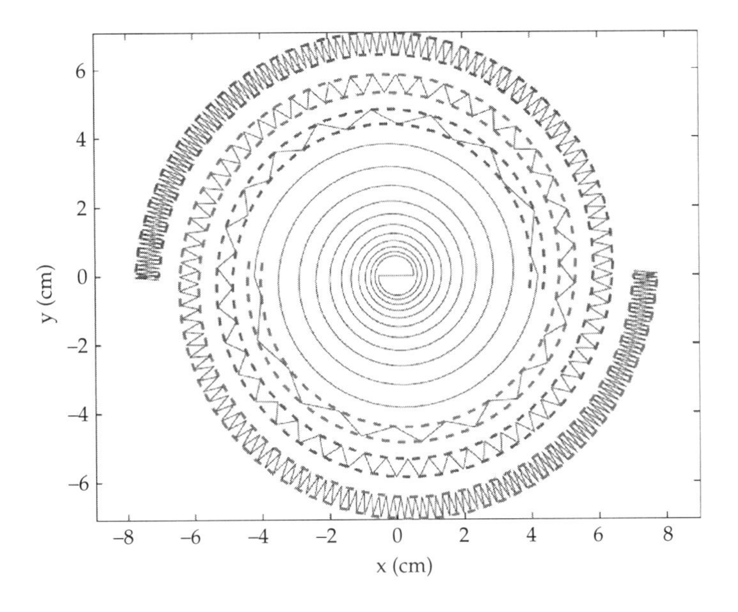

FIGURE 5.26 Top view of the 6″ inductively loaded wire spiral antenna having a constant coil height and an exponentially tapered pitch.

fabricated 6″ spiral antenna, and Fig. 5.30 plots the measured return loss and total realized gain of the 6″ spiral in Fig. 5.29. The measured realized gain is compared with FEKO simulations for a spiral in free space (without dielectric material) and printed on an infinite 0.25″ thick dielectric layer. As compared to the ideal infinite substrate

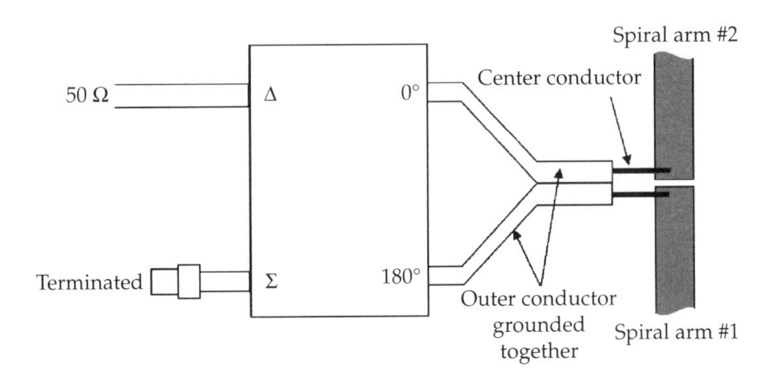

FIGURE 5.27 Balanced feeding of the spiral antenna using a 0° to 180° hybrid.

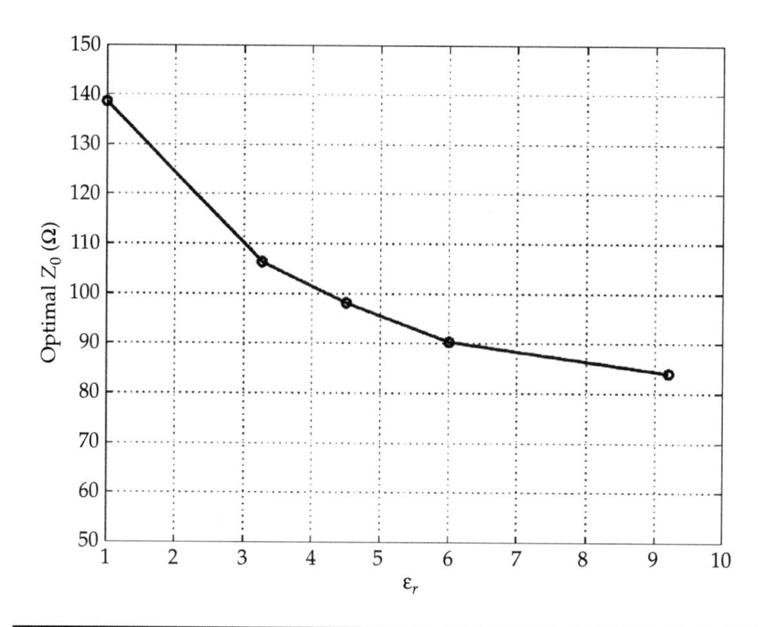

FIGURE 5.28 Effect of the dielectric substrate ε_r value on the spiral's input impedance at high frequencies.

FIGURE 5.29 Miniature spiral antenna fabricated to realize the 3D inductive coiling using a 0.25″ thick TMM-4 PCB board (only one side of the traces are shown here, see also Fig. 3.22) [23, 46].

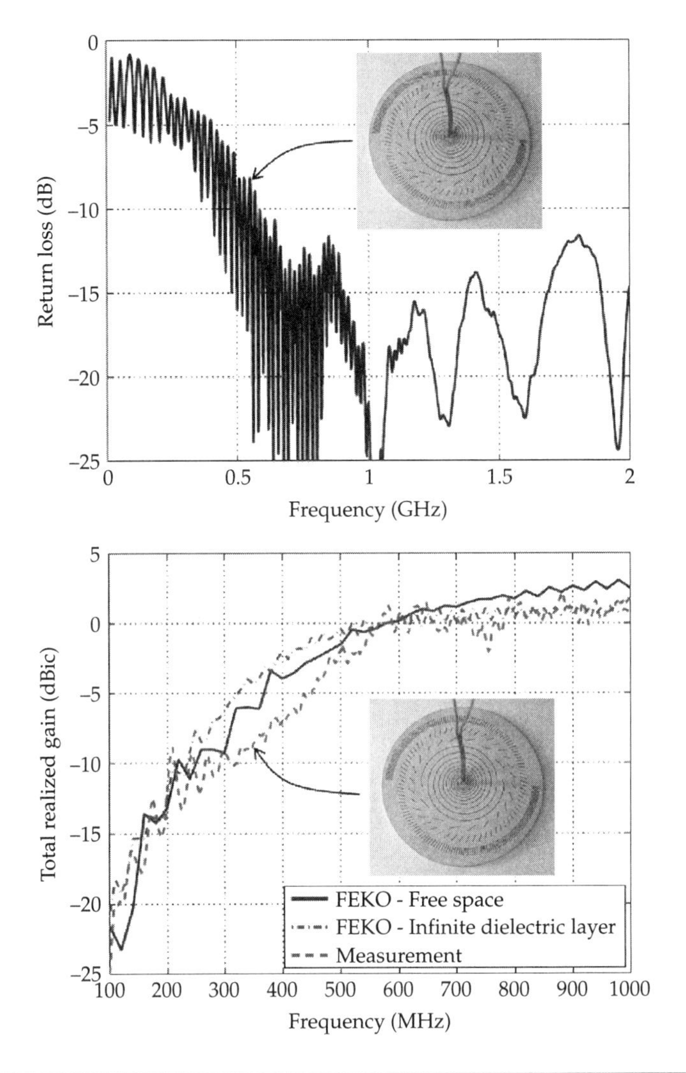

FIGURE 5.30 VSWR and gain performance of a free-standing 6″ spiral.

calculations, the measured gain does show lower levels between 250 MHz and 500 MHz due to diffractions from the truncated dielectric substrate. Nevertheless, the measured gain levels are very similar with and without dielectric layers below −10 dBi gain point. Note that the rapid oscillations in the return loss curve, below 1 GHz, are caused by reflections from cable connections in the measurement setup.

The previous 6″ design can be scaled to 12″ for operating down to lower frequencies. However, this would lead to poor performance at high frequencies due to coarser spiral arms in terms of wavelengths. To avoid this and maintain proper operation up to 2 GHz as in the original 6″ design, the spiral windings were tightened near the center. Figure 5.31 shows the new 12″ spiral design fabricated on a 0.25″ TMM-4 board. The antenna was placed 2.25″ above a conducting ground 24″ in diameter. A layer of ferrite absorbing tile (TDK IB-015) of diameter 12″ and 0.25″ thick was also placed on the metallic ground plane. The ferrite-reduced, adverse ground plane effects were discussed in [49].

In addition to the 6″ and 12″ design discussed in the preceding paragraph, we also show in Fig. 5.32 a miniaturized 18″ spiral. This design uses optimized conical coils and was measured in absence of ground plane. It was measured with a 4:1 transformer balun to match the antenna port impedance of 200 Ω. However, the actual antenna impedance was greater than 200 Ω, resulting in the oscillatory gain behavior below 300 MHz depicted in Fig. 5.32.

In summary, this chapter discussed the fundamental characteristics and operational principles of spiral antennas. We increased the electrical size of the spirals by introducing loading to increase the series inductance and shunt capacitance along the arms and thus reduce wave

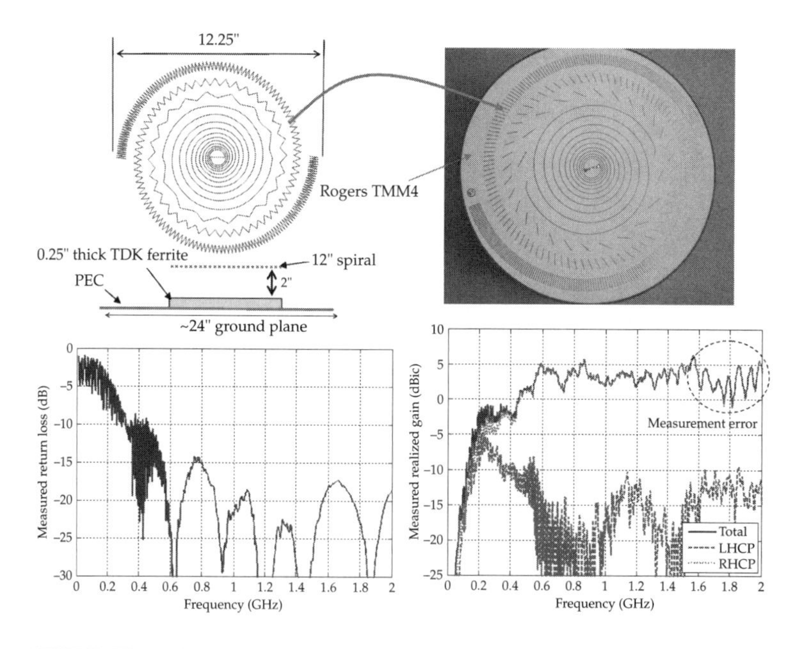

FIGURE 5.31 Measured return loss and realized gain of a fabricated 12″ miniaturized spiral antenna packaged in a PVC enclosure.

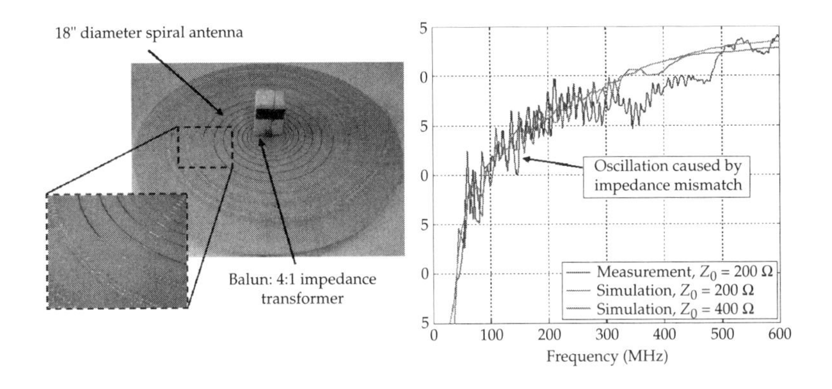

FIGURE 5.32 Measured and simulated free-standing 18″ spiral antenna with conical coiled arms [46].

velocity. As such different levels of miniaturization were achieved. It was shown that low-loss dielectric material loading serves as an effective means to increase shunt capacitance. Planar meandered arms and 3D coiling of the arms serve to increase series inductance. Strategically applying both of this loading allows miniaturization while maintaining antenna impedance. Several examples of the fabricated miniaturized spirals were presented. The measured gain performance of these antennas was shown to agree with full-wave simulations and demonstrated the capability to operate at lower frequencies (after miniaturization). We may comfortably note that although these miniaturization techniques were applied to spirals only, they are applicable to other antenna types as well. Of course there are many different ways to implement shunt capacitance and series inductance for miniaturization.

References

1. V. Rumsey, "Frequency independent antennas," *IRE National Convention Record*, pt. 1, March 1957, pp. 114–118.
2. J. Donnellan and R. Close, "A spiral-grating array," *IRE Transactions on Antennas and Propagation*, vol. AP-9, May 1961, pp. 291–294.
3. J. Dyson "The equiangular spiral antenna," *IRE Transactions on Antennas and Propagation*, vol. AP-7, April 1959, pp. 181–187.
4. J. Dyson "The unidirectional equiangular spiral antenna," *IRE Transactions on Antennas and Propagation*, vol. AP-7, October 1959, pp. 239–334.
5. J. D. Dyson, R. Bawer, P. E. Mayes, and J. I. Wolfe, "A note on the difference between equiangular and archimedes spiral antennas," *IEEE Transactions on Microwave Theory and Techniques*, vol. 9, no. 2, March 1961, pp. 203–205.
6. J. Dyson and P. Mayes, "New circular-polarized frequency-independent antennas with conical beam or omnidirectional patterns," *IEEE Transactions on Antennas and Propagation*, vol. 9, no. 4, July 1961, pp. 334–342.

7. W. Curtis, "Spiral antennas," *IRE Transactions on Antennas and Propagation*, vol. AP-8, May 1960, pp. 293–306.
8. J. Kaiser, "The Archimedean two-wire spiral antenna," *IRE Transactions on Antennas and Propagation*, vol. AP-8, May 1960, pp. 312–323.
9. J. Donnellan "A spiral-dublet scanning array," *IRE Transactions on Antennas and Propagation*, vol. AP-9, May 1961, pp. 276–279.
10. W. L. Stutzman, "Wide bandwidth antenna array design," *Proceedings of the IEEE Southeastern Regional Meeting*, Raleigh, NC, April 1985, pp. 92–96.
11. A. A. Chatzipetros, "Planar wideband arrays with variable element size," Independent Study Report, Virginia Technical University, March 1993.
12. D. G. Shively, *Wideband Planar Array Antennas: Theory and Measurements*, Master's Thesis, Virginia Technical University, May 1988.
13. D. B. Shively and W. L. Stutzman, "Wideband arrays with variable element sizes," *Proceedings of the IEEE*, vol. 137, pt. H, no. 4, August 1990.
14. M. Nurnberger and J. L. Volakis, "A new planar feed for slot spiral antennas," *IEEE Transactions on Antennas and Propagation*, vol. 44, January 1996, pp. 130–131.
15. T. Ozdemir and J. L. Volakis, "A new planar feed for slot spiral antennas," *IEEE Transactions on Antennas and Propagation*, vol. 44, January 1996, pp. 130–131.
16. M. W. Nurnberger and J. L. Volakis, "Extremely broadband slot spiral antennas with shallow reflecting cavities," *Electromagnetics*, vol. 20, no. 4, 2000, pp. 357–376.
17. J. L. Volakis, M. W. Nurnberger, and D. S. Filipović, "A Broadband cavity-backed slot spiral antenna," *IEEE Antennas and Propagation Magazine*, vol. 43, no. 6, December 2001, pp. 15–26.
18. M. Nurnberger and J. L. Volakis, "New termination for ultra wide-band slot spirals," *IEEE Transactions on Antennas and Propagation*, vol. 50, January 2002, pp. 82–85.
19. D. S. Filipovic and J. L. Volakis, "Novel slot spiral antenna designs for dual-band/multi-band operation," *IEEE Transactions on Antennas and Propagation*, vol. 51, March 2003, pp. 430–440.
20. D. S. Filipovic and J. L. Volakis, "A flush mounted multifunctional slot aperture (Combo-antenna) for automotive applications," *IEEE Transactions on Antennas and Propagation*, vol. 52, no. 2, February 2004, pp. 563–571.
21. D. S. Filipovic and J. L. Volakis, "Broadband meanderline slot spiral antenna," *IEE Proceedings—Microwaves, Antennas and Propagation*, vol. 149, no. 2, April 2002, pp. 98–105.
22. B. A. Kramer, M. Lee, C. C. Chen, and J. L. Volakis, "Design and performance of an ultra wideband ceramic-loaded slot spiral," *IEEE Transactions on Antennas and Propagation*, vol. 53, no. 7, June 2004, pp. 2193–2199.
23. B. A. Kramer, C. C. Chen, and J. L. Volakis, "Size reduction and a low-profile spiral antenna using inductive and dielectric loading," *IEEE Transactions on Antennas and Wireless Propagation Letters*, vol. 7, 2008, pp. 22–25.
24. B. A. Kramer, M. Lee, C. C. Chen, and J. L. Volakis, "Fundamental limits and design guidelines for miniaturizing ultra-wideband antennas," *IEEE Antennas and Propagation Magazine*, vol. 51, August 2009, pp. 57–69.
25. J. J. H. Wang, "The spiral as a traveling wave structure for broadband antenna applications," *Electromagnetics*, vol. 20, no. 4, 2000, pp. 323–342.
26. R. Corzine and J. Mosko, *Four-Arm Spiral Antennas*, Artech House Norwood, Mass., 1990.
27. W. L. Curtis, "Spiral antennas," *IRE Transactions on Antennas and Propagation*, vol. 8, May 1960, pp. 298–306.
28. M. S. Wheeler, "On the radiation from several regions in spiral antennas," *IRE Transactions on Antennas and Propagation*, vol. AP-9, January 1961, pp. 100–102.
29. C. C. Chen and J. L. Volakis, "Spiral Antennas: Overview, Properties and Miniaturization Techniques," in book R. Waterhouse, ed., *Printed Antennas for Wireless Communications*, John Wiley & Sons, Hoboken, New Jersey, 2007.

30. B. R. S. Cheo, V. H. Rumsey, and W. J. Welch, "A solution to the frequency-independent antenna problem," *IRE Transactions on Antennas and Propagation*, vol. AP-9, November 1961, pp. 527–534.
31. M. Lee, B. A. Kramer, C. C. Chen, and J. L. Volakis, "Distributed lumped loads and lossy transmission line model for wideband spiral antenna miniaturization and characterization," *IEEE Transactions on Antennas and Propagation*, vol. 55, no. 10, October 2007, pp. 2671–2678.
32. J. Galejs, "Dielectric loading of electric dipole antennas," *Journal of Research of the National Bureau of Standards*, vol. 66D, September–October 1962, pp. 557–562.
33. C. Y. Ying, "Theoretical study of finite dielectric-coated cylindrical antennas," *Journal of Mathematical Physics*, vol. IO, 1969, pp. 480-483.
34. J. R. James and A. Henderson, "Electrically short monopole antennas with dielectric or ferrite coatings," *Proceedings of the Institute of Electrical Engineers*, vol. 125, September 1978, pp. 793–803.
35. J. R. James, A. J. Schuler, and R. F. Binham, "Reduction of antenna dimensions by dielectric loading," *Electronics Letters*, vol. 10, no. 13, June 1974, pp. 263–265.
36. D. I. Kaklamani, C. N. Capsalis, and N. K. Uzunoglu, "Radiation from a monopole antenna covered by a finite height dielectric cylinder," *Electromagnetics*, vol. 12, April–June 1992, pp. 185–216.
37. Y. P. Zhang, T. K. Lo, Y. Hwang, "A dielectric-loaded miniature antenna for microcellular and personal communications," *Proceedings of the IEEE APS-Symposium*, Newport Beach, Calif., June 1995, pp. 1152–1155.
38. C. W. Harrison, "Monopole with inductive loading," *IEEE Transactions on Antennas and Propagation*, vol. 11, July 1983, pp. 394–400.
39. W. P. Czerwinski, "On optimizing efficiency and bandwidth of inductively loaded antennas," *IEEE Transactions on Antennas and Propagation (Communication)*, vol. AP-13, September 1965, pp. 811–812.
40. R. C. Hansen, "Efficiency transition point for inductively loaded monopole," *Electronics Letters*, vol. 9, March 8, 1973, pp. 117–118.
41. B. A Kramer, M. Lee, C. C. Chen, and J. L. Volakis, "Miniature UWB Antenna with Embedded Inductive Loading," Technology Small Antennas and Novel Metamaterials, 2006 IEEE International Workshop, 2006, pp. 289–292.
42. S. Koulouridis, G. Kiziltas, Y. Zhou, D. J. Hansford, and J. L. Volakis, "Polymer-ceramic composites for microwave applications: fabrication and performance assement," *IEEE Transactions on Microwave Theory and Techniques*, vol. 54, no. 12, pt. 1, December 2006, pp. 4202–4208.
43. G. J. Burke, E. K. Miller, and A. J. Poggio, "The Numerical Electromagnetics Code (NEC) - a brief history," *Antennas and Propagation Society International Symposium*, 2004 IEEE, vol. 3, June 2004, pp. 2871–2874.
44. S. R. Best, "A discussion on the properties of electrically small self-resonant wire antennas," *IEEE Antennas and Propagation Magazine*, vol. 46, no. 6, December 2004, pp. 9–22.
45. J. Rashed and C. T. Tai, "A new class of wire antennas," *IEEE Transactions on Antennas and Propagation*, vol. 39, no. 9, September 1991, pp. 1428–1430.
46. B. A. Kramer, "Size reduction of an ultrawideband low-profile spiral antenna," Ph.D. dissertation, The Ohio State University, 2007.
47. J. E. Rowe, *Nonlinear Electron-Wave Interaction Phenomena*, Academic Press Inc., New York, 1965.
48. B. A. Kramer, S. Koulouridis, C. C. Chen, and J. L. Volakis, "A novel reflective surface for an UHF spiral antenna," *IEEE Antennas and Wireless Propagation Letters*, vol. 5, 2006, pp. 32–34.
49. F. Erkmen, C. C. Chen, and J. L. Volakis, "Impedance matched ferrite layers as ground plane treatments to improve antenna wideband performance," *IEEE Transactions on Antennas and Propagation*, 2008.

CHAPTER 6

Negative Refractive Index Metamaterial and Electromagnetic Band Gap Based Antennas

Gokhan Mumcu and John L. Volakis

6.1 Introduction

Reducing antenna size and profile via traditional miniaturization methods (such as shorting pins [1], meandering [2], and dielectric loading [3]; see Chap. 3) often result in low radiation efficiencies, narrow bandwidth, and undesired radiation patterns. It is not therefore surprising that strong interest exists in engineered materials (referred to as metamaterials) as a method to optimize or enhance antenna performance.

Metamaterials are typically constructed from periodic arrangements of available materials (isotropic or anisotropic dielectrics, magnetic materials, conductors, etc.) and exhibit electromagnetic properties not found in any of their bulk individual constituents. Over the last decade, theory and practical applications of metamaterials at microwave frequencies became an extensive research area. An early boost to this research was the initial experimental verifications of electromagnetic band gap structures (EBGs) [4] and negative refractive index (NRI or equivalently $\varepsilon < 0$ and $\mu < 0$) media [5,6]. Since then, NRI metamaterials were studied to demonstrate sub-wavelength focusing for greater sensitivity [7] in lens systems. Printed circuit realizations of NRI media led to smaller radio frequency (RF) devices such as phase shifters, couplers, and antennas [8,9]. Similarly, band structures of EBGs were exploited to realize novel waveguides, resonators, and filters [10]. The strong resonances provided by defect modes in EBGs were used to transform small radiators into directive antennas [11]. In addition, forbidden propagation bands of EBGs were exploited as high impedance ground planes to improve antenna radiation properties such as gain, conformability, and coupling reduction [4]. Likewise, composite substrates assembled from different dielectric materials were used to improve performance of printed antennas [12].

Indeed, the novel propagation properties found in NRI and EBG metamaterials are promising for various antenna applications, including miniaturization, coupling reduction, gain increase, and scanning. In this chapter, we begin by presenting an overview of wave propagation in NRI metamaterials using their equivalent transmission line circuit models [8,9]. Subsequently, antenna examples from the recent literature are presented to demonstrate how NRI properties lead to novel and smaller antennas. The last section of the chapter is devoted to the high-gain antennas based on EBG structures.

6.2 Negative Refractive Index Metamaterials

Veselago was the first in the 1960s to examine media having negative dielectric (ε) and magnetic (μ) constitutive parameters ($\varepsilon < 0$, $\mu < 0$) [13]. In such media, electric ($\boldsymbol{E}$) and magnetic ($\boldsymbol{H}$) fields of plane waves along with the wavevector ($\mathbf{k}$) form a "left-handed" (LH) triad. As such, the group ($\mathbf{v}_g$) and phase ($\mathbf{v}_p$) velocities have opposite directions. That is, the media can support backward wave propagation but forward energy flow. Moreover, this LH behavior is associated with the reversal of Snell's law and implies a negative index of refraction ($n < 0$) with reversed (negative) refraction angle.

6.2.1 Propagation in ($\varepsilon < 0$, $\mu < 0$) Media

To better understand the LH nature of plane wave propagation in a homogenous ($\varepsilon < 0$, $\mu < 0$) medium, we start with the time harmonic

Maxwell's equations

$$\nabla \times E = -j\omega B - M_s \tag{6.1}$$

$$\nabla \times H = j\omega D + J_s \tag{6.2}$$

$$\nabla \cdot D = \rho_e \tag{6.3}$$

$$\nabla \cdot B = \rho_m \tag{6.4}$$

where $M_s(\mathrm{V/m^2})$ and J_s $(\mathrm{A/m^2})$ are the magnetic and electric current densities, respectively. Also, ρ_m $(\mathrm{Wb/m^3})$ and $\rho_e(\mathrm{C/m^3})$ are, respectively, magnetic and electric charge densities. As usual, E stands for electric field intensity (V/m), H denotes the magnetic field intensity (A/m), D is the electric flux density $(\mathrm{C/m^2})$, and B is the magnetic flux density $(\mathrm{Wb/m^2})$. In above equations and throughout the rest of this chapter, an $e^{j\omega t}$ time dependence was assumed and suppressed.

For a linear and nondispersive medium (although LH media must be always dispersive to satisfy entropy conditions [1], their dispersion may be assumed as weak), the vector pairs (D, E) and (B, H) are related to each other via the constitutive relations as

$$D = \varepsilon_0 \varepsilon_r E = \varepsilon E \tag{6.5}$$

$$H = \mu_0 \mu_r H = \mu H \tag{6.6}$$

where ε_0 and μ_0 are the permittivity and permeability of free space, respectively. Likewise, ε_r and μ_r represent relative permittivity and permeability constants, respectively. A solution of Eqs. (6.1) and (6.2) are the plane waves

$$E = E_0 e^{-j\mathbf{k}\cdot\mathbf{r}} \tag{6.7}$$

$$H = \frac{E_0}{\eta} e^{-j\mathbf{k}\cdot\mathbf{r}} \tag{6.8}$$

where $\eta = \sqrt{\mu/\varepsilon}$ denotes the intrinsic media impedance. Since plane waves can be used to express any arbitrary electromagnetic field distribution (via Fourier transforms), Eqs. (6.7) and (6.8) can be used to study the electromagnetic properties of materials.

To examine LH materials, let us set $\varepsilon = s|\varepsilon|$ and $\mu = s|\mu|$ with $s = \pm 1$. With substituting Eqs. (6.7) and (6.8) in Eqs. (6.1) and (6.2), Maxwell's equations become

$$\mathbf{k} \times E = s\omega\,|\mu|\, H \tag{6.9}$$

$$\mathbf{k} \times H = -s\omega\,|\varepsilon|\, E \tag{6.10}$$

assuming that $M_s = J_s = 0$. For $s = -1$; E, H, and $\mathbf{k}$ then form a LH triad. Concurrently, the Poynting vector $S = E \times H$ is in the opposite direction (antiparallel) of $\mathbf{k}$. That is, the outgoing (along $+r$) propagation is associated with a negative wavenumber $\mathbf{k} = k\hat{\mathbf{r}}, k < 0$ in Eqs. (6.7) and (6.8). Consequently, the refractive index becomes negative ($n = s\sqrt{\varepsilon_r \mu_r}$). An important property of the LH media is that the group velocity

$$\mathbf{v}_g = \frac{\partial \omega}{\partial k}\hat{\mathbf{k}} = \frac{\partial}{\partial k}\left(\frac{ck}{n}\right)\hat{\mathbf{k}} = -\frac{c}{|n|}\hat{\mathbf{k}} \tag{6.11}$$

is also opposite to the direction of the phase velocity

$$\mathbf{v}_p = \frac{\omega}{|\mathbf{k}|}\hat{\mathbf{k}}, \quad (\hat{\mathbf{k}} = \mathbf{k}/|\mathbf{k}|) \tag{6.12}$$

The fact that $\mathbf{v}_g$ is coincidant with the direction of S (in the direction of $-\mathbf{k}$) corresponds to energy transfer from the source to infinity in spite of the reversed phase velocity. More discussions on the properties and electromagnetic theory of NRI ($\varepsilon < 0$, $\mu < 0$) media can be found in recently published books [8, 9].

Naturally available materials do not exhibit simultaneously negative dielectric and magnetic constitutive relations. Methodologies to "artificially" realize such effective media (using periodic material inclusions) have been recently developed. These types of engineered materials are often referred to as LH metamaterials to denote the left handed plane wave propagation and antiparallel phase and group velocities. Another common terminology is "negative refractive index (NRI)" media to stress that $n < 0$. Other terminologies include "double negative (DNG)" metamaterials to emphasize simultaneously negative constitutive parameters, and "backward wave (BW)" metamaterials to stress the antiparallel group and phase velocities.

Several techniques have been proposed to implement artificial media that support the NRI phenomena. For instance, at certain microwave frequencies, periodic arrangements of split ring resonators (SRRs) and metallic thin wires were shown to exhibit negative permeability and permittivity, respectively [5, 6]. Another common method is to employ reactively loaded periodic transmission line (TL) grids [8, 9]. As compared to the resonator-based NRI structures, loaded one-dimensional (1D) and two-dimensional (2D) TLs exhibit broader bandwidth and can be conveniently realized using standard microstrip fabrication techniques. As a result, the concept of NRI-TLs found applications in various printed antenna and RF device designs (such as sub-wavelength focusing [7], phase shifters [14], couplers [8], miniature [15,16], and leaky wave antennas [17]). To better understand the design and operational principles of these antennas, it is beneficial to first consider NRI-TL models and their dispersion properties.

6.2.2 Circuit Model of ($\varepsilon < 0$, $\mu < 0$) Media

In general, a standard lossless TL can be modeled as a periodic structure with its unit cell (smallest element of periodicity) consisting of a circuit having series inductance (L_R) and shunt capacitor (C_R). This circuit is assumed to represent a small section of the TL (much less than $\lambda_g/4$, λ_g = guided wavelength. For an NRI-TL ($\varepsilon < 0$, $\mu < 0$), the circuit model is shown in Fig. 6.1*a* and relies on flipping the inductor and capacitor locations.

Wave propagation in 1D periodic media can be effectively analyzed using transfer matrix ($\bar{\mathbf{T}}$) formalism and making use of Bloch's theorem (i.e., periodic boundary conditions) [18]. In accordance with Bloch's theorem, electromagnetic fields propagating along the z direction can be represented as a superposition of Bloch eigenmodes satisfying the periodic relation

$$\begin{bmatrix} V(z+p) \\ I(z+p) \end{bmatrix} = \begin{bmatrix} V(z) \\ I(z) \end{bmatrix} e^{-jkp} \tag{6.13}$$

where p is the physical length of the unit cell. Since the transfer matrix $\bar{\mathbf{T}}$ relates the fields at z to the fields at $z+p$ via

$$\begin{bmatrix} V(z) \\ I(z) \end{bmatrix} = \bar{\mathbf{T}} \begin{bmatrix} V(z+p) \\ I(z+p) \end{bmatrix} \tag{6.14}$$

substituting Eq. (6.13) into Eq. (6.14) leads to the eigenvalue equation

$$\begin{bmatrix} V(z) \\ I(z) \end{bmatrix} = \bar{\mathbf{T}} \begin{bmatrix} V(z) \\ I(z) \end{bmatrix} e^{-jkp} \Rightarrow \left[\bar{\mathbf{T}} - \bar{\mathbf{I}}e^{jkp}\right] \begin{bmatrix} V(z) \\ I(z) \end{bmatrix} = 0 \tag{6.15}$$

We can now identify e^{jkp} as the eigenvalues of the transfer matrix $\bar{\mathbf{T}}$ having unity magnitude (i.e., k real) for propagating waves. We also remark that nonunity magnitude eigenvalues are associated with evanescent modes. Also, Eq. (6.15) leads to the same eigenvectors for k and $k + 2\pi n/p$, $n = \pm 1, \pm 2, \pm 3 \ldots$ Thus, it is convenient to restrict k within the range $-\pi/p < k < \pi/p$ without loss of information. This region is referred to as the first Brillouin zone and the eigenvalues are represented by the Bloch wavenumber $\mathrm{K} = kp$ in the horizontal axis scaled from $-\pi$ to π.

Figure 6.1*b* depicts the dispersion (K-ω) diagram of an NRI unit cell when the equivalent lumped loads of the unit cell TL are $C_L = 5$ pF and $L_L = 1$ nH. As would be expected, the medium supports two Bloch modes, one propagating along the +z direction and the other along −z. The power flow direction is dented by group velocity $v_g = \partial\omega/\partial k$, and therefore positive slope of the Bloch mode curves implies +z propagating modes and negative slope refers to propagation along −z.

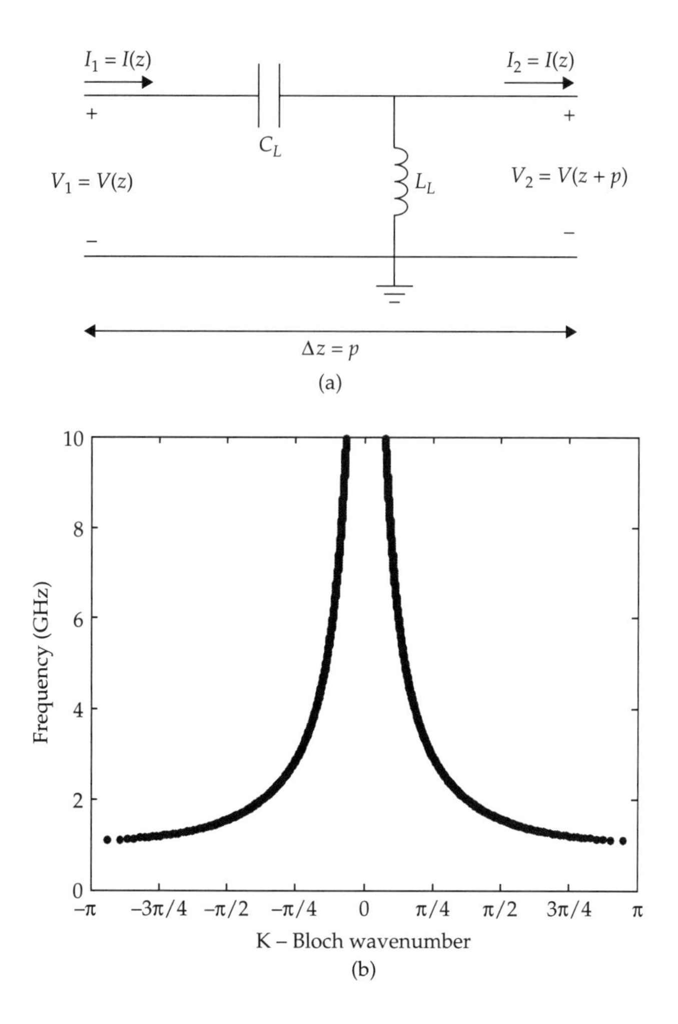

FIGURE 6.1 (a) Unit circuit to model propagation in NRI media; (b) K-ω diagram with $C_L = 5\,\text{pF}$ and $L_L = 1\,\text{nH}$.

It is clear that the propagating waves have antiparallel group and phase velocities ($v_p = \omega/k^L$) as in the case with an unbounded ($\varepsilon < 0$, $\mu < 0$) media. We do note however that the NRI-TL does not allow propagation at very low frequencies due to the high-pass nature of the unit cell. That is, a forbidden propagation band (i.e., a band gap) occurs for frequencies less than 1.2 GHz in the case of Fig. 6.1*b*.

6.2.3 Composite Circuit Model for the NRI-TL Medium

As noted, periodic media with unit cells composed of series capacitors and shunt inductors support backward waves and constitute the artificial NRI metamaterial. However, in a practical implementation, the reactive loads responsible for the NRI effects must be placed within an ordinary host TL or material. Due to the parasitic effects of the host medium, NRI metamaterials also support ordinary right handed (RH) wave propagation at higher frequencies. Therefore, the complete circuit model for the NRI-TL (see Fig. 6.2) must also include distinct sets of reactive loads to represent both LH and RH propagation. For this reason, the NRI-TL circuit model in Fig. 6.2 is also referred to as the "composite right left handed" (CRLH) metamaterial unit cell. As can be surmised from Fig. 6.2, the series C_L and shunt L_L elements are responsible for LH propagation, whereas the shunt C_R and series L_R elements cause RH propagation at higher frequencies. Of course, the LH and RH propagation properties can be tuned to some desired frequency range by simply controlling the unit cell parameters (i.e., amount of the reactive loads).

Using the transfer matrix formalism, the dispersion relation of the NRI-TL is given by

$$K = \cos^{-1}\left(1 - \frac{\chi}{2}\right) \tag{6.16}$$

in which

$$\chi = \left(\frac{\omega_L}{\omega}\right)^2 + \left(\frac{\omega}{\omega_R}\right)^2 - \omega_L^2\alpha \qquad \alpha = L_R C_L + L_L C_R \tag{6.17}$$

and $\omega_L = 1/\sqrt{L_L C_L}$, $\omega_R = 1/\sqrt{L_R C_R}$ are the resonance frequencies of the loads responsible for LH and RH behaviors, respectively. Likewise, $\omega_{se} = 1/\sqrt{L_R C_L}$ and $\omega_{sh} = 1/\sqrt{L_L C_R}$ denote the resonance frequencies of the series and shunt loads. Figure 6.3*a* gives a representative

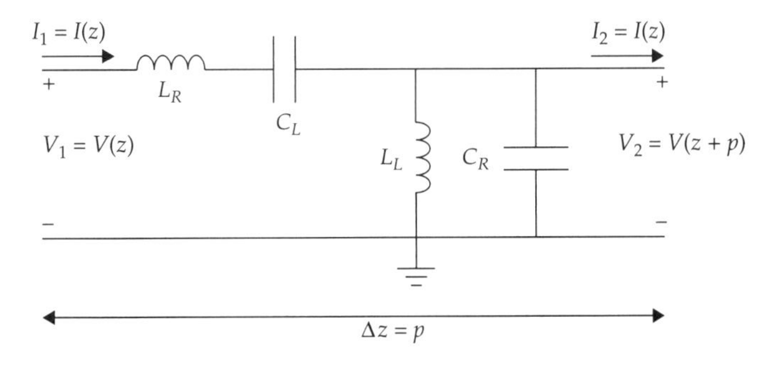

FIGURE 6.2 Complete circuit model of NRI-TL media.

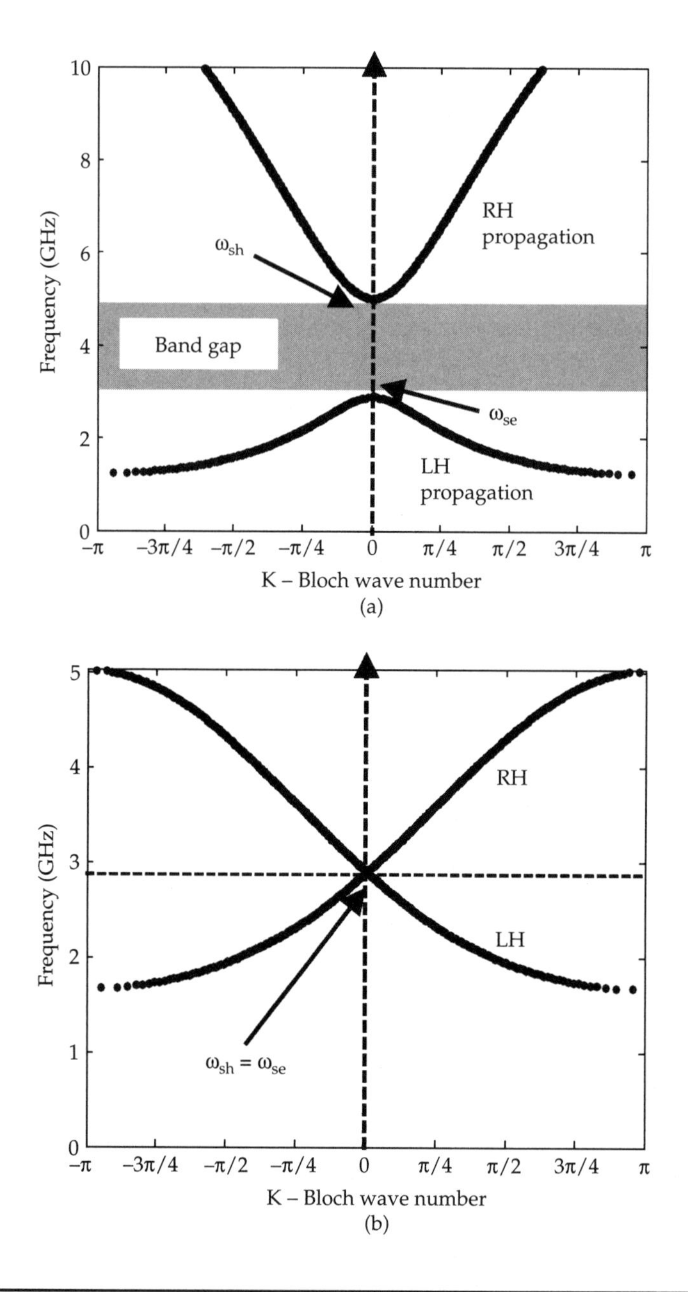

FIGURE 6.3 K-ω diagram of the NRI unit cell shown in Fig. 6.2 with the parameter values (a) $L_L = L_R = 1\,\text{nH}$, $C_L = 3\,\text{pF}$, and $C_R = 1\,\text{pF}$; (b) $L_R = 3\,\text{nH}$, $L_L = 1\,\text{nH}$, $C_L = 1\,\text{pF}$, and $C_R = 3\,\text{pF}$.

K-ω diagram in the 0 to 10 GHz band computed for $L_L = L_R = 1\,\text{nH}$, $C_L = 3\,\text{pF}$, and $C_R = 1\,\text{pF}$. As seen, the circuit acts as a LH medium in the frequency range 1.27 to 2.91 GHz by supporting opposite group and phase velocities. Between 2.91 and 5.03 GHz no propagation occurs (band gap), and above 5.03 GHz the circuit behaves as a RH TL.

It can also be shown that the K = 0 frequencies defining the upper and lower frequencies of the band gap are equal to ω_{se} and ω_{sh} [8]. Thus, it is possible to remove the band gap completely (to obtain a continuous propagation band around the K = 0 frequency) by equating the series and shunt resonance frequencies, viz. by setting $\omega_{se} = \omega_{sh}$. Such NRI-TL (or CRLH) structures are referred to as "balanced" TLs and satisfy the load condition $L_R C_L = L_L C_R$. It is interesting to observe that the dispersion relation of balanced NRI-TLs can further be simplified to read $K = \omega\sqrt{L_R C_R} - 1/\omega\sqrt{L_L C_L}$ around the K = 0 frequency with the modes having corresponding impedances $Z = Z_L = \sqrt{L_L/C_L} = Z_R = \sqrt{L_R/C_R}$. In this case, the overall phase attained through the unit cell is simply the sum of positive and negative phases of the distinct RH and LH sections (as expected from the composite structure). To illustrate the concept of balanced NRI-TLs, an example dispersion diagram (computed for $L_R = 3\,\text{nH}$, $L_L = 1\,\text{nH}$, $C_L = 1\,\text{pF}$, and $C_R = 3\,\text{pF}$) is shown in Fig. 6.3*b*.

6.3 Metamaterial Antennas Based on NRI Concepts

The unit cell model of the NRI metamaterials offers design flexibilities to reduce the operational frequency by employing reactive loads within the regular transmission lines. Also, NRI metamaterials can conveniently exhibit multiple modes (in the LH and RH regime) when a finite number of unit cells is used to form microwave resonators. As such, printed NRI metamaterials were extensively studied to realize a variety of miniature and multiband antennas over the recent years. Further, the backward and forward waves realized on printed NRI metamaterials were found especially attractive to realize back-fire to end-fire scanning leaky wave antennas [17,19]. Finally, we remark that since the balanced NRI-TL unit cell exhibits continuous propagation band across the K = 0 frequency, they can also provide broadband and broadside leaky wave radiation. In the following, we present antenna examples from the recent literature and demonstrate how propagation properties of NRI metamaterials can be harnessed for antenna applications similar to the traveling wave antennas.

6.3.1 Leaky Wave Antennas

For LWAs, guided power gradually leaks away in the form of coherent radiation as it propagates along the periodic waveguide structure (see Fig. 6.4*a*) [20]. In general, the rate of energy leakage is small along the waveguide and therefore the guided mode is not much perturbed. LWAs take advantage of this mechanism to achieve a desirable radiating structure exhibiting high gain, broadband impedance matching, and pattern scanning with frequency.

Critical to the radiation characteristics of a LWA is the complex propagation constant $\gamma = \beta - j\alpha$. As shown in Fig. 6.4*a*, the attenuation constant α is responsible for the radiation of the supported traveling wave over the aperture, whereas the wavenumber β (smaller than the free space wavenumber k_0) defines the direction of radiation. As can be understood, the traveling wave structure must be kept long enough for most of the power to be radiated prior the wave reaching to the end of the antenna. By ensuring that minimal amount of the traveling wave reaches the antenna end, the LWA can exhibit broadband impedance match. Also, the LWA radiation patterns can be conveniently scanned with frequency over a large bandwidth due to the frequency dependent propagation constant.

The K-ω diagram for a typical NRI-LWA is shown in Fig. 6.4*b*. The radiating region (i.e., $\mathrm{K} = \mathrm{K} = \mathrm{K}_x < \mathrm{K}_0$) is bounded by the light lines (K_0). Since the LH waves at lower frequencies (2.5–2.91 GHz) exhibit $\mathrm{K} < 0$ for +x propagating modes (i.e., $v_g > 0$), the LWA radiates in the backward (i.e., −x) direction. On the other hand, the LWA radiates in the forward direction (i.e., +x) when the operational frequency is in the higher RH band (5.03–10.6 GHz). As expected, the LWA seizes to radiate from 2.91 to 5.03 GHz due to the band gap. An important property of the NRI-LWAs is their capability to scan from backward to forward directions. In addition, a continuous backward to forward scan via frequency can be achieved when a balanced NRI-TL is employed. In contrast, conventional LWAs scan only a small range of angles either in the forward or backward directions. Moreover, conventional LWAs operate in higher order modes and require complicated feeding mechanisms to suppress the nonradiating dominant modes [19].

One of the earliest printed layouts used to realize artificial NRI-TLs is shown in Fig. 6.5*a* [17]. As seen, a unique aspect of the NRI-TL is the interdigital capacitors and shorted inductive stubs formed along its path. These serve to slowdown the wave and realize the LH propagation. The dispersion diagrams or K-ω curves for this layout can be calculated via full-wave electromagnetic solvers by employing periodic boundary conditions. Alternatively, an equivalent circuit can be extracted via full-wave circuit solvers for usage in a transfer matrix analysis [8]. As noted, the reactive loading realized with interdigital capacitor and shorted microstrip stub gives rise to the LH behavior,

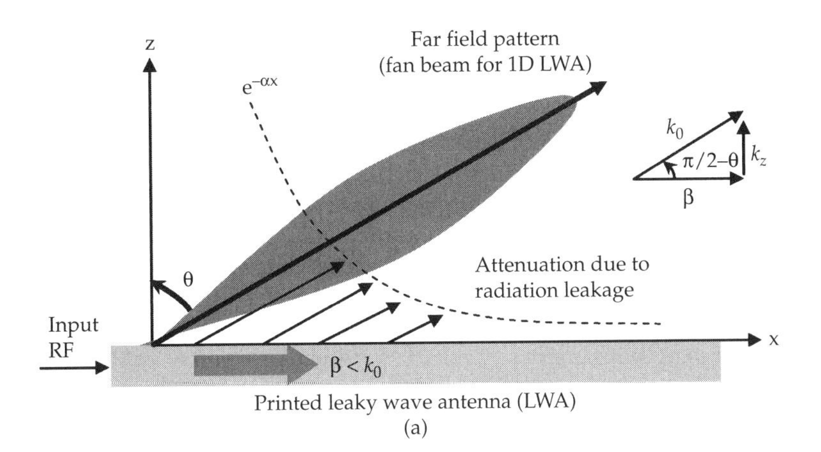

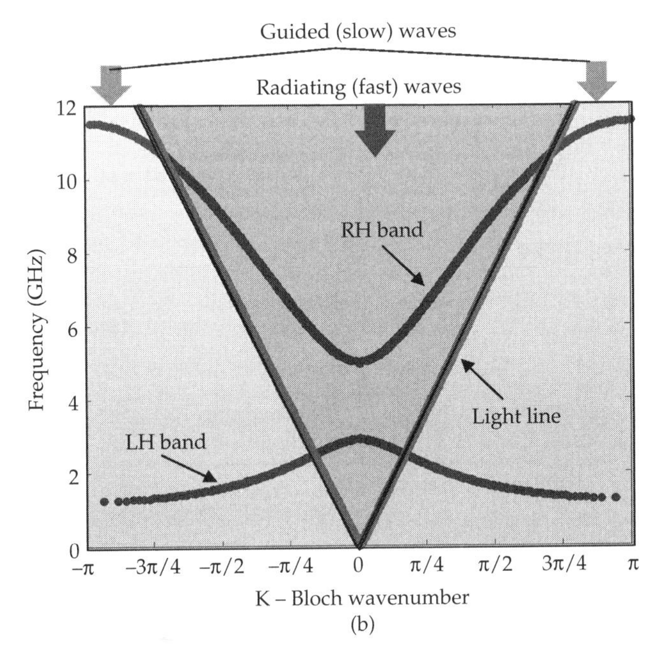

FIGURE 6.4 (a) Operation principle of LWAs; (b) K-ω diagram of the NRI-TL forming the LWA for $L_L = L_R = 1\,\text{nH}$, $C_L = 3\,\text{pF}$, $C_R = 1\,\text{pF}$, and assumed periodicity $p = 1$ cm. Light line is defined by $\text{K} = k_0 p$ (*See* [21]).

and the parasitic effects due to metallization realize the RH behavior. The LWA in [17] consisted of N = 24 unit cells and operated from 3.1 to 6.3 GHz, implying 61% bandwidth. The transition from LH to RH behavior (i.e., K = 0 point) occurred at 3.9 GHz. This structure was reported to exhibit more than 40% radiation efficiency with gain values above 6 dB (>10 dB side lobe level) over the operational bandwidth. Fig. 6.5*b* gives the measured E-plane radiation patterns, showing the backward to forward scan as the operating frequency is increased.

NRI-LWAs can have limited applicability in modern communication systems due to their frequency dependent nature. To overcome this, pattern scanning capability at a fixed frequency is desired. To do so, [21] modified the NRI-TL unit cell to incorporate varactor diodes as shown in Fig. 6.6. Conceptually (see Fig. 6.6), the bias voltages of the varactor diodes (3 per unit cell) are used to shift the dispersion diagram around the same operational frequency (ω_0) to create

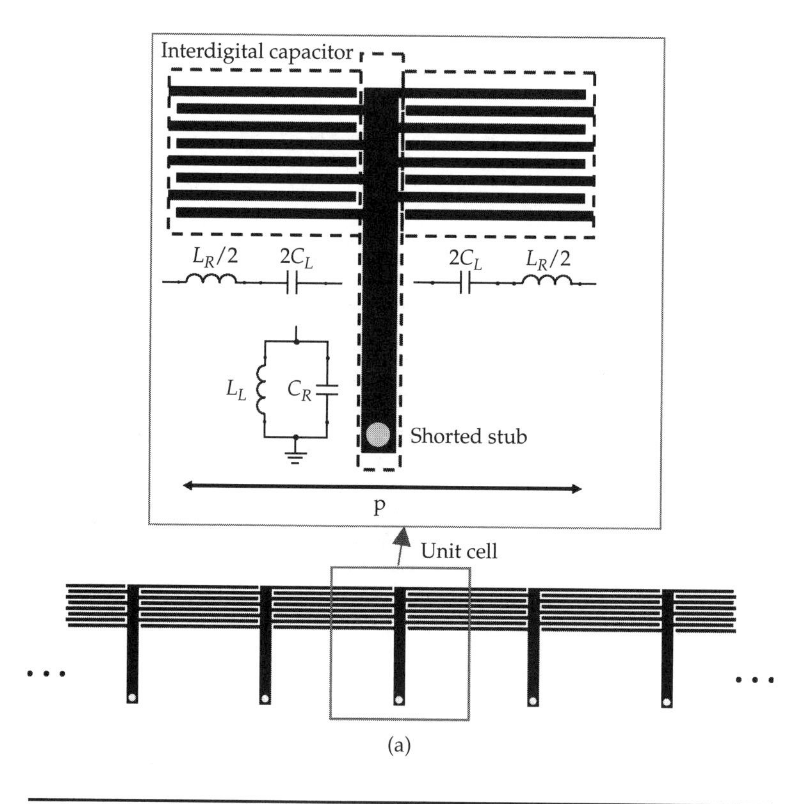

FIGURE 6.5 (a) Microstrip implementation of NRI-TL unit cell; (b) Radiation patterns of N = 24 unit cell 1D NRI-LWA. Unit cells have the layout shown on the left. (*After Liu et al. [17], ©IEEE 2002.*)

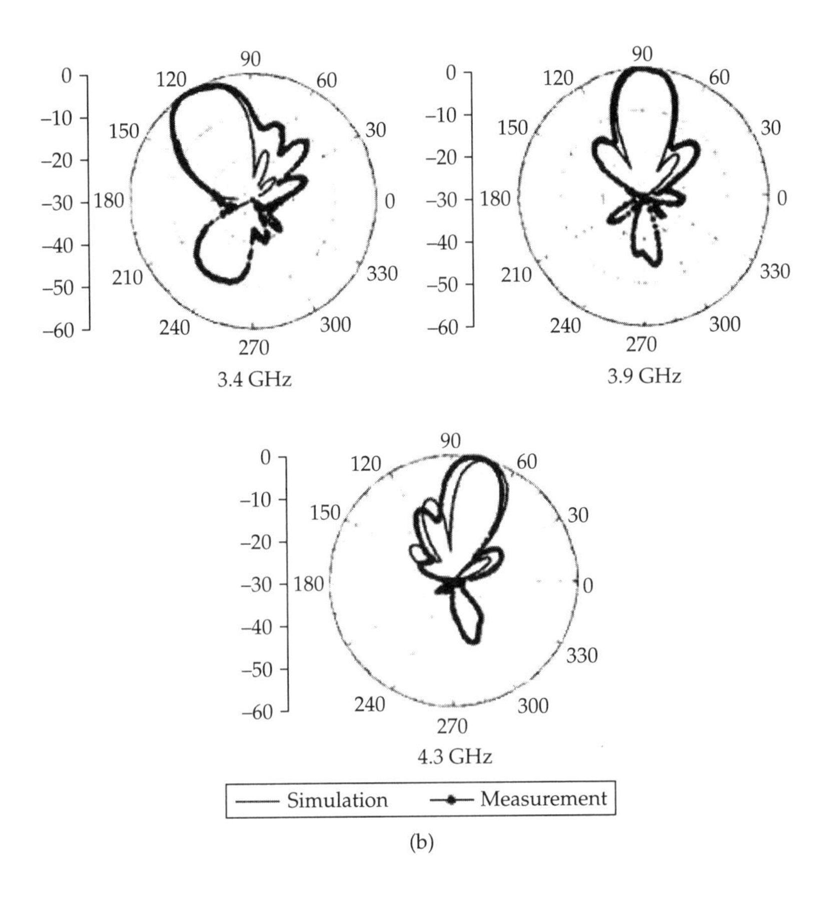

FIGURE 6.5 *Continued.*

an electrically controllable K-ω diagram. This implies different propagation constants (β) for each bias setting are obtained at the same ω_0. Thus the LWA pattern is steered toward the angle $\sin^{-1}(\beta/k_0)$ as illustrated in Fig. 6.4*a*. The LWA shown in Fig. 6.6 consisted of N = 30 unit cells and was reported to [21] allow scanning from −50° to +50° at 3.33 GHz as the varactor bias was varied from 0 V to 21 V. The observed maximum gain at broadside was around 18 dB. It was also reported that the intermodulation effects due to the nonlinear varactors was negligible for such antenna applications.

The maximum effective aperture (and gain) of an LWA occurs at a length where 90% of the input power is radiated. In order to obtain broadside radiators capable of arbitrary gain levels, NRI-TL leaky wave sections with RF amplifiers were proposed [22]. In this approach,

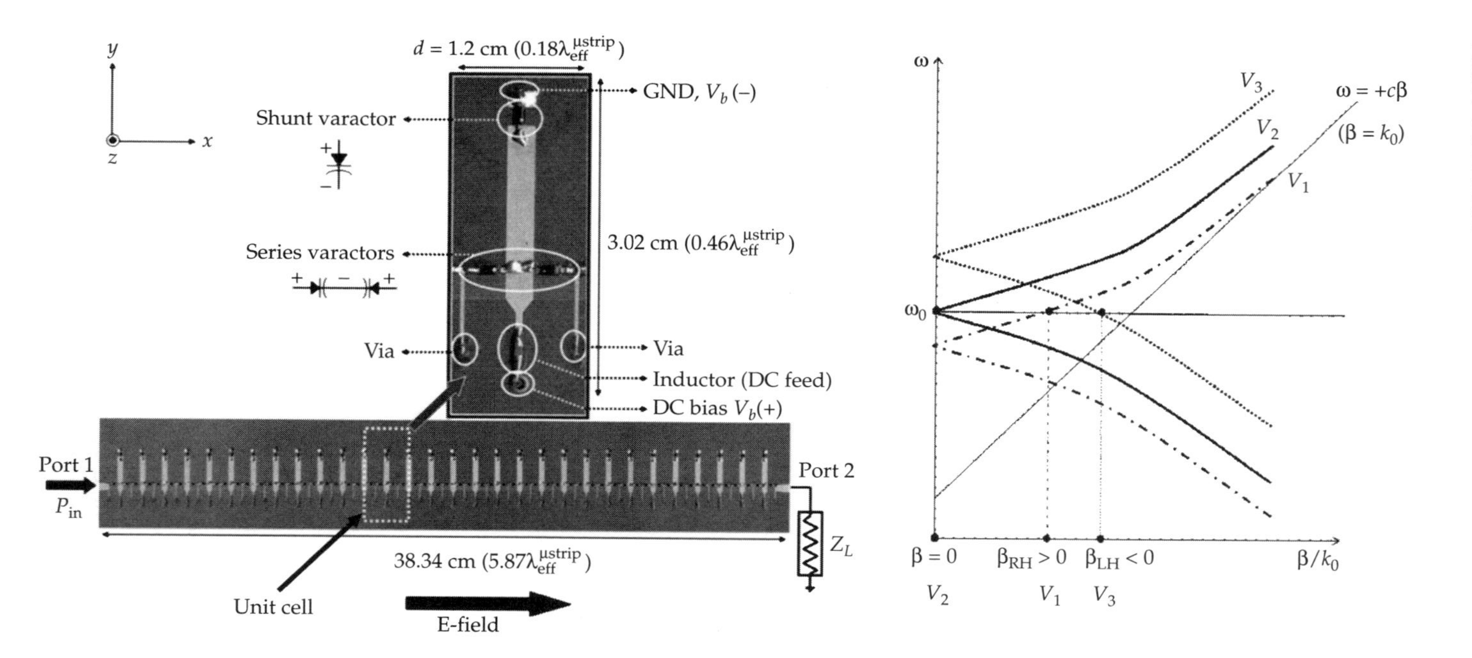

FIGURE 6.6 LWA consisting of N = 30 varactor diode loaded microstrip NRI-TL unit cells. The bias voltage of the varactors is used to vary the dispersion diagram as illustrated to the right. Various radiation patterns are obtained at a fixed ω_0 by controlling the β values. (*After Lim et al. [21], ©IEEE 2005.*)

RF amplifiers are used to pump power to each NRI-TL section to realize much larger effective aperture. Arbitrary gain and side lobe levels can then be conveniently achieved by adjusting various design parameters such as leakage factor, excitation coefficients, amplifier repetition, and overall length. An active LWA prototype consisting of 8×6 unit cell NRI-TL-LWA sections connected with seven amplifiers is depicted in Fig. 6.7. The LWA was printed on the top substrate and the amplifiers were located at the bottom substrate being connected to the NRI-TL sections with via holes. As compared to the same length passive LWA exhibiting 8.9 dB gain, it was reported [22] that the active structure had 17.8 dB broadside gain, that is, an 8.9 dB gain enhancement. This gain enhancement was due to larger effective aperture, power generation, and better impedance matching.

A 2D version of a NRI-TL-LWA is depicted in Fig. 6.8 [23]. Dependent on the excitation edge, this antenna uses the frequency dependent nature of the leaky wave radiation to scan in one principal plane. To provide simultaneous scanning capability in the other plane, phase-shift tuning can also be employed [19]. The shown antenna prototype has dimensions of 9×9 cm (4×4 unit cells) and can scan continuously from $-24°$ to $47°$. A maximum gain of 18 dB was measured for broadside radiation. Also, $|S_{11}| < -10$ dB bandwidth was 5.76 to 7.4 GHz. As expected, the gain dropped at the band ends, and was reported to be 4 and 2.5 dB at the lower and upper frequency ends, respectively.

There are a variety of printed circuit realizations of the NRI-TL circuit that were proposed for different RF applications and needs. For example, the microstrip layout shown in Fig. 6.5*a* was subsequently

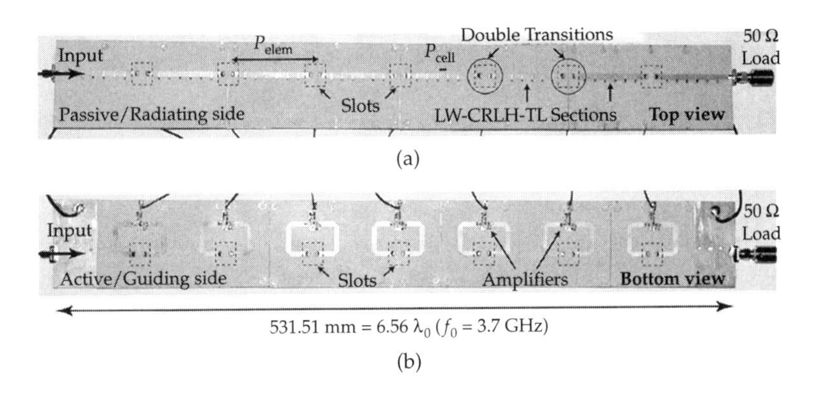

FIGURE 6.7 Active LWA prototype consisting of 8×6 unit cell NRI-TL-LWA sections connected with seven amplifiers. (a) Radiating side; (b) Active side. (*After Casares-Miranda et al. [22], ©IEEE 2006.*)

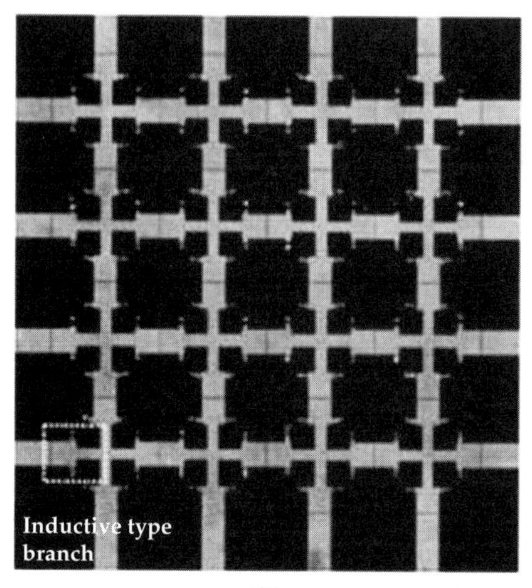

2D

FIGURE 6.8 4 × 4 2D NRI-TL-LWA. Unit cell layout is closely related to the NRI-TL microstrip implementation shown in Fig. 6.5*a*. (*After Allen et al. [23], ©IET 2007.*)

combined with circuit elements (such as varactors in [22]) to achieve added design flexibility. Figure 6.9*a* displays an implementation employing coplanar strip (CPS) technology for removing the need for via holes. The CPS NRI-TL unit cell was employed to develop a 1D LWA with reduced beam squinting [24]. By adjusting the slope and shape of the K-ω diagram, the LWA shown in Fig. 6.9*b* was designed to achieve 56° scanning over 1.8 GHz bandwidth, implying an average beam squint of 0.031°/MHz. At the center frequency (5 GHz), the LWA was adjusted to radiate toward 45°.

Another possible NRI-TL unit cell implementation is shown in Fig. 6.10 [25]. Since conventional microstrip implementations exhibit large conduction losses at higher millimeter wave frequencies, this unit cell was designed using open-walled rectangular waveguides by incorporating a dielectric resonator to achieve the NRI properties. Unit cell parameters were adjusted to achieve a balanced NRI structure for continuous backward to forward pattern scan. More specifically, the N = 30 unit cell antenna in Fig. 6.10 achieved 100° beam steering from 10.9 to 12 GHz. The gain values were measured to be in the range of 5.2 to 8.7 dB.

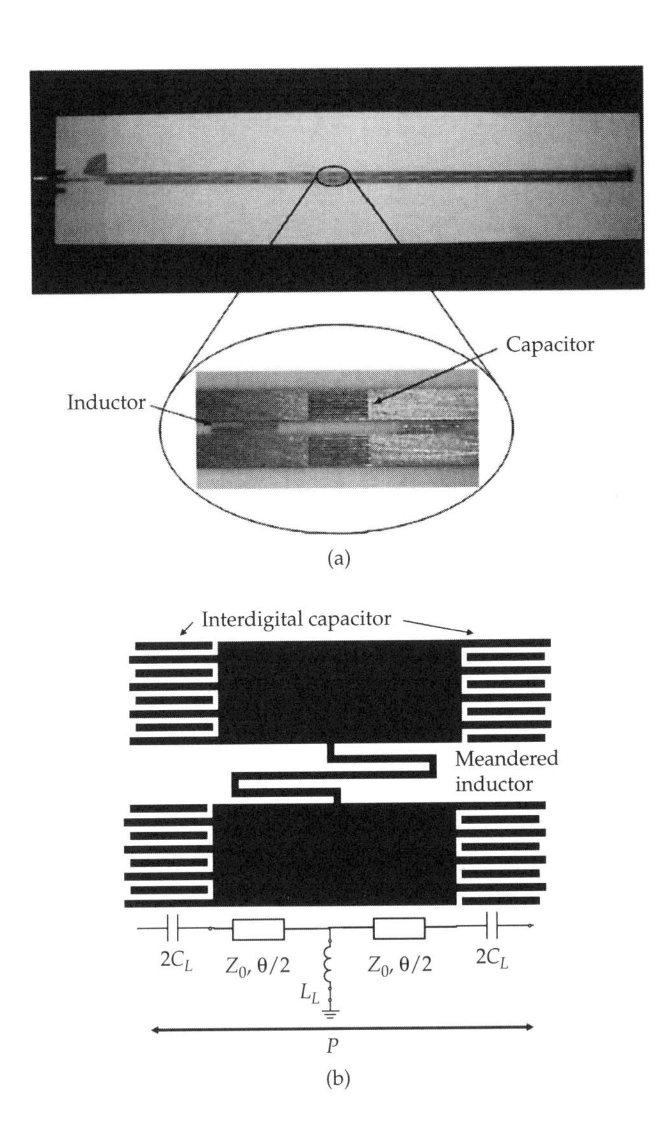

FIGURE 6.9 CPS technology realization of the NRI-TL to form a LWA; (a) N = 20 unit cell LWA designed for reduced beam squinting; an average of 0.031°/MHz beam scan was measured over 1.8 GHz bandwidth; (b) NRI-TL unit cell implemented in CPS technology. (*After Antoniades et al. [24], ©IEEE 2008.*)

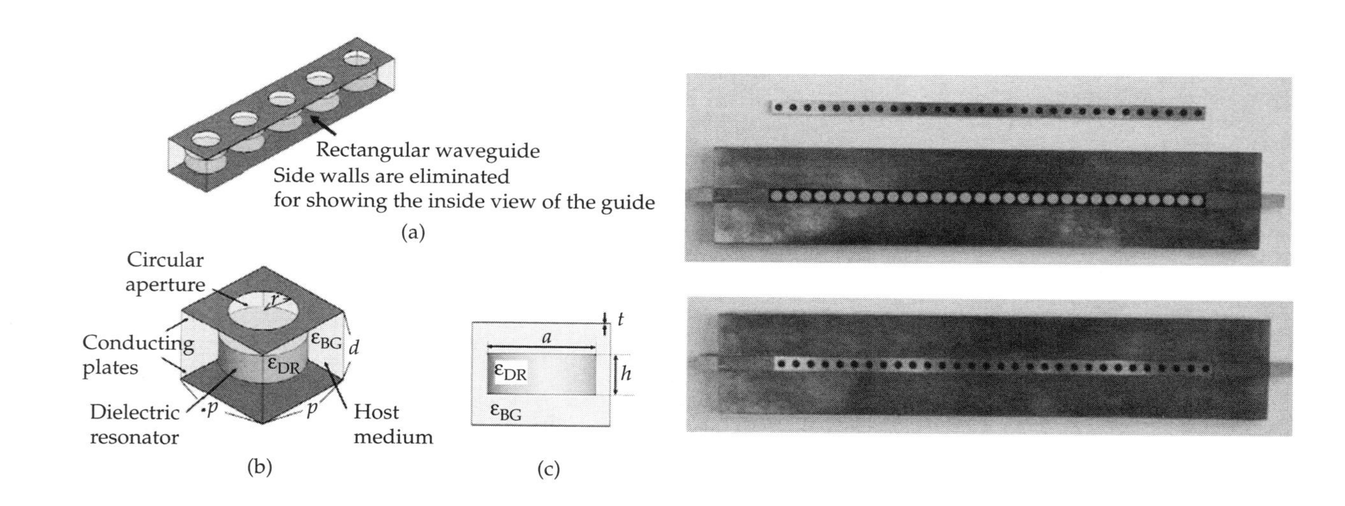

FIGURE 6.10 NRI-TL unit cell using a dielectric resonator; the N = 30 unit cell LWA achieved 100° beam steering from 10.9 to 12.0 GHz with gain values in the 5.2 to 8.7 dB range. (*After Ueda et al. [25], ©IEEE 2008.*)

6.3.2 Miniature and Multi-band Patch Antennas

Similar to conventional resonators, the ones based on NRI-TL unit cells must be multiples of half wavelength (i.e., $\ell = n\lambda_g/2 \rightarrow \beta = n\pi$, where λ_g = guided wavelength). Therefore, resonant modes of an N unit cell NRI-TL occurs at $K = \pi n/N$ frequencies (n being an integer smaller than N). As illustrated in Fig. 6.11*a*, the LH branch of the NRI-TL allows for negative resonances ($n < 0$) whereas the RH branch gives rise to the traditional ($n > 0$) resonances. Since the dual resonances ($\pm n$) exhibit similar field distributions and impedance characteristics (due to the identical K magnitude), the NRI-TL structures were exploited for a variety of multi-band antennas. Also, NRI-TL structures provide a convenient antenna miniaturization technique as their unit cells can be kept electrically small by employing interdigital or lumped reactive circuit elements. An interesting situation relates to the zeroth order resonance ($n = 0$) as the operational frequency at $K = 0$ becomes independent of the overall antenna physical length. In this case, to improve radiation efficiency and gain, the size of a zeroth order resonator (i.e., number of unit cells) can be increased without changing the bandwidth and impedance characteristics [19].

Figure 6.11*b* demonstrates a zeroth order antenna using $N = 4$ unit cells [16]. The unit cell layout is almost identical to that shown in Fig. 6.4*a*. However, a virtual ground capacitor is used to achieve a via-free design. The actual antenna [16] was implemented on a 31 mil thick $\varepsilon_r = 2.68$ substrate and had 10 mm ($p = 2.5$ mm) length. The measured return loss was -11 dB at the resonant frequency of 4.88 GHz. As compared to a patch antenna (having a maximum length of 20.6 mm) printed on the same substrate, this zeroth order antenna had 75% smaller footprint. The dual band ring antenna shown in Fig. 6.11*c* [8] also relied on a similar unit cell layout. However, the circularly periodic condition of such loop configurations support only even modes. Further, the common connection of the virtual grounds implied a floating potential for the zeroth order resonance. Hence, the $n = 0$ resonance was not supported using the shown antenna layout. Instead, the shown antenna supported $n = \pm 2$ resonances at 1.93 and 4.16 GHz using a straightforward single-feed mechanism. The antenna footprint diameter (30.6 mm) was $\lambda_0/5.1$ at 1.93 GHz and the corresponding measured gains were about 6 and -4 dB at 4.16 and 1.93 GHz, respectively.

To achieve further footprint reduction, antennas based on NRI-TL unit cells incorporating lumped capacitors were also reported [26]. For the antenna example in Fig. 6.12*a* [26], the inductive posts of the mushroom-shaped unit cell provided the needed LH loading, whereas series LH edge capacitances were increased with the lumped MIM

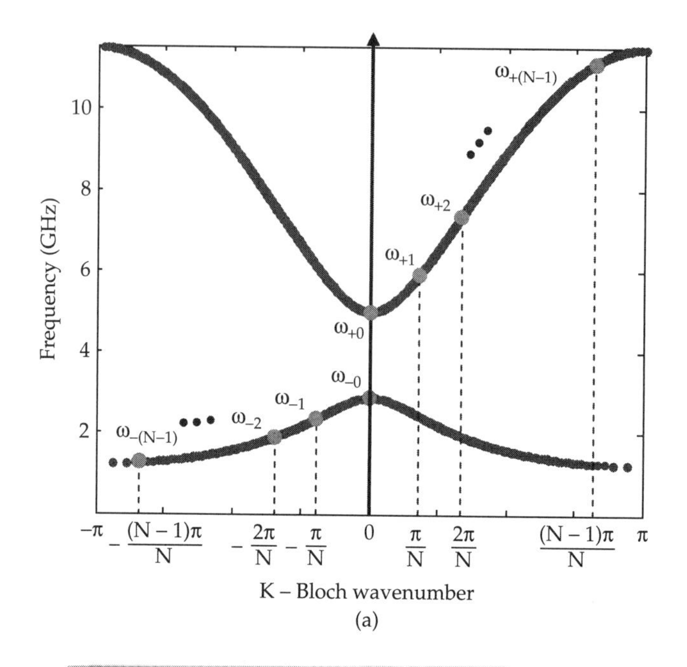

(a)

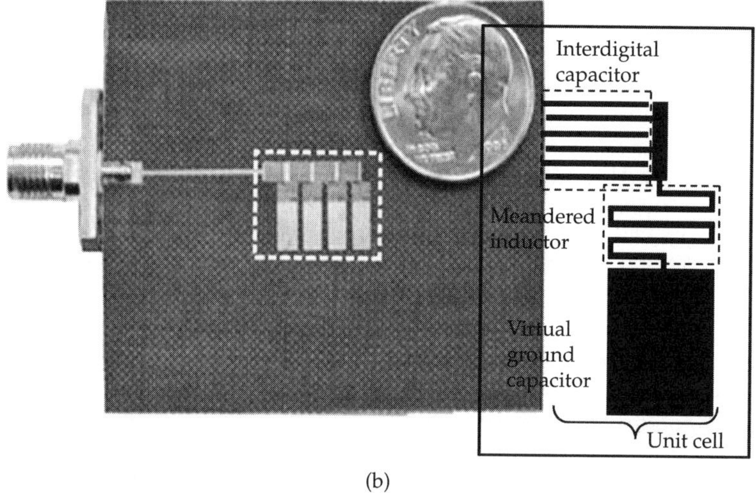

(b)

FIGURE 6.11 (a) Resonant modes of an unbalanced N unit cell NRI-TL structure; (b) N = 4 unit cell zeroth order NRI-TL antenna and its unit cell, (*After Sanada et al. [16], ©IEEE 2004.*); (c) Dual mode NRI-TL ring antenna. (*After Caloz and Itoh, ©J. Wiley & Sons, 2005.*)

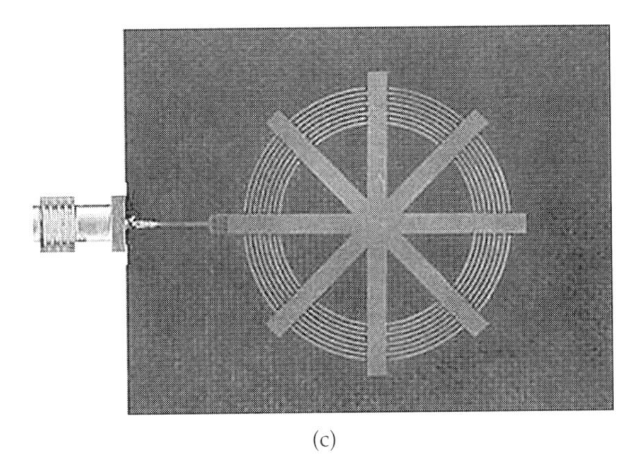

(c)

FIGURE 6.11 *Continued.*

capacitors. This unit cell configuration achieved miniaturization by shifting the K-ω branches to lower frequencies. The shown antenna in Fig. 6.12*a* was designed to work at the $n = -1$ resonance for broadside radiation. Although 3×1 unit cells were satisfactory to have the $n = -1$ resonance, a 3×3 unit cell configuration was considered to improve radiation aperture. The physical size of the footprint was $12.4 \times 12.4 \times 3.414$ mm corresponding to an electrical size of $\lambda_0/10 \times \lambda_0/10 \times \lambda_0/36$ at the resonance frequency of 2.42 GHz. This antenna had 1% $|S_{11}| < -10$ dB bandwidth, 3.3 dB gain, and 44% radiation efficiency.

Another interesting multi-band antenna design was carried out by partially filling a typical patch antenna with an NRI-TL structure [27]. The inclusion of the NRI-TL unit cells (RH + NRI-TL + RH resonator) within the regular patch allowed for design flexibility and tuning to realize the $n = \pm1$ and $n = 0$ resonances at desired frequencies. In addition, the antenna electrical size was reduced at the lowest frequency due to the presence of NRI-TL unit cells. As shown in Fig. 6.12*b*, this tri-band antenna included two mushroom-like NRI-TL unit cells within the patch metallization. The antenna resonated at 1.06 GHz (for GSM), 1.45 GHz (for navigation), and 2.16 GHz (for UMTS). The resonances exhibited 3% ($n = -1$), 3% ($n = 0$), and 15% ($n = 1$)$|S_{11}| < -6$ dB bandwidths, respectively. At the $n = \pm1$ resonances the patterns were dipole-like, whereas at the $n = 0$ mode the pattern was monopole-like. Measured gains and efficiencies were -3 dB and 17.8% for the $n = -1$ mode, 1 dB and 39% for the $n = 0$ mode, 6.5 dB and 82% for the $n = 1$ mode. The associated physical size of the antenna footprint was $42 \times 42 \times 10$ mm, corresponding to an electrical length of $\lambda_0/6.7$ at the lowest resonance frequency.

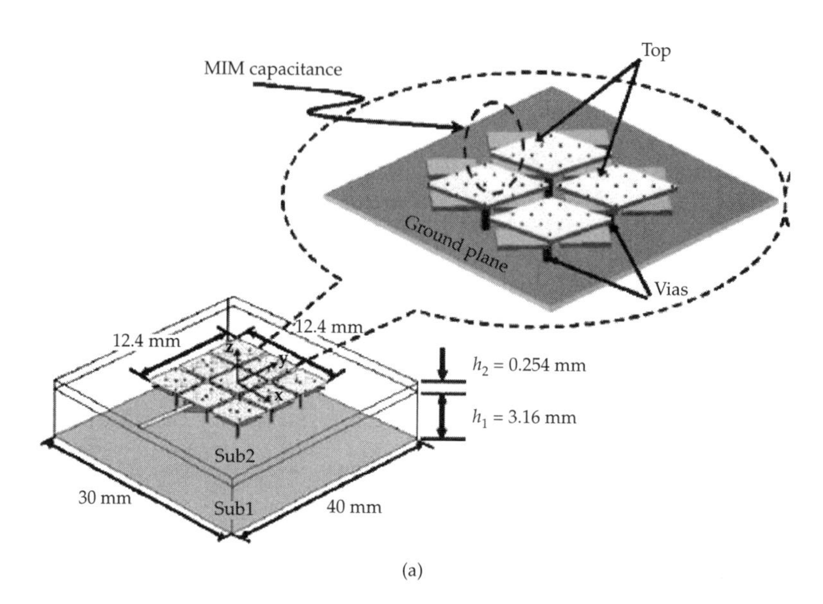

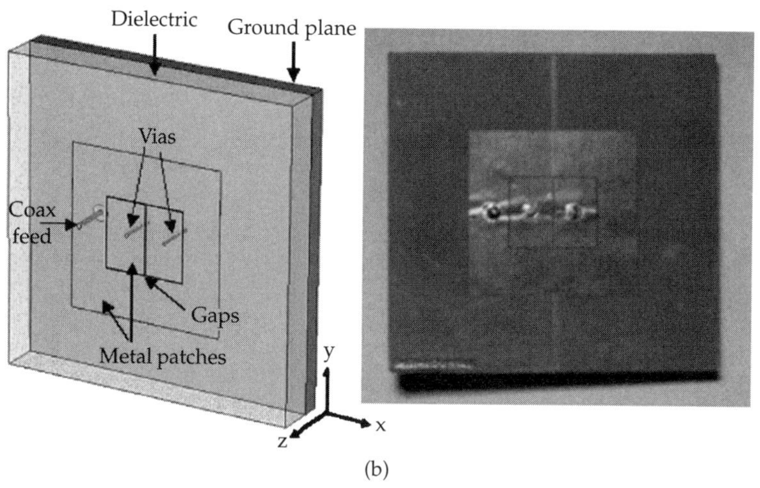

FIGURE 6.12 (a) Illustration of the 3 × 3 unit cell $n = -1$ NRI-TL antenna (*After Leong et al. [26], ©IEEE 2007*); (b) Tri-band patch antenna partially filled with NRI-TL unit cells (*After Herraiz-Martinez et al. [27], ©IEEE 2008.*) (c) Dual-band patch antenna partially filled with NRI-TL unit cells (*After Herraiz-Martinez et al. [27], ©IEEE 2008.*)

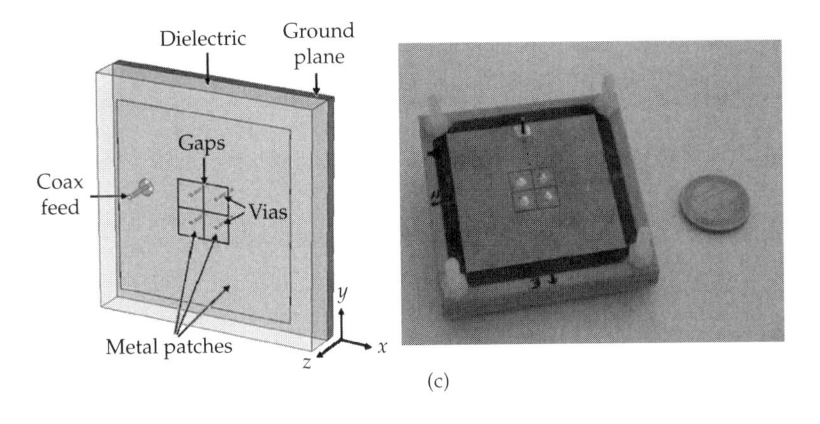

FIGURE 6.12 *Continued.*

In addition to the tri-band antenna, a dual mode patch operating at $n = \pm 1$ modes was also designed as shown in Fig. 6.12*c*. This dual band antenna employed a 2×2 unit cell arrangement to misalign the coax feed with the vias of the NRI-TL unit cells, and thus avoid excitation of the undesired $n = 0$ mode. It was noted that the antenna operated at 1.81 GHz (for DCS) and 2.2 GHz (for UMTS) with 7.2% and 5% $|S_{11}| < -6$ dB bandwidths, respectively. The corresponding measured gain and efficiency was 4.5 dB and 60% for the $n = -1$ mode, 6.8 dB and 83% for the $n = 1$ mode. The associated antenna physical size (footprint) was $48.2 \times 48.2 \times 8$ mm, implying an electrical length of $\lambda_0/3.4$ at the lowest resonance frequency.

6.3.3 Compact and Low-Profile Monopole Antennas

NRI metamaterials have also been considered for developing compact and low-profile electrically small monopole-like radiators. To demonstrate how monopoles can benefit from metamaterial structures, we consider the resonant mechanism of a folded monopole antenna. As depicted in Fig. 6.13, a traditional monopole usually has a $\lambda_0/4$ height to achieve in-phase radiating currents. However, a low-profile monopole (see Fig. 6.13) can be instead considered provided it is excited to support the common mode. Since a compact footprint is also desired, the $\sim\lambda/2$ phase shifter required for bringing the currents in-phase must be kept physically small. Using metamaterial based compact phase shifters, the monopoles can be made with smaller footprints and lower profiles.

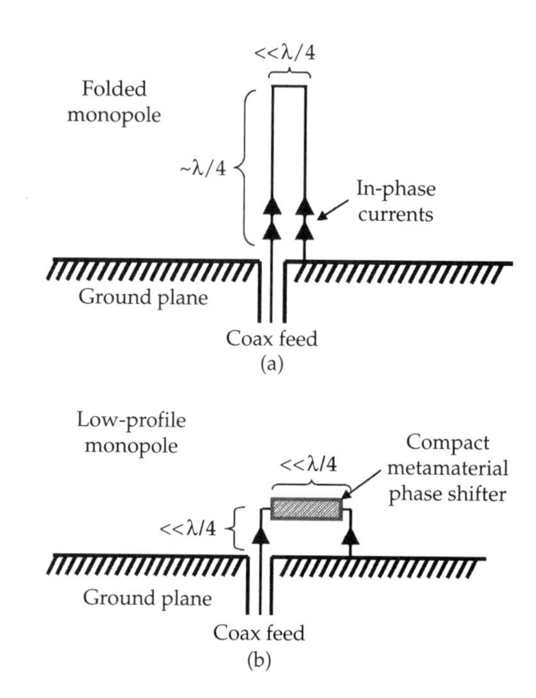

FIGURE 6.13 Traditional folded monopole and a low-profile version.

Folded monopoles supporting the common mode can be modeled through the equivalent circuit shown in Fig. 6.14*a* [28]. In this model, the electrically short vias carrying the in-phase currents are represented by radiation resistances R_R. The vias are also associated with lumped inductive loads represented by elements of value $2L_L$. The RH loads C_R and L_R are inherent to the host TL, whereas the LH capacitor C_L is specific to the unit cell implementation (e.g., edge capacitance in the mushroom-like layout) and inserted lumped loads. A straightforward even-odd mode analysis reveals that the current magnitudes are much stronger (larger than 100 times) in the even mode (i.e., in-phase) as compared to the odd-mode when a balanced NRI-TL unit cell structure (i.e., $L_R/C_R = L_L/C_L$) is used to feed the vias [28]. This implies that the NRI-TL structure acts as a good low-profile radiator. Unfortunately, as usual, short monopoles have very small input resistances (i.e., $R_R = 160\pi^2(h/\lambda_0)^2$, where h stands for height [29]). To increase the input resistance, a common approach is to employ multiple (M) folded arms and improve impedance matching through the relation $R_{in} \approx M^2 R_R$ [30].

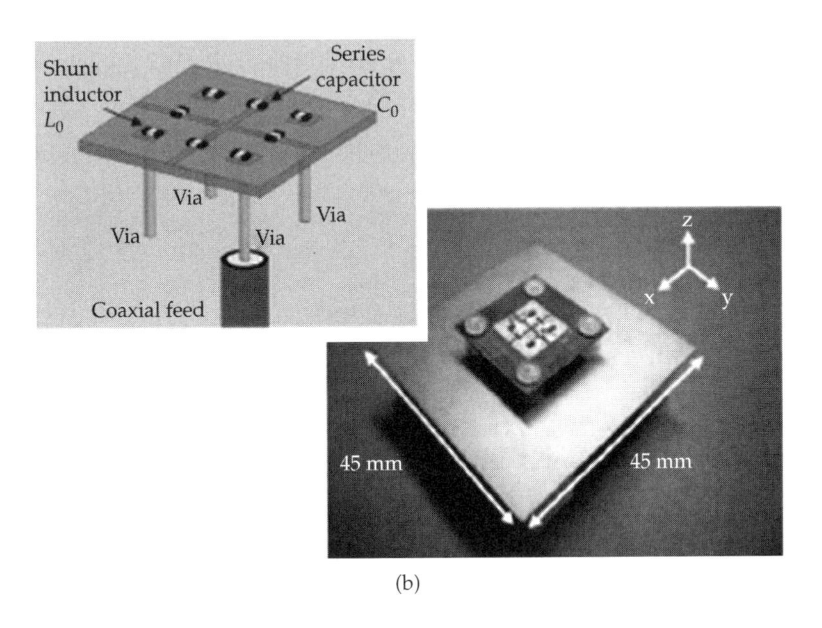

FIGURE 6.14 (a) NRI-TL loaded folded monopole circuit model; (b) Example of four-arm folded NRI-TL monopole; footprint is $\lambda_0/9.6 \times \lambda_0/9.6 \times \lambda_0/17.9$ at 3.1 GHz; the $|S_{11}| < -10$ dB bandwidth is 1.7%; the gain is 0.9 dB and radiation efficiency is 69.1% (*After Antoniades et al. [28], ©IEEE 2008.*)

Figure 6.14*b* demonstrates a compact and low-profile NRI-TL ring antenna consisting of four unit cells [28]. For this, the footprint was 10 mm ($\lambda_0/9.6$) × 10 mm($\lambda_0/9.6$) × 5.4 mm($\lambda_0/17.9$) at the resonance frequency of 3.1 GHz. The corresponding measured $|S_{11}| < -10$ dB bandwidth was reported to be 1.7% with 0.9 dB gain and 69.1% radiation efficiency. Another compact ring antenna is shown in Fig. 6.15*a* and took advantage of the NRI-TL structures to realize a $\lambda_0/11 \times \lambda_0/13 \times \lambda_0/28$ footprint area at 1.77 GHz [15]. Since this antenna included only two very short ($\lambda_0/28$) radiating vias, the impedance match was achieved by employing an asymmetric NRI-TL structure. The measured $|S_{11}| < -10$ dB bandwidth was reported [15] to be 6.8% with 0.95 dB gain and 54% radiation efficiency.

Alternatively, the low-profile monopoles introduced in [31] and [32] made use of compact 180° phase shifters to achieve a common mode in low-profile monopole realization. Specifically, the antenna shown in Fig. 6.15*b* consisted of a standard transmission line connected to two shorting posts loaded with lumped inductors [31]. The inductor choices of 10 nH and 5 nH resulted in a $\lambda_0/4.9 \times \lambda_0/45.6 \times \lambda_0/12.2$ footprint antenna operating at 2.19 GHz. Larger inductive loads were shown to further lower the resonance frequency. The antenna in Fig. 6.15*b* had −2.9 dB gain and 32% radiation efficiency with 4.7% $|S_{11}| < -10$ dB bandwidth. Another monopole-like miniature antenna is shown in Fig. 6.15*c*. It realized the 180° phase shift using an embedded spiral [32]. For this, the measured $|S_{11}| < -10$ dB bandwidth was about 1% and the antenna footprint was $\lambda_0/4.9 \times \lambda_0/45.6 \times \lambda_0/16$ footprint at about 2.4 GHz.

Typically, compact and low-profile monopoles suffer from radiation efficiency and narrow bandwidths. A NRI-TL based low-profile monopole with two nearby resonances was considered in [33]. As shown in Fig. 6.16*a*, the antenna is composed of two arms, each consisting of five unit cells. The unit cell layout of each arm is, however, slightly different to achieve resonance at two closely spaced frequencies. This antenna had 3.1% $|S_{11}| < -10$ dB bandwidth at 3.28 GHz with a $\lambda_0/4 \times \lambda_0/7 \times \lambda_0/29$ footprint. The corresponding antenna gain was 0.79 dB with 66% radiation efficiency. Another interesting design was reported in [34] and shown in Fig. 6.16*b*. This NRI-TL monopole resonates at a lower frequency (5.5 GHz) as compared to the unloaded one operating at 6.3 GHz. The NRI-TL loading also introduces a resonance around 3.55 GHz by exciting the ground plane. These two resonances allow for a wideband $|S_{11}| < -10$ dB impedance match from 3.15 to 6.99 GHz. The overall size of the antenna is 30 × 22 × 1.59 mm.

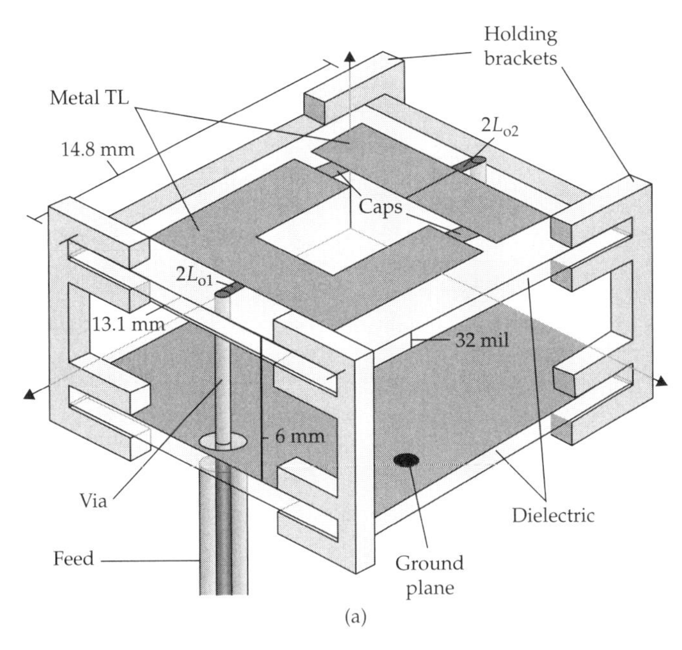

(a)

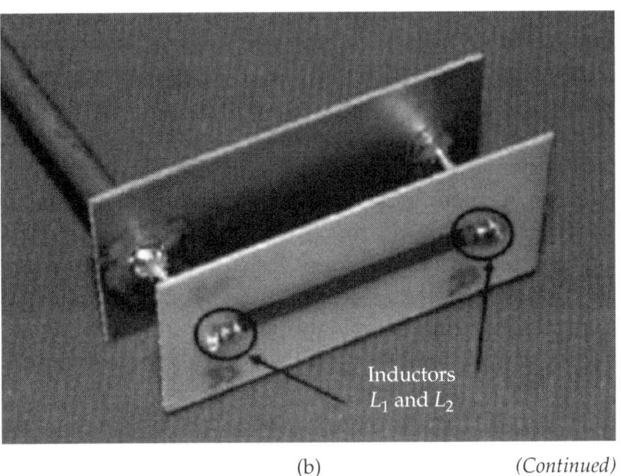

(b) *(Continued)*

FIGURE 6.15 (a) Low-profile NRI-TL ring antenna consisting of two asymmetric unit cells; footprint is $\lambda_0/11\times\lambda_0/13\times\lambda_0/28$ at 1.77 GHz. $|S_{11}| < -10$ dB bandwidth is 6.8%, gain is 0.95 dB, and radiation efficiency is 54% (*After Qureshi et al. [15], ©IEEE 2005.*); (b) Inductor loaded antenna having footprint of $\lambda_0/4.9\times\lambda_0/45.6\times\lambda_0/12.2$ at 2.19 GHz, $|S_{11}| < -10$ dB bandwidth is 4.7%, gain is −2.9 dB, and radiation efficiency is 32% (*After Lee et al. [31], ©IEEE 2007.*); (c) Miniature antenna with embedded spiral loading; footprint is $\lambda_0/4.9\times\lambda_0/45.6\times\lambda_0/16$ at 2.4 GHz. $|S_{11}| < -10$ dB bandwidth is 1% (*After Kokkinos et al. [32], ©IEEE 2007.*)

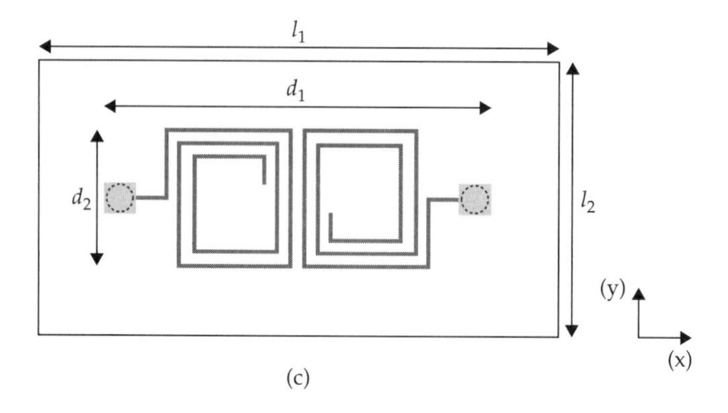

FIGURE 6.15 *Continued.*

6.3.4 Metamaterial-Inspired Antennas

As is well known, small dipoles are insufficient radiators due to high impedance match and their small radiation resistance. The small dipole must therefore be conjugately matched by means of a (possibly) bulky microwave network. To mitigate matching network requirements, a common approach is to form a self-resonant system by employing traditional miniaturization techniques such as meandering [2, 35] or material loading [3, 12] (see Chap. 3). Recently, it was demonstrated in [36] that electrically small resonant antennas can also be realized by enclosing small dipoles within metamaterial shells made up from negative permittivity ($\varepsilon < 0$) materials. It was shown that when properly designed, the $\varepsilon < 0$ shell acts as a distributed inductive element that matches to the highly capacitive impedance of the enclosed miniature antenna. The shell system in turn provides efficient radiation due to the purely real input impedance which can match to a specified source resistance.

Unfortunately, there are certain issues associated with practical realizations of such metamaterial shells. For example, tiny and highly conformal unit cells are required to form an electrically small metamaterial shell. In addition, the unit cells must have very low loss to achieve the high-radiation efficiency. Hence, for a more practical implementation, electrically small antennas were placed in close proximity of a metamaterial (e.g., $\varepsilon < 0$, or $\mu < 0$, or NRI) unit cell [37] (referred to as "metamaterial-inspired").

Figure 6.17*a* demonstrates an example of electrically small metamaterial-inspired antennas [37]. A small rectangular loop antenna was placed within close proximity of a parasitic capacitively

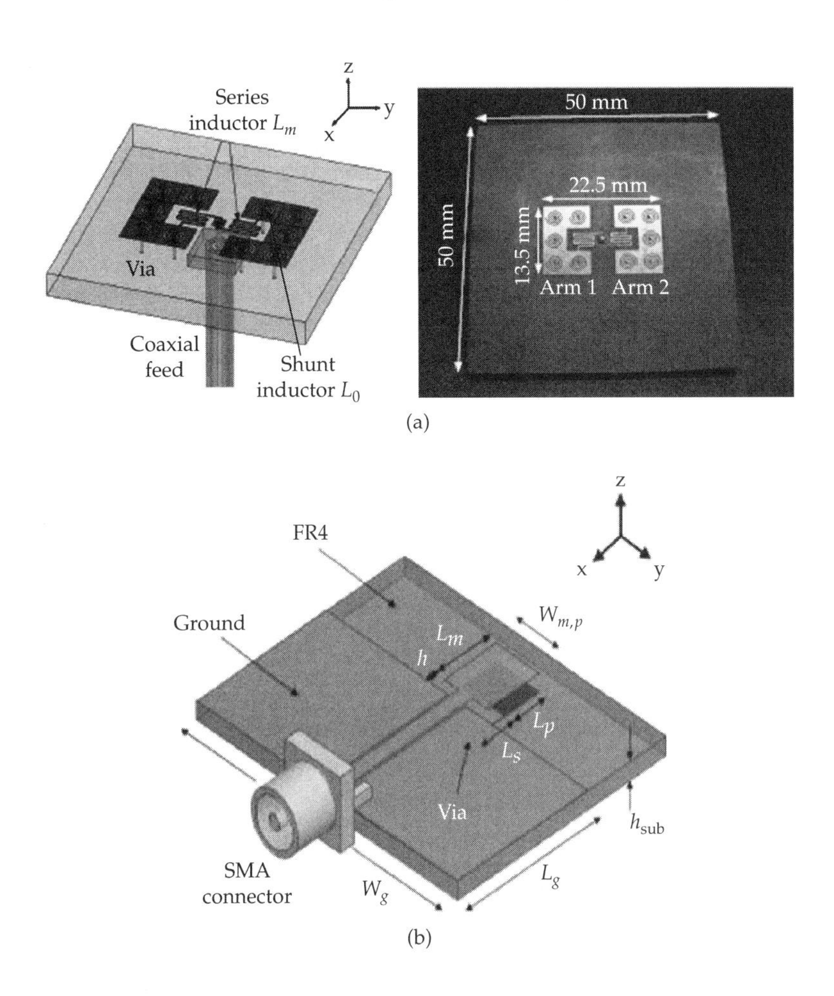

FIGURE 6.16 (a) NRI-TL monopole with each arm supporting slightly different resonances to enhance bandwidth; footprint is $\lambda_0/4 \times \lambda_0/7 \times \lambda_0/29$ at 3.28 GHz; 3.1% $|S_{11}| < -10$ dB bandwidth; 0.79 dB gain; and 66% radiation efficiency, (*After Zhu et al. [33], ©IEEE 2009.*); (b) Broadband dual-mode printed monopole antenna; 30 × 22 × 1.59 mm in size; $|S_{11}| < -10$ dB bandwidth from 3.15 to 6.99 GHz (*After Antoniades et al. [34], ©IEEE 2009.*)

loaded loop element to achieve 50 Ω impedance match. The capacitive load was realized with an interdigital capacitor. The antenna resonated at 438.1 MHz with a 0.43% $|S_{11}| < -10$ dB bandwidth. The antenna dimensions (excluding the ground plane) were $\lambda_0/9.4 \times \lambda_0/18 \times \lambda_0/867$. The computed gain value was 4.8 dB with 79.5%

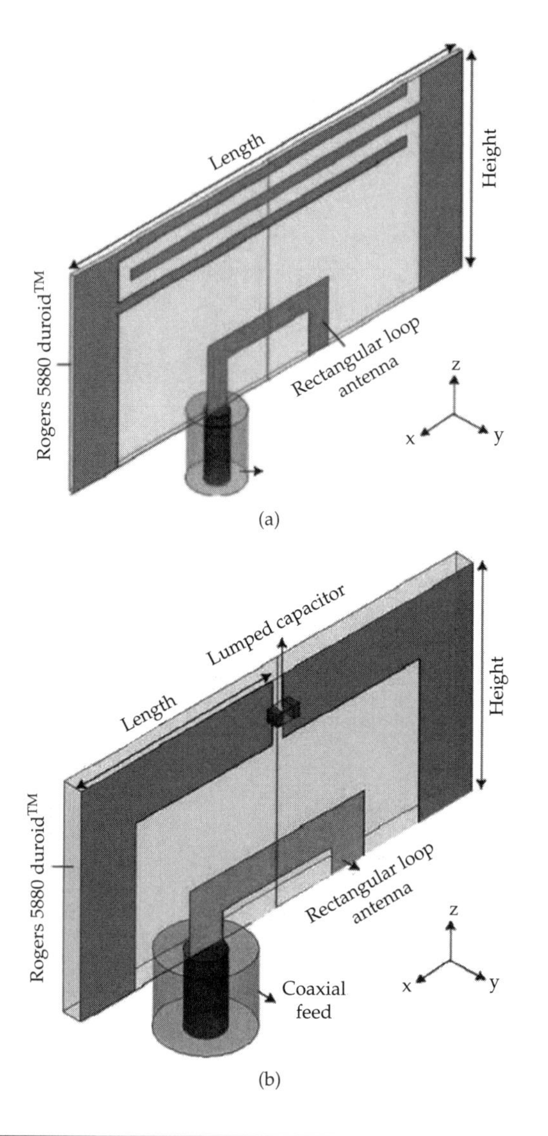

FIGURE 6.17 Metamaterial-inspired antennas: (a) Capacitively loaded loop based antenna; footprint is $\lambda_0/9.4\times\lambda_0/18\times\lambda_0/867$ at 438.1 MHz; 0.43% $|S_{11}| < -10$ dB bandwidth; 4.8 dB gain; and 79.5% radiation efficiency (*After Erentok et al. [37], ©IEEE 2008.*); (b) Capacitively loaded loop based antenna; footprint is $\lambda_0/9.2\times\lambda_0/17.9\times\lambda_0/833$ at 455.8 MHz; 0.5% $|S_{11}| < -10$ dB bandwidth; 5.36 dB gain; and 89.8% radiation efficiency (*After Erentok et al. [37], ©IEEE 2008.*); (c) Meander line based antenna; footprint is $\lambda_0/12\times\lambda_0/15.3\times\lambda_0/276$ at 1373 MHz; 1.3% $|S_{11}| < -10$ dB bandwidth; 1.03 dB gain; and 88% radiation efficiency (*After Erentok et al. [37], ©IEEE 2008.*); (d) Dipole loaded with multiple split ring resonators to achieve multi-band operation at 1.3 GHz and 2.8 GHz; footprint size is 42.05 × 16.4 × 0.5 mm. (*After Herraiz-Martinez et al. [38], ©IEEE 2008.*)

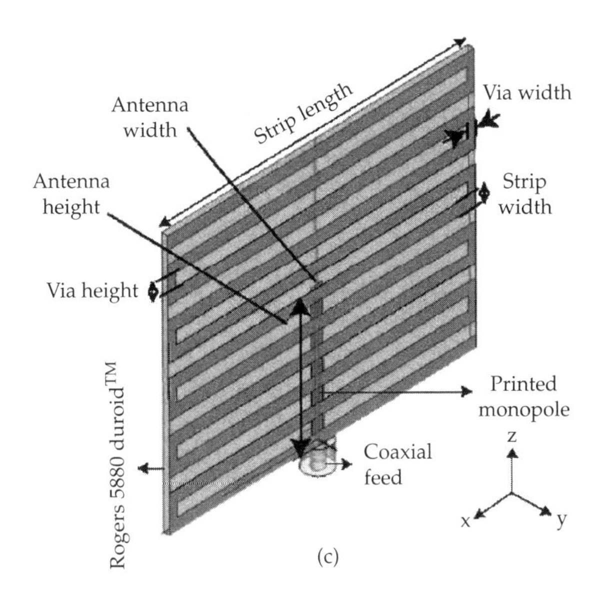

(d)

FIGURE 6.17 *Continued.*

efficiency. Figure 6.17*b* shows a slightly different version of the antenna when the interdigital capacitor was replaced with a lumped circuit element. In this case, electrical size and bandwidth of the antenna remained about the same with an improved 5.36 dB gain and 89.8% efficiency [37]. A different metamaterial-inspired antenna configuration that was introduced in [37] is shown in Fig. 6.17*c*. This design had dimensions of $\lambda_0/12 \times \lambda_0/15.3 \times \lambda_0/276$ at 1373 MHz. The computed $|S_{11}| < -10$ dB bandwidth was 1.3% with 1.03 dB gain and 88% efficiency. We note, however, that the antennas in Fig. 6.17*a* to *c* must be considered as whole structure that includes the large ground plane backing. As another example, we mention the dipole antenna shown in Fig. 6.17*d* [37]. This antenna was loaded with multiple split ring resonators to achieve multi-band operation at 1.3 and 2.8 GHz.

6.4 High-Gain Antennas Utilizing EBG Defect Modes

Electromagnetic band gap structures (EBGs) are artificial periodic media that can prevent electromagnetic wave propagation in specified directions and frequency bands (e.g., see the band gap regime of the 1D NRI-TL medium shown in Fig. 6.3*a*). Depending on the structural complexity (i.e., 1D, 2D, or 3D periodicity) and contrast difference of the constituent materials, EBGs can be designed to block propagation in single or multiple directions with relatively narrow or wide bandwidths [10]. Due to these exotic propagation properties and design flexibilities, EBGs have attracted considerable interest from the electromagnetic community even since their initial introduction [39]. Consequently, EBGs have found a wide range of RF applications and have been successfully employed to enhance performance of various devices such as filters, waveguides, and antennas [10, 40, 41].

EBGs are generally harnessed in several ways for antenna applications. Their band gap feature is utilized as high impedance ground planes [4] to enhance radiation performance (gain, pattern, reduced side/back lobes, etc.) by suppressing undesired surface waves and mutual couplings. In addition, EBGs provide in-phase reflection at their band gap frequencies and form artificial magnetic ground planes that lead to low-profile antenna realizations [42] (see Chap. 3 for EBG based antennas).

An important characteristic of EBGs are their capability of supporting localized electromagnetic modes within band gap frequencies when defects are introduced in their periodic structure (e.g., by replacing one or several unit cells with a different material) [10]. These localized modes are associated with narrow transmission regimes occurring within the band gap frequencies. Since transmission is sensitive to the propagation direction, an antenna embedded within the defect can radiate to free space only in particular directions. Therefore, very high-gain levels can be obtained with simpler structures. This approach can be an alternative to high gain 2D antenna arrays requiring complex and lossy feed networks [11,43–46].

More specifically, [45] and [46] investigated directivity enhancements of a dipole or patch antenna placed under multiple layer dielectric slabs and explained the concept as a leaky wave [47]. Although the multilayered structure was not referred to as a defect mode EBG at that time, it can be considered as one of the earliest high-gain EBG antennas. The fact that small antennas embedded in defect mode EBGs get converted into directive radiators was experimentally demonstrated in [11]. Figure 6.18*a* demonstrates a patch antenna located inside a defect mode EBG [48]. The EBG was constructed from rows of dielectric (alumina) rods separated by air. When the separation between the ground plane and rods are properly adjusted (according to the reflection phase of the EBG superstrate), the overall structure can be considered as having four rows of rods: two above the ground plane and two below (as a consequence of image theory). The defect is created by enlarging the separation between the second and third rows (i.e., larger gap between the second row and ground plane. A $\lambda_0/3 \times \lambda_0/3 \times \lambda_0/19.9$ patch antenna placed in the defect area resonates at 4.75 GHz and delivers 19 dB gain with the EBG size of $4.05\lambda_0 \times 4.05\lambda_0 \times 0.97\lambda_0$ and 50% aperture efficiency. The gain bandwidth (frequency interval over which gain stays within 3 dB of the peak value) was 3% and a minimum $|S_{11}|$ of -8 dB was measured.

To provide more control over the radiation properties, it was proposed in [49] to employ an EBG structure capable of exhibiting band gap in all propagation directions (i.e., 3D). As depicted in Fig. 6.18*b*, a woodpile EBG was placed above a patch or double slot antenna to enhance directivity. An EBG aperture of 15.2 × 15.2 cm (with total antenna thickness of 11.7 mm and a ground plane size of 30 × 30 cm) was employed to maximize the directivity of the patch feed operating near 12.5 GHz. The authors found that the double slot antenna exhibits about 1 dB better gain than the patch at the expense of a more complex feed mechanism. The measured gain was 19.8 dB with 1% $|S_{11}| < -10$ dB bandwidth. The structure had 63% radiation efficiency and 19% aperture efficiency.

Partially reflecting surfaces (PRS) or frequency selective surfaces (FSS) [50–52] are other alternatives of EBGs in forming resonator based high-gain antennas. In such resonators, confined EM energy between PRS and ground plane gradually leaks away and thus, a large radiation aperture supporting high levels of directivity is realized. Since PRS exhibit a reflection coefficient near to -1, it must be

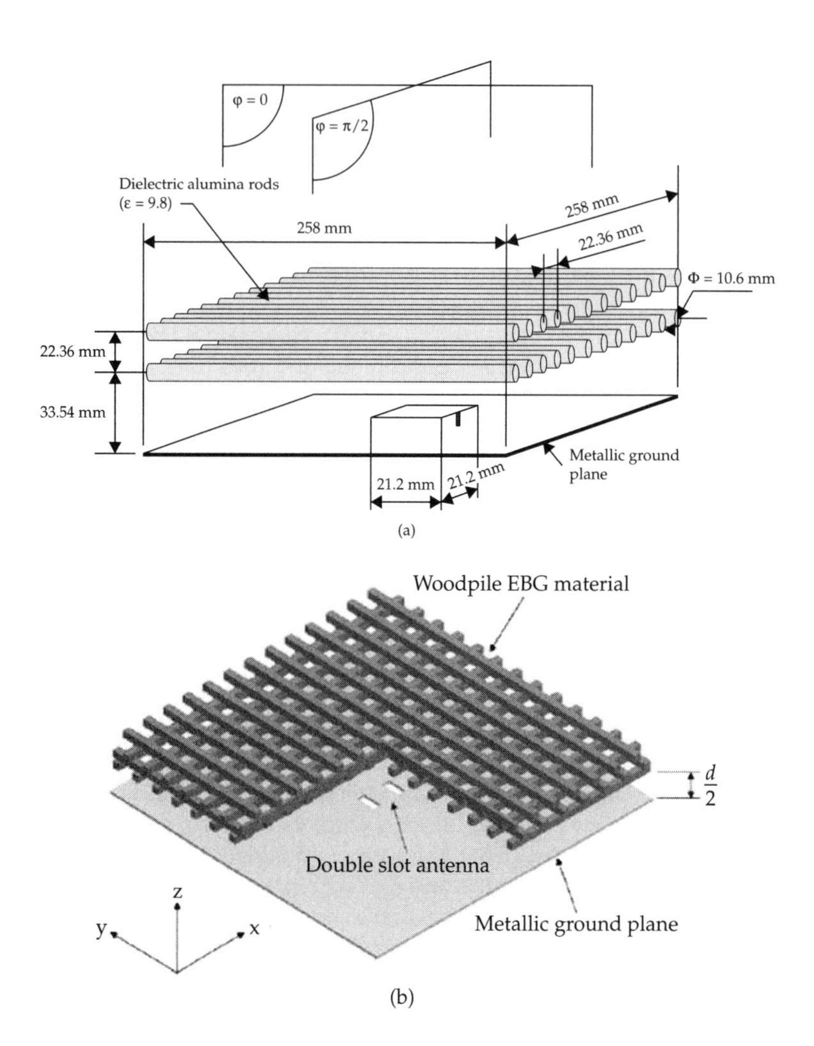

FIGURE 6.18 (a) Antenna placed inside a defect mode EBG made up from alumina rods delivering 19 dB gain, 50% aperture efficiency (*After Thevenot et al. [48], ©IEEE 1999.*); (b) Woodpile EBG resonator fed by a double slot antenna operating near 12.5 GHz; antenna has 19.8 dB gain, 63% radiation efficiency, and 19% aperture efficiency; EBG size: $6.3\lambda_0 \times 6.3\lambda_0 \times 0.49\lambda_0$ (*After Weily et al. [49], ©IEEE 2005.*)

placed about half wavelength ($\sim\lambda_0/2$) above the ground plane to form the resonator (see Fig. 6.19*a*). To lower the profile, a concept is to replace the ground plane with one that exhibit properties of a magnetic ground plane, that is, one that has a reflection coefficient near to +1. With this in mind, the paper in [53] proposed to employ an artificial magnetic ground plane (AMC) under the PRS. As shown in Fig. 6.19*b*, a patch antenna placed in the middle of the AMC excites the PRS-AMC cavity. For a simpler design, the AMC was designed from a via-free unit cell layout [54, 55] instead of a mushroom-like EBG structure. To form the cavity, the PRS in Fig. 6.19*c* was placed about a quarter wavelength (5.9 mm, $0.28\lambda_0$) above the AMC. The patch resonated at about 14.2 GHz with an electrical length of $0.116\lambda_0$ and 2% $|S_{11}| < -10$ dB bandwidth. The size of the AMC metallization was 15 cm ($7.1\lambda_0$) × 15 cm ($7.1\lambda_0$), whereas the PRS had 10.6 cm ($5\lambda_0$) × 10.6 cm ($5\lambda_0$) footprint. The measured directivity was 19.5 dB with a side lobe level below −15 dB.

EBG, PRS, or FSS based antennas exhibit high levels of gain as a consequence of their high Q resonance mechanisms. However, these antennas inherently suffer from narrow bandwidths since high Q resonances are very sensitive to frequency variations. On the other hand, multiband designs can be proposed to partially address the narrow bandwidth issue. For example, [56] proposed to insert an additional FSS surface inside an PRS-PEC resonator to create dual band operation at two closely spaced frequencies (5.15 and 5.7 GHz). As illustrated in Fig. 6.20*a*, an FSS can be designed to act as an AMC at a certain frequency (f_1), whereas it can be mostly transparent at another frequency (f_2). To realize dual resonance at closely spaced frequencies, [56] employed a highly selective FSS surface built from Hilbert curve inclusions [57]. At the experiment stage, the FSS was redesigned to reside right under the PRS in order to facilitate a simpler fabrication through the use of a single circuit board with both sides metalized. The antenna prototype shown in Fig. 6.20*b* employed 24.2 × 24.2 cm PRS/FSS superstrate and had a total height of 2.6 cm. The computed gain was 18 dB for the shown configuration.

Reference [58] presented a detailed analysis of dual band directivity enhancement using two different frequency selective surfaces printed on different sides of a thin dielectric layer. The patch-fed antenna configuration shown in Fig. 6.20*c* resonated at 8.2 and 11.7 GHz with 2.4% and 2.8% $|S_{11}| < -10$ dB bandwidth, respectively. The overall size of the structure was 12 × 12 × 1.6 cm. The antenna had 19.7 dB gain at 8.2 GHz and 21.4 dB gain at 11.7 GHz.

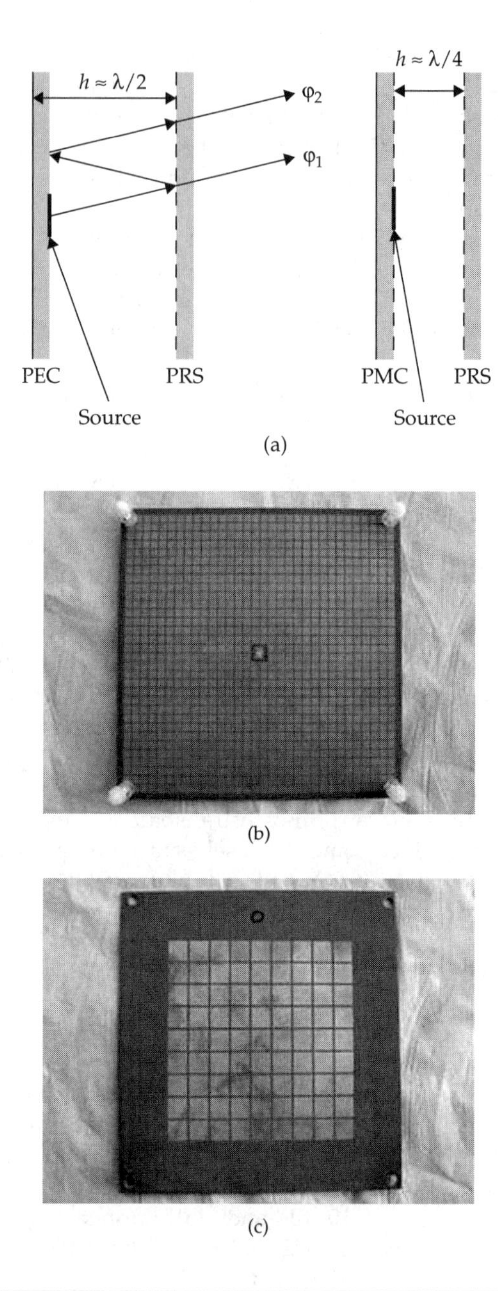

FIGURE 6.19 (a) Lowering the resonant cavity profile with PMC ground plane; (b) 15 cm ($7.1\lambda_0$) $\times$ 15 cm ($7.1\lambda_0$) AMC with a patch antenna feed; (c) 10.6 cm ($5\lambda_0$) $\times$ 10.6 cm ($5\lambda_0$) footprint PRS placed 5.9 mm ($0.28\lambda_0$) above the AMC. The antenna resonates at 14.2 GHz with 2% $|S_{11}| < -10$ dB bandwidth, 19.5 dB directivity, and < -15 dB side lobe level (*After Feresidis et al. [53], ©IEEE 2005.*)

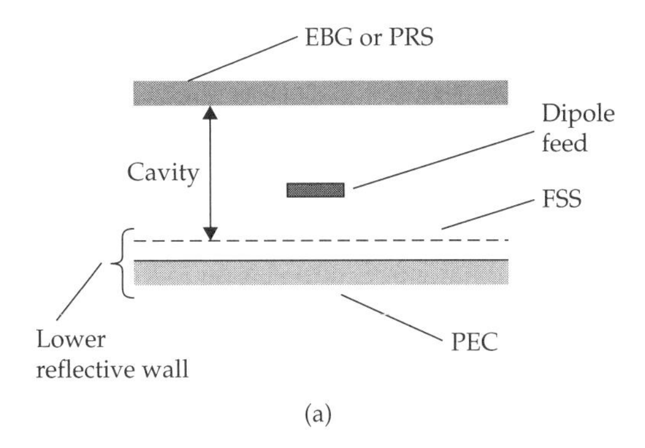

(a)

(b)

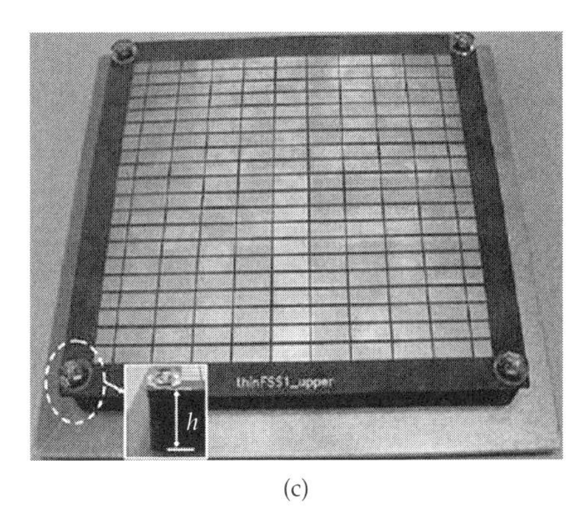

(c)

FIGURE 6.20 (a) Dual band resonant cavity antenna with an additional FSS layer (*After Rodes et al. [56], ©IEEE 2007.*); (b) 24.2 × 24.2 × 2.6 cm antenna resonating at 5.15 GHz and 5.7 GHz; 18 dB simulated gain 18 dB (*After Rodes et al. [56], ©IEEE 2007.*); (c) 12 × 12 × 1.6 cm dual band high gain antenna, 19.7 dB gain at 8.2 GHz with 2.4% $|S_{11}| < -10$ dB bandwidth, 21.4 dB gain at 11.7 GHz with 2.8% bandwidth (*After Lee et al. [58], ©IET 2007.*)

References

1. R. Porath, "Theory of miniaturized shorting-post microstrip antennas," *IEEE Transactions on Antennas and Propagation*, vol. 48, no. 1, January 2000, pp. 41–47.
2. J. H. Lu and K. L. Wong, "Slot-loaded meandered rectangular microstrip antenna with compact dual frequency operation," *Electronic Letters*, vol. 34, May 1998, pp. 1048–1050.
3. D. H. Schaubert, D. M. Pozar, and A. Adrian, "Effect of microstrip antenna substrate thickness and permittivity," *IEEE Transactions on Antennas and Propagation*, vol. 37, no. 6, June 1989, pp. 677–682.
4. D. Sievenpiper, L. Zhang, R. F. J. Broas, N. G. Alexopolous, and E. Yablonovitch, "High-impedance electromagnetic surfaces with a forbidden frequency band," *IEEE Transactions on Microwave Theory and Techniques*, vol. 47, no. 11, November 1999, pp. 2059–2074.
5. R. A. Shelby, D. R. Smith, and S. Schultz, "Experimental verification of a negative index of refraction," *Science*, vol. 292, 2001, pp. 77–79.
6. J. B. Pendry, A. J. Holden, D. J. Robbins, and W. J. Stewart, "Magnetism from conductors and enhanced nonlinear phenomena," *IEEE Transactions on Microwave Theory and Techniques*, vol. 47, no. 11, 1999, pp. 2075–2084.
7. A. Grbic and G. V. Eleftheriades, "Overcoming the diffraction limit with a planar left handed transmission line lens," *Physical Review Letters*, vol. 92, no. 11, March 2004, pp. 117403 1–117403 4.
8. C. Caloz and T. Itoh, *Electromagnetic Metamaterials: Transmission Line Theory and Microwave Applications*, John Wiley & Sons, New Jersey, 2006.
9. G. V. Eleftheriades and K. G. Balmain, *Negative–Refraction Metamaterial*, IEEE Press–John Wiley & Sons, New York, 2005.
10. J. Joannopoulos, R. Meade, and J. Winn, *Photonic Crystals—Molding the Flow of Light*, Princeton University Press, Princeton, N.J., 1995.
11. B. Temelkuran, M. Bayindir, E. Ozbay, R. Biswas, M. M. Sigalas, G. Tuttle, and K. M. Ho, "Photonic crystal based resonant antenna with a very high directivity," *Journal of Applied Physics*, vol. 87, no. 1, January 2000, pp. 603–605.
12. G. Kiziltas, D. Psychoudakis, J. L. Volakis, and N. Kikuchi, "Topology design optimization of dielectric substrates for bandwidth improvement of a patch antenna," *IEEE Transactions on Antennas and Propagation*, vol. 51, no. 10, October 2003, pp. 2732–2743.
13. V. G. Veselago, "The electrodynamics of substances with simultaneously negative values of permittivity and permeability," *Soviet Physics Uspekhi*, vol. 10, no. 4, 1968, pp. 509–514.
14. M. A. Antoniades and G. V. Eleftheriades, "Compact linear lead/lag metamaterial phase shifters for broadband applications," *IEEE Antennas and Wireless Propagation Letters*, vol. 2, no. 7, July 2003, pp. 103–106.
15. F. Qureshi, M. A. Antoniades, and G. V. Eleftheriades, "A compact and low-profile metamaterial ring antenna with vertical polarization," *IEEE Antennas and Wireless Propagation Letters*, vol. 4, 2005, pp. 333–336.
16. A. Sanada, M. Kimura, I. Awai, C. Caloz, and T. Itoh, "A planar zeroth-order resonator antenna using a left-handed transmission line," *34th European Microwave Conference*, Amsterdam, 2004.
17. L. Liu, C. Caloz, and T. Itoh, "Dominant mode (DM) leaky-wave antenna with backfire-to-endfire scanning capability," *Electronics Letters*, vol. 38, no. 23, November 2002, pp. 1414–1416.
18. R. E. Collin, *Field Theory of Guided Waves*. IEEE Press, New York, 1991.
19. C. Caloz, T. Itoh, and A. Rennings, "CRLH metamaterial leaky-wave and resonant antennas," *IEEE Antennas and Propagation Magazine*, vol. 50, no. 5, October 2008, pp. 25–39.
20. A. A. Oliner and D. R. Jackson, "Leaky-wave antennas," in *Antenna Engineering Handbook*, J. L. Volakis, ed. McGraw Hill, New York, 2007, ch. 11.
21. S. Lim, C. Caloz, and T. Itoh, "Metamaterial-based electronically controlled transmission-line structure as a novel leaky-wave antenna with tunable

radiation angle and beamwidth," *IEEE Transactions on Microwave Theory and Techniques*, vol. 53, no. 1, January 2005.
22. F. P. Casares-Miranda, C. Camacho-Penalosa, and C. Caloz, "High-gain active composite right/left-handed leaky-wave antenna," *IEEE Transactions on Antennas and Propagation*, vol. 54, no. 8, August 2006.
23. C. A. Allen, K. M. K. H. Leong, and T. Itoh, "Design of a balanced 2D composite right-/left-handed transmission line type continuous scanning leaky-wave antenna," *IET Microwaves, Antennas and Propagation*, vol. 1, no. 3, June 2007, pp. 746–750.
24. M. A. Antoniades and G. V. Eleftheriades, "A CPS leaky-wave antenna with reduced beam squinting using NRI-TL metamaterials," *IEEE Transactions on Antennas and Propagation*, vol. 56, no. 3, March 2008, pp. 708–721.
25. T. Ueda, N. Michishita, M. Akiyama, and T. Itoh, "Dielectric-resonator-based composite right/left-handed transmission lines and their application to leaky wave antenna," *IEEE Transactions on Microwave Theory and Techniques*, vol. 56, no. 10, October 2008, pp. 2259–2269.
26. K. M. K. H. Leong, C. J. Lee, and T. Itoh, "Compact metamaterial based antennas for MIMO applications," *International Workshop on Antenna Technology (IWAT)*, March 2007, pp. 87–90.
27. F. J. Herraiz-Martinez, V. Gonzalez-Posadas, L. E. Garcia-Munoz, and D. Segovia-Vargas, "Multifrequency and dual-mode patch antennas partially filled with left-handed structures," *IEEE Transactions on Antennas and Propagation*, vol. 56, no. 8, August 2008.
28. M. A. Antoniades and G. V. Eleftheriades, "A folded monopole model for electrically small NRI-TL metamaterial antennas," *IEEE Antennas and Wireless Propagation Letters*, vol. 7, 2008.
29. S. A. Schelkunoff and H. T. Friis, *Antennas: Theory and Practice*, John Wiley and Sons, Inc., New York, 1952, pp. 309–309.
30. S. R. Best, "The performance properties of electrically small resonant multiple-arm folded wire antennas," *IEEE Antennas and Propagation Magazine*, vol. 47, no. 4, August 2005, pp. 13–27.
31. D. H. Lee, A. Chauraya, J. C. Vardaxoglou, and W. S. Park, "Low frequency tunable metamaterial small antenna structure," *The Second European Conference on Antennas and Propagation (EuCap)*, Edinburgh, UK, November 2007.
32. T. Kokkinos, A. P. Feresidis, and J. L. Vardaxoglou, "A low-profile monopole-like small antenna with embedded metamaterial spiral-based matching network," *The Second European Conference on Antennas and Propagation (EuCap)*, Edinburgh, UK, November 2007.
33. J. Zhu and G. V. Eleftheriades, "A compact transmission-line metamaterial antenna with extended bandwidth," *IEEE Antennas and Wireless Propagation Letters*, vol. 8, 2009, pp. 295–298.
34. M. A. Antoniades and G. V. Eleftheriades, "A broadband dual-mode monopole antenna using NRI-TL metamaterial loading," *IEEE Antennas and Wireless Propagation Letters*, vol. 8, 2009, pp. 258–261.
35. S. R. Best, "The radiation properties of electrically small folded spherical helix antennas," *IEEE Transactions on Antennas and Propagation*, vol. 52, no. 4, April 2004, pp. 953–960.
36. R. Ziolkowski and A. Erentok, "Metamaterial-based efficient electrically small antennas," *IEEE Transactions on Antennas and Propagation*, vol. 54, no. 7, July 2006, pp. 2113–2129.
37. A. Erentok and R. W. Ziolkowski, "Metamaterial-inspired efficient electrically small antennas," *IEEE Transactions on Antennas and Propagation*, vol. 56, no. 3, March 2008, pp. 691–707.
38. F. J. Herraiz-Martinez, L. E. Garcia-Munoz, V. Gonzalez-Posadas, and D. Segovia-Vargas, "Multi-frequency printed dipoles loaded with metamaterial particles," *14th Conference on Microwave Techniques (COMITE)*, Prague, Czech Republic, April 2008.
39. E. Yablonovich, "Inhibited spontaneous emission in solid-state physics and electronics," *Physical Review Letters*, vol. 58, 1987, p. 2059.

40. Y. J. Lee, J. Yeo, R. Mittra, and W. S. Park, "Application of electromagnetic bandgap (EBG) superstrates with controllable defects for a class of patch antennas as spatial angular filters," *IEEE Transactions on Antennas and Propagation*, vol. 53, no. 1, January 2005, pp. 224–235.
41. A. Mekis, J. C. Chen, I. Kurland, S. Fan, P. R. Villeneuve, and J. D. Joannopoulos, "High transmission through sharp bends in photonic crystal waveguides," *Physical Review Letters*, vol. 77, no. 18, October 1996, pp. 3787–3790.
42. F. Yang and Y. Rahmat-Samii, "A low-profile circularly polarized curl antenna over an electromagnetic bandgap (EBG) surface," *Microwave and Optical Technology Letters*, vol. 31, no. 4, November 2001, pp. 264–267.
43. R. Biswas, E. Ozbay, B. Temelkuran, M. Bayindir, M. M. Sigalas, and K. M. Ho. "Exceptionally directional sources with photonic–bandgap crystals," *Journal of the Optical Society of America B*, vol. 18, no. 6.11, November 2001, pp. 1684–1689.
44. C. Cheype, C. Serier, M. Thevenot, T. Monediere, A. Reineix, and B. Jecko, "An electromagnetic bandgap resonator antenna," *IEEE Transactions on Antennas and Propagation*, vol. 50, no. 9, September 2002, pp. 1285–1290.
45. D. R. Jackson and N. G. Alexopoulos, "Gain enhancement methods for printed circuit antennas," *IEEE Transactions on Antennas and Propagation*, vol. 33, no. 9, September 1985, pp. 976–987.
46. H. Y. Yang and N. G. Alexopoulos, "Gain enhancement methods for printed circuit antennas through multiple substrates," *IEEE Transactions on Antennas and Propagation*, vol. 35, no. 7, July 1987, pp. 860–863.
47. D. R. Jackson and A. A. Oliner, "Leaky-wave analysis of the high-gain printed antenna configuration," *IEEE Transactions on Antennas and Propagation*, vol. 36, 1988, pp. 905–910.
48. M. Thevenot, C. Cheype, A. Reineix, and B. Jecko, "Directive photonic-bandgap antennas," *IEEE Transactions on Microwave Theory and Techniques*, vol. 47, no. 11, 1999, pp. 2115–2122.
49. A. R. Weily, L. Horvath, K. P. Esselle, B. C. Sanders, and T. S. Bird, "A planar resonator antenna based on a woodpile EBG material," *IEEE Transactions on Antennas and Propagation*, vol. 53, no. 1, January 2005, pp. 216–222.
50. B. A. Munk, *Frequency Selective Surfaces: Theory and Design*. John Wiley & Sons, Inc., Hoboken, New Jersey, 2000.
51. B. A. Munk, *Finite Antenna Arrays and FSS*. John Wiley & Sons, Inc., Hoboken, New Jersey, 2000.
52. R. Sauleau, "Fabry-Perot resonators," *Encyclopedia of RF and Microwave Engineering*, John Wiley, Hoboken, N. J., 2008.
53. A. F. Feresidis, G. Goussetis, S. Wang, and J. C. Vardaxoglou, "Artificial magnetic conductor surfaces and their application to low-profile high-gain planar antennas," *IEEE Transactions on Antennas and Propagation*, vol. 53, no. 1, January 2005, pp. 209–215.
54. Y. E. Erdemli, K. Sertel, R. A. Gilbert, D. E. Wright, and J. L. Volakis, "Frequency-selective surfaces to enhance performance of broad-band reconfigurable arrays," *IEEE Transactions on Antennas and Propagation*, vol. 50, no. 12, December 2002, pp. 1716–1724.
55. Y. Zhang, J. von Hagen, M. Younis, C. Fischer, and W. Wiesbeck, "Planar artificial magnetic conductors and patch antennas," *Special Issue on Metamaterials, IEEE Transactions on Antennas and Propagation*, vol. 51, no. 10, October 2003, pp. 2704–2712.
56. E. Rodes, M. Diblanc, E. Arnaud, T. Monédière, and B. Jecko, "Dual-band EBG resonator antenna using a single-layer FSS," *IEEE Antennas and Wireless Propagation Letters*, vol. 6, 2007, pp. 368–371.
57. J. McVay, N. Engheta, and A. Hoorfar, "High impedance metamaterial surfaces using Hilbert-curve inclusions," *IEEE Microwave and Component Letters*, vol. 14, no. 3, March 2004.
58. D. H. Lee, Y. J. Lee, J. Yeo, R. Mittra, and W. S. Park, "Design of novel thin frequency selective surface superstrates for dual-band directivity enhancement," *IET Microwaves, Antennas and Propagation*, vol. 1, no. 1, February 2007, pp. 248–254.

CHAPTER 7

Antenna Miniaturization Using Magnetic Photonic and Degenerate Band Edge Crystals

Gokhan Mumcu, Kubilay Sertel, and John L. Volakis

7.1 Introduction

It is well recognized that extraordinary propagation properties offered by engineered metamaterials is the new frontier for achieving the optimal performance in RF devices [1]. Undoubtedly, such engineered material mixtures and modifications also provide unique and highly sought advantages in antenna applications. For example, ferrites [2,3], loaded ferrites, and ferroelectrics [4] have already been exploited for phase shifters [5], antenna miniaturization, and beam control. As already discussed in Chap. 6, an extensive literature exists on negative refractive index (NRI) and electromagnetic band gap (EBG) metamaterials. NRI metamaterials has allowed for sub-wavelength focusing and greater sensitivity in lens applications [6–8]. Zeroth order resonances found in printed circuit NRI realizations have been utilized to develop various miniature devices such as phase shifters and antennas [9–11]. Similarly, controllable dispersion properties of EBG structures [12,13] have been employed to form shorter waveguide, smaller resonator, and higher Q filter designs [9–11]. It was already discussed in Chap. 6 that defect mode EBG structures can transform small radiators into directive antennas [14,15]. Also, their forbidden propagation bands have been exploited to form high impedance ground planes to improve antenna radiation properties [16–19]. Further, antenna substrates formed by mixing different dielectric materials and frequency selective surfaces have been found successful in improving the gain of printed antennas and arrays [20,21].

A key goal with metamaterials is to reduce wave velocity or even reverse it in some frequency regions to achieve an electrically small RF structure (see Chap. 6, [9–11],[22–25]). In this context, recently introduced metamaterial concepts based on periodic assemblies of layered anisotropic media (and possibly magnetic materials) hold a great promise for novel devices. Such assemblies forming magnetic photonic (MPC) and degenerate band edge (DBE) crystals display higher order dispersion (K-ω) diagrams. Therefore, they support new resonant modes and degrees of freedom (in essence volumetric anisotropy) to optimize antenna performance [26,27]. For instance, MPCs display one-way transparency (unidirectionality) and wave slowdown due to the stationary inflection points (SIP) within their K-ω diagrams [28]. In addition, they exhibit faster vanishing rate of the group velocity around the vicinity of SIP frequency [as compared to the regular band edges (RBE) of ordinary crystals where zero group velocities occur]. This leads to stronger resonances suitable for high-gain antenna apertures. DBE crystals are realized by removing the magnetic layers of MPCs and lead to higher order band edge resonances (fourth vs. second in RBEs). This provides thinner resonators suitable for high-gain conformal antennas [27]. It is shown in this chapter that utilizing the MPC and DBE dispersions on uniform microwave substrates (in contrast to layered media) may enable their unique properties for novel printed devices.

In this chapter, we first describe the fundamental electromagnetic properties of DBE and MPC crystals. Subsequently, we demonstrate that small antennas embedded in MPC and DBE crystals are converted into directive radiators due to the presence of higher order K-ω resonances. Specifically, a finite DBE assembly is realized and shown to exhibit large aperture efficiency for conformal high-gain antenna applications.

The second half of the chapter presents a coupled transmission line concept capable of emulating DBE and MPC dispersions on traditional microwave (printed circuit board) substrates. Lumped circuit models are also given to provide design flexibilities via lumped element inclusions. Several printed antenna examples are given to verify miniaturization using the coupled line concept. These miniaturized DBE and MPC antennas are designed to be comparable or better than recently published metamaterial antennas.

7.2 Slow Wave Resonances of MPC and DBE Crystals

A key property of any photonic crystal (i.e., EBG structure) is their support for slow wave modes, that is, modes associated with vanishing group velocities ($\partial\omega/\partial K \rightarrow 0$). Typically, photonic crystals display these slow waves as the operational frequency approaches the forbidden propagation bands. Characteristics of MPC and DBE crystals [26–28] is the intercoupling of modes to generate new modes having slower group velocities (see Fig. 7.1). These novel slow wave modes can then be harnessed for miniature and directive antennas [29–32]. The behavior of the K-ω curve (see Chap. 6 for the definition of K-ω diagram) can conveniently be approximated with polynomial relations around the K-ω points (see Fig. 7.1) displaying the slow group velocity modes [33]. In general, photonic crystals constructed of periodic layers of isotropic materials exhibit second order band edges. In contrast, as depicted in Fig. 7.1, a DBE mode is associated with a fourth order K-ω curve $[\omega'(K) = \omega''(K) = \omega'''(K) = 0]$ and a MPC medium is associated with an asymmetric dispersion relation [i.e., $\omega(K) \neq \omega(-K)$]. In addition, the MPC medium can exhibit a third order polynomial K-ω curve having a stationary inflection point (SIP) $[\omega'(K) = \omega''(K) = 0, \omega'''(K) \neq 0]$ [28].

In contrast to the slow mode resonances observed in RBE or DBE mode supporting crystals, SIPs occur away from band edge frequencies. Hence, semi-infinite MPC slabs can exhibit a large transmittance ($> 90\%$) for incoming RF pulses. As illustrated in Fig. 7.1*c*, once the incoming RF pulse penetrates into the MPC medium; it significantly slows down, gets spatially compressed, and concurrently attains large field amplitude. Due to these phenomena, SIP region of the band diagram is also referred to as the frozen mode regime [26–28, 35, 36].

DBE crystals achieve better transmittance, and amplitude increase (equivalently higher Q resonances) as opposed to RBE crystals due to their flatter higher order band edge [27]. Specifically, semi-infinite DBE slabs are better matched at the dielectric interface experiencing a transmittance decay rate of $\Delta\omega_d^{1/4}$. In comparison, the transmittance decay rate around the RBE is on the order of $\Delta\omega_d^{1/2}$($\Delta\omega_d$ is the difference between the band edge and actual frequency of operation). Also, the group velocity decreases more quickly in the DBE ($\Delta\omega_d^{3/4}$)

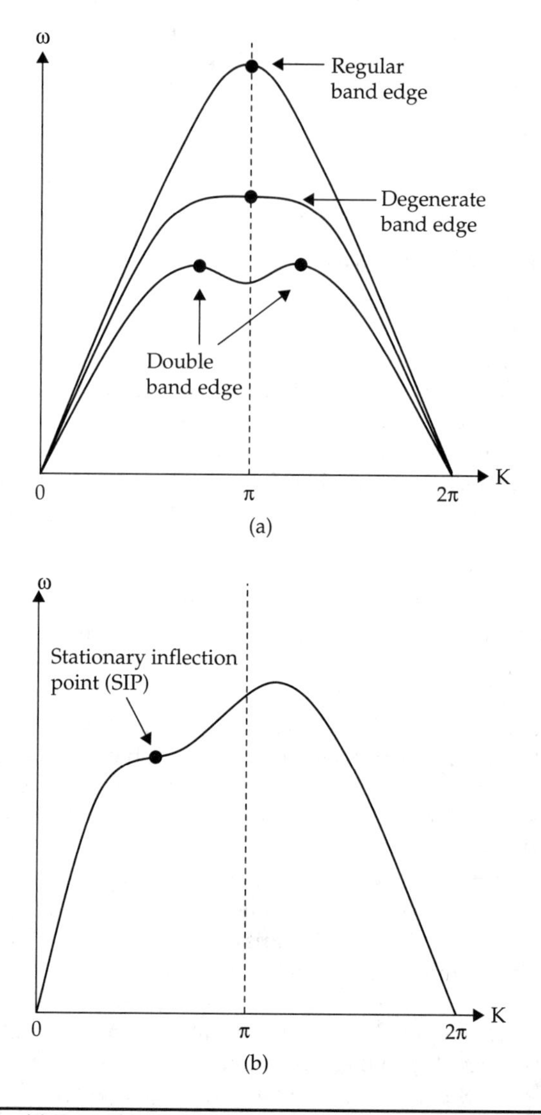

FIGURE 7.1 (a) Typical band edge and K-ω curves realized by layered anisotropic media (bottom K-ω branches are not shown for simplicity); DBE mode is associated with a fourth order maximally flat K-ω curve, whereas the RBE and double band edge (DbBE) modes are second order band edges; (b) MPC media exhibiting a third order K-ω curve with an SIP; (c) Illustration of a wave impingent on the MPC, almost 100% of the wave is converted into a frozen mode with near zero group velocity. (*After Mumcu et al. [34,35], ©IEEE 2009.*)

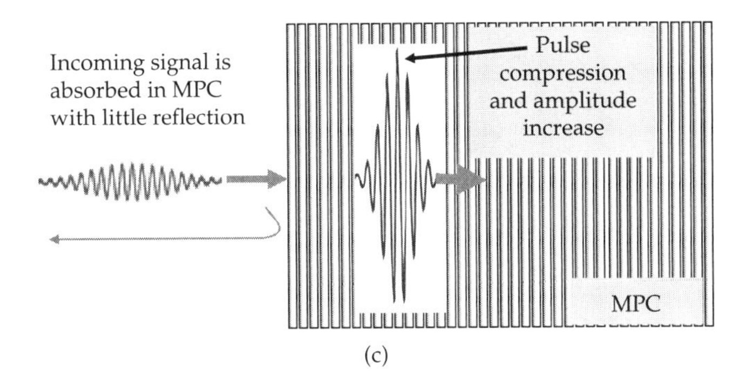

FIGURE 7.1 *Continued.*

medium as compared to the RBE crystal ($\Delta\omega_d^{1/2}$). A consequence of this behavior is a higher amplitude increase in the vicinity of the DBE and SIP modes. Equally important is that the DBE and MPC crystals can be thinner to achieve the same amplitude increase (as compared to RBE). They can therefore be potentially exploited for improved antenna gain, matching, and miniaturization [36]. In the following sections, we demonstrate how higher order K-ω resonances of the MPC and DBE crystals can be harnessed to develop high gain and miniature antennas.

7.3 High Gain Antennas Embedded Within Finite Thickness MPC and DBE Crystals

7.3.1 Transmission Characteristics of Finite Thickness MPCs

The simplest possible MPC structure displaying an asymmetric dispersion relation is shown in Fig. 7.2*a* [26]. The unit cell is composed of two identical misaligned anisotropic dielectric layers (the A layers) and a ferromagnetic layer (the F layer). The A layers can be conveniently realized using naturally available uniaxial crystals, such as "rutile" [36]. Engineered anisotropic layers [29,37] or metallic textures [38] have also been shown to provide cost-effective A layer realizations. The F layers must be formed from gyroelectric or gyromagnetic media to provide some amount of Faraday rotation. In the microwave band, properly biased narrow line-width ferrites such as yttrium iron garnet (YIG) or calcium vanadium garnet (CVG) can be used to construct low loss F layers.

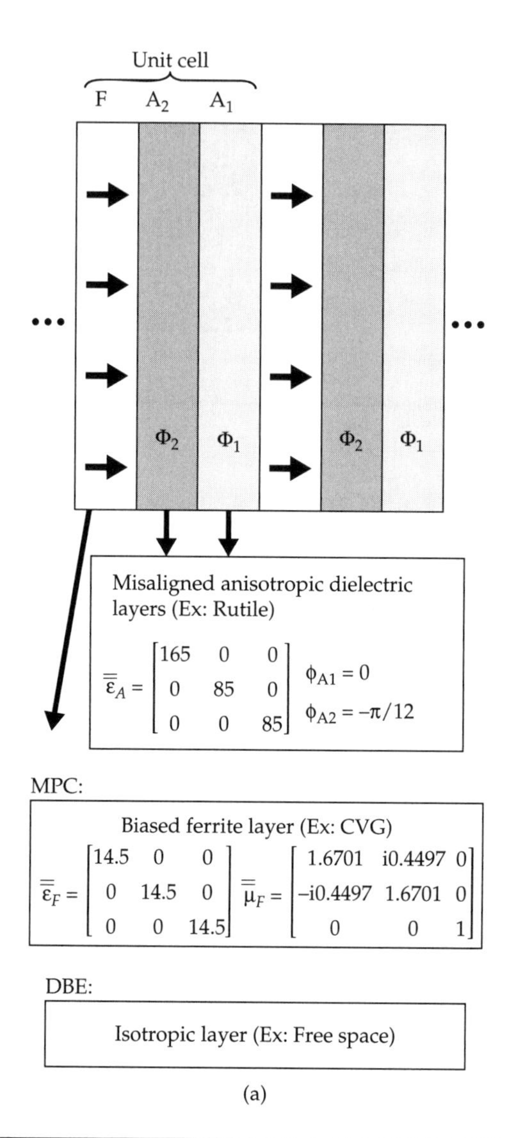

FIGURE 7.2 (a) MPC and DBE unit cell configurations (*After Mumcu et al. [35], ©IEEE 2005.*); (b) MPC dispersion diagram when 0.5695 mm thick A and 0.2161 mm thick F layers are used [with the material parameters in (a)]; (c) Band diagram as a function of incidence angle; (d) Transmittance of semi-infinite and finite (N = 107) MPC slab; the latter showing Fabry–Perot resonances; (e) $|E_x|$ inside the MPC for normal incidence (*After Mumcu et al. [39], ©IEEE 2006.*)

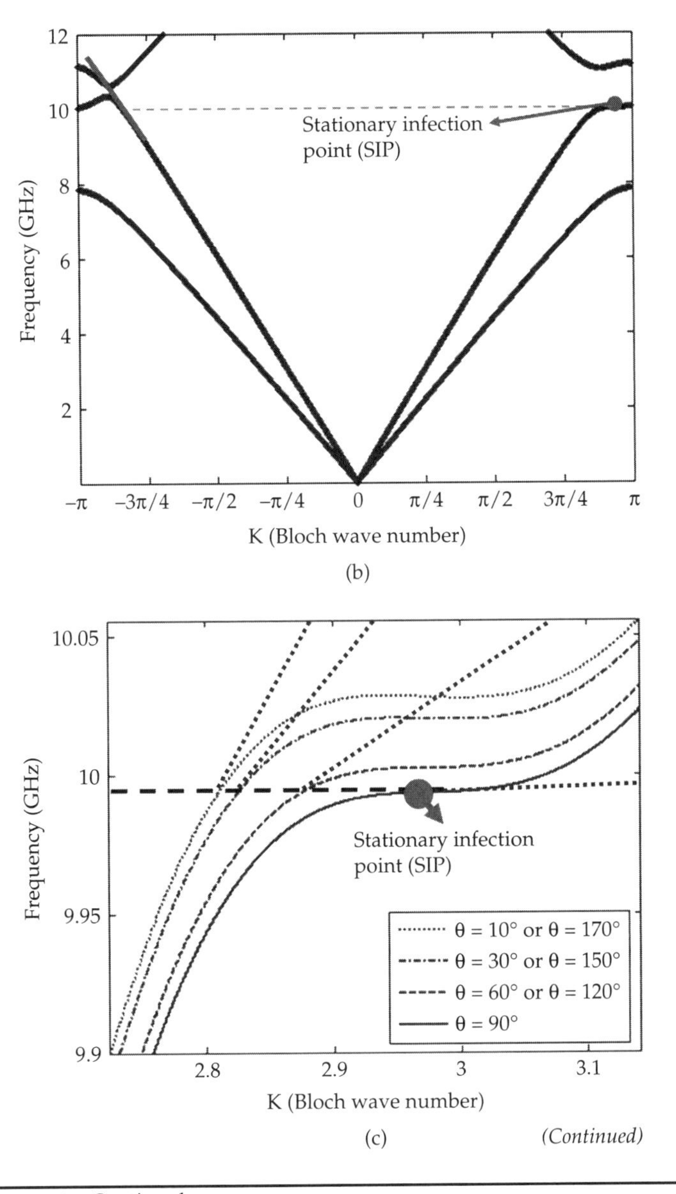

FIGURE 7.2 *Continued.*

Figure 7.2*b* demonstrates the K-ω diagram when the A layers are realized using rutile (TiO_2) and the F layers with calcium vanadium garnet (CVG—having 1 Oe linewith 2.01 gyromagnetic ratio, 1950 G saturation magnetization, and 5.295 kOe bias field). As mentioned

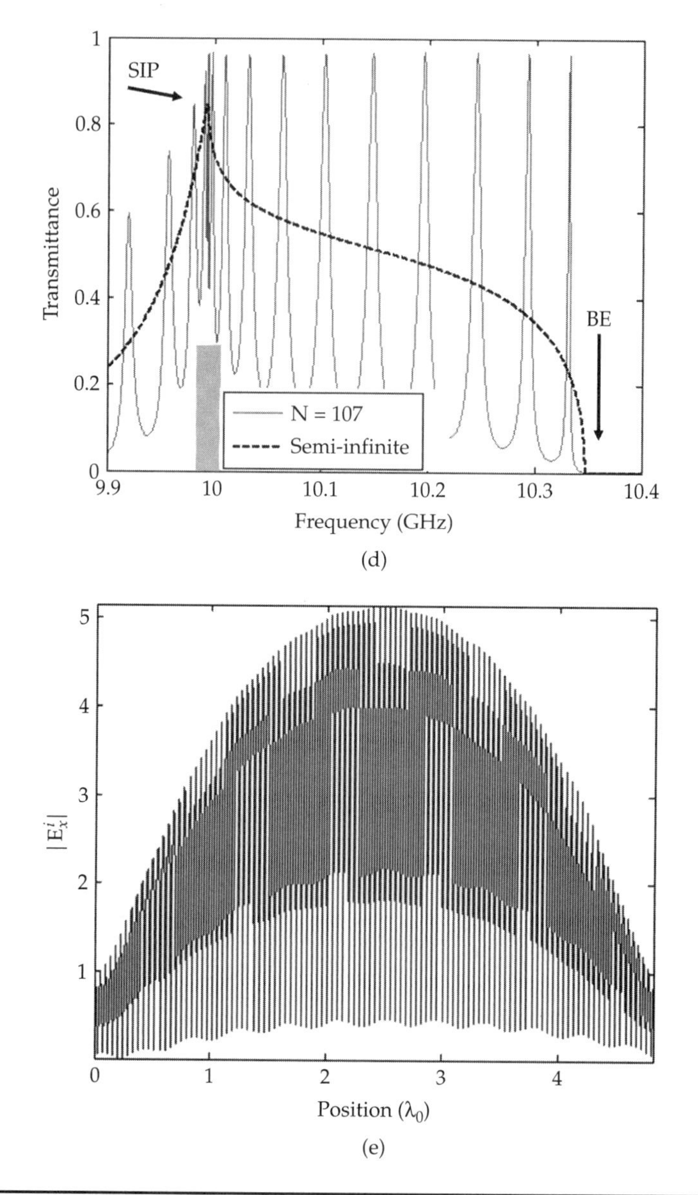

FIGURE 7.2 *Continued.*

earlier, the key characteristic of this diagram is the presence of the SIP giving rise to amplitude increases much like that of an enclosed resonator. However, since the SIP location in the K-ω diagram is sensitive to the direction of incidence (see Fig. 7.2*c*), wave velocity slowdown and amplitude increase will be a function of incidence angle. Also, for finite thickness slabs (rather than semi-infinite media), transmittance occurs only at the Fabry–Perot resonances. These resonances are depicted in Fig. 7.2*d* for a slab composed of N = 107 unit cells. It is necessary that the Fabry–Perot resonances overlay the SIP resonance (associated with the infinite medium) to realize large field amplitude growth. An advantage of the SIP is its substantially large bandwidth as it occurs away from the band edge. In contrast, for RBE media, it is only possible to obtain a Fabry–Perot peak near the band edge (rather than on top of the band edge). One can, of course, increase the thickness of the RBE slab to bring the Fabry–Perot resonances closer to the band edge, but this may not be practical.

MPCs can be suitable as a host medium for high-gain antennas by taking advantage of the amplitude growth. Concurrently, the high contrast dielectrics used to form the MPC medium can result in significant miniaturization. Before discussing high-gain MPC antennas, it is important to point out the difference of the SIP resonance from previously employed gain enhancement techniques such as substrate–superstrate [40], multiple substrates [41], and the defect mode EBGs [14]. In all of these approaches, higher gains were obtained by harnessing a strong volumetric resonance tailored to a specific location of the antenna element. In contrast, the SIP resonance occurs within the propagation band and can give a larger bandwidth beyond a single resonance by merging several Fabry–Perot resonances. In addition, since amplitude growth inside the crystal is more uniformly distributed throughout the crystal (due to slow mode propagation, see Fig. 7.2*e*), miniature arrays that take advantage of the entire volume can be realized.

7.3.2 Dipole Performance Within Magnetic Photonic Crystal

To demonstrate the MPC benefits for antenna gain enhancements, a dipole of length $\lambda_0/20$ (λ_0 = free-space wavelength) was modeled inside an N = 107 unit cell MPC with a total thickness of $4.8\lambda_0$ (see Fig. 7.2*a* for material parameters, Fig. 7.3*a* for problem setup) [39]. The strip dipole resonated at the SIP frequency of 9.99342 GHz. The received power by this dipole was compared to another resonant $\lambda_0/20$ ($\lambda_g/2$) dipole placed in a uniform high contrast $\varepsilon_r = 100$ slab. For a fair comparison, a similar slab thickness $4\lambda_0$ was chosen to minimize the reflections from the air–dielectric interface (based on the $\lambda/4$ transformer concept). The dipole was specifically placed at the maximum field location within the crystal, corresponding to the front of the forty-ninth unit cell as depicted in Fig. 7.3*a*.

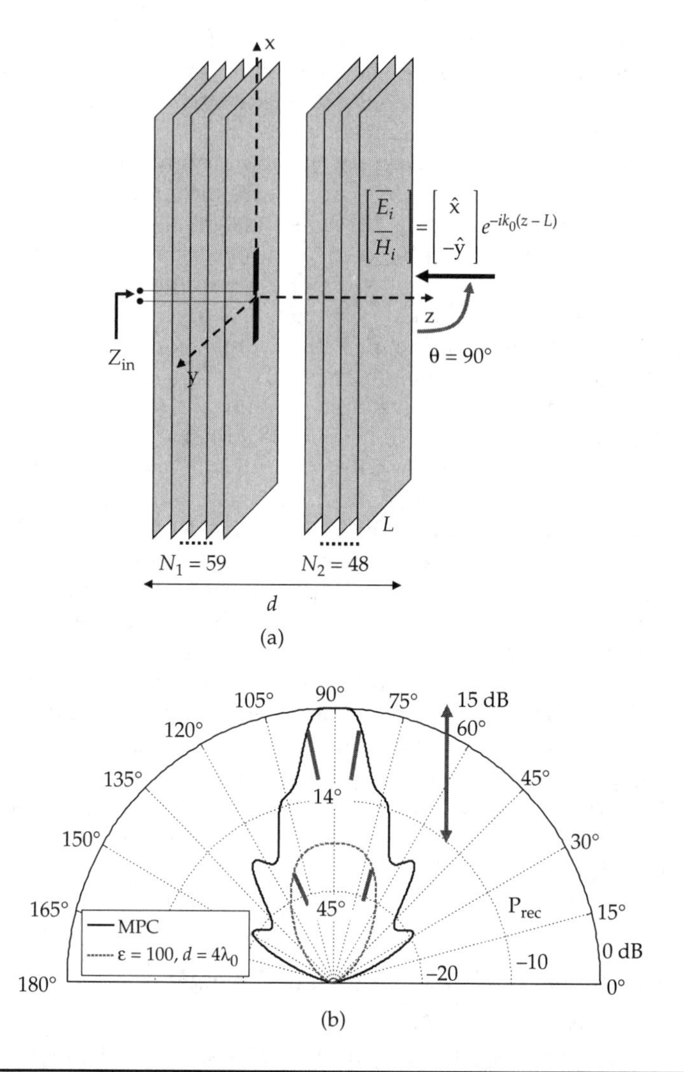

FIGURE 7.3 (a) Problem setup of a small dipole embedded in the MPC medium; (b) Received power comparison between the dipole in the MPC and that in the simple medium; (c) Receiving patterns of a dipole within the MPC medium for varying frequencies (*After Mumcu et al. [39], ©IEEE 2006.*)

As expected from the SIP properties, the $\lambda_0/20$ dipole was found to receive 15 dB more power in the MPC as compared to the same dipole in a simple dielectric slab (see Fig. 7.3*b*). In addition, the half-power beamwidth reduced from 45° to 14° when the dipole is placed in

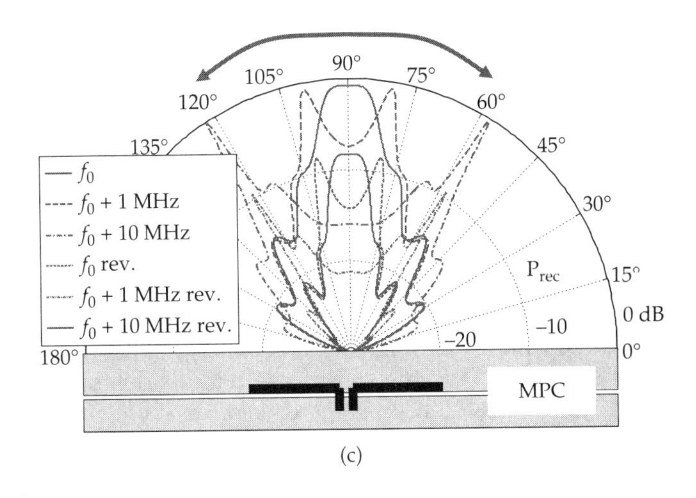

FIGURE 7.3 *Continued.*

MPC. This significant power enhancement and pattern narrowing is a direct consequence of the MPC's sensitivity to the incidence angle. As mentioned earlier in this chapter, the K-ω diagram shift as a function of incidence angle, leading to a more confined pattern even if the antenna is small. The sensitivity of the inflection point to incidence angle can be also advantageous for beam shaping without changing orientation or phasing of the antenna element. From Fig. 7.3*c*, we observe that the pattern splits into two peaks that move apart due to the occurrence of the frozen mode at different angles of incidence. As would be expected, once the frequency is increased beyond a certain value (where no frozen mode exists), the pattern seizes to display the large peaks.

Another interesting property of the MPCs pertains to the bias direction. Reverse–biasing can disable or change the direction of reception by modifying the Fabry–Perot peaks and flipping the K-ω diagram about the K = 0 axis. For example, when the F layers are reverse biased, the received power becomes 8 dB less as compared to the case when the F layer is forward–biased, and this is depicted in Fig. 7.3*c*. To consider the issues associated with a practical realization of the MPCs, material losses and slab/crystal thickness need to be considered. Typically, a minimum thickness is required to observe the maximum achievable amplitude at the inflection point of an MPC [35]. Therefore, the received power drops as the thickness of the MPC becomes smaller (from N = 107 down to N = 30). Nevertheless, even in the case of N = 30 (1.5 λ_0 thick), the MPC embedded dipole can receive 6 dB more power as compared to being embedded in a simple medium. A similar drop in the received power was also observed when the material loss tangent was included in the calculations. Indeed, to fully make use of the MPC in high-gain antenna applications, a low loss tangent of $\tan\delta = 1 \times 10^{-4}$ is needed.

7.3.3 Resonance and Amplitude Increase Within DBE Crystals

Although MPCs allow for design flexibility and gain, they are understandably difficult to fabricate and prone to losses due to the need for biased ferrite layers. In this regard, DBE crystals offer a more convenient unit cell structure by replacing the magnetic layers of MPCs with ordinary isotropic materials. As shown in Fig. 7.2*a*, one of the simplest DBE unit cell configurations [27] includes two identical anisotropic dielectric layers misaligned with respect to each other (A layers) and an isotropic layer (instead of the magnetic F layer of the MPC). Figure 7.4*a* presents the band diagram when the A layers are realized using rutile and the F layers are replaced with free space. As was previously discussed in the introduction of this chapter, the DBE crystal has a relatively flatter band edge as compared to the RBE crystals. Therefore, the group velocity vanishes faster (with better transmittance) when the operational frequency approaches the band edge. This implies higher amplitude increase that can deliver substantial gain over regular crystals. It is therefore equally important to discuss gain enhancements within the DBE crystals as was done for the MPCs.

To demonstrate the amplitude increase and effect of finite thickness, we consider here two different crystal configurations consisting of N = 20 and N = 10 unit cells (see Fig. 7.4 for material parameters) [36,42]. Figure 7.4*b* depicts the Fabry–Perot resonances in the 8 to 15 GHz range when the N = 20 unit cell crystal is illuminated with an x–polarized incident field. We observe that the closest Fabry–Perot peak to the DBE resonance occurs at 11.196 GHz, being almost totally transparent to the incident polarization. Specifically, inside the crystal, $|E_x|$ and $|E_y|$ reach to amplitudes of 4.1 and 4.3, respectively, as depicted in Fig. 7.4*c*. Due to the fewer number of unit cells, the second crystal exhibits a Fabry–Perot resonance further away from the band edge. The closest Fabry–Perot peak occurs at 11.036 GHz and the maximum amplitude inside the crystal becomes $|E_x| = 1.8$ and $|E_y| = 2.2$. Clearly, two to four fold amplitude increase coupled with high K-ω diagram sensitivity to incidence angles (similar to the MPCs) can make the DBE crystals attractive for high-gain antenna applications.

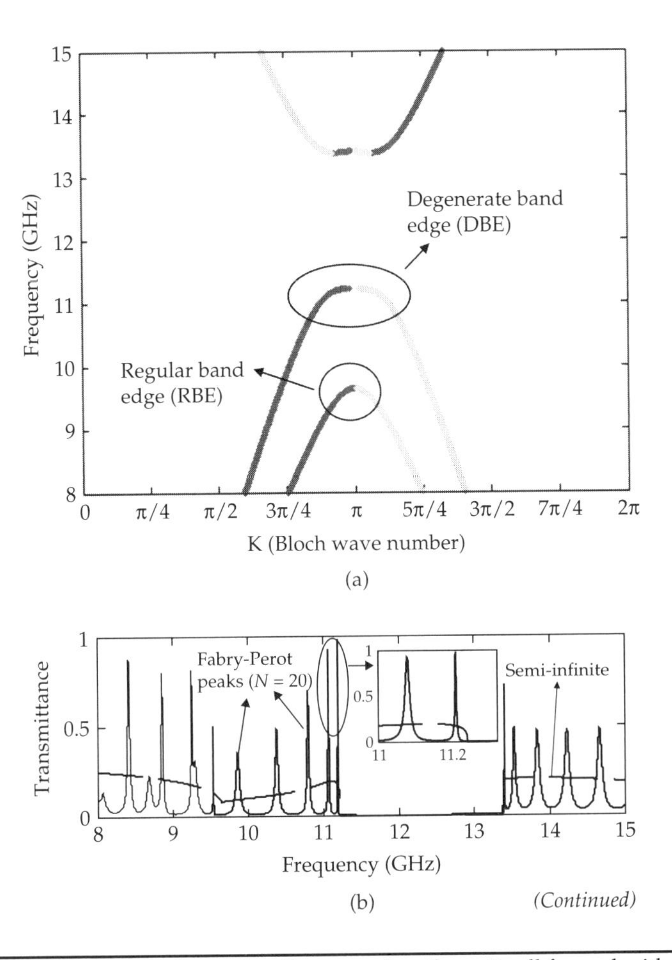

(Continued)

FIGURE 7.4 (a) DBE dispersion diagram when the unit cell formed with 0.5 mm thick A and 0.25 mm thick F layers (material parameters are given in Fig. 7.2*a*; also $\varphi_{A1} = 0$, $\varphi_{A2} = \pi/4$); (b) Transmittance for a N = 20 unit cell crystal with x-polarized incidence. The overall crystal structure is rotated by $-3\pi/20$ about its center axis (i.e., $\varphi_{A1} = -3\pi/20$, $\varphi_{A2} = -\pi/20$) to achieve best transmittance for x-polarization; (c) $|E_x|$ and $|E_y|$ inside the crystal for x-polarized normal incidence (*After Mumcu et al. [39], ©IEEE 2006.*) [42].

7.3.4 Dipole Performance Within Degenerate Band Edge Crystal

To demonstrate antenna gain enhancement using DBE crystals, a dipole of length $\lambda_0/20$ is modeled inside the crystal configurations of the previous section. The dipole operates at the Fabry–Perot peaks closest to the DBE resonance (11.196 GHz for N = 20 and 11.036 GHz

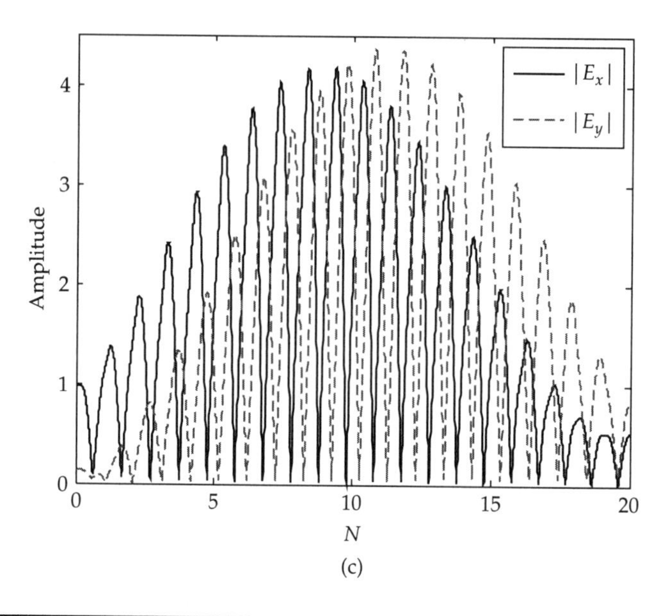

FIGURE 7.4 *Continued.*

for N = 10). Since it is more practical to place the dipole at the layer boundaries (rather than inside the individual layers), the dipole is placed on top of the layer where the field attains its maximum value. Specifically, the location of the dipole for the first case is the end of the F layer at the tenth unit cell. In the second case, it is placed at the end of the F layer of the fifth unit cell. Since the average dielectric constant inside the crystal is around $\varepsilon_r = 100$, the dipole performance is compared with the same antenna embedded within an isotropic homogenous dielectric slab having $\varepsilon_r = 100$. Also, the slab thicknesses are chosen to be very close to that of the DBE crystals, being $1\lambda_0$ and $0.5\lambda_0$, respectively (for N = 20 and N = 10).

As expected, the amplitude increase inside the DBE crystal manifests itself in higher received power. Specifically, the dipole inside the first DBE crystal receives 11.7 dB more power than that in a simple medium when the crystal is excited by an x–polarized incident field (see Fig. 7.5*a*).

We also observe that the half-power beamwidth of the dipole is greatly reduced inside the first DBE crystal (from 76° down to 8°). When the loss factors ($\tan\delta = 1\times10^{-4}$) of the dielectric layers are taken into account, the dipole received power is reduced by only 1.9 dB, implying that it displays almost the same performance as in a low loss DBE crystal. As expected, when the dipole is placed inside a thin

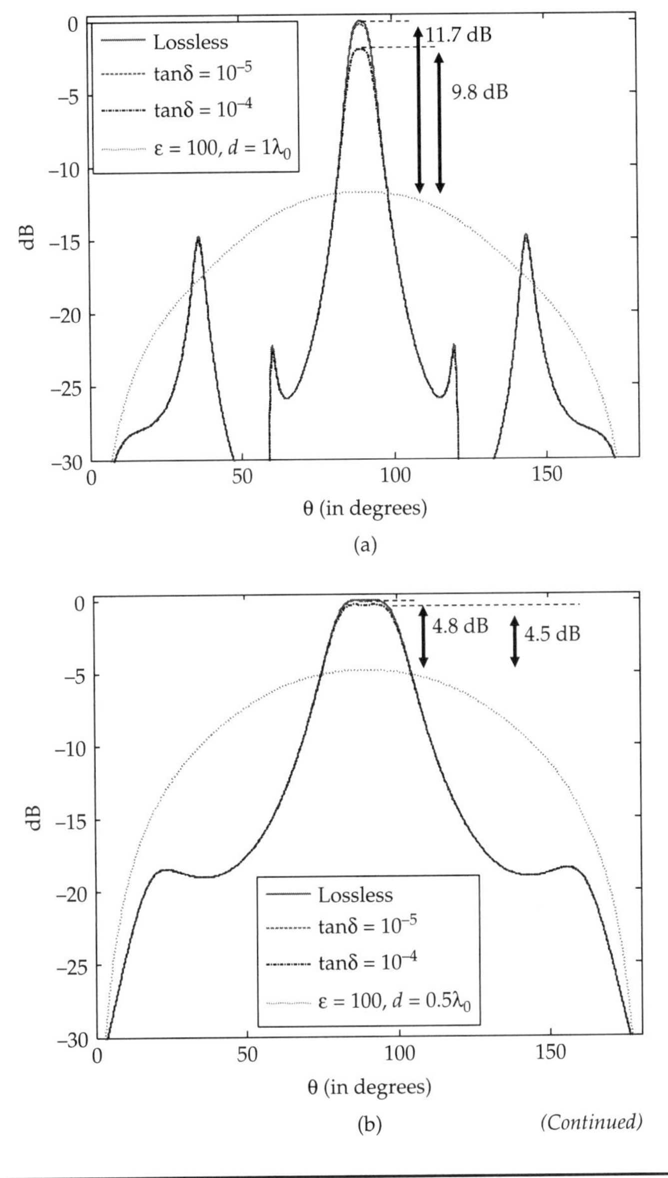

(Continued)

FIGURE 7.5 Received power by a dipole in DBE medium consisting of (a) N = 20; (b) N = 10 unit cells; (c) Received power by an 11 element end-fire dipole array in the N = 20 unit cell DBE crystal. (*After Volakis et al. [36], ©IEEE 2006.*) [42].

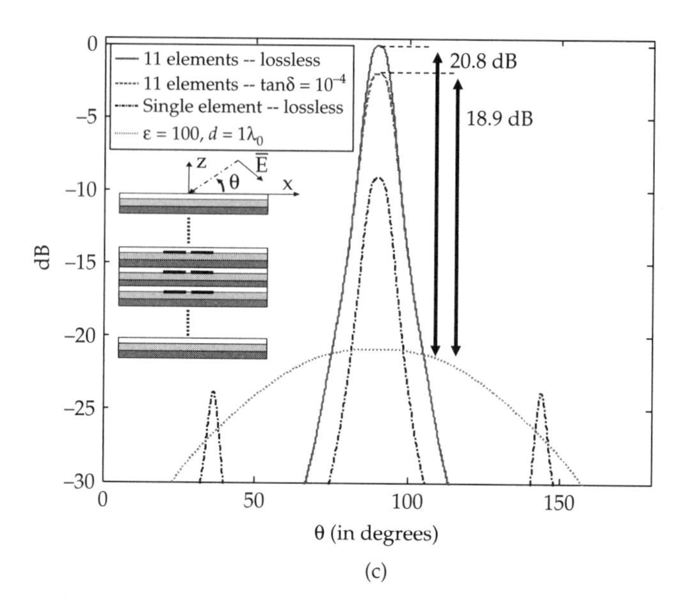

FIGURE 7.5 *Continued.*

DBE crystal, the received power (relative to the reference antenna) drops (see Fig. 7.5*b*). Nevertheless, the dipole still receives 4.8 dB more power (4.5 dB more when lossy) over the reference antenna. Further, the half-power beamwidth of the dipole inside the DBE is also much narrower than the reference antenna (26°–76°).

It is also important to mention the performance of possible array configurations within these crystals. Since a large propagation constant is observed in the z–direction (due to the large effective dielectric constant), end-fire array configurations are attractive. Figure 7.5*c* presents an example of an 11 element end-fire dipole array within the DBE crystal (each dipole being $\lambda_0/20$).

The 11 elements are placed in front and back of the original location of the single dipole considered earlier, that is, at the end of the F layer of the tenth unit cell. The array elements are separated a unit cell from each other with a linear phasing of $-\pi$ to direct the beam along the $+$z direction (at the band edge, the phase difference across a unit cell is approximately π). As seen from Fig. 7.5*c*, the end-fire array allows for an additional 9.1 dB increase in received power over the single element in the DBE crystal. Also, the side lobes of the receiving pattern are greatly suppressed.

7.3.5 Practical Degenerate Band Edge Antenna Realizations

The flatness of the DBE band structure is entirely attributed to the introduced anisotropy of the dielectric layers since their isotropic counterparts can only yield RBE behavior (see Fig. 7.1). Therefore, the anisotropic layers (needed in a practical realization) must exhibit a sufficient degree of anisotropy. Another issue to be addressed relates to losses. Crystal losses must be kept at a minimum since slowly propagating modes experience higher losses per unit cell.

Above, we considered a naturally available material "rutile" for the A layers of the DBE crystals. Indeed, rutile exhibits a very large permittivity with an anisotropy ratio of 1.9 ($\varepsilon_{xx} = 165$, $\varepsilon_{yy} = \varepsilon_{zz} = 85$). It also exhibits a very low loss tangent ($\tan\delta = 1 \times 10^{-4}$) in the X-band (8–12 GHz) [43]. However, rutile samples are costly. Therefore, anisotropic layers of DBE crystals have so far been utilized via engineered periodic assemblies that effectively yield uniaxial permittivity tensors. A way to introduce anisotropy using low-cost techniques is to mix isotropic dielectric materials [44] and suitable arrangements of metallic inclusions [45]. Due to implementation ease, a realization of DBE crystals was constructed by periodically stacking patterned printed circuit boards (PCBs) [38]. The proposed unit cell is depicted in Fig. 7.6*a* with the first layer (leftmost) consisting of two dielectric slabs with short metallic strips printed on each side.

Periodic arrangement of metal strips gives rise to polarization sensitivity [46–48] and serves to generate small dipole moments, especially when the electric field is polarized in the direction of the strips. Therefore, when the strips are small (as compared to the wavelength), the medium emulates an anisotropic dielectric material from a macroscopic point of view. One can control the effective permittivity (anisotropic) by altering dimensions of the strips, their separation distance, and permittivity of the host medium. As an example, the dipole strips on the second layer of the DBE crystal unit cell are rotated 45° to introduce the necessary misalignment. The third layer is chosen as air for simplicity. The other parameters of the unit cell were optimized (using a full-wave periodic finite element method [49] to account for surface depolarization effects of thin PCB layers) to obtain a maximally flat DBE K-ω diagram (see Fig. 7.6*b*).

To ensure presence of the DBE mode inside 3D finite size crystals, a crystal made of N = 8 unit cells (a total of 32 PCBs with 0.05 cm thick Rogers RO4350 substrate having $\varepsilon_r = 3.48$ and $\tan\delta = 0.0037$) was manufactured (see Fig. 7.7). The DBE crystal had a total thickness of 5.84 cm and was 10.92 × 10.92 cm. As depicted in Fig. 7.7*a*, transmittance and probing experiments were conducted to verify the existence of the DBE mode predicted computationally. Figure 7.7*b*

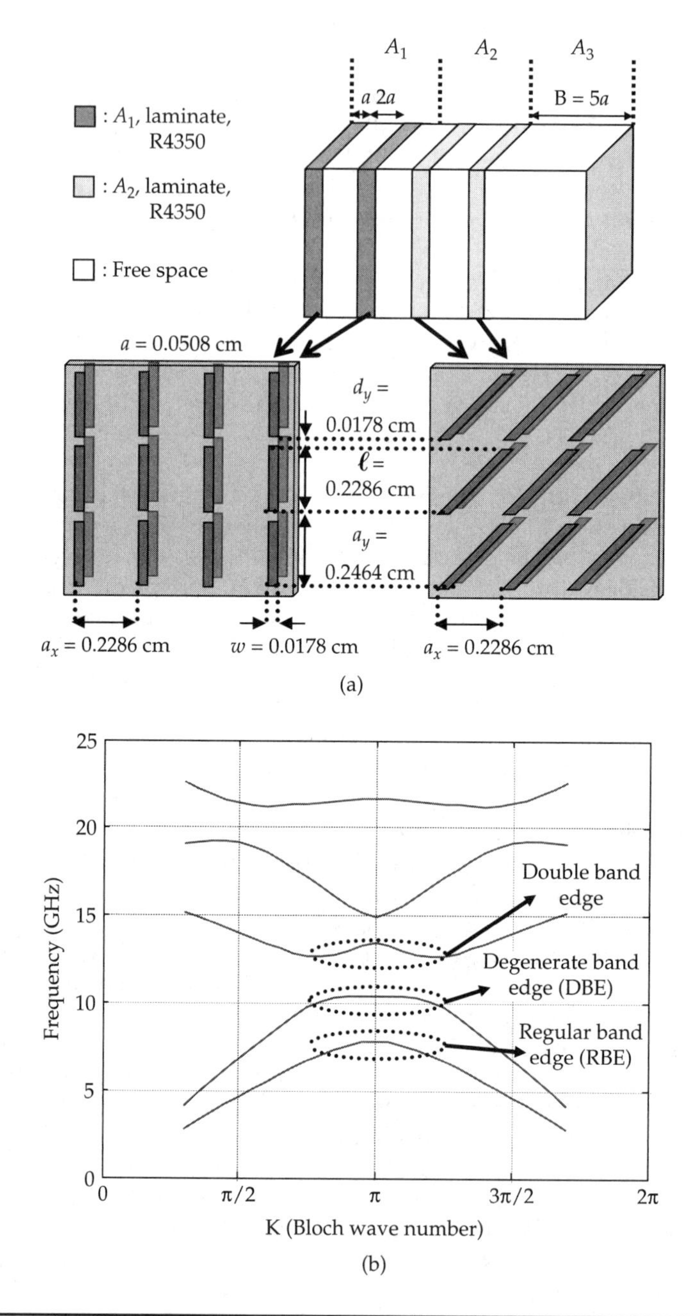

FIGURE 7.6 (a) DBE unit cell configuration; (b) Band diagram. (*After Yarga et al. [38], ©IEEE 2008.*)

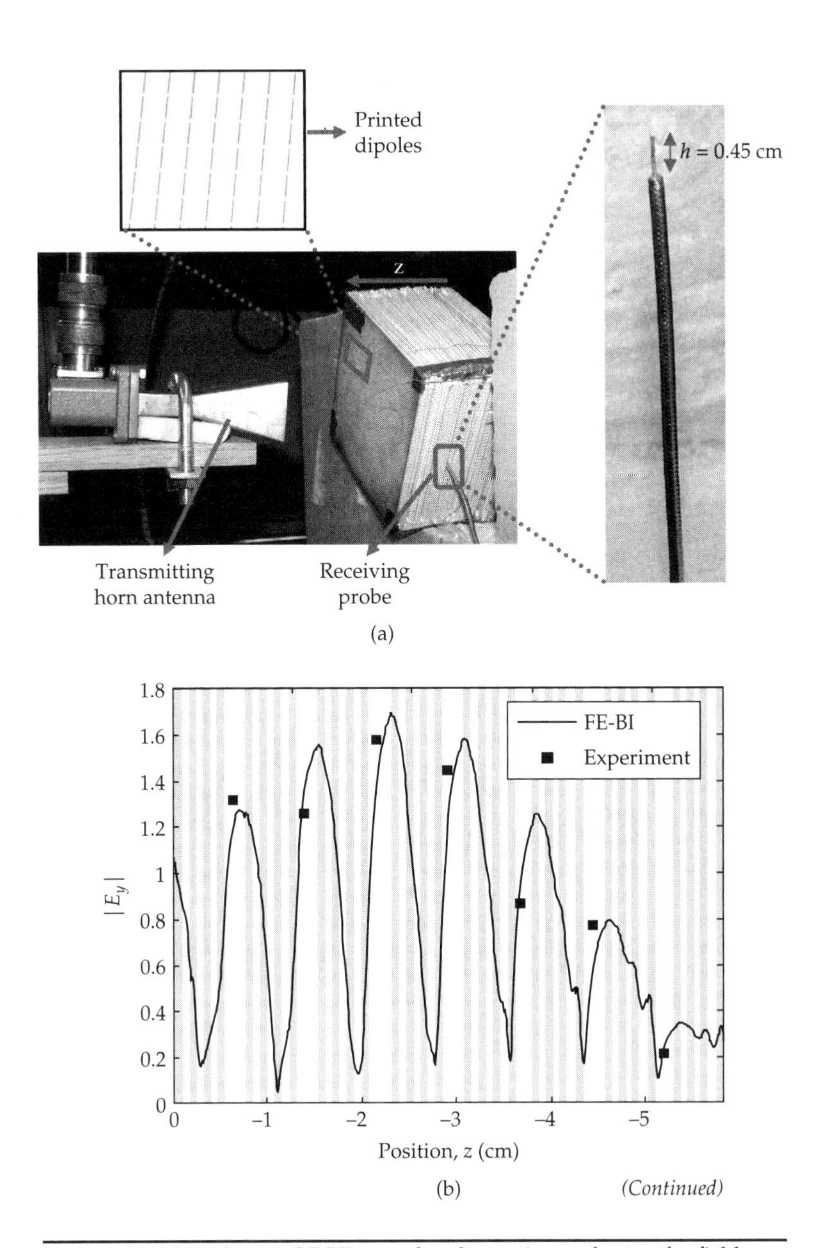

FIGURE 7.7 (a) Fabricated DBE crystal and experimental setup for field probing; (b) Measured field amplitude profile inside the DBE crystal; (c) Experimental setup for measuring the pattern of a dipole embedded inside the DBE crystal; (d) Receiving pattern of the dipole (*After Yarga et al. [38], ©IEEE 2008.*)

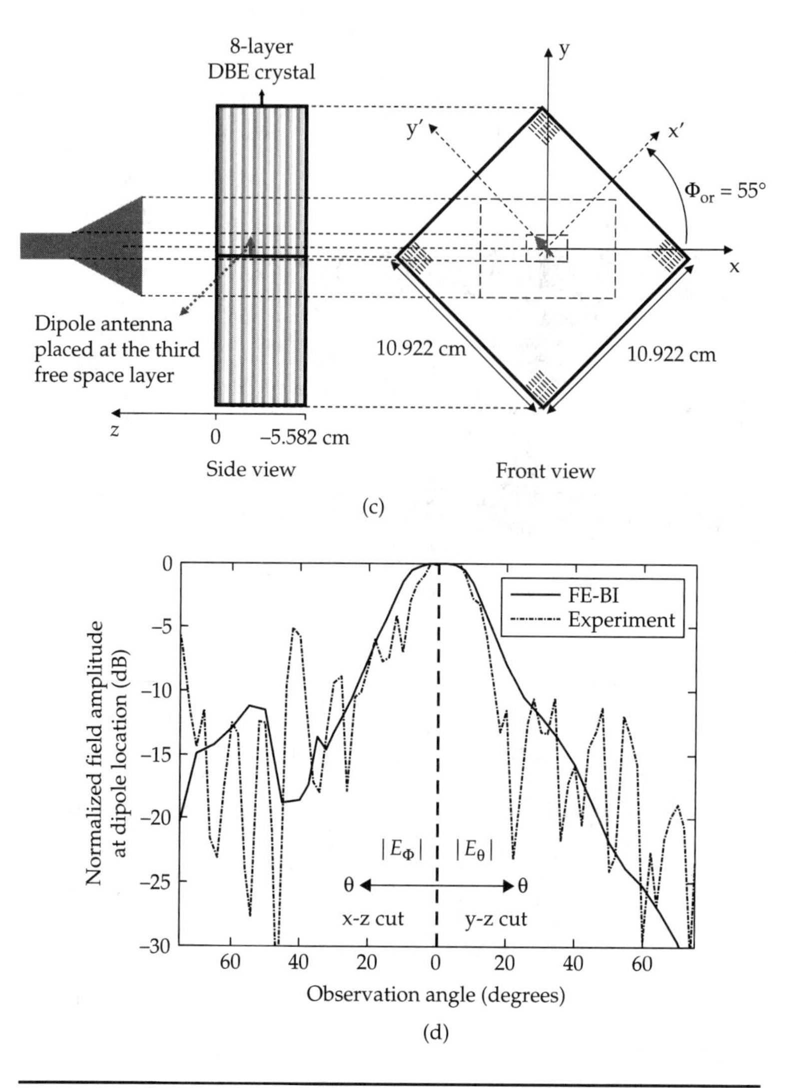

FIGURE 7.7 *Continued.*

demonstrates the amplitude increase inside the crystal at the DBE frequency. Maximum field amplitude of 1.65 at 10 GHz is seen corresponding to the flat portion of the band diagram, and occurs at the third free-space layer. To demonstrate directivity enhancement, a dipole antenna (2 cm in total length) was placed to this maximum field location. Figure 7.7*c* depicts the orientation of the crystal and the embedded dipole. The measured receive pattern is shown in Fig. 7.7*d* and as seen, the agreement with the finite element-boundary integral (FE-BI) calculations is quite good. We note that since the designed crystal was rotated to achieve optimum illumination at the DBE mode, the receiving pattern of the embedded dipole was measured along the two principal cuts shown in Fig. 7.7*c*. As seen, the measured pattern has large ripples for illumination angles larger than 30°. This is due to diffractions from the finite aperture, not taken into account by the 1-D numerical analysis. We remark that the received power was 5 dB higher in presence of the crystal, and had about 25° in half-power beamwidth (HPBW). This implies a directivity of 18 dB.

The higher directivity exemplifies the utility of the DBE modes. However, the measured 38% aperture efficiency is not the best attainable from an aperture of the same size. Also, the crystal was fabricated with a large lateral extent to demonstrate transmission/reflection characteristics of 1-D DBE crystals. Further, the prototype is rather lossy due to metallic inclusions. Therefore, an alternative and smaller DBE crystal constructed from mixtures of very low loss ceramic materials was presented in [29, 37]. As seen in Fig. 7.8*a*, the 1 mm thick anisotropic layers of the unit cell was realized by alternating square-cross-section rods of barium titanate ($BaTiO_3$, $\varepsilon_r \approx 80$, $\tan\delta = 3.7 \times 10^{-4}$ @ 2 GHz) and alumina (Al_2O_3 $\varepsilon_r \approx 10$, $\tan\delta = 2.8 \times 10^{-4}$ @ 3 GHz). When the individual rod cross sections are much smaller than a wavelength, the alternating rods exhibit a high-contrast equivalent permittivity tensor (see Fig. 7.8*a*) which leads to antenna size reduction. To achieve the best performance in terms of dielectric loss, the rods were glued using M-bond 610 resulting in $\tan\delta = 1.9 \times 10^{-3}$ at 7.59 GHz (the intrinsic loss tangent of the assembled layers without adhesive was $\tan\delta = 0.9 \times 10^{-3}$ GHz). When the unit cell parameters were optimized (i.e., misalignment angles and air layer thicknesses), a DBE mode at 9.25 GHz was obtained. Placement on the ground plane effectively doubled the thickness of the crystal. As shown in Fig. 7.8*b*, a three unit cell structure (as thin as $0.29\lambda_0$ at 8.71 GHz) supported a large magnetic field intensity (on the ground plane) when the crystal was illuminated around the sharp Fabry–Perot peak closest to the DBE mode frequency.

To harness the amplitude increase in the crystal of Fig. 7.8, a slot-fed DBE resonator antenna having an aperture of $0.8\lambda_0 \times 0.8\lambda_0$, was designed and manufactured. As shown in Fig. 7.8*c*, the feeding slot was oriented to match the polarization of the magnetic field under plane wave illumination at the Fabry–Perot resonance. Figure 7.8*d* and 7.8*e* demonstrates the simulated and measured far-field gain patterns of the antenna assembly. Disagreements between measurements and numerical simulations can be attributed to dielectric losses in the feed structure and the uniaxial layers, as well as the external foam used to

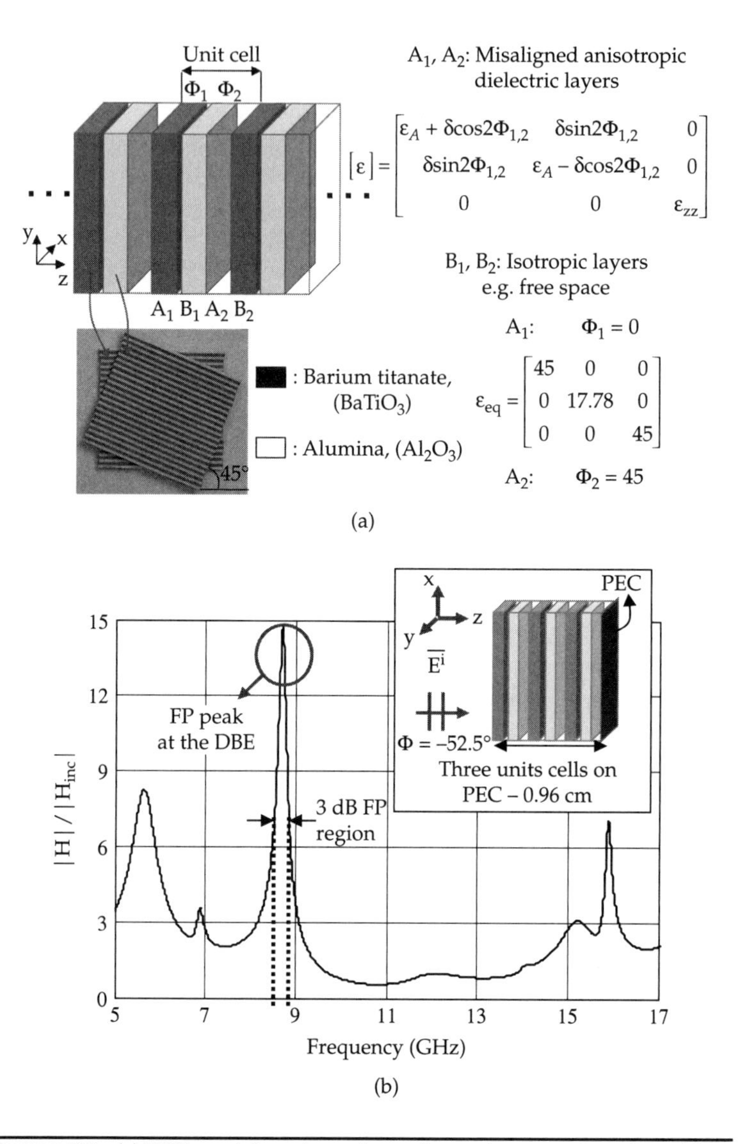

FIGURE 7.8 (a) DBE crystal made up from alternating ceramic rods and its dispersion diagram; (b) Magnetic field amplitude enhancement at the Fabry–Perot near the DBE frequency. Note that three layer DBE assembly is backed by a PEC to create an effectively thick crystal; (c) Fabricated slot fed DBE crystal antenna; (d) to (e) Measured gains in two principal cuts (*After Yarga et al. [37], ©IEEE 2009.*)

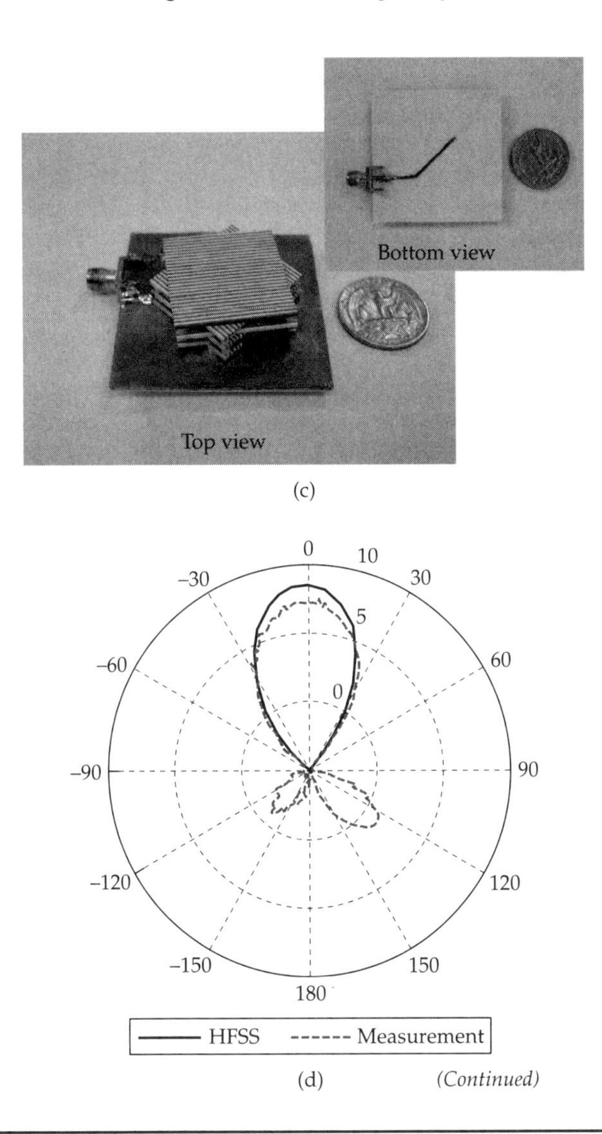

(Continued)

FIGURE 7.8 *Continued.*

hold the sample in place during chamber measurements. This DBE antenna delivered a peak (computed) directivity of 10.16 dB (measured realized gain is about 8 dB at 8.67 GHz), at $S_{11} < -10$ dB bandwidth of 3.85% and a 3 dB directivity bandwidth of 3.97%. Also, a remarkable aperture efficiency of 126.55% (referred to the top surface of the crystal) was achieved.

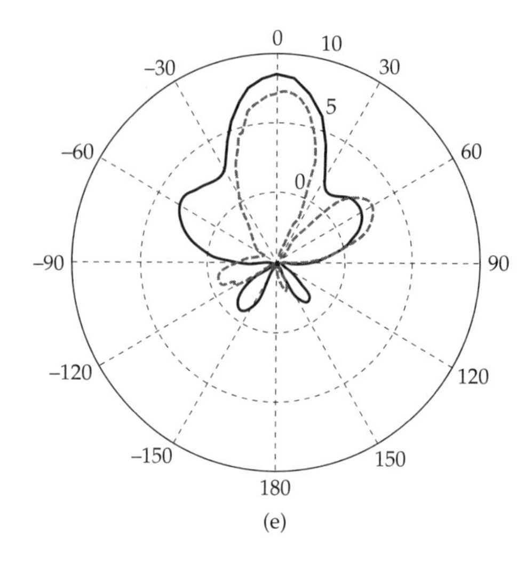

FIGURE 7.8 *Continued.*

The concept of DBE crystals made up from engineered anisotropic dielectrics was also extended to realize dielectric resonator antennas (DRAs) having small lateral dimensions [50]. As depicted in Fig. 7.9*a*, maximally flat DBE K-ω diagram was utilized for shifting the resonance points to smaller frequencies without increasing the antenna size. Another important advantage of DBE-DRAs was the tunable operation frequency with the misalignment angle (without requiring practically unavailable permittivity values). The manufactured antenna assembly shown in Fig. 7.9*b* to 7.9*d* consists of three uniaxial layers and an isotropic layer placed over a finite ground plane. The antenna is excited by using a slot coupled microstrip feed. The DBE-DRA resonated at 2.74 GHz with $|S_{11}| < -10$ dB bandwidth of 2.73%. The measured far-field pattern and realized gain of the DBE-DRA is shown in Fig. 7.9*e*. Specifically, a gain of 6.51 dB was measured. The corresponding directivity was calculated to be 6.86 dB, implying a measured efficiency of 92.3%.

7.4 Printed Antenna Miniaturization via Coupled Lines Emulating Anisotropy

7.4.1 Antenna Miniaturization Using Degenerate Band Edge Dispersion

For miniaturization, one goal is to lower the resonance frequency without resorting to any antenna size increments. It is therefore essential to understand resonance conditions of simple printed antennas.

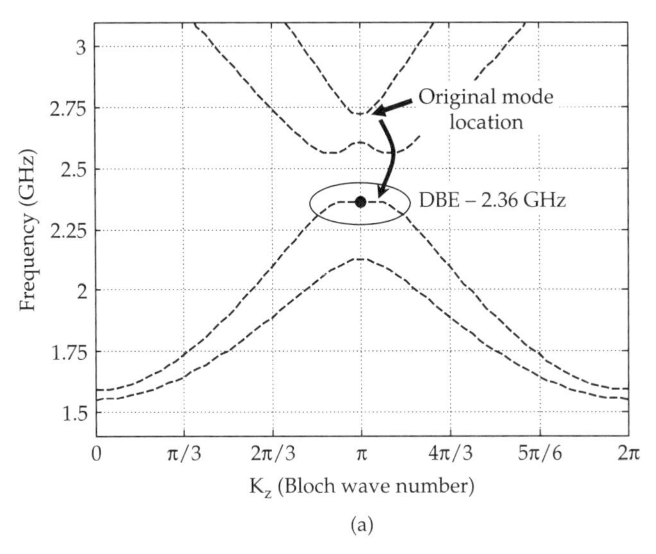

(a)

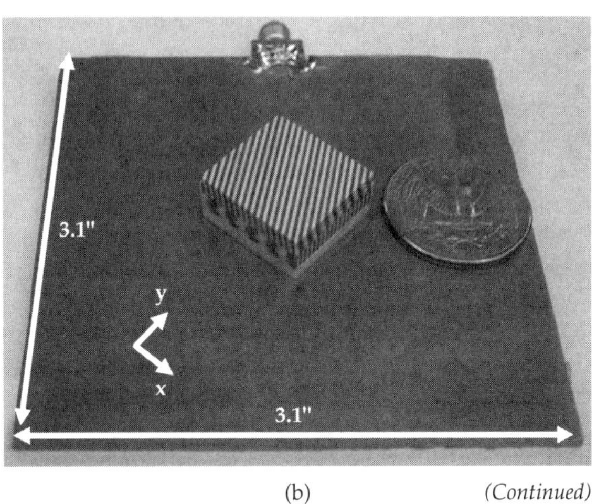

(b) *(Continued)*

FIGURE 7.9 (a) DBE-DRA dispersion diagram; Realized DBE-DRA antenna; (b) Top view; (c) Side view; (d) Bottom and top view showing the slot coupled microstrip line feed; (e) Simulated and measured gains in principle cuts. (*After Yarga et al. [50], ©IEEE 2009.*)

As an example, Fig. 7.10*a* considers a printed loop that can be thought as a circularly periodic structure involving two unit cells (top and bottom half loops) [32]. The corresponding band diagram shown in Fig. 7.10*b* indicates two linear K-ω curves (identical to a uniform transmission line) corresponding to propagating waves in opposite

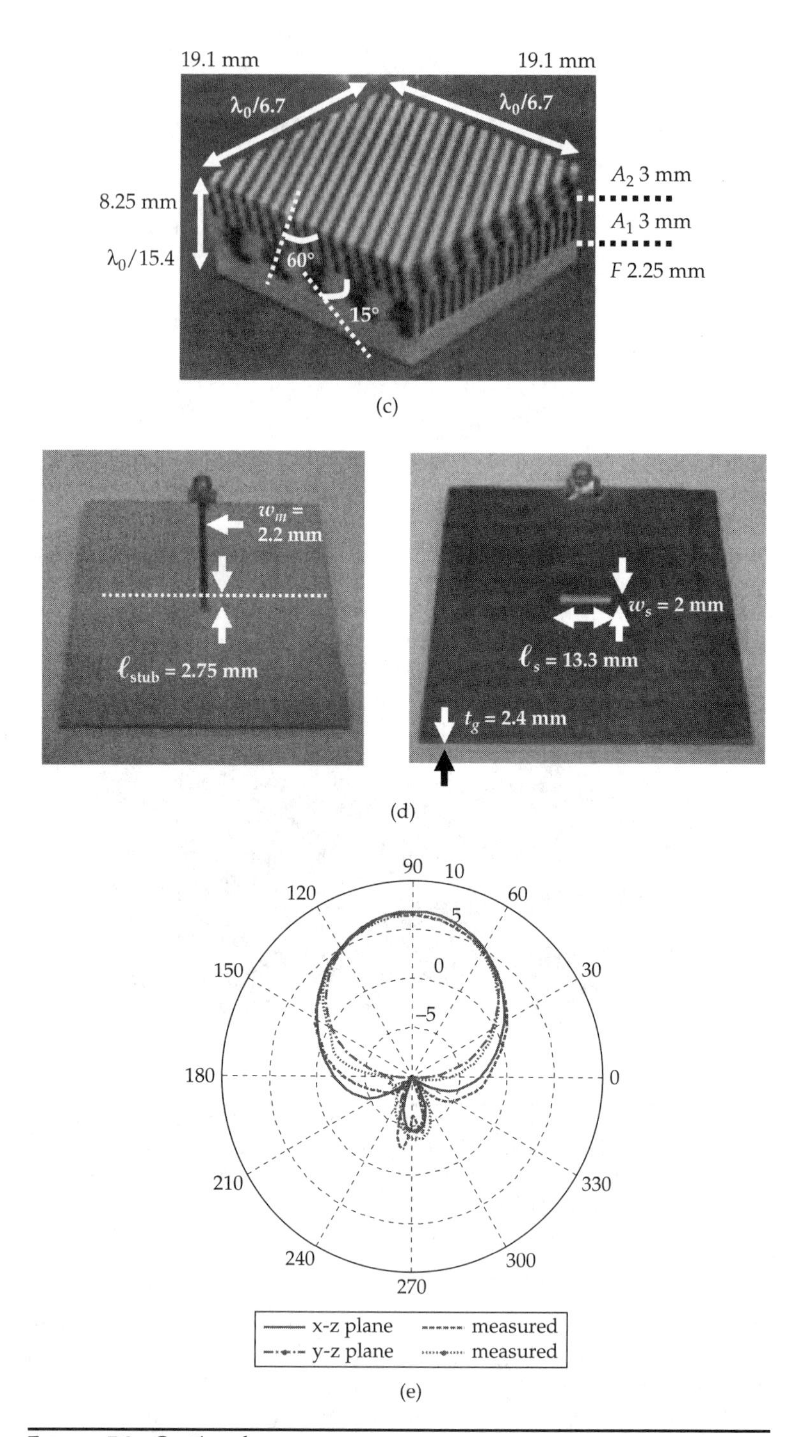

FIGURE 7.9 *Continued.*

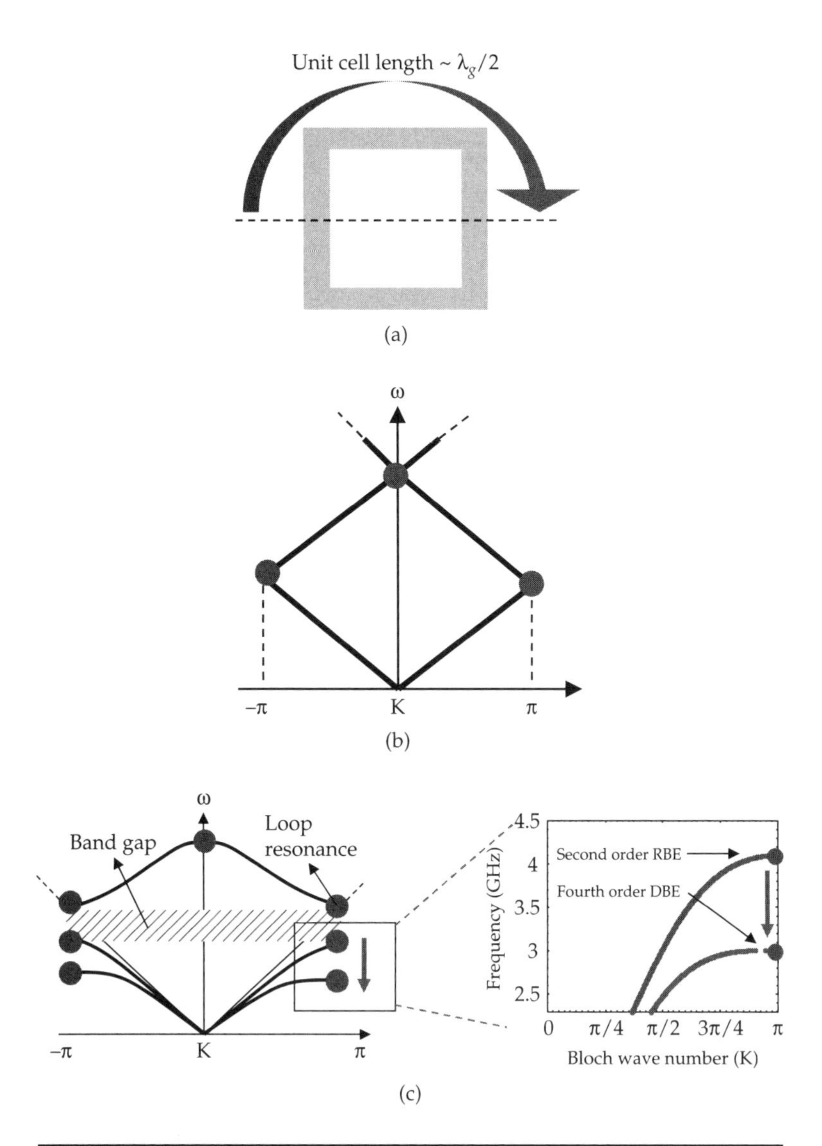

FIGURE 7.10 (a) Simple printed loop antenna; (b) Dispersion diagram of a unit cell forming the rectangular printed loop antenna; resonances for this circularly periodic structure (two unit cells) are marked by dots; (c) Bending of the K-ω diagram to shift resonance to lower frequencies; magnified view of the dispersion diagram around the band edge. (*After Mumcu et al. [32], ©IEEE 2009.*)

directions ($K < 0$ and $K > 0$). The lowest order resonance occurs at $K = \pm\pi$ due to matching phases of the two propagating waves on the loop. To reduce the frequency of the resonances, it is necessary to lower the K-ω curve as shown in Fig. 7.10*c*. This can be done by inserting reactive elements (capacitive or inductive) within the transmission lines. In general, such a structure forms a second order RBE curve at frequencies where $K = \pm\pi$. Although the frequency where $K = \pm\pi$ can be further pushed down (by increasing reactive loading), the parabolic relation still remains and limits the amount of curvature at the band edge. On the other hand, DBE crystals exhibit fourth order K-ω curves at the band edges. As seen in Fig. 7.10*c*, a maximally flat K-ω curve pushes the resonance further down in frequency, improving miniaturization. Improved flexibility of the dispersion diagram can also be used to design reconfigurable loop antennas with larger frequency tuning ranges.

7.4.2 Realizing DBE Dispersion via Printed Circuit Emulation of Anisotropy

As already noted, the simplest volumetric DBE unit cell is composed of two anisotropic homogenous layers having two different misaligned permittivity tensors [27]. However, it is possible to realize the DBE behavior on otherwise uniform substrates using a pair of transmission lines as depicted in Fig. 7.10*a* [32, 51]. Each transmission line is thought as carrying one polarization component of the electric field propagating within the DBE crystal. By using different line lengths, a phase delay is introduced between the two polarizations to emulate diagonal anisotropy. In a similar fashion, even–odd mode impedances and propagation constants on the coupled lines can be used to emulate a general anisotropic medium (i.e., nondiagonal anisotropy tensor). Finally, by cascading the uncoupled and coupled transmission line sections as in Fig. 7.11*a*, an equivalent printed circuit is realized that emulates the volumetric DBE crystal. The circuit can then be tuned to achieve the DBE mode by appropriately selecting the line thicknesses and coupled line separations.

The emulation of anisotropy using printed circuits is, of course, very attractive as several new variables can be adjusted for design and propagation control. Moreover low-cost manufacturing of the designed printed circuit unit cells is a major advantage. However, printed DBE designs (i.e., tuning of line lengths, widths, and coupling ratio) usually requires many layout iterations using full wave numerical solutions. Furthermore an understanding of various loading effects (i.e., capacitive, inductive) is not as easy due to the geometrical and computational challenges. To alleviate these issues, a four-port lumped element circuit model can be introduced to represent the

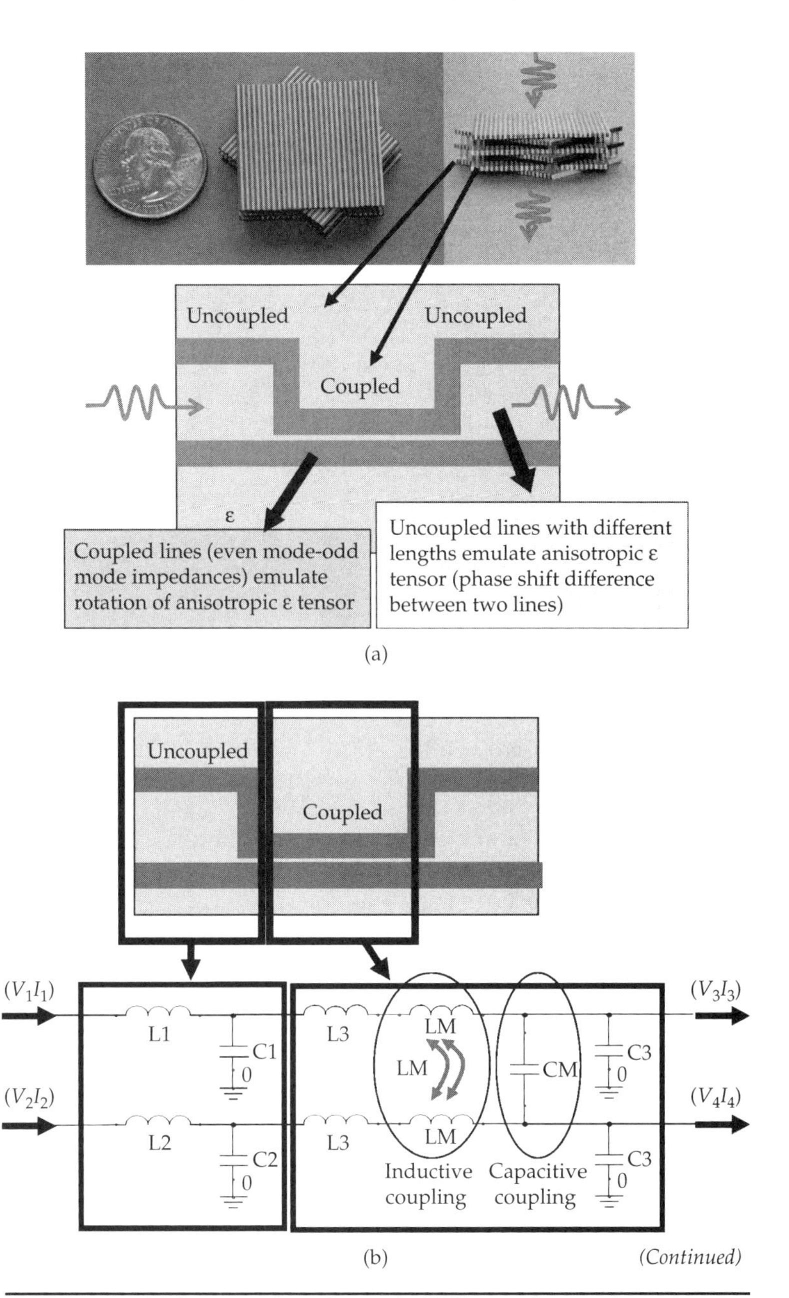

(Continued)

FIGURE 7.11 (a) Concept of emulating the DBE crystal on a microstrip substrate using cascaded coupled and uncoupled transmission line pairs; (b) Lumped circuit model of partially coupled lines; (c) Different band edges obtained by changing the capacitive coupling between the transmission lines (represented by C_M) in case uncoupled section is capacitively loaded ($L_1 = L_2 = L_3 = 1$ nH, $C_1 = 10$ pF, $C_2 = C_3$). (*After Mumcu et al. [32, 34, 52, 53], ©IEEE.*)

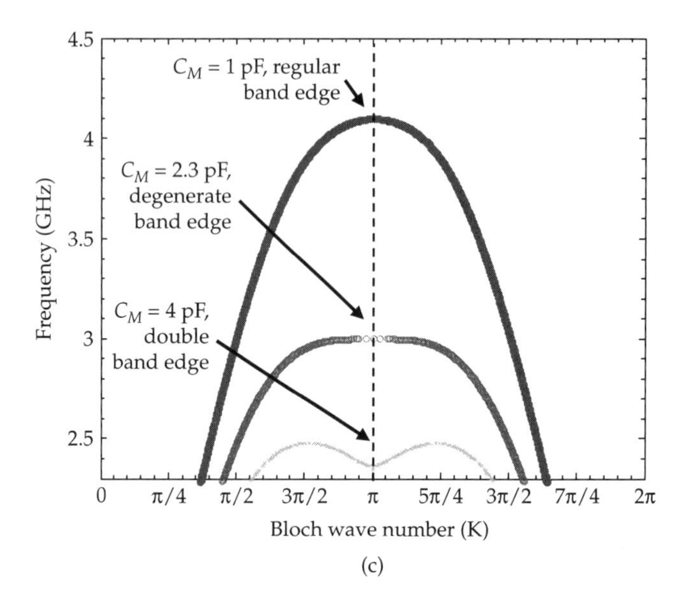

FIGURE 7.11 *Continued.*

partially coupled microstrip lines [34, 52, 53]. Such a circuit can then be used to find the lumped element value to be in turn translated into printed sections.

A possible four-port lumped circuit model of the printed unit cell in Fig. 7.11*a* is depicted in Fig. 7.11*b*. Specifically, the uncoupled section of the circuit is modeled using the lumped loads (L_1, C_1) and (L_2, C_2) per line. For the coupled sections, mutual inductances and/or a capacitor connecting the two circuit branches can be introduced to create controlled coupling to emulate nondiagonal anisotropy. Cascading these uncoupled and coupled sections then forms the DBE unit cell. As an example, we consider the dispersion diagram (calculated by transfer matrix method—see Chap. 6) of a unit cell whose top branch in the uncoupled section is capacitively loaded. From Fig. 7.11*c*, it is clear that the circuit's dispersion diagram can be significantly altered by varying the value of the coupling capacitor C_M. Specifically, for $C_M = 1$ pF the circuit displays a regular band edge (RBE) behavior. However, for $C_M = 2.3$ pF the resulting K-ω curve displays a DBE behavior. If $C_M = 4$ pF, a double band edge (DbBE) is obtained [27].

An important reason for using lumped circuit model is to provide guidelines for designing the DBE K-ω diagram. For instance, in the above circuit example, changing the value of the mutual inductance (L_M) for constant C_M does not affect the dispersion diagram. However, when the uncoupled branch is inductively loaded (i.e.,

$L_1 = 10$ nH, $C_1 = 1$ pF) the same DBE behavior is obtained for $L_M = 2.3$ nH. That is, an appropriate coupling mechanism must be present to realize the DBE. Regardless, the circuit model allows for a convenient approach to develop a printed DBE antenna. In addition, introduction of anisotropy increases the overall number of $K = \pi$ or $K = 0$ resonances by creating additional K-ω branches (i.e., one branch per transmission line). This mode diversity and concurrent tuning flexibility offered by coupling may play a key role in designing planar and small footprint antennas operating at multiple frequencies.

7.4.3 DBE Antenna Design Using Dual Microstrip Lines

In this subsection, we proceed to harness the maximally flat DBE band diagram to realize reduced size printed antennas. To do so, we cascade the unit cells in Fig. 7.11 in a circular periodic fashion as in Fig. 7.12*a*. We note that the transmission line sections consisting of longer uncoupled lines are bent toward the center of the structure to keep the footprint as small as possible. To achieve greater control in bending the K-ω diagram, the circuit layout is enhanced with additional lumped loads. Specifically, one of the uncoupled line branches is capacitively loaded whereas a capacitive coupling mechanism between the circuit branches is employed for K-ω bending. Overall, the lumped element circuit model allows for a convenient way to implement DBE unit cells on uniform microwave substrates using partially coupled and loaded microstrip lines [34].

Figure 7.12*b* demonstrates a lumped circuit element loaded microstrip line layout that implements the proposed DBE circuit on a 125 mil thick Duroid ($\varepsilon_r = 2.2$, $\tan\delta = 0.0009$) substrate. The corresponding dispersion diagram is computed using the transfer matrix approach once the ABCD matrix [22] of the unit cell has been extracted via a full-wave electromagnetic solver (such as Ansoft HFSS). As expected, from the design guidelines, the bending of the K-ω curve and the location of the $K = \pi$ resonance is conveniently controlled by increasing the coupling amount. For $C_M = 0.65$ pF, the layout exhibits a DBE behavior around 2.25 GHz and provides > 20% reduction of the $K = \pi$ resonance frequency.

The fabricated DBE antenna and its measured performance is depicted in Fig. 7.13*a* to 7.13*c* [53]. The antenna is printed on a 2″ × 2″, 125 mil thick Duroid substrate ($\varepsilon_r = 2.2$, $\tan\delta = 0.0009$). As usual, the back side of the substrate is used as the ground plane. For impedance matching, the antenna is fed with a capacitively coupled 50 Ω coaxial cable. The dual transmission lines of the antenna were also loaded with S series R05 high Q capacitors obtained from Johanson Technology, CA, USA (www.johansontechnology.com). From Fig. 7.13*b*, a

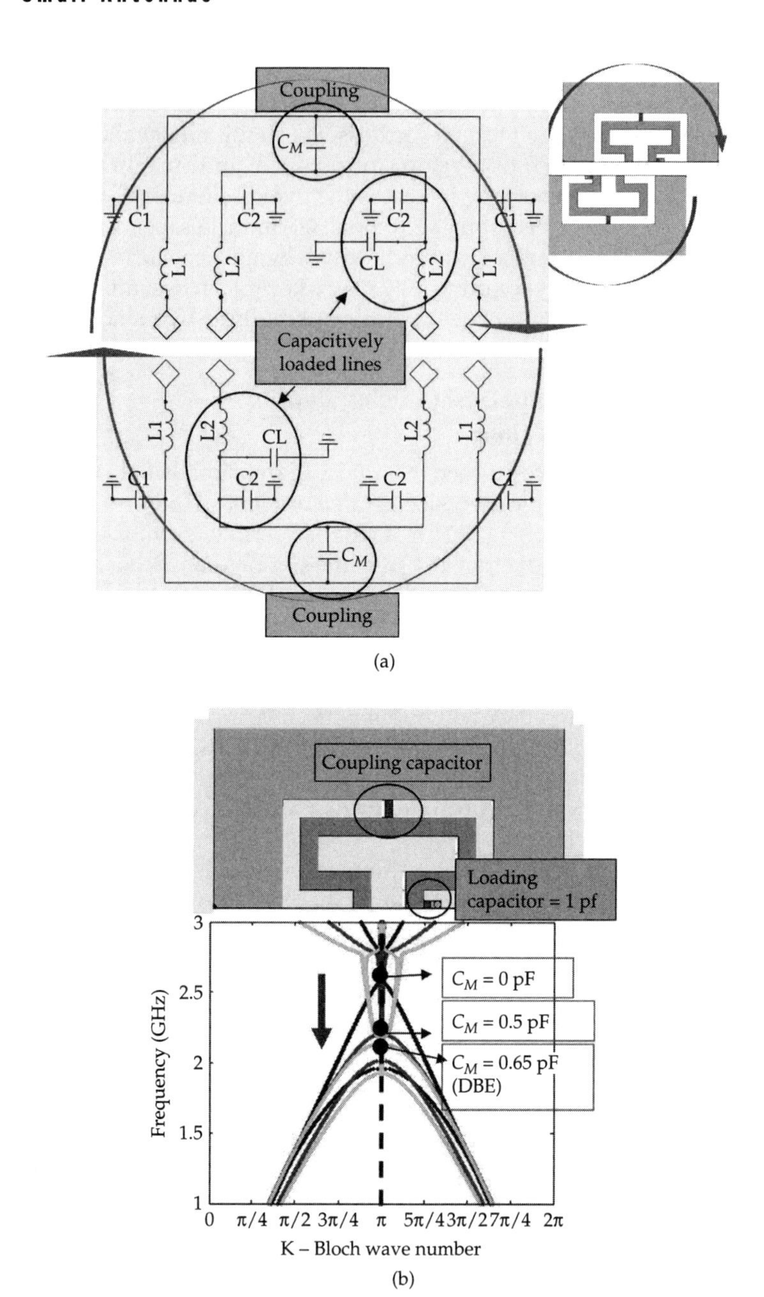

FIGURE 7.12 (a) Lumped element model of the DBE antenna composed of circularly cascaded two unit cells; (b) Loaded dual transmission lines implementing DBE unit cell on a 125 mil thick Duroid ($\varepsilon_r = 2.2$, $\tan\delta = 0.0009$) substrate and corresponding dispersion diagram. $K = \pi$ resonance is taken to lower frequencies by switching to a DBE. (*After Mumcu et al. [53], ©IEEE 2009.*)

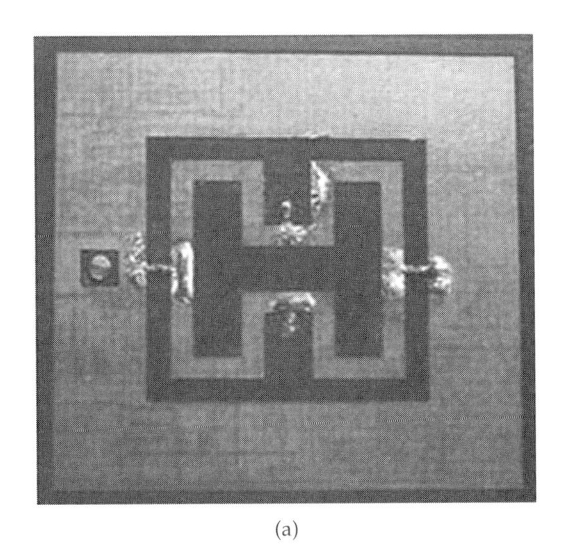

(a)

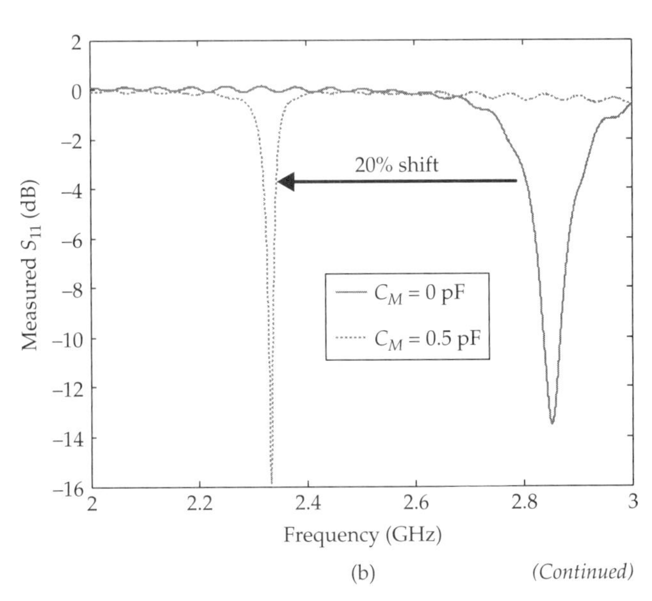

(b) *(Continued)*

FIGURE 7.13 (a) 1″ × 1″ DBE antenna layout on 2″ × 2″, 125 mil thick Duroid substrate ($\varepsilon_r = 2.2$, $\tan\delta = 0.0009$); (b) Measured $|S_{11}| < -10$ dB resonances for different coupling capacitors; (c) Measured gain pattern at 2.33 GHz. The antenna has 0.4% bandwidth and 2.6 dB realized gain corresponding to 38% radiation efficiency. (*After Mumcu et al. [53], ©IEEE 2009.*)

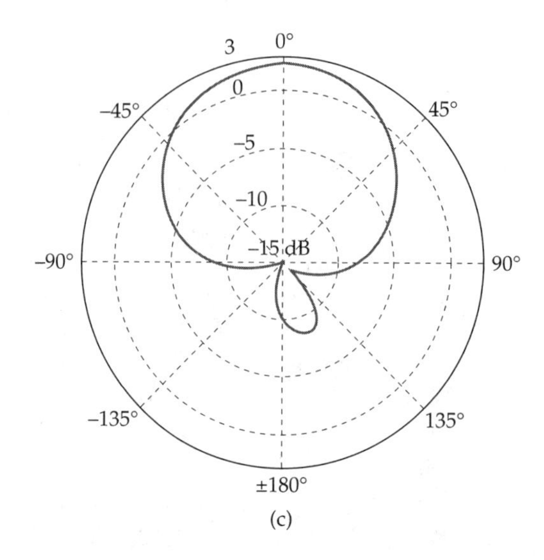

FIGURE 7.13 *Continued.*

coupling capacitor of $C_M = 0.5$ pF shifts the antenna resonance from 2.85 to 2.35 GHz (as expected from the dispersion diagram). On the other hand, a smaller $|S_{11}| < -10$ dB bandwidth of 0.4% is measured at 2.35 GHz as compared to a 0.9% bandwidth at 2.85 GHz. However, we note that the substrate thickness can also be adjusted to compensate bandwidth reduction (without compromising other antenna parameters). For instance, increasing the substrate thickness to 375 mil results in a considerably larger bandwidth of 1.9% for $C_M = 0.5$ pF. The measured broadside directivity in Fig. 7.13*c* agrees well with the 6.78 dB directivity computed using Ansoft HFSS. However, due to the high losses associated with the chip capacitors, the realized gain is only 2.6 dB as compared to a 6.3 dB gain for the unloaded antenna. This rather low 38% antenna efficiency can be substantially improved using higher Q lumped elements or interdigital microstrip capacitors. Of importance is that in contrast to a patch antenna operating at the same frequency and on the same substrate, the DBE antenna in Fig. 7.13*a* provides 47% footprint size reduction. In addition, it occupies 36% less area as compared to a simple loop antenna.

To further reduce DBE antenna size, a high-contrast substrate can be used. Among readily available high-contrast dielectric materials, alumina (A_2O_3) having $\varepsilon_r = 9.6$ and $\tan\delta = 0.0003$ was selected to keep substrate losses at a minimum. The design followed the method described above. However, lumped element circuit inclusions were not considered for this antenna example [32]. The resulting antenna footprint was 0.85″ × 0.88″ and the DBE resonance occurred at a much lower frequency (~1.45 GHz) due to the higher dielectric constant. As expected, due to the higher dielectric constant substrate, this DBE antenna had a very small bandwidth (much less than 1%) on a 100 mil thick substrate. Thus, the substrate thickness was increased to 500 mil without changing the layout to increase bandwidth. It was also observed that decreasing the outer line widths resulted in wider bandwidths.

Figure 7.14*a* depicts the fabricated DBE antenna design using AD–995 substrate layers (from CoorsTek Inc.). The AD–995 substrate is 99.5% pure alumina with electrical properties $\varepsilon_r = 9.7$ and $\tan\delta = 0.0001$. Since the thickest (readily available) AD–995 substrate is only 50 mil, 10 such layers were stacked to realize the 500 mil thick substrate. These 10 layers were held together using two plastic straps (instead of glue) to minimize losses. The antenna was again fed through a capacitively coupled coaxial cable, realizing a probe. This design occupied a remarkably small $\lambda_0/9 \times \lambda_0/9 \times \lambda_0/16$ footprint at 1.48 GHz. The back surface of the substrate was metal coated to form the ground plane. The measured gain and bandwidth plots, given in Fig. 7.14*b* and 7.14*c*, show a good agreement with the calculated curves. Specifically, a 4.5 dB broadside gain with 3.0% $|S_{11}| < -10$ dB bandwidth was measured. This 4.5 dB gain also implies a radiation efficiency as high as 95%.

The small electrical size and large radiation efficiency place the above antenna among the smallest in the recently reported literature. Although other metamaterial antennas [23, 24] achieved a small electrical size (on the order of $\lambda_0/10 \times \lambda_0/10$), the included capacitive and inductive loadings resulted in radiation efficiencies of less than 60%. Specifically, the zeroth order planar patch in [23] has a very small bandwidth and the miniature antenna in [24] suffers from small bandwidth and a low 44% radiation efficiency. On the other hand, although a radiation efficiency of 97% was reported for the metamaterial-inspired small antennas in [54], these designs had relatively small bandwidth.

To better evaluate antenna performance, Fig. 7.15 compares the gain-bandwidth products of various antennas from recent literature. For a more fair comparison, the footprint size of the DBE antenna and the one in [24] was used in computing the horizontal axis *ka*. That is $ka = 2\pi a/\lambda$, where a is the radius of the smallest semi-sphere that can enclose the printed antenna. The optimal performance is calculated in accordance to [55–57] (see Chap. 1). However, since small antennas can exceed the gain limit defined in [56], another curve with a fixed gain of 3 is also provided. As seen, the DBE antenna provides a rather large G/Q ratio better than other metamaterial antennas [24, 54].

(a)

FIGURE 7.14 (a) Fabricated DBE antenna on 2″ × 2″, 500 mil thick alumina substrate ($\varepsilon_r = 9.6$, $\tan\delta = 0.0003$); the footprint size $\lambda_0/9 \times \lambda_0/9 \times \lambda_0/16$ at 1.48 GHz; (b) Measured gain pattern; (c) Measured $|S_{11}| < -10$ dB bandwidth. The antenna has 3% bandwidth and 4.5 dB realized gain corresponding to > 90% radiation efficiency. (*After Mumcu et al. [32], ©IEEE 2009.*)

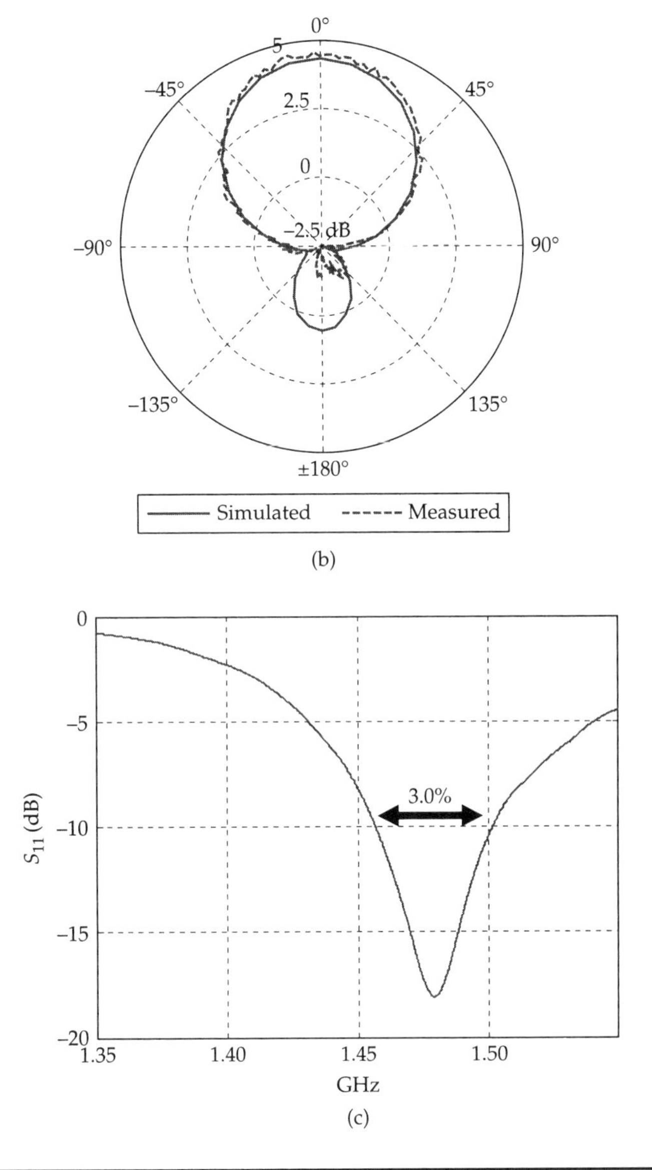

FIGURE 7.14 *Continued.*

Specifically, the footprint size of the MS-DBE antenna is slightly larger than the metamaterial designs; however, it has a performance close to the optimum limit. The spherical folded helix proposed in [58] performs better by utilizing the spherical volume. By comparison, the

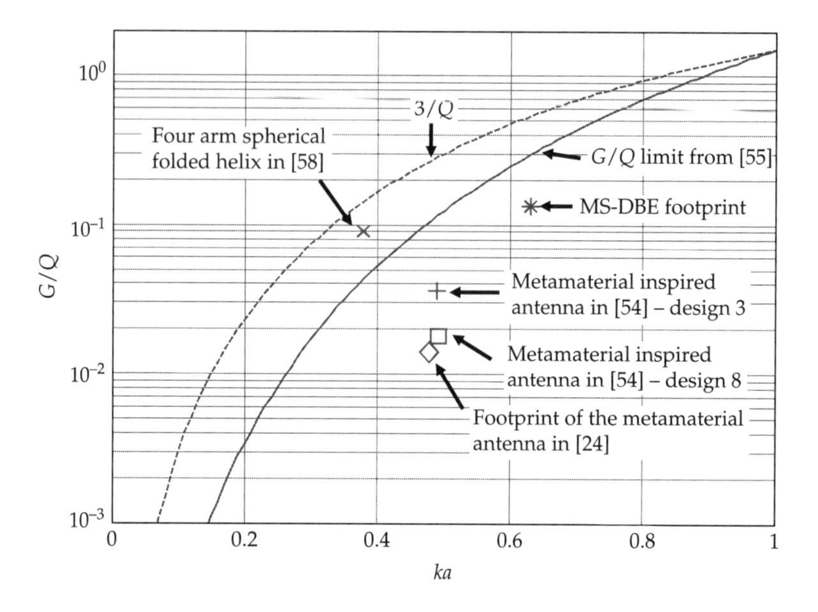

FIGURE 7.15 Comparison of G/Q ratio of the antenna in Fig. 7.14 (*After Mumcu et al. [32], ©IEEE 2009.*)

DBE antenna is planar and therefore appropriate for conformal installations. Inclusion of lumped elements within the design is expected to provide further size reduction. In addition, the broadside radiation pattern of the DBE antenna may be advantageous as it relates to conformal phased antenna arrays.

7.4.4 Coupled Double Loop Antennas

Although capacitively loaded DBE antennas provide a significant footprint reduction with respect to a patch antenna, their radiation efficiency suffers from conductor and capacitor losses. A closer observation of the resonance field distribution (see Fig. 7.16*a*) shows that the inner line capacitors are located at the field nulls and therefore not decisive on affecting resonance frequency. Hence, removal of inner line capacitors may lead to more efficient antenna layouts without affecting other radiation parameters. For example, Fig. 7.17*a* demonstrates a coupled double loop (CDL) antenna that incorporates interdigital capacitors for coupling. The antenna resonates at 2.26 GHz and exhibits almost identical 0.4% impedance bandwidth as compared to the DBE antenna presented in Fig. 7.13. On the other hand, the measured 3.9 dB gain (51% efficiency) is 1.3 dB higher than that of

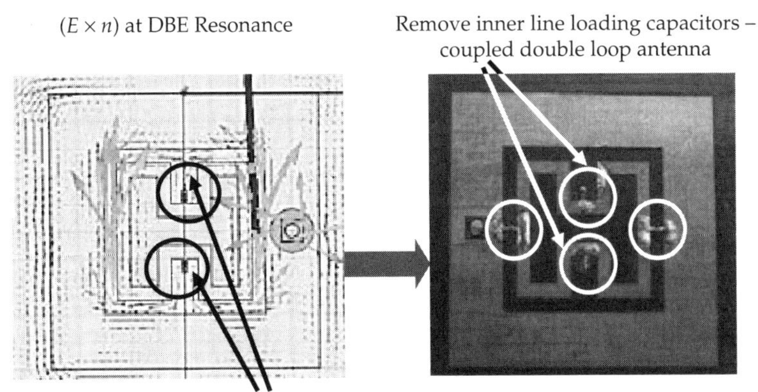

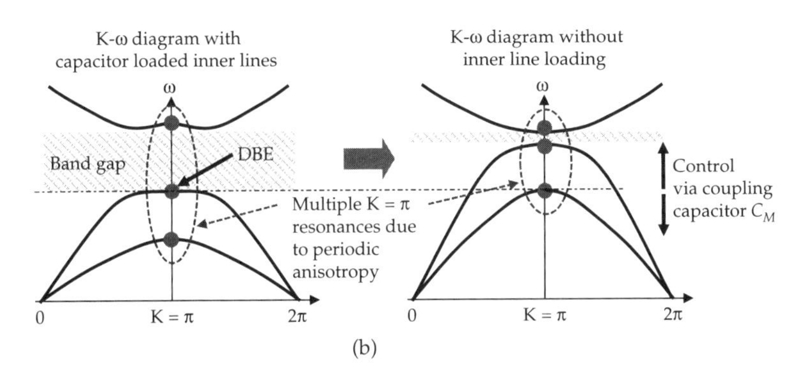

FIGURE 7.16 (a) Resonant field distribution on the top surface of the lumped capacitor loaded DBE antenna (see also Fig. 7.13). Inner line capacitors are located at the field nulls; (b) Removal of inner line capacitors and their effect on the dispersion diagrams.

the original DBE antenna (i.e., 2.6 dB gain with 38% efficiency). An alternative CDL antenna implemented using high Q lumped capacitors was also measured to provide a gain as high as 4.6 dB with 60% radiation efficiency at 2.35 GHz. It should also be remarked that the maximum radiation efficiency to be achieved by the outer loop (on the same substrate) is only 80% at 2.85 GHz.

Figure 7.16*b* illustrates how the K-ω diagram gets modified when the inner line capacitors are removed from the printed DBE unit cell. As expected, the unit cell seizes to exhibit a DBE behavior. In addition, the K $= \pi$ resonances of different K-ω branches shift to other frequencies. Nevertheless, a controllable (via capacitive loading)

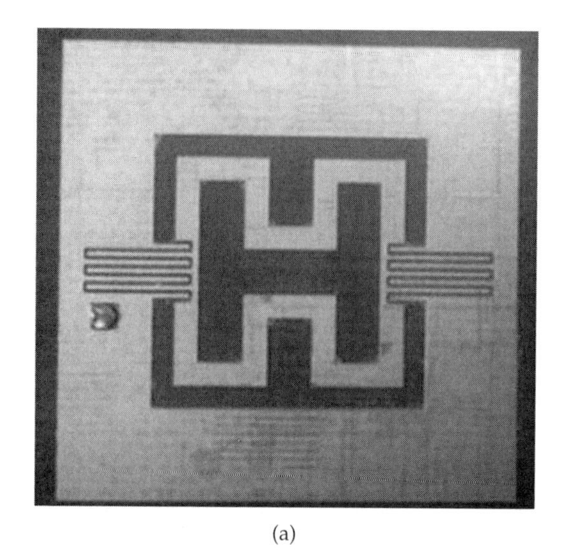

(a)

0.4 pF lumped capacitor for impedance matching

0.8 pF lumped capacitors

y

x

(b)

FIGURE 7.17 (a) Capacitively loaded 1″ × 1″ double loop (CDL) antenna layout printed on a 2″ × 2″, 125 mil thick Duroid substrate ($\varepsilon_r = 2.2$, $\tan\delta = 0.0009$); this antenna has improved realized gain of 3.9 dB with 51% efficiency (as opposed to 2.6 dB gain for the DBE antenna in Fig. 7.13) at 2.26 GHz; (b) Fabricated double loop antenna on a 250 mil thick 1.5″ × 1.5″ Rogers TMM 10i ($\varepsilon_r = 9.8$, $\tan\delta = 0.002$) substrate. An additional 0.4 pF capacitor is connected between the coaxial probe and the outer microstrip line to improve $S_{11} < -10$ dB matching; (c) Comparison of simulated and measured return loss; (d) 4.34 dB x−pol gain is measured at 2.65 GHz on the y−z plane. Antenna footprint is $\lambda_0/9.8 \times \lambda_0/9.8 \times \lambda_0/19.7$ at 2.4 GHz.

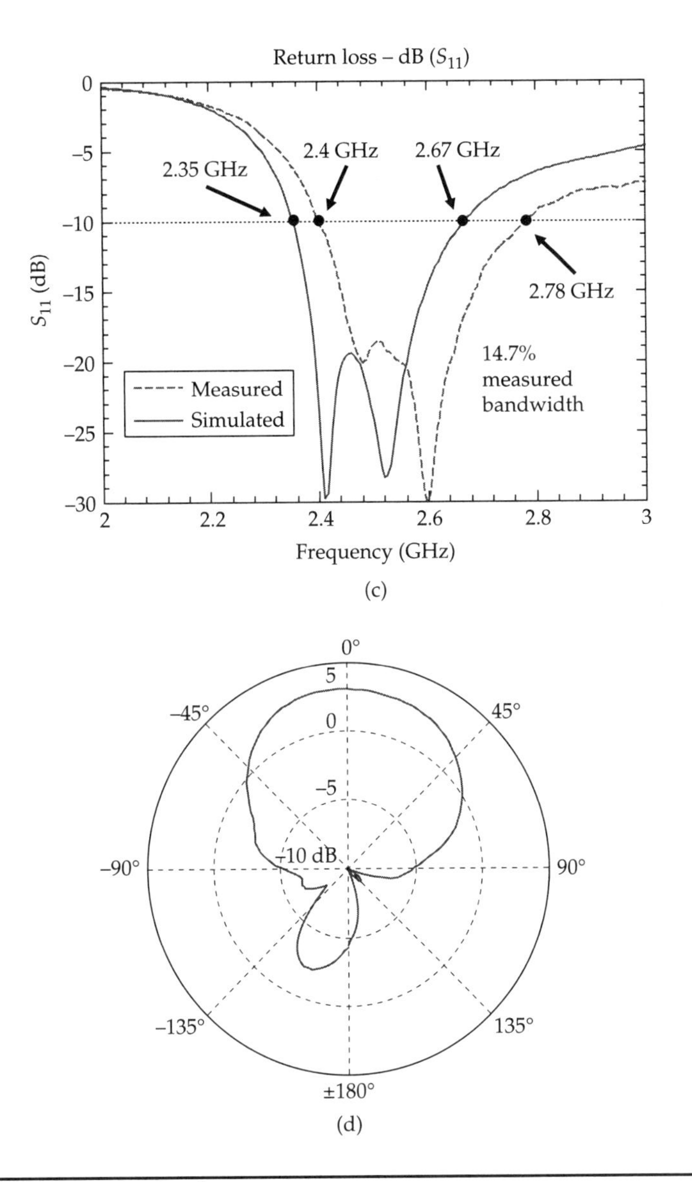

FIGURE 7.17 *Continued.*

K = π resonance associated with an RBE behavior continues to exist at the former DBE frequency. This modified layout can also be used to develop efficient miniature antennas with multiband or extended bandwidth performance. This can be achieved by simultaneously exciting and combining multiple K = π resonances.

To demonstrate bandwidth enhancement by exciting two nearby $K = \pi$ resonances, a miniature CDL antenna is presented in Fig. 7.17*b*. The antenna is implemented on a 250 mil thick 1.5″ × 1.5″ Rogers TMM 10i ($\varepsilon_r = 9.8$, $\tan\delta = 0.002$) substrate. The inserted 0201 size code capacitors are from Johanson Technology (S series) and have an equivalent series resistance of 0.35 Ω. Similar to previous designs, the antenna is fed by a capacitively coupled coaxial probe. To improve matching, a lumped capacitor was connected between the probe and the outer transmission line as shown in Fig. 7.17*b*. We note that the value of the capacitor (0.4 pF) was determined experimentally via trial and error. For the 0.8 pF coupling capacitors and the chosen layout, the calculated antenna resonances are at 2.52 GHz and 2.32 GHz (see Fig. 7.17*c*) and provide a 12.7% bandwidth. From Fig. 7.17*c* and 7.17*d*, it is observed that the measured gain and bandwidth curves are in reasonable agreement with the simulated ones. Specifically, a bandwidth of 14.7% from 2.4 to 2.78 GHz was measured. Also, this double loop antenna exhibited 4.34 dB x-pol. gain at 2.65 GHz with corresponding co-pol. gain measured to be larger than 3.5 dB across the entire bandwidth. The maximum cross-pol. gain level was determined to be 6.5 dB lower than the co-pol. gain. But this cross polarization level can be further reduced by employing a better feeding mechanism and resorting to a more symmetric unit cell. The radiation efficiency was found to be above 80% over the entire bandwidth. Overall, the antenna footprint was only $\lambda_0/9.8 \times \lambda_0/9.8 \times \lambda_0/19.7$ at 2.4 GHz.

7.4.5 Printed MPC Antennas on Biased Ferrite Substrates

As observed in above sections, DBE and CDL antennas suffered from narrower bandwidths as their footprints were miniaturized via capacitive loadings. It is, nevertheless, possible to increase bandwidth by employing magnetic substrate (i.e., $\varepsilon_r > 1$, $\mu_r > 1$) inserts that will allow the realization of the MPC modes using the coupled line concept. Specifically, a ferrite substrate section placed under the coupled lines allows emulating spectral asymmetry of the MPC modes. Changing the bias of the ferrite adjusts the nonreciprocal current displacement effects and bends the K-ω branch to exhibit the SIP behavior [59].

To design a printed MPC antenna, the partially coupled microstrip line layout was implemented on a composite substrate having ferrite insertions. Specifically, to realize the nonreciprocal current displacement effect, the coupled lines were placed on calcium vanadium garnet blocks (from TCI Ceramics, $4\pi M_s = 1000$ G, $\Delta H = 6$Oe, $\varepsilon_r = 15$, $\tan\delta = 0.00014$). The specific design is shown in Fig. 7.18*a*. An external bias field (1000 G) was applied normal to the ground plane to saturate

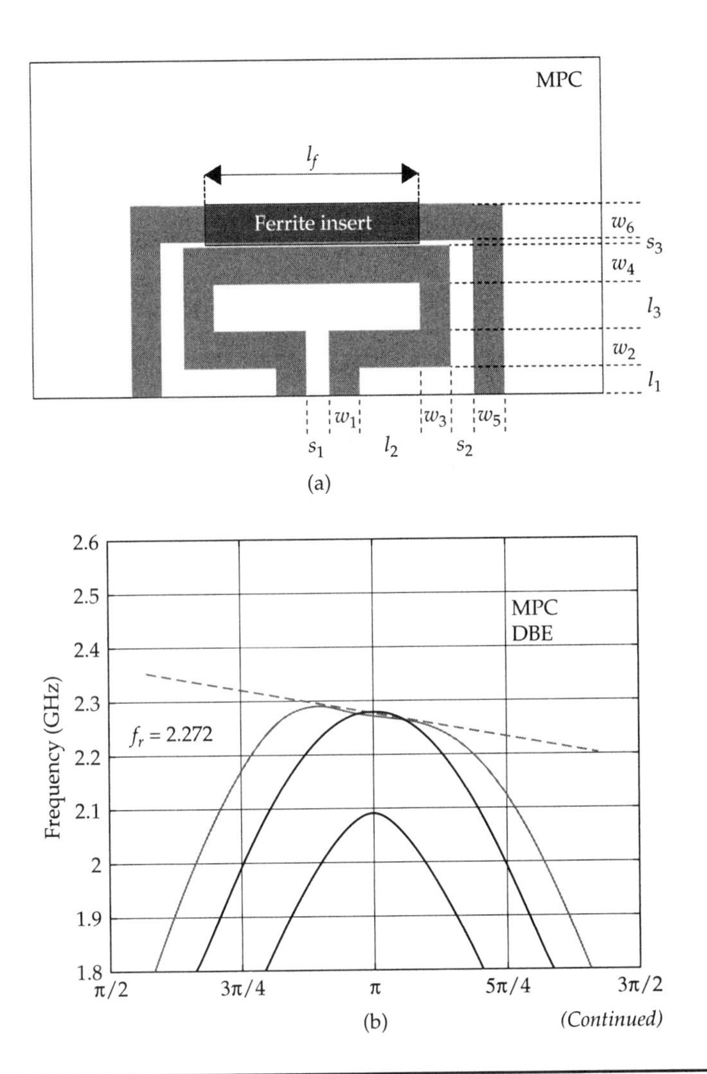

FIGURE 7.18 (a) MPC unit cell with design parameters (in mils): $w_1 = w_2 = w_3 = 100$, $w_4 = 20$, $w_5 = 80$, $w_6 = 120$, $s_1 = 50$, $s_2 = 70$, $s_3 = 10$, $l_1 = 50$, $l_2 = 60$, $l_3 = 100$, $l_f = 500$; (b) corresponding dispersion diagram of the unit cell; (c) Fabricated MPC antenna on composite substrate. Calcium vanadium garnet (CVG, $4\pi M_s = 1000\ G$, $\Delta H = 6\,\text{Oe}$, $\varepsilon_r = 15$, $\tan\delta = 0.00014$) sections are inserted into the low-contrast Duroid substrate ($\varepsilon_r = 2.2$, $\tan\delta = 0.0009$). Bottom view of the antenna with magnets is shown on the bottom right inset; (d) Comparison of simulated and measured gains; (e) Miniature MPC antenna performance. The substrate is formed by inserting CVG sections into the high-contrast Rogers RT/Duroid 6010 laminate ($\varepsilon_r = 10.2$, $\tan\delta = 0.0023$) (After Apaydin et al. [60] ©IET.)

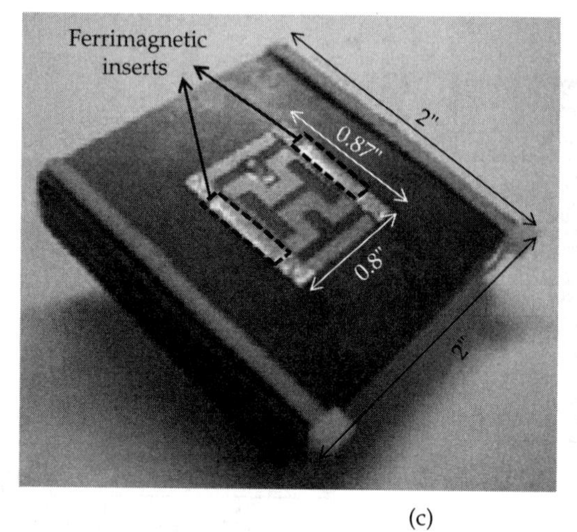

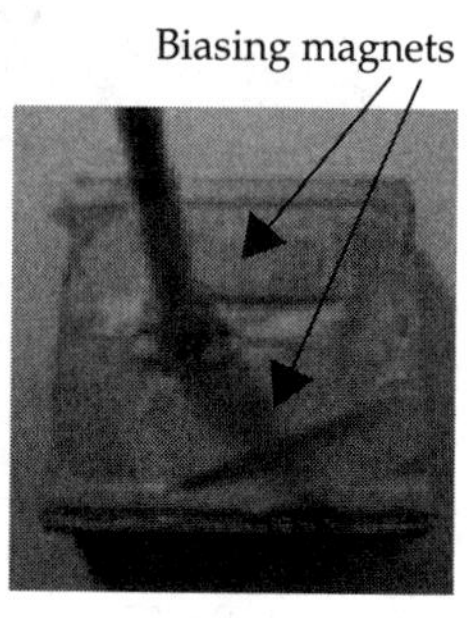

(c)

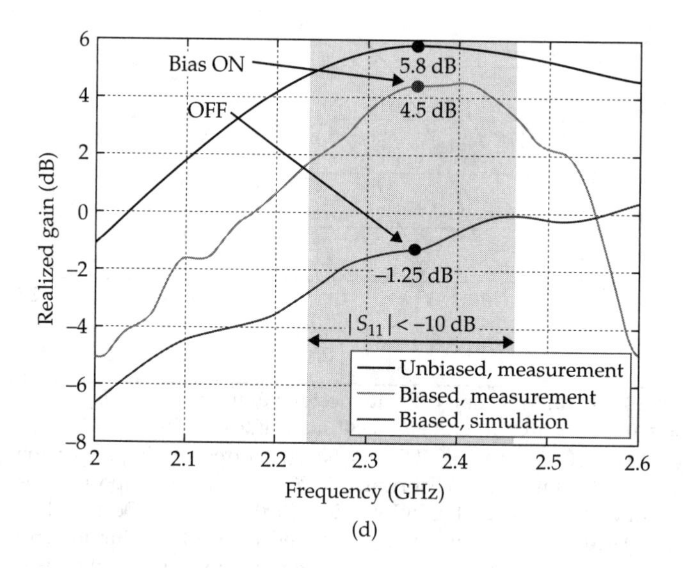

(d)

FIGURE 7.18 *Continued.*

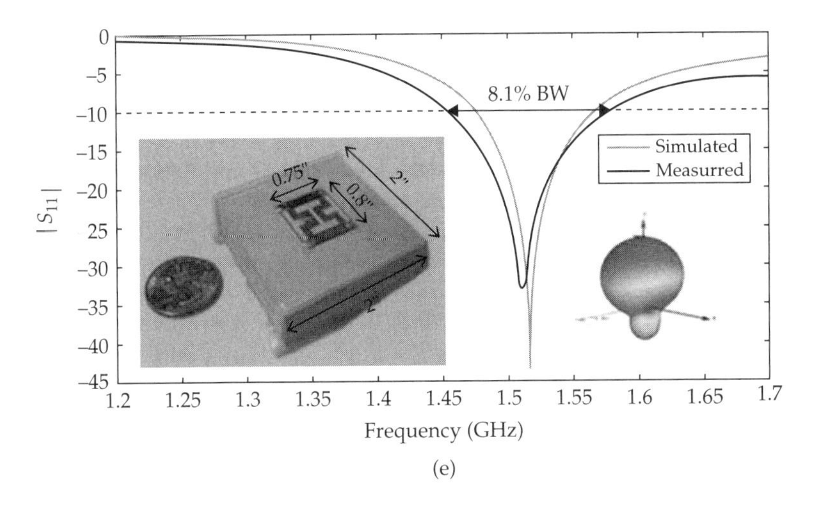

(e)

FIGURE 7.18 *Continued.*

the magnetic inserts. The ferrite blocks were inserted into the low contrast 2″ × 2″, 500 mil thick Duroid ($\varepsilon_r = 2.2$, $\tan\delta = 0.0009$) substrate. We note that for this design the uncoupled lines were printed on the earlier Duroid substrate, whereas the coupled lines were placed on the ferrite blocks using copper tapes. The layout was then tuned to achieve the MPC dispersion (see Fig. 7.18*b*) by appropriately selecting the line widths, lengths, and coupled line separations [60]. The final fabricated design is depicted in Fig. 7.18*c* and its gain/bandwidth performance is demonstrated in Fig. 7.18*d*. As seen, the antenna has 4.5 dB realized gain at 2.35 GHz with 9.1% impedance bandwidth. The corresponding efficiency is 67%, and this lower value may be attributed to the nonuniformity of the biasing field through the ferrite blocks.

To achieve further miniaturization, a similar MPC antenna design (see Fig. 7.18*e*) was carried out on a 2″ × 2″, 500 mil thick high contrast Rogers RT/Duroid 6010 substrate ($\varepsilon_r = 10.2$, $\tan\delta = 0.0023$). This antenna delivered 3.1 dB realized gain with 8.1% bandwidth at 1.51 GHz. The measured 73% radiation efficiency was again lower than expected due to the losses associated with nonuniformly biased ferrite sections. Nevertheless, the antenna had a remarkably small $\lambda_0/9.8 \times \lambda_0/10.4$ footprint on $\lambda_0/16$ thick substrate, making it near optimal in terms of gain-bandwidth product with respect to Chu-Harrington limit.

7.5 Platform/Vehicle Integration of Metamaterial Antennas

When installing antennas on vehicles, one concern is the integration of antennas into the body of platforms. Conformal integration of antenna structures is especially crucial for airborne targets to reduce drag and observability. During integration, a challenge relates to the limited available surface for antenna placement. This is even more acute for ground vehicles, where the hood and most of the roof cannot be utilized. That is, although large conducting surfaces are available, the antenna needs to be placed (still conformably) at the edges of vehicle roof, doors, bumpers, and window frames. These restrictions place major design challenges, and imply antennas that are less susceptible to nearby structure effects. Miniaturization can play a critical role in reducing loading effects without producing pronounced degradation in antenna bandwidth, gain, and radiation pattern.

The observed footprint reduction in DBE, CDL, and MPC antennas also implies that the substrate size can concurrently be reduced without affecting much performance (as compared to a patch antenna). For example, in Fig. 7.19*a*, we consider a smaller $1.5'' \times 1.5'' \times 0.5''$ substrate (as compared to MPC antenna in Fig. 7.18*c*) and choose to recess the MPC antenna below the ground plane by placing it in a cavity. The recessed (embedded) MPC antenna, in addition to being flush, is also less susceptible to interference from nearby edges and corners when placed above a vehicle platform. It is well known that when antennas are placed close to corners or edges of platforms, their resonance frequency gets detuned due to the location dependent ground plane size and geometry. A robust antenna system, on the contrary, mitigate such platform effects and have the flexibility to be used almost anywhere on the vehicle, independent from its location and ground plane size with little or no degradation in performance.

To demonstrate the advantages of the embedded MPC antenna (as compared to the more protruding and larger patch antennas) we considered experiments with different ground plane sizes. Return losses of these antennas are shown in Fig. 7.19*b*. It is observed that antennas recessed in a ground plane larger than $3'' \times 3''$ stay tuned to a 2.7% bandwidth. Another advantage of MPC antennas is their tunability with applied magnetic bias. In fact, the antennas in Fig. 7.19*b* can be tuned so that they all resonate at 2.45 GHz by slightly changing the bias field. From Fig. 7.19*c*, we observe that all recessed MPC antennas share (stay tuned to) the same 3% bandwidth. Specifically, retuning the magnetic bias allows the recessed antennas to be placed very close to or just at vehicle corners/edges.

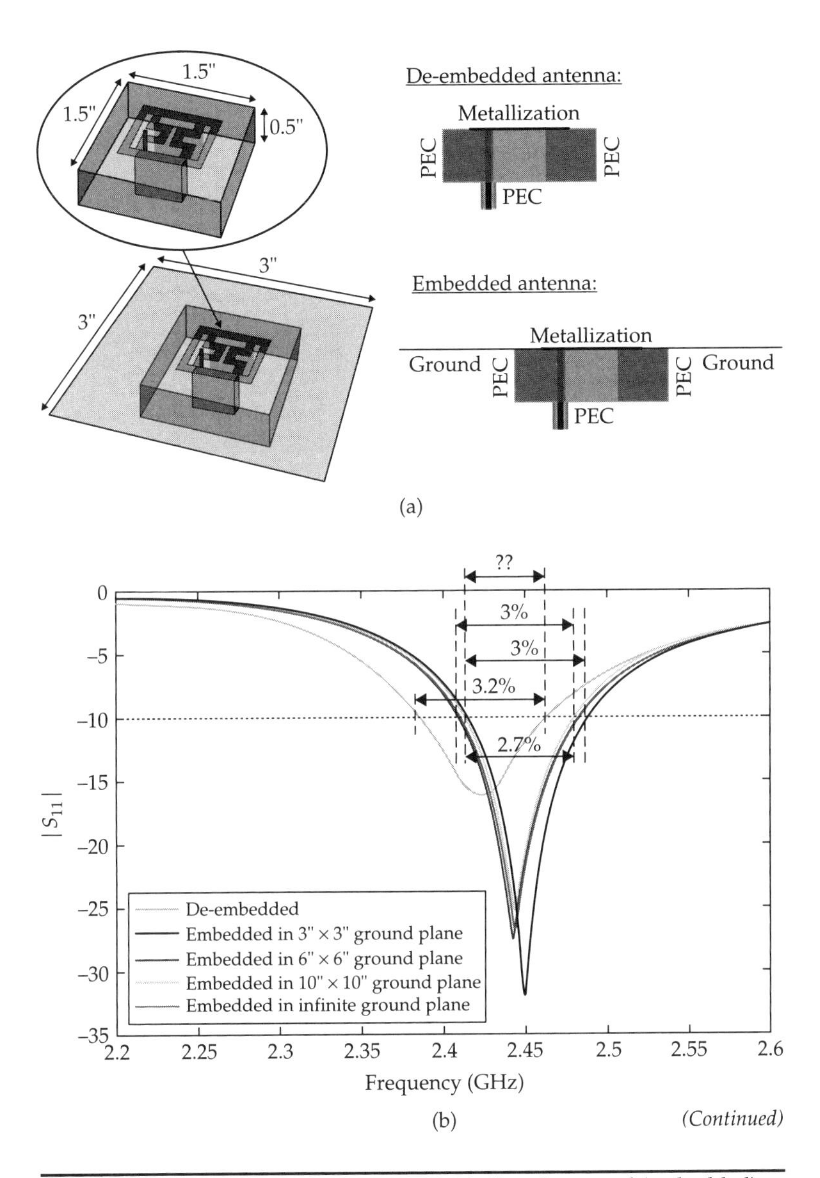

FIGURE 7.19 (a) Cavity backed (de-embedded) and recessed (embedded) MPC antenna geometries; (b) Return loss of the cavity-backed MPC antenna when recessed in different ground plane sizes; (c) Return loss of the cavity-backed MPC antenna when magnetic bias is adjusted to resonate at 2.45 GHz. (*After Irci et al. [61], ©Elsevier.*)

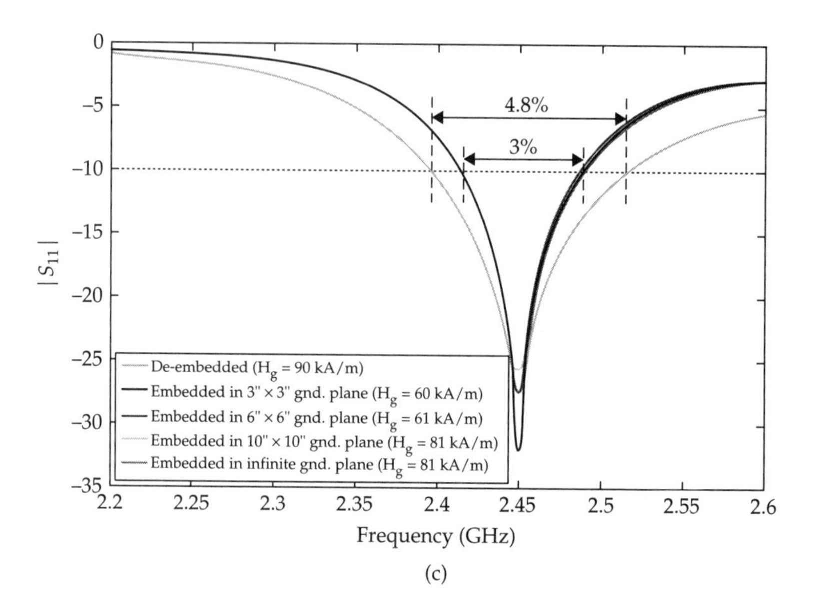

FIGURE 7.19 *Continued.*

To investigate the in situ performance of the cavity-backed MPC antenna recessed in 3″ × 3″ ground plane, we considered the case where magnetic biasing was fixed 80 kA/m. From Fig. 7.19*b*, it is understood that center of the MPC antenna should be at least 1.5″ away from any vehicle edge/corner to guarantee a 2.7% tuned bandwidth. Indeed, we considered several such locations at several locations on a real size high mobility multipurpose wheeled vehicle (HMMWV) mock-up for antenna placement at the 2.45 GHz design frequency.

Since the vehicle is electrically very large, we used the uniform theory of diffraction (UTD) package of the FEKO for analysis. Far-field radiation pattern of the MPC antenna was obtained using Ansoft HFSS commercial package. Then, this pattern was utilized in FEKO as excitation at the selected antenna locations to illuminate the mock-up HMMWV ground vehicle. Figure 7.20 summarizes the resulting radiation patterns for several antenna placement scenarios. The radiation patterns reveal that the cavity backed MPC retains the pattern shape with a broadside gain higher than 5.4 dB. We only observe some pattern ripple variations, particularly for the MPC antenna placed near the center of the rooftop. This ripple is, obviously, due to diffraction contribution from the roof edges and tips. As the antenna location changes, the phase of the diffraction terms also changes, causing ripple variations that are location dependent. These ripples can be avoided by placing the antennas closer to corners/edges of the vehicle [61].

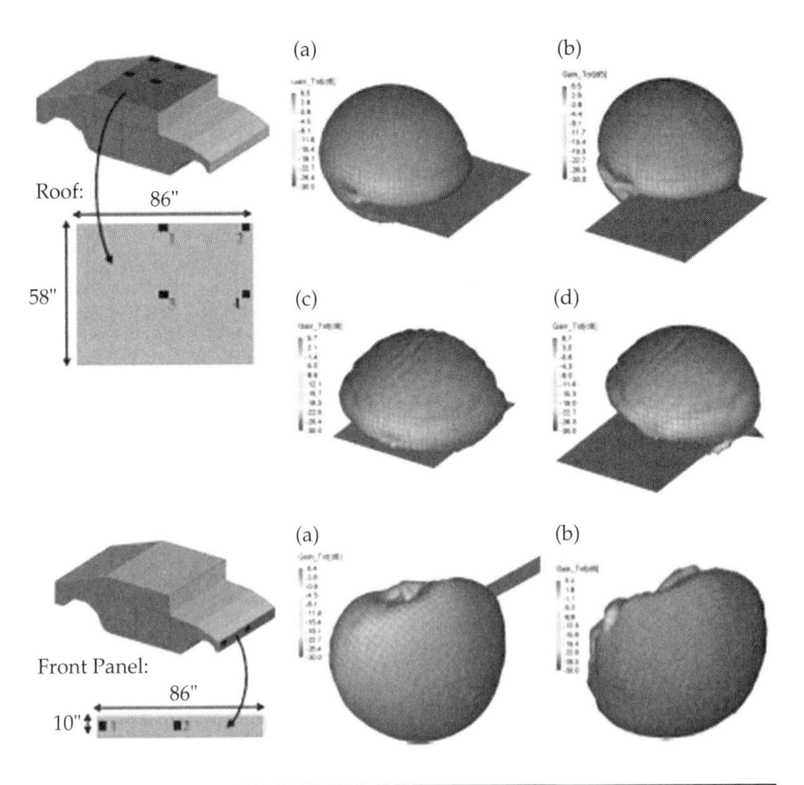

FIGURE 7.20 Antenna placement on HMMWV roof and front panel. (*After Irci et al. [61], ©Elsevier.*)

References

1. *IEEE Transactions on Antennas and Propagation, Special Issue on Metamaterials*, vol. 51, 2003.
2. D. Polder, "On the theory of ferromagnetic resonance," *Philosophical Magazine*, vol. 40, January 1949, pp. 99–115.
3. G. Tyras, "The permeability matrix for a ferrite medium magnetized at an arbitrary direction and its eigenvalues," *IRE Transactions on Microwave Theory and Techniques*, vol. 7, no. 1, January 1959, pp. 176–177.
4. H. Gundel, H. Riege, E.J.N. Wilson, J. Handerek, and K. Zioutas, "Fast polarization changes in ferroelectrics and their application in accelerators," *Nuclear Instruments and Methods in Physics Research. Section A*, vol. 280, 1989, pp. 1–6.
5. X. Zuo, H. How, S. A. Oliver, and C. Vittoria, "Development of high frequency ferrite phase-shifter," *IEEE Transactions on Magnetics*, vol. 37, no. 4, July 2001, pp. 2395–2397.
6. R. A. Shelby, D. R. Smith, and S. Schultz, "Experimental verification of a negative index of refraction," *Science*, vol. 292, 2001, pp. 77–79.
7. A. Grbic and G. V. Eleftheriades, "Overcoming the diffraction limit with a planar left handed transmission line lens," *Physical Review Letters*, vol. 92, no. 11, March 2004, pp. 117403 1–117403 4.

8. A. Grbic and G. V. Eleftheriades, "Practical limitations of subwavelength resolution using negative-refractive-index transmssion-line lenses," *IEEE Transactions on Antennas and Propagation*, vol. 53, no. 10, October 2005.
9. M. A. Antoniades and G. V. Eleftheriades, "Compact linear lead/lag metamaterial phase shifters for broadband applications," *IEEE Antennas and Wireless Propagation Letters*, vol. 2, no. 7, July 2003, pp. 103–106.
10. G. V. Eleftheriades and K. G. Balmain, *Negative–Refraction Metamaterial*. IEEE Press-John Wiley & Sons, New York, USA, 2005.
11. C. Caloz and T. Itoh, *Electromagnetic Metamaterials: Transmission Line Theory and Microwave Applications*, Wiley-IEEE Press, 2005.
12. E. Yablonovich. "Inhibited spontaneous emission in solid-state physics and electronics," *Physical Review Letters*, vol. 58, 1987, p. 2059.
13. J. Joannopoulos, R. Meade, and J. Winn, *Photonic Crystals—Molding the Flow of Light*, Princeton University Press, Princeton, N.J., 1995.
14. B. Temelkuran, M. Bayindir, E. Ozbay, R. Biswas, M. M. Sigalas, G. Tuttle, and K. M. Ho, "Photonic crystal based resonant antenna with a very high directivity," *Journal of Applied Physics*, vol. 87, no. 1, January 2000, pp. 603–605.
15. R. Biswas, E. Ozbay, B. Temelkuran, M. Bayindir, M. M. Sigalas, and K. M. Ho. "Exceptionally directional sources with photonic–bandgap crystals," *Journal of the Optical Society of America B*, vol. 18, no. 11, November 2001, pp. 1684–1689.
16. R. F. J. Broas, D. F. Sievenpiper, and E. Yablonovitch, "A high-impedance ground plane applied to a cellphone handset geometry," *IEEE Transactions on Microwave Theory and Techniques*, vol. 49, no. 7, July 2001, pp. 1262–1265.
17. F. Yang and Y. Rahmat-Samii, "A low profile circularly polarized curl antenna over an electromagnetic bandgap (EBG) surface," *Microwave and Optical Technology Letters*, vol. 31, no. 4, November 2001, pp. 264–267.
18. R. W. Ziolkowski and A. D. Kipple, "Application of double negative materials to increase the power radiated by electrically small antennas," *IEEE Transactions on Antennas and Propagation*, vol. 51, no. 10, October 2003, pp. 2626–2640.
19. A. Erentok, P. L. Luljak, and R. W. Ziolkowski, "Characterization of a volumetric metamaterial realization of an artificial magnetic conductor for antenna applications," *IEEE Transactions on Antennas and Propagation*, vol. 53, no. 1, January 2005, pp. 160–172.
20. G. Kiziltas, D. Psychoudakis, J. L. Volakis, and N. Kikuchi, "Topology design optimization of dielectric substrates for bandwidth improvement of a patch antenna," *IEEE Transactions on Antennas and Propagation*, vol. 51, no. 10, October 2003, pp. 2732–2743.
21. Y. E. Erdemli, K. Sertel, R. A. Gilbert, D. E. Wright, and J. L. Volakis, "Frequency-selective surfaces to enhance performance of broad-band reconfigurable arrays," *IEEE Transactions on Antennas and Propagation*, vol. 50, no. 12, December 2002, pp. 1716–1724.
22. R. Ziolkowski and A. Erentok, "Metamaterial-based efficient electrically small antennas," *IEEE Transactions on Antennas and Propagation*, vol. 54, no. 7, July 2006, pp. 2113–2129.
23. A. Sanada, M. Kimura, I. Awai, C. Caloz, and T. Itoh, "A planar zeroth-order resonator antenna using a left-handed transmission line," *34th European Microwave Conference*, Amsterdam, 2004.
24. K. M. K. H. Leong, C. J. Lee, and T. Itoh, "Compact metamaterial based antennas for MIMO applications," *International Workshop on Antenna Technology (IWAT)*, March 2007, pp. 87–90.
25. F. Qureshi, M. A. Antoniades, and G. V. Eleftheriades, "A compact and low-profile metamaterial ring antenna with vertical polarization," *IEEE Antennas and Wireless Propagation Letters*, vol. 4, 2005.
26. A. Figotin and I. Vitebsky, "Nonreciprocal magnetic photonic crystals," *Physical Review E*, vol. 63, no. 066609, May 2001, pp. 1–20.
27. A. Figotin and I. Vitebsky, "Gigantic transmission band-edge resonance in periodic stacks of anisotropic layers," *Physical Review E*, vol. 72, no. 036619, September 2005, pp. 1–12.

28. A. Figotin and I. Vitebsky, "Electromagnetic unidirectionality in magnetic photonic crystals," *Physical Review B*, vol. 67, no. 165210, April 2003, pp. 1–20.
29. L. Zhang, G. Mumcu, S. Yarga, K. Sertel, J. L. Volakis, and H. Verweij, "Fabrication and characterization of anisotropic dielectrics for low-loss microwave applications," *Journal of Materials Science*, vol. 43, no. 5, March 2008, pp. 1505–1509.
30. G. Mumcu, K. Sertel, and J. L. Volakis, "Printed coupled lines with lumped loads for realizing degenerate band edge and magnetic photonic crystal modes," *IEEE Antennas and Propagation Society Symposium*, San Diego, Calif., July 2008.
31. G. Mumcu, K. Sertel, and J. L. Volakis, "Partially coupled microstrip lines for antenna miniaturization," *IEEE International Workshop on Antenna Technology: Small Antennas and Novel Metamaterials (IWAT)*, Santa Monica, CA, March 2009.
32. G. Mumcu, K. Sertel, and J. L. Volakis, "Miniature antenna using printed coupled lines emulating degenerate band edge crystals," *IEEE Transactions on Antennas and Propagation*, vol. 57, no. 6, June 2009, pp. 1618–1624.
33. B. E. A. Saleh and M. C. Teich, *Fundamentals of Photonics*, Ch. 15, John Wiley & Sons Inc., New York, USA, 1991.
34. G. Mumcu, K. Sertel, and J. L. Volakis, "Lumped Circuit models for degenerate band edge and magnetic photonic crystals," *IEEE Microwave and Wireless Components Letters*, vol. 20, no. 1, January 2010, pp. 4–6.
35. G. Mumcu, K. Sertel, J. L. Volakis, I. Vitebskiy, and A. Figotin, "RF propagation in finite thickness unidirectional magnetic photonic crystals," *IEEE Transactions on Antennas and Propagation*, vol. 53, no. 12, December 2005, pp. 4026–4034.
36. J. L. Volakis, G. Mumcu, K. Sertel, C. C. Chen, M. Lee, B. Kramer, D. Psychoudakis, and G. Kiziltas, "Antenna miniaturization using magnetic photonic and degenerate band edge crystals," *IEEE Antennas and Propagation Magazine*, vol. 48, no. 5, October 2006, pp. 12–28.
37. S. Yarga, K. Sertel, and J. L. Volakis, "A directive resonator antenna using degenerate band edge crystals," *IEEE Transactions on Antennas and Propagation*, vol. 57, no. 3, March 2009, pp. 799–803.
38. S. Yarga, K. Sertel, and J. L. Volakis, "Degenerate band edge crystals for directive antennas," *IEEE Transactions on Antennas and Propagation*, vol. 56, no. 1, January 2008, pp. 119–126.
39. G. Mumcu, K. Sertel, and J. L. Volakis, "Miniature antennas and arrays embedded within magnetic photonic crystals," *IEEE Antennas and Wireless Propagation Letters*, vol. 5, no. 1, December 2006, pp. 168–171.
40. D. R. Jackson and N. G. Alexopoulos, "Gain enhancement methods for printed circuit antennas," *IEEE Transactions on Antennas and Propagation*, vol. 33, no. 9, September 1985, pp. 976–987.
41. H. Y. Yang and N. G. Alexopoulos, "Gain enhancement methods for printed circuit antennas through multiple substrates," *IEEE Transactions on Antennas and Propagation*, vol. 35, no. 7, July 1987, pp. 860–863.
42. J. L. Volakis, G. Mumcu, and K. Sertel, "Miniature antennas within degenerate band edge crystals," *Proceedings of Joint 9th International Conference on Electromagnetics in Advanced Applications, ICEAA, and 11th European Electromagnetic Structures Conference, EESC*, Torino, Italy, September 2005.
43. G. Mumcu, K. Sertel, and J. L. Volakis, "A measurement process to characterize natural and engineered low-loss uniaxial dielectric materials at microwave frequencies," *IEEE Transactions on Microwave Theory and Techniques*, vol. 56, no. 1, January 2008, pp. 217–223.
44. R. E. Collin, "A simple artificial anisotropic dielectric medium," *IEEE Transactions on Microwave Theory and Techniques*, vol. 6, April 1958, pp. 206–209.
45. M. M. Z. Kharadly and W. Jackson, "The properties of artificial dielectrics comprising arrays of conducting elements," *Proceedings of the IEE*, vol. 100, July 1953, pp. 199–212.
46. W. E. Kock, "Metallic delay lenses," *Bell System Technical Journal*, vol. 27, May 1948, pp. 58–82.

47. A. Munir, N. Hamanaga, H. Kubo, and I. Awai, "Artificial dielectric rectangular resonator with novel anisotropic permittivity and its TE modewaveguide filter application," *IEICE Transactions on Electronics*, vol. E88-C, no. 1, January 2005, pp. 40–46.
48. I. Awai, H. Kubo, T. Iribe, D.Wakamiya, and A. Sanada, "An artificial dielectric material of huge permittivity with novel anisotropy and its application to a microwave BPF," *IEICE Transactions on Electronics*, vol. E88-C, no. 7, July 2005, pp. 1412–1419.
49. S. Yarga, K. Sertel, and J. L. Volakis, "Finite element method for periodic structures," *Presented at the 8th International Workshop on Finite Elements for Microwave Engineering*, Stellenbosch, South Africa, May 25–26, 2006.
50. S. Yarga, K. Sertel, and J. L. Volakis, "Multilayer dielectric resonator antenna operating at degenerate band edge modes," *IEEE Antennas and Wireless Propagation Letters*, vol. 8, 2009, pp. 287–290.
51. C. Locker, K. Sertel, and J. L. Volakis, "Emulation of propagation in layered anisotropic media with equivalent coupled microstrip lines," *IEEE Microwave and Wireless Component Letters*, vol. 16, no. 12, December 2006, pp. 642–644.
52. G. Mumcu, K. Sertel, and J. L. Volakis, "Printed coupled lines with lumped loads for realizing degenerate band edge and magnetic photonic crystal modes," *2008 IEEE Antennas and Propagation Society Symposium*, San Diego, CA, July 2008.
53. G. Mumcu, K. Sertel, and J. L. Volakis, "Partially coupled microstrip lines for antenna miniaturization," *IEEE International Workshop on Antenna Technology: Small Antennas and Novel Metamaterials (IWAT)*, Santa Monica, CA, March 2009.
54. A. Erentok and R. W. Ziolkowski, "Metamaterial-inspired efficient electrically small antennas," *IEEE Transactions on Antennas and Propagation*, vol. 56, no. 3, March 2008, pp. 691–707.
55. A. K. Skrivervik, J. F. Zurcher, O. Staub, and J. R. Mosig, "PCS antenna design: the challenge of miniaturization," *IEEE Antennas and Propagation Magazine*, vol. 43, no. 4, August 2001, pp. 12–27.
56. R. F. Harrington, *Time-Harmonic Electromagnetic Fields*. IEEE Press/Wiley, Piscataway, N.J./New York, 2001.
57. A. D. Yaghjian and S. R. Best, "Impedance, bandwidth, and Q of antennas," *IEEE Transactions on Antennas and Propagation*, vol. 53, April 2005, pp. 1298–1324.
58. S. R. Best, "The radiation properties of electrically small folded spherical helix antennas," *IEEE Transactions on Antennas and Propagation*, vol. 52, April 2004, pp. 953–960.
59. M. B. Stephanson, K. Sertel, J. L. Volakis, "Frozen modes in coupled microstrip lines printed on ferromagnetic substrates," *IEEE Microwave and Wireless Components Letters*, vol. 18, no. 5, 2008, pp. 305–307.
60. N. Apaydin, E. Irci, G. Mumcu, K. Sertel, and J. L. Volakis, "Miniature antennas based on printed coupled lines emulating anisotropy," submitted to *IET Microwaves, Antennas and Propagation*.
61. E. Irci, K. Sertel, and J. L. Volakis, "Antenna miniaturization for vehicular platforms using printed coupled lines emulating magnetic photonic crystals," submitted to *Elsevier Metamaterials Journal*.

CHAPTER 8

Impedance Matching for Small Antennas Including Passive and Active Circuits

Stavros Koulouridis

8.1 Introduction

Antenna miniaturization at VHF and UHF frequencies is of particular interest due to the larger size of these antennas. Chapters 3 and 4 discussed several miniaturization concepts, including material loading. Indeed, material usage for miniaturization has been increasingly used to achieve antenna size reduction [1, 2]. Often, this amounts to canceling the antenna's reactance. But materials add to the weight and are difficult to realize in practice. In this chapter, we focus on using lumped loads (inductors or capacitors) to develop matching circuits. These are intended to tune the antenna and bring it to resonance. Indeed, any antenna (regardless of its size) can be retuned to bring it to resonance. Of course, the Chu limits must be obeyed, implying a compromise in bandwidth and gain.

Matching circuits are typically placed between the source and antenna (load), as shown in Fig. 8.1. They are designed to transform the antenna's input impedance to that of the feed network. Secondary

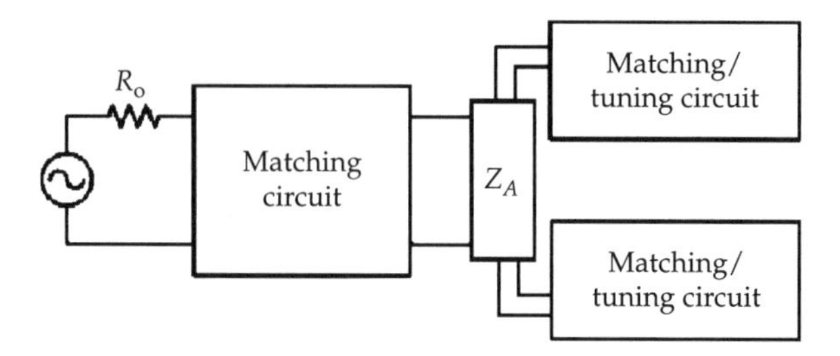

FIGURE 8.1 Matching/tuning networks at the antenna input and/or within antenna to transform its impedance Z_A into the characteristic impedance Z_0 of the feeding network.

matching circuits can be also placed within antenna's structure for additional tuning.

To obtain a broadband match, one must cancel the reactance over a broad frequency band. This is challenging and two approaches can be followed:

1. Passive impedance matching (governed by the Bode-Fano limit; see Chap. 2 and Fig. 8.2).
2. Non-Foster matching using active circuit elements to negate the reactive part of the antenna impedance. Such negative elements can be realized in practice using solid state devices. However, losses and instability are key issues to be overcome.

In this chapter, we consider matching circuits employing both passive and active elements. We present several examples and impedance matching techniques using Foster and non-Foster elements.

8.2 Passive Narrowband Matching

Impedance matching at a given frequency is carried out by placing a matching circuit between the feed network and antenna port. The matching circuit cancels the reactive part of the antenna and enables a resonance shift, allowing the antenna to radiate at lower frequencies.

As shown in Fig. 8.3, two configurations are possible for a network to match the antenna impedance, $Z_A = R_A + jX_A$. Circuits of type (a) can generally achieve matching at a chosen frequency if $R_A < Z_0$ (characteristic impedance of the feed network). For $R_A > Z_0$, circuits of type (b) must be used. The values of the lumped elements,

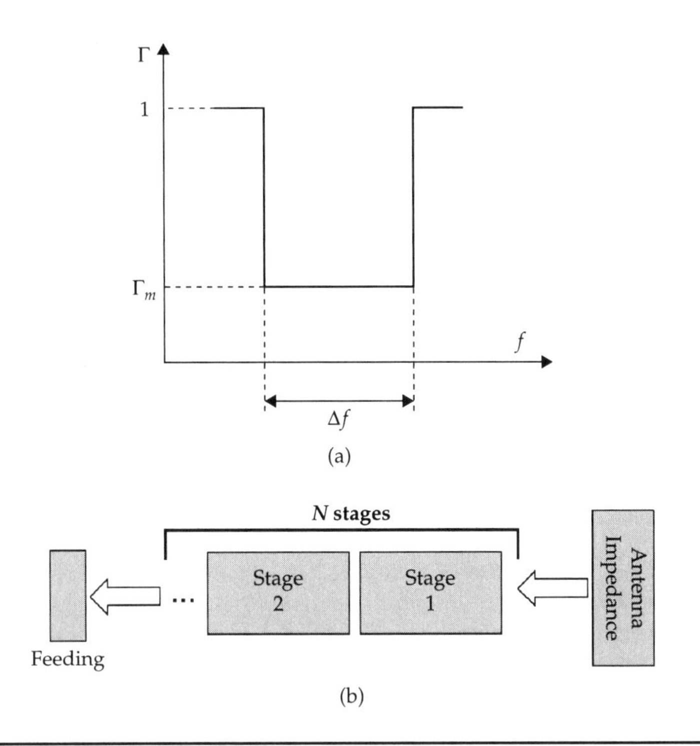

FIGURE 8.2 Bode-Fano limit criterion. The product bandwidth (Δf) - return loss (Γ_m) is restricted and finite even if an infinite number of lossless passive matching stages are used.

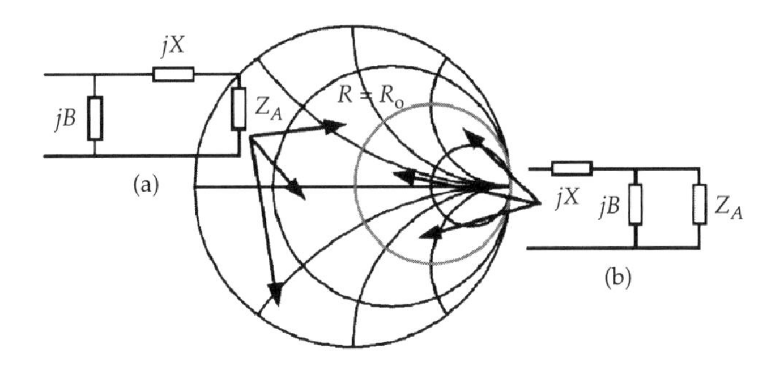

FIGURE 8.3 Smith chart with the $R = Z_0$ circle (red). The two matching circuits can match the impedance $Z_A = R_A + jX_A$ to Z_0 either for $R_A < Z_0$ [points outside $r = 1$ circle. (a) circuit] or $R_A > Z_0$ [points inside $r = 1$ circle; (b) circuit].

B (susceptance) and X (reactance), are given [3] from Eqs. (8.1) and (8.2) for circuits (a) and (b), respectively. Note that both positive and negative solutions are physically possible for B and X ($B < 0$ and/or $X > 0$ implies an inductor whereas $B > 0$ and/or $X < 0$ implies a capacitor).

$$X = \pm\sqrt{R_A(Z_o - R_A)} - X_A \qquad B = \pm\frac{\sqrt{(Z_o - R_A)/R_A}}{R_o} \tag{8.1}$$

$$B = \frac{X_A \pm \sqrt{(R_A/Z_o)\left(R_A^2 + X_A^2 - Z_o R_A\right)}}{R_A^2 + X_A^2} \qquad X = \frac{1}{B} + \frac{X_A Z_o}{R_A} - \frac{Z_o}{B R_A} \tag{8.2}$$

Matching circuits typically use lossless reactive elements (capacitors and inductors) placed in series or parallel. In the following sections, we discuss several matching applications.

8.2.1 Dipole

Let us consider a 6″-long dipole antenna resonating at 1 GHz. The goal is to lower its resonant frequency by adding matching circuits. To do so, we begin by calculating the real (R_A) and imaginary (X_A) impedance components of the 6″ dipole from 50 to 400 MHz. These calculations are shown in Fig. 8.4, and since the impedance's real part is lower than $Z_0 = 50\,\Omega$ (characteristic line impedance), a matching circuit of type (a) can be used (see Fig. 8.4*c*). As depicted in Fig. 8.4, the dipole has large reactance part at lower frequencies. To cancel it, we introduce the shown LC circuit* of Fig. 8.4*c*, such that (L, C) values of the circuit depend on the frequency of operation. Thus, for 65 MHz, 140 MHz, 280 MHz, etc., different (L, C) matching circuits must be chosen, ($L_1 = 9.36\,\mu$H, $C_1 = 721$ pF), ($L_2 = 1.99\,\mu$H, $C_2 = 153$ pF), and ($L_3 = 0.47\,\mu$H, $C_3 = 36.5$ pF). Figure 8.4*c* shows the gain at each of the aforementioned frequencies after matching. As desired, resonance is achieved at these lower frequencies. However, as the frequency is reduced, the gain and bandwidth are lowered. This is expected, and discussed in several of the earlier chapters. That is, miniaturization comes at the "price" of gain and bandwidth compromise.

A two-stage passive matching circuit can also be considered as in Fig. 8.5*a*. To do so, a second matching circuit is introduced next to the one defined by Eq. (8.1). The topology of the second matching circuit

*Alternative solution of type (a) is possible with two inductors placed in series and in parallel. However, this circuit has no significant difference in the achieved bandwidth (see Fig. 8.4*d*).

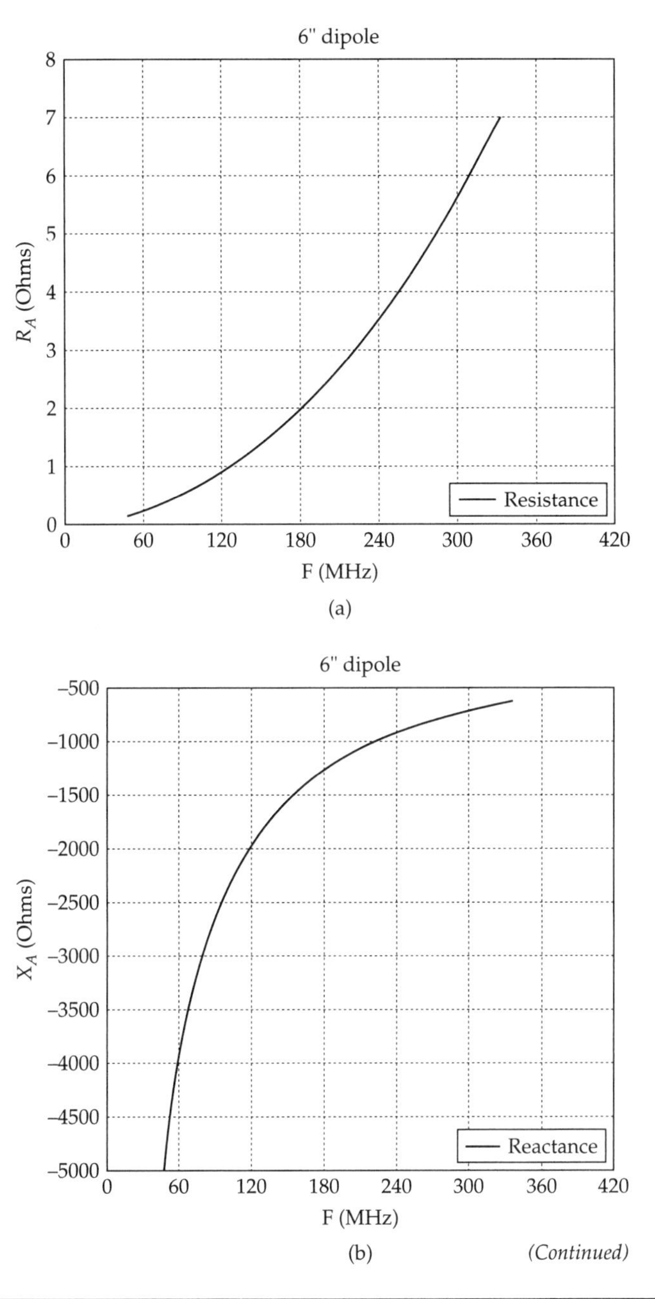

(Continued)

FIGURE 8.4 Narrowband matching for a 6″ linear dipole (resonance at 1 GHz) at low frequencies values of L and C elements from Eq. (8.1). (a) Resistance; (b) Reactance; (c) Gain performance using LC matching circuit shown (with inductor in series and capacitor in shunt); (d) Gain performance using alternate matching circuit with inductors only.

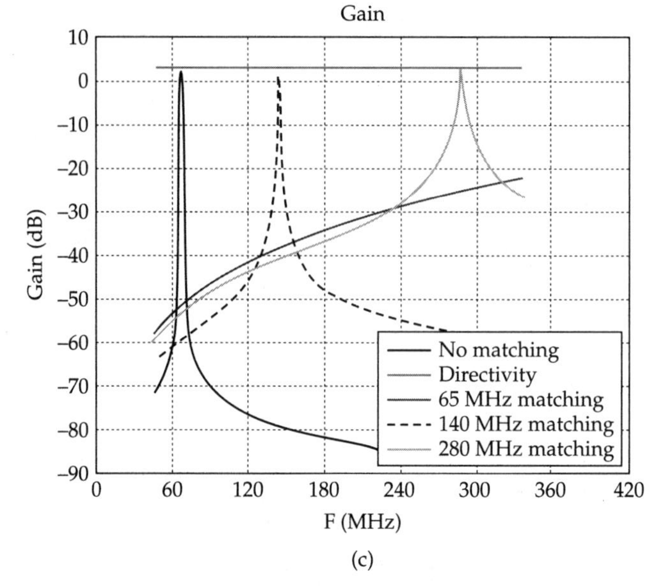

(c)

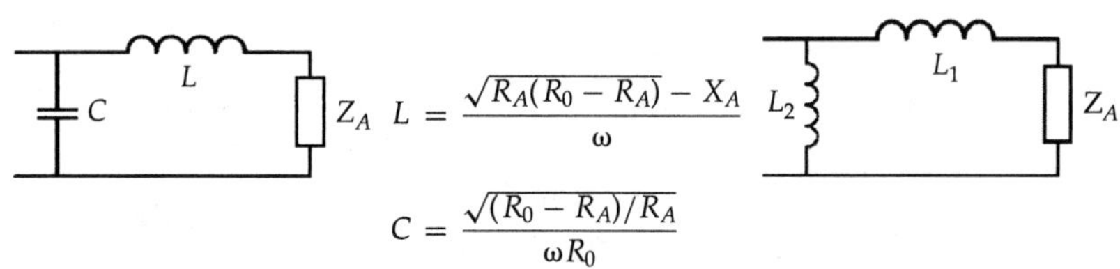

$$L = \frac{\sqrt{R_A(R_0 - R_A)} - X_A}{\omega}$$

$$C = \frac{\sqrt{(R_0 - R_A)/R_A}}{\omega R_0}$$

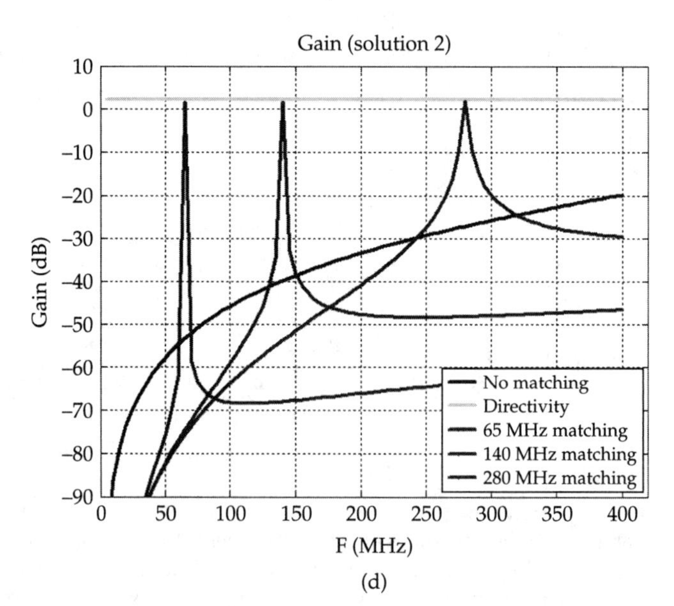

(d)

FIGURE 8.4 *Continued.*

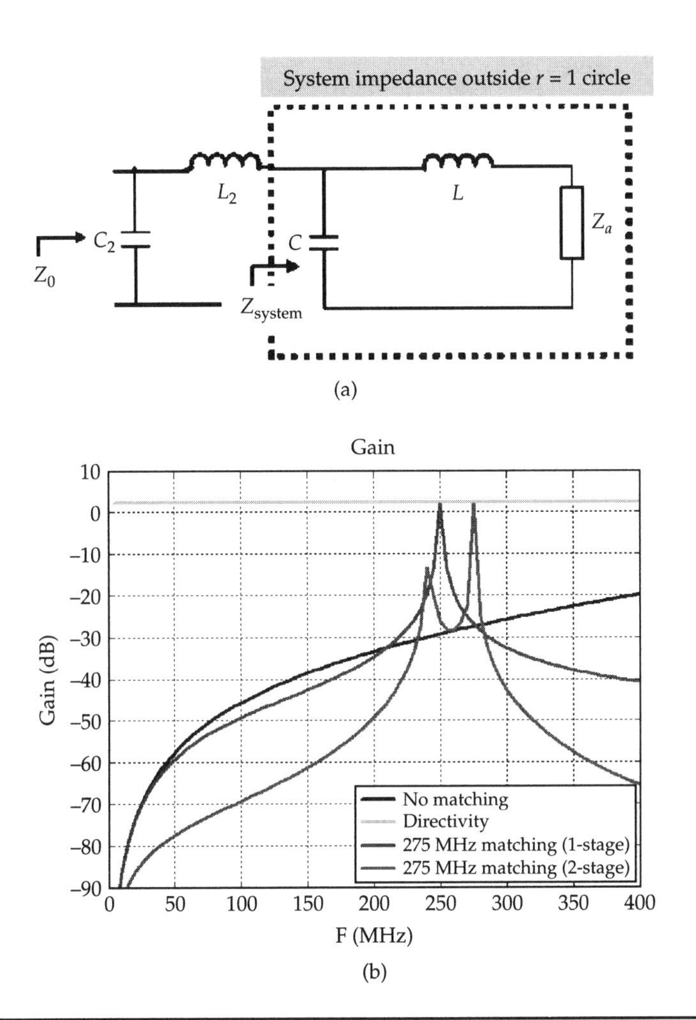

FIGURE 8.5 Two-stage matching circuit for a 6″-long dipole. (a) Circuit topology; (b) Gain performance using a two-stage matching circuit as compared to a one-stage matching circuit.

can be obtained using the same procedure. As would be expected, this second circuit needs to be tuned at a slightly different, but adjacent frequency to increase bandwidth. Also, the $Z_A = R_A + jX_A$ in Eqs. (8.1) or (8.2) needs to be replaced with $Z_{\text{system}} = R_{\text{system}} + jX_{\text{system}}$ as depicted in Fig. 8.5. Here, Z_{system} is the combined impedance of the antenna and the first matching circuit as defined in Fig. 8.4. Based on this, the (L, C) parameters for the second stage matching circuit in

Fig. 8.5 is given by

$$Z_0 = \frac{\dfrac{1}{j\omega C_2}(j\omega L_2 + jX_{\text{system}} + R_{\text{system}})}{\dfrac{1}{j\omega C_2} + (j\omega L_2 + jX_{\text{system}} + R_{\text{system}})}$$

$$L_2 = \frac{\sqrt{R_{\text{system}}(Z_0 - R_{\text{system}})} - X_{\text{system}}}{\omega}$$

$$C_2 = \frac{1}{\omega Z_0}\sqrt{\frac{Z_0 - R_{\text{system}}}{R_{\text{system}}}} \tag{8.3}$$

As an example, at 275 MHz, if $L = 1.85\ \mu\text{H}$, $C = 143$ pF for the first matching circuit, we find that $L_2 = 8.21$ nH, $C_2 = 612$ pF for the second matching circuit. Figure 8.5*b* shows the gain response of the same dipole as in Fig. 8.4 at 275 MHz. As can be observed, the two-stage matching circuit provides a significant improvement in bandwidth.

8.3 Passive Broadband Matching

Passive broadband matching networks are typically built with successive matching circuits (or L segments). Each segment consists of two lumped elements (capacitive and/or inductive). Usually, 3 to 5 segments are sufficient to obtain most of the benefit from a matching network. In practice, the matching circuit segment can introduce inevitable losses that must also be considered.

Extensive work exists in formal design of lossless multiple stage matching circuits [4–7]. As already noted, each stage is an LC network of L-shape with the formulae referring to the load modeled as a single RLC. Since it is not always trivial to model the load, computed aided design approaches have been developed [4]. Other robust techniques have been proposed [5–7] recently, but they are more complex to apply and implement.

A very simple and yet highly efficient optimization algorithm to obtain an optimum number of matching circuit stages (with optimum L and C values) for broadband matching was developed by Koulouridis and Volakis [10]. As an example, Fig. 8.6 shows N circuits (networks) connected in series to match the antenna's impedance to the feed. Each network has two elements (A and B), that can be inductors or capacitors. The topology of each network and the values of its elements are defined by five parameters (a, b, c, d, e) as depicted in Fig. 8.6. For optimization, we start with $N = 1$ and successively increase the number of matching circuits until optimum matching is found.

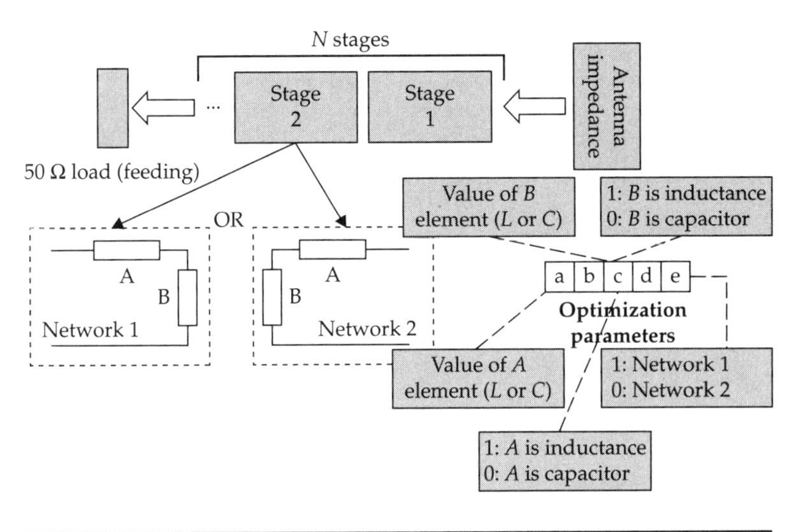

FIGURE 8.6 Multistage matching circuit for broadband antennas via optimization.

The optimization process stops when two consecutive additions lead to a VSWR, which is worse or no better than the previous. A global optimizer called NOMADm, a suite of MATLAB that functions for nonlinear and mixed variable optimization, was employed for the example given here. This suite is based on the mesh adaptive direct search (MADs) algorithm [8,9].

8.3.1 Broadband Planar Dipole

As an example, let us apply the N-network matching circuit (see Fig. 8.6) approach to a planar broadband dipole as shown in Fig. 8.7*a* [10]. The dipole is printed on a very thin, flexible FR4 sheet having $\lambda/1000$ thickness. Using matching circuits, the goal is to obtain a realized gain greater than 0 dBi at the lowest possible frequency without affecting the dipole's high-frequency behavior. Indeed, the optimization algorithm (see Fig. 8.6) yields an optimum topology with passive elements values for the proposed three-stage lossless network. This three-stage matching network lowered the operational frequency from around 195 to 150 MHz, implying a 25% minimization.

Nevertheless, the realization of the actual matching circuit values is an issue. For example, when the lumped elements are mounted on a PCB board, the distance between the lumped elements may alter the matching circuit's equivalent values. Hence, the matching network was also simulated on a realistic PCB board using Agilent ADS. A symmetric design was also employed (see Fig. 8.7*c*) to ensure a more

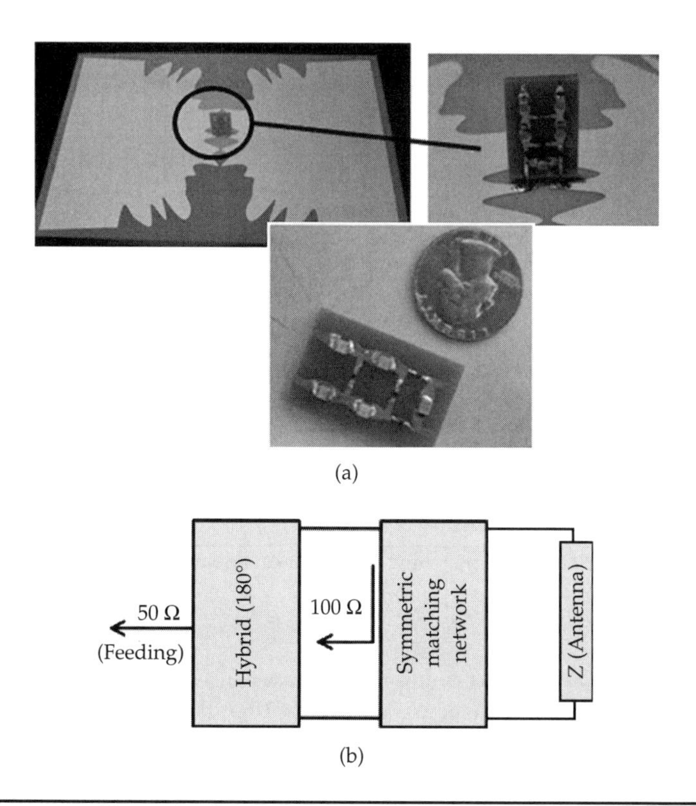

FIGURE 8.7 Planar dipole matching data. (a) Return Loss; (b) Efficiency predicted via Agilent's ADS; (c) Three-stage symmetric matching network modeled in ADS and taking into account printed line lengths.

balanced feed. As shown in Fig. 8.7*b*, when the proposed matching network is included, even after tuning, the antenna efficiency is not as good. It is, however, above 75% in the 150-MHz to 1-GHz region, making it good enough for a gain greater than 0 dBi over the return loss bandwidth in Fig 8.7*a*. We note that measured data for the spline-optimized flare dipole's impedance were employed in the optimization to find the matching circuit values.

Measurements of the final three-stage matching circuit and antenna were carried out (Fig. 8.8). Indeed, the gain reached 0 dBi at 165 MHz. However, there was an efficiency decrease in the 300 to 350 MHz band for the matched dipole (further left than predicted by ADS). This was likely due to imperfections in the elements. Above 400 MHz, the matched dipole performance followed the original dipole (prior to matching) behavior but it dropped fast bellow 0 dBi at 900 MHz.

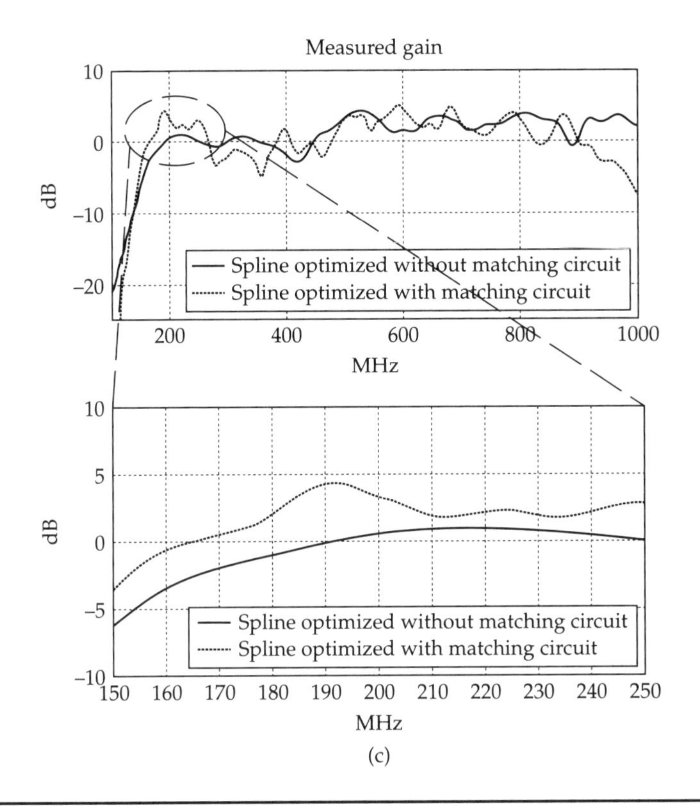

FIGURE 8.7 *Continued.*

As would be expected, there were inherent circuit element conduction losses from soldering. Referring to [11], these losses can greatly affect antenna performance when a large number of elements are used. Certainly, better results can be obtained when boards are fabricated in a controlled environment. However, the delivered antenna has better bandwidth (i.e., 190–1000 MHz bandwidth of the pre-matched shape optimized dipole compared to 165 to 900 MHz bandwidth of the matched antenna). That is, a 15% size-reduction was obtained. Nevertheless, more impressive results can be obtained using negative matching elements as restrictions in bandwidth and gain can be partially overcome. This has been discussed in Sec. 8.4.

8.3.2 Inverted Hat Antenna

Another example considered is the inverted hat antenna (IHA) in Fig. 8.9. A single L-matching circuit was designed for this broadband IHA antenna [12] to improve its low-frequency performance at

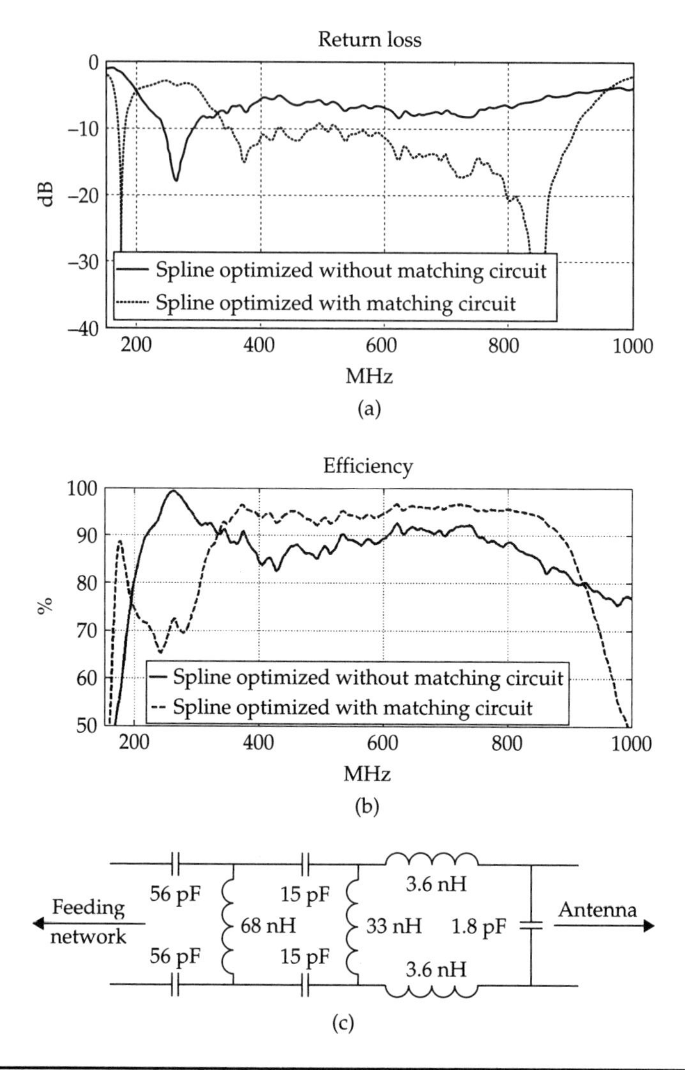

FIGURE 8.8 Symmetrical three-stage matching network realized for the planar dipole. (a)Photo of the three-stage matching network in Fig. 8.6 applied to the spline-optimized planar dipole; (b) dipole feeding using a 180° hybrid; (c) Measured gain with and without insertion of the matching circuit for the flare dipole.

approximately 36 MHz. After the optimization algorithm was employed in MATLAB, a circuit as shown in Fig. 8.9 is applied at the antenna input. The values of the lumped elements obtained after the optimization served as the initial point to a tuning procedure in ADS, aiming to maximize antenna performance around that frequency.

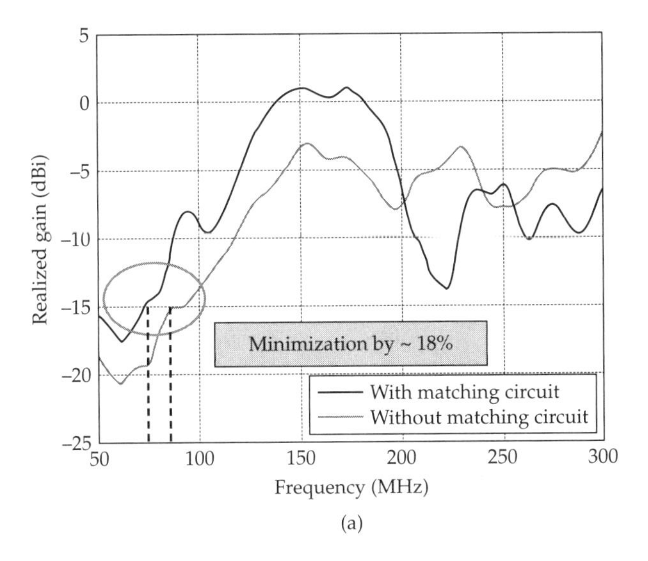

(a)

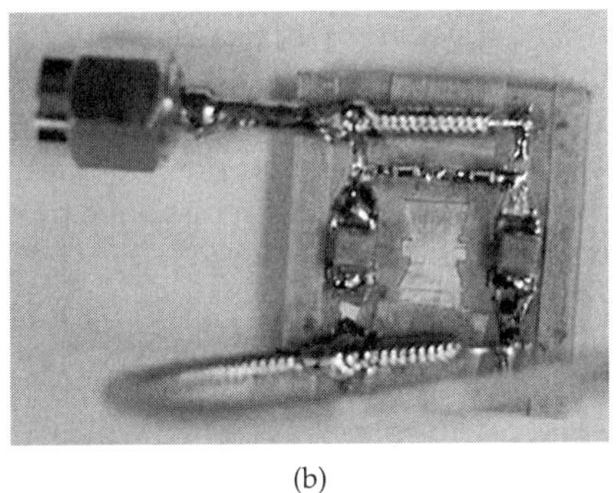

(b)

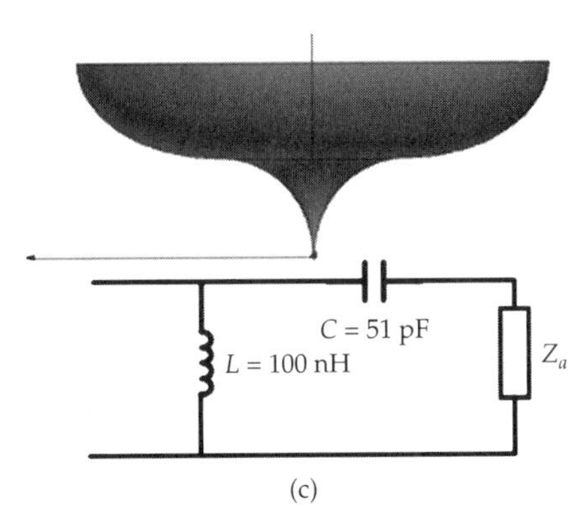

(c)

FIGURE 8.9 Single-stage broadband matching circuit for the "inverted hat" monopole. The fabricated circuit provides 18% minimization (at the −15 dB point). It generally improves antenna performance in the 50 to 200 MHz region. A frequency switch can be used for the higher frequencies.

Indeed, after optimization, the realized gain improved, resulting in 18% minimization. However, the matching circuit is low-pass and affects high frequencies negatively. In this case, a bypass frequency switch may be used for higher frequencies.

8.4 Negative Matching

A non-Foster network can be used to cancel the reactive part of the antenna impedance. Such a network realizes negative impedance, $Z = -kZ_E$, for $k > 0$ with Z_E being the impedance of the element. Non-Foster circuits have been considered since 1950s when Linvill [13] and Yanagisawa [14] introduced negative impedance converters (NIC). A NIC is a two-port network realized via a combination of active devices (amplifiers) and lumped loads (capacitors and inductors)*. Several configurations of negative impedance converters can be found in [15–17], with some simple ones shown in Fig. 8.10. A key disadvantage of NIC's is the need to power them externally. Also, biasing with a DC voltage [18] needs special handling. To build stable circuits, basic rules must be followed (see Fig. 8.11) as well. Bahr [19] introduced negative matching for antennas in 1970, but inherent bandwidth and noise issues with solid state devices limited their use. Recent developments of integrated circuits have provided interest in non-Foster matching.

Among recent examples in using NICs, negative capacitance was realized through a lumped capacitor and a NIC to cancel the dipole's reactance [20]. Negative matching was also designed for a monopole in [18]. In [21], a microstrip patch was loaded with a negative capacitor to match it at 9.5 MHz. Doing so, a bandwidth increase from 12% to 24% was achieved and the radiation characteristics improved for the compensated antenna via negative matching.

Overall, non-Foster matching can improve bandwidth and reduce antenna size as it is not limited by the Chu and Fano limits. Hence, active matching can allow for very small broadband antennas. In the following sections, we discuss some examples.

8.4.1 Loop Antenna

As an example, let us consider a 6″ loop antenna (see Fig. 8.12*a*), with the given input impedance and return loss. The loop has a narrow resonance at $f = 0.67$ MHz and our goal is to improve its bandwidth and reduce its size using non-Foster matching whose circuit element

*Negative matching implies the use of active devices. Active positive matching is also possible, but it is not discussed in this chapter.

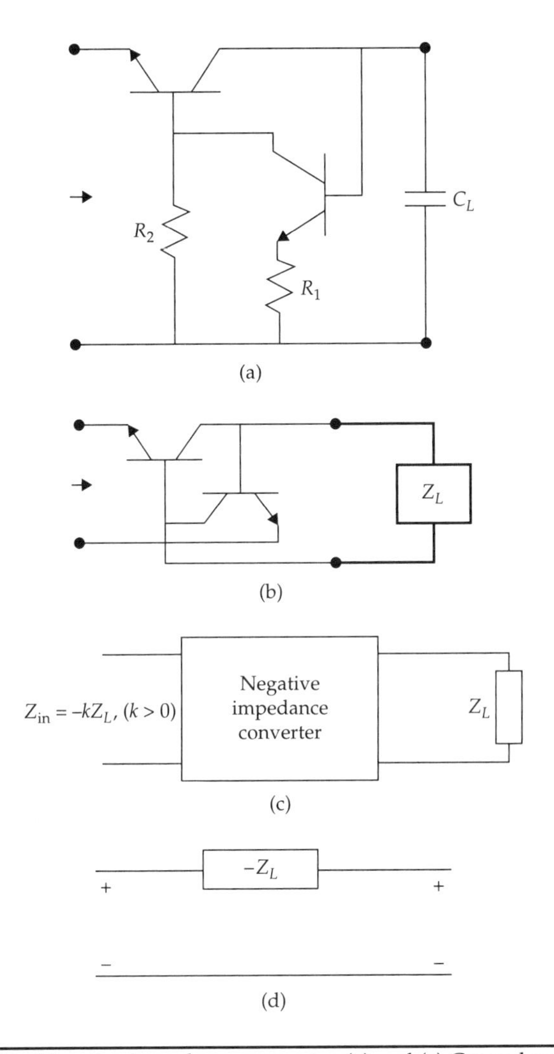

FIGURE 8.10 Negative impedance converter. (a) and (c) Ground negative impedance converter ($C_{in} = -C_L$). *(Linvill 1953 [13].)*; (b) and (d) Floating negative impedance converter ($Z_{in} = -Z_L$). *(Linvill 1953 [13].)*

values are obtained via optimization. Indeed, on introducing the negative matching circuit (see Fig. 8.12*b*), the network delivers a 320 MHz impedance bandwidth (−10 dB return loss). This is a bandwidth increase from 290 to 610 MHz as seen in Fig. 8.12*c*. That is, a narrow resonance antenna can be easily transformed to a smaller broadband antenna using NICs.

To see how an active matching element affects antenna efficiency, a floating negative impedance converter circuit (Fig. 8.10*b*) is used

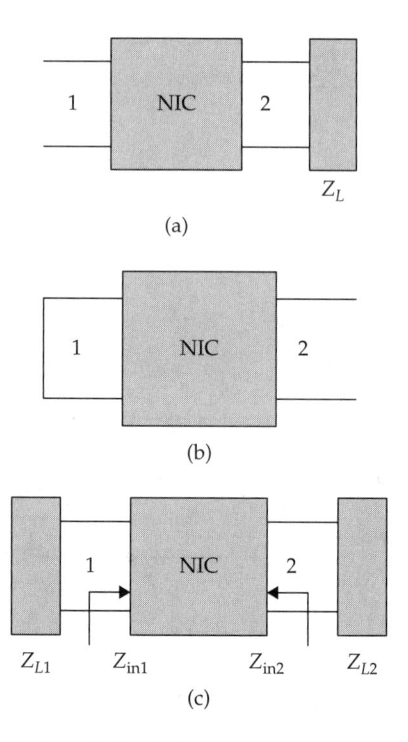

FIGURE 8.11 Negative impedance converters for stability. (a) Open circuit stable as seen from port 1; (b) Short circuit stable as seen from port 2; and (c) Stable NIC subject to conditions $|Z_{L1}| > |Z_{\text{in1}}|$, $|Z_{L2}| < |Z_{\text{in2}}|$.

to realize the negative series elements. Also, a ground negative impedance converter (Fig. 8.10*a*) is used for the shunt elements. These elements were employed to realize the matching circuit shown in Fig. 8.12 using Agilent ADS. Figure 8.13 shows the ADS calculated return loss for the matched loop antenna, and it is seen to be in reasonable agreement to that in Fig. 8.12. However, the efficiency of the overall antenna has considerably dropped. Obviously, some of the input power is lost in the active elements. We also note that the peak in efficiency and return loss is around 650 MHz implying circuit instability around that frequency. Further, the lumped elements values must be retuned (see Fig. 8.13) with the addition of active elements.

Negative elements can also be used inside the antenna structure. Of course, employing matching circuits within the antenna can affect input impedance. In [22], an antenna for handheld devices (see Fig. 8.14) was loaded with a negative inductance as shown in Fig. 8.15. For analysis, this negative inductance was modeled as a single passive inductor with the appropriate NIC topology. Figure 8.16 shows that

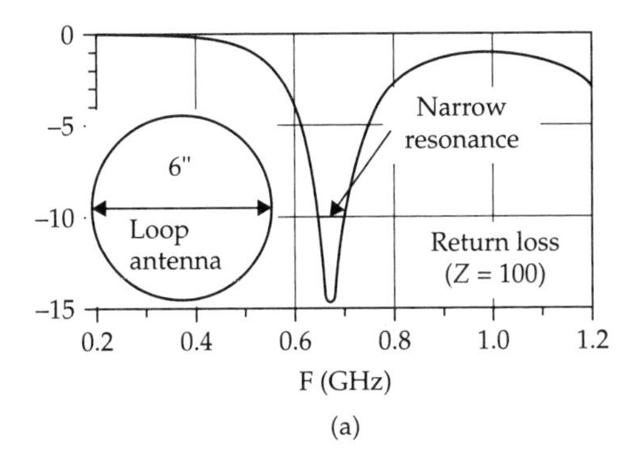

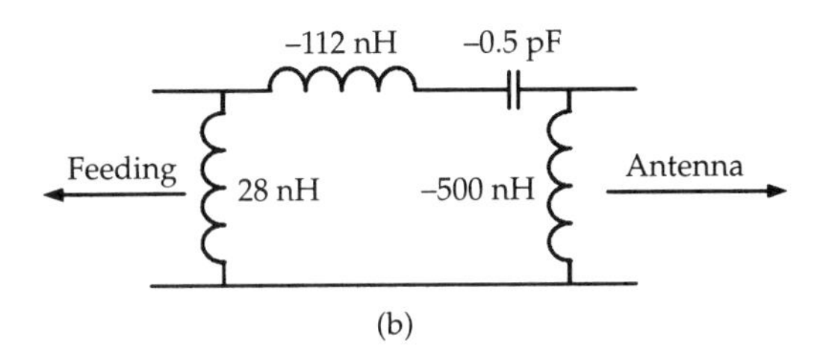

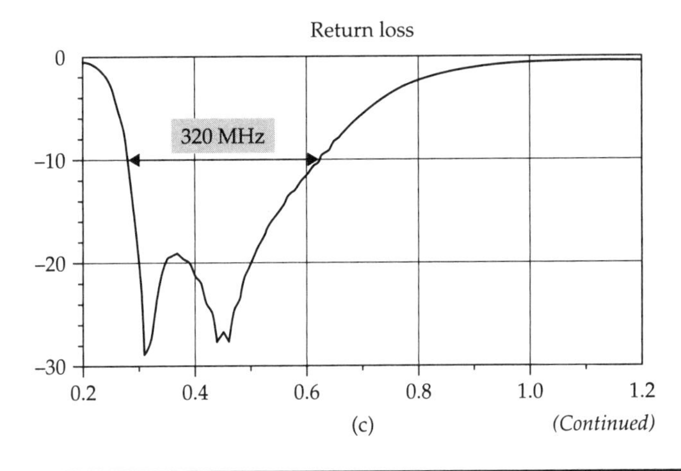

(Continued)

FIGURE 8.12 Performance for a NIC matched antenna loop. (a) Resonance of a 6″ loop at $f = 0.67$ GHz prior to NIC matching; (b) Negative circuit used for matching; (c) Return loss after matching (320 MHz impedance bandwidth; (d) Efficiency performance after NIC matching (600-MHz bandwidth with 50% efficiency).

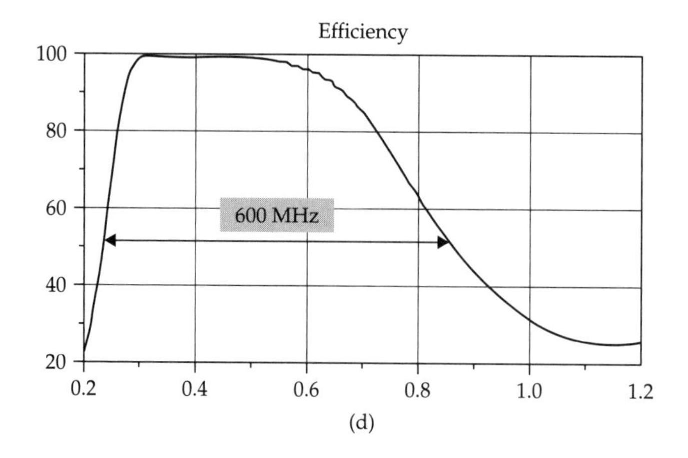

FIGURE 8.12 *Continued.*

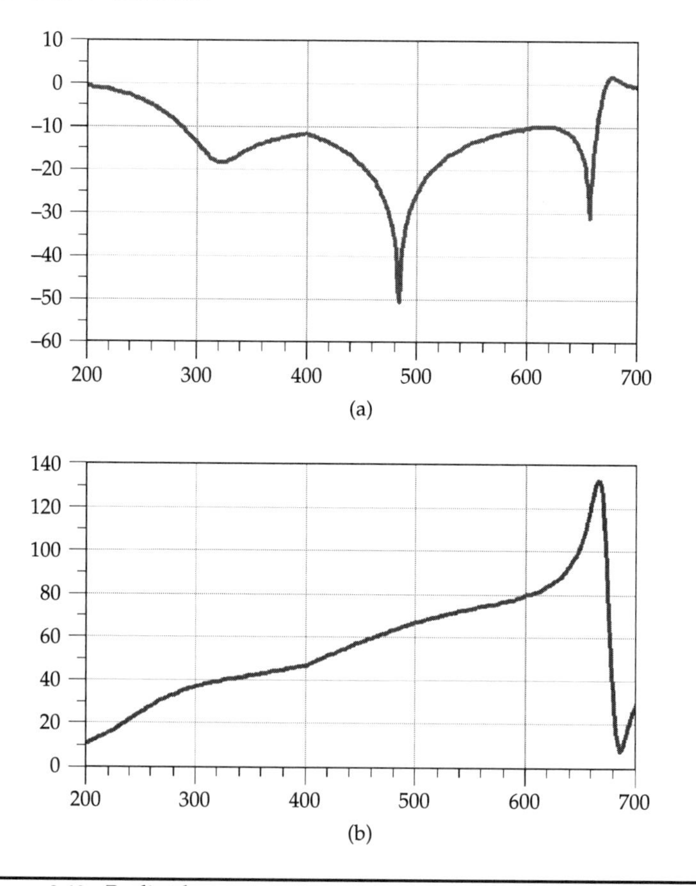

FIGURE 8.13 Realized active network (for the network in Fig. 8.6) to match the loop antenna in Fig. 8.12.

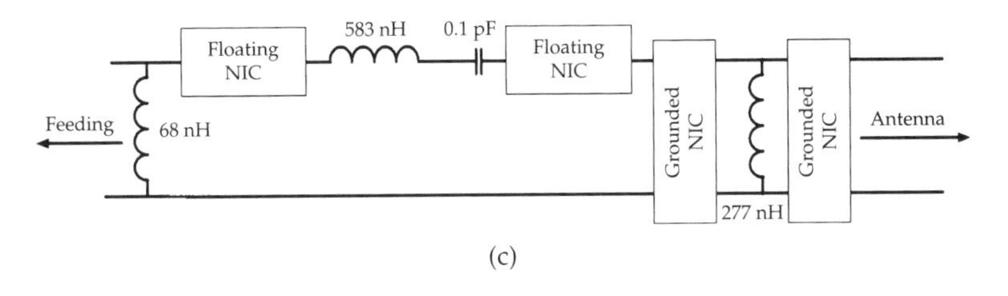

FIGURE 8.13 *Continued.*

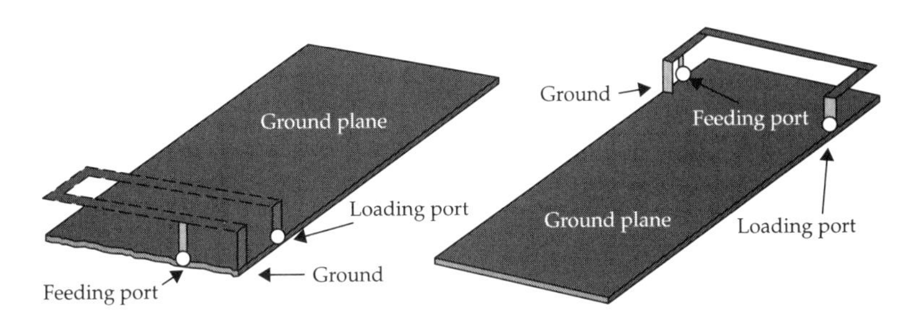

FIGURE 8.14 Antenna with negative elements for handheld device. Upon consideration of the real and imaginary parts of antenna impedance, the antenna can be matched using a single negative inductor. (*After Bit-Babik et al. ©IEEE, 2007 [22].*)

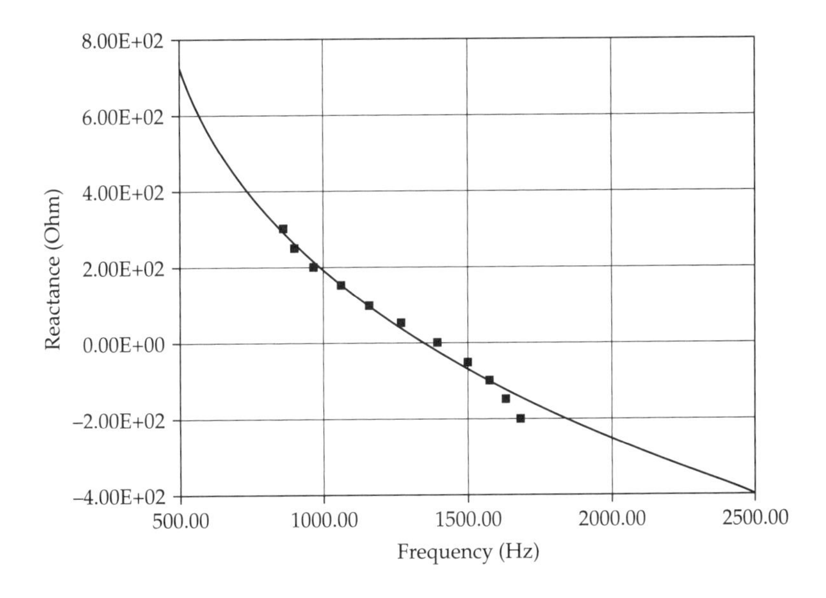

FIGURE 8.15 Ideal inductance (dots) and synthesized negative inductance (continuous line) from an inductor connected to an impedance converter for the matching port in Fig. 8.14. (*After Bit-Babik et al. ©IEEE, 2007 [22].*)

this inductor delivers a −10 dB return loss over a 1-GHz bandwidth (1–2 GHz). Also shown in Fig. 8.16 is a comparison of the matched return losses when passive inductors are used. That is, negative inductance gives continuous bandwidth but the passive elements give narrowband performance.

8.4.2 Flare Dipole

As the antenna size is reduced, we expect its resistance to become very small and its reactance to increase significantly. Therefore, impedance matching becomes critical in such cases. In this chapter, we demonstrated that passive matching can attain 15% size reduction for the flare dipole (without affecting bandwidth or gain). It is, however, possible to further reduce antenna size using non-Foster circuits. In this subsection, we again consider the flare dipole [8] before shape optimization (shown in Fig. 8.17) and discuss matching circuits to improve its performance at lower frequencies. For reference, its original impedance and gain are also given in Fig. 8.18. As seen in Fig. 8.18*a* and 8.18*b*, at lower frequencies (below 200 MHz), the flare dipole reactance reaches extreme negative values, and the resistance practically vanishes. Consequently, the flare dipole's efficiency becomes very low.

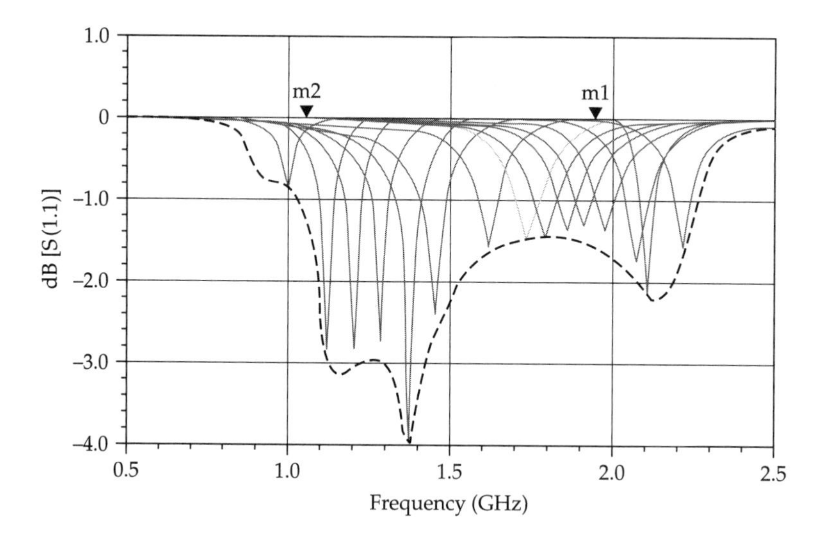

FIGURE 8.16 Return loss after matching using negative inductance for the antenna in Fig. 8.14 (dotted line) and return loss when passive elements are employed (continuous lines). *(After Bit-Babik et al. ©IEEE, 2007 [22].)*

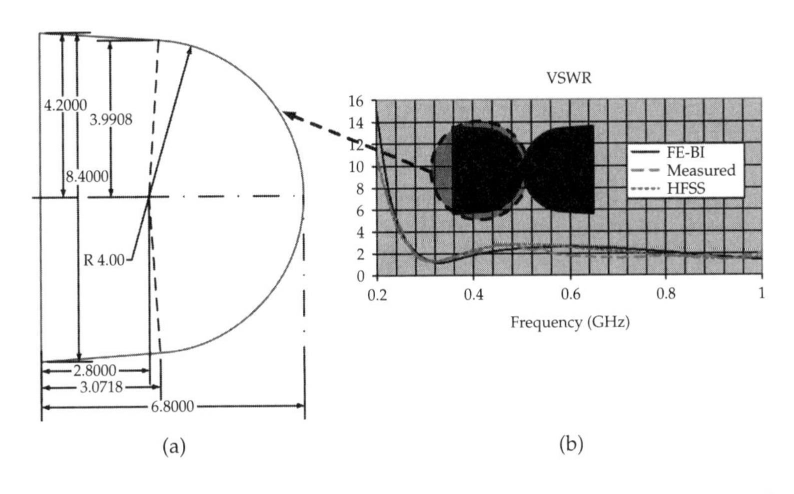

FIGURE 8.17 Flare dipole geometry and VSWR. (a) Flare dipole (dimensions in inches) geometry; (b) Measured and simulated VSWR for the shown dipole; simulations were carried out using HFSS and a finite element boundary integral (FE-BI) method [23]. (See also Koulouridis and Volakis [8,10]).

The matching procedure amounts to adding a fictitious port within the antenna to introduce suitable loads computed via the S matrix values as depicted in Fig. 8.19. If we apply a load Z_L at the introduced fictitious port, the corresponding antenna input impedance, Z_{in}, can then be controlled by adjusting Z_L. Indeed, as shown in Fig. 8.19*b* and 8.19*c*, we can introduce a load Z_L that generates a nearly ideal input impedance. Following these steps, we can control the antenna's efficiency as well.

Concerning the flare dipole, if we can increase the low radiation resistance and cancel the reactive part of the antenna impedance, it will radiate down to 50 MHz. To do so, we define two additional ports (A and B) within the flare dipole structure as depicted in Fig. 8.20*a*. Obviously, because of symmetry, a two-port system can be studied. For instance, a capacitor C at port A will be equivalent to applying two capacitors of value $C/2$ at ports A and B. Thus, the application of the formula in Fig. 8.19*c* becomes straightforward.

Ideally, to match the input impedance to 50 Ω in the 50 to 400 MHz band, we need to introduce a complex load at ports A and B as that in Fig. 8.20*a* (calculated from the formula in Fig. 8.19*c*). However, if we only focus on matching the real part of the input impedance to 50 Ω we find that an inductive load (see Fig. 8.20*b*) can

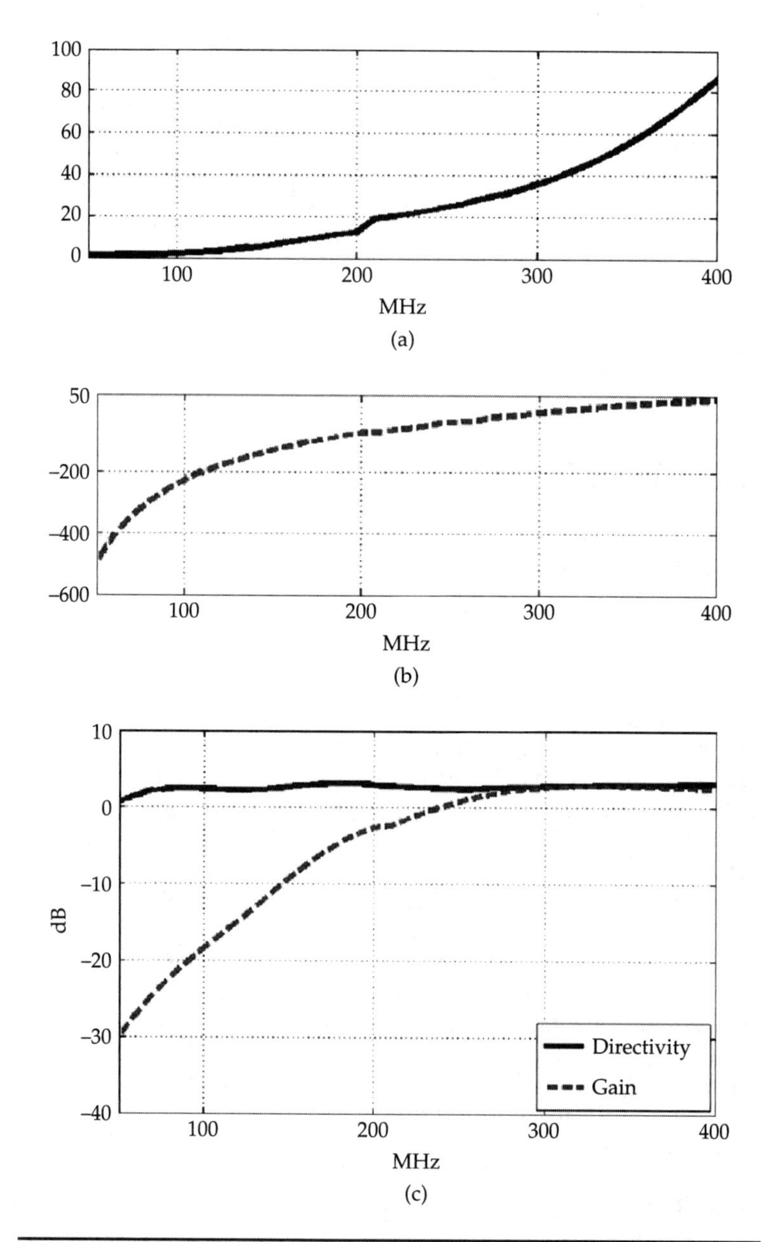

FIGURE 8.18 Input impedance and gain of the original flare dipole before matching. (a) Real; (b) Imaginary part of the input impedance; (c) Directivity and realized gain.

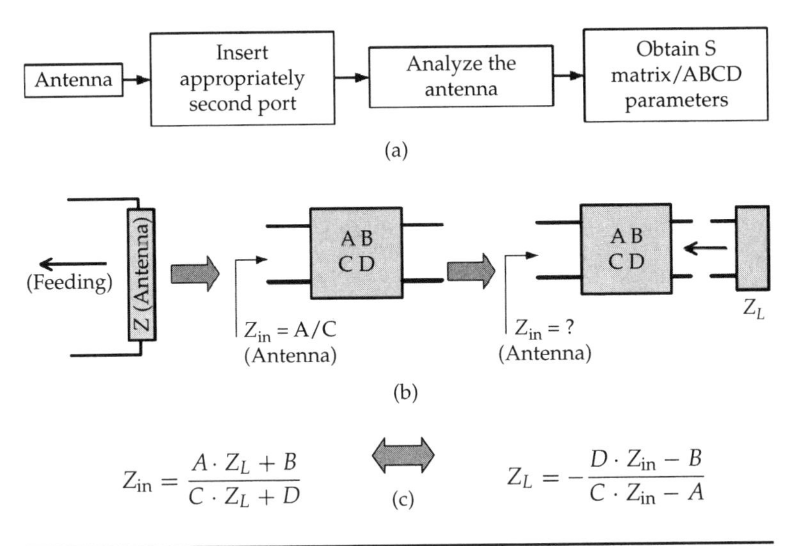

$$Z_{in} = \frac{A \cdot Z_L + B}{C \cdot Z_L + D} \qquad Z_L = -\frac{D \cdot Z_{in} - B}{C \cdot Z_{in} - A}$$

FIGURE 8.19 Introduction of a fictitious port to accommodate impedance matching. (a) Addition of a fictitious port to calculate the S matrix (and equivalent ABCD parameters); (b) Z_{in} controlled by adding a load Z_L; (c) Z_L calculated from the ABCD parameters to achieve a desired Z_{in}.

considerably improve the antenna's input resistance. Indeed, when two non-Foster capacitors $C_A = C_B = C/2 = -2.76$ pF (that "mimic" the needed inductive load) are applied at ports A and B, respectively, can generate the input resistance of Figs. 8.20*c* and 8.21*c*. As seen, the input resistance is very close to the ideal 50 Ω value for the 50 to 400 MHz region of interest. However, the reactive part of the input impedance worsened as depicted in Fig. 8.20*d*, and needs to be compensated.

To compensate for the reactance in Fig. 8.20*d*, we proceed to introduce a series non-Foster network at the antenna input terminals. (It is remarked that application of the series non-Foster reactive matching network at the feeding port does not affect the input resistance.) Upon optimization, we find that the series non-Foster network of Fig. 8.21*b* (see Fig. 8.21*a* for part location) leads to the input resistance and reactance depicted in Fig. 8.21*c*. As seen from Fig. 8.21*c*, the non-Foster circuit cancels the imaginary part of the input impedance from approximately 80 to 270 MHz. With this finalized matching circuit and negative capacitors at ports A and B, the resulting bandwidth is 3.5:1 for a flare dipole at one-third of its original size. Also, of importance is that the gain is mostly above 0 dBi (see Fig. 8.20*d*).

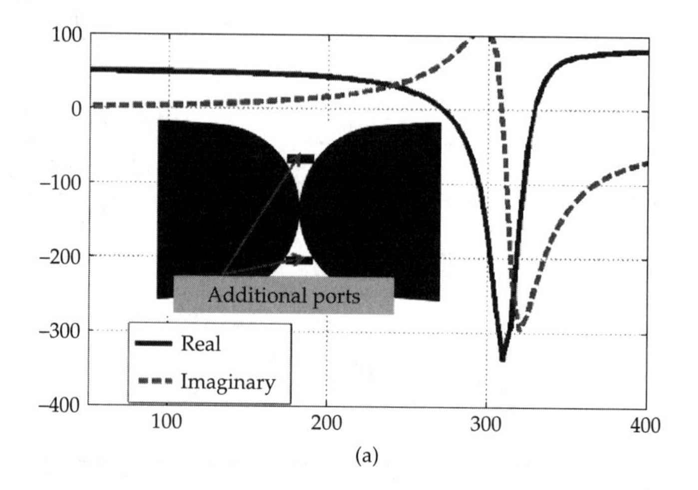

(a)

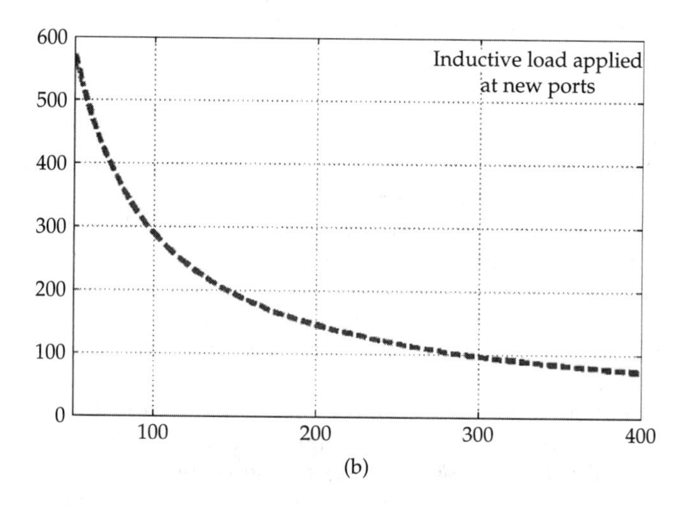

(b)

FIGURE 8.20 Evaluation of the reactive load at the fictitious ports A and B in Fig. 8.20 to obtain 50 Ω. (a) ideal port loads as a function of frequency (MHz); (b) Reactance using a negative capacitor at the new ports to transform the input impedance to a value close to 50 Ω (the actual load placed at ports A and B is $C/2 = -2.76\,\text{pF}$); (c) Input resistance at the antenna input after the load in (b) is applied; (d) input reactance at the antenna port after the load in (b) is applied.

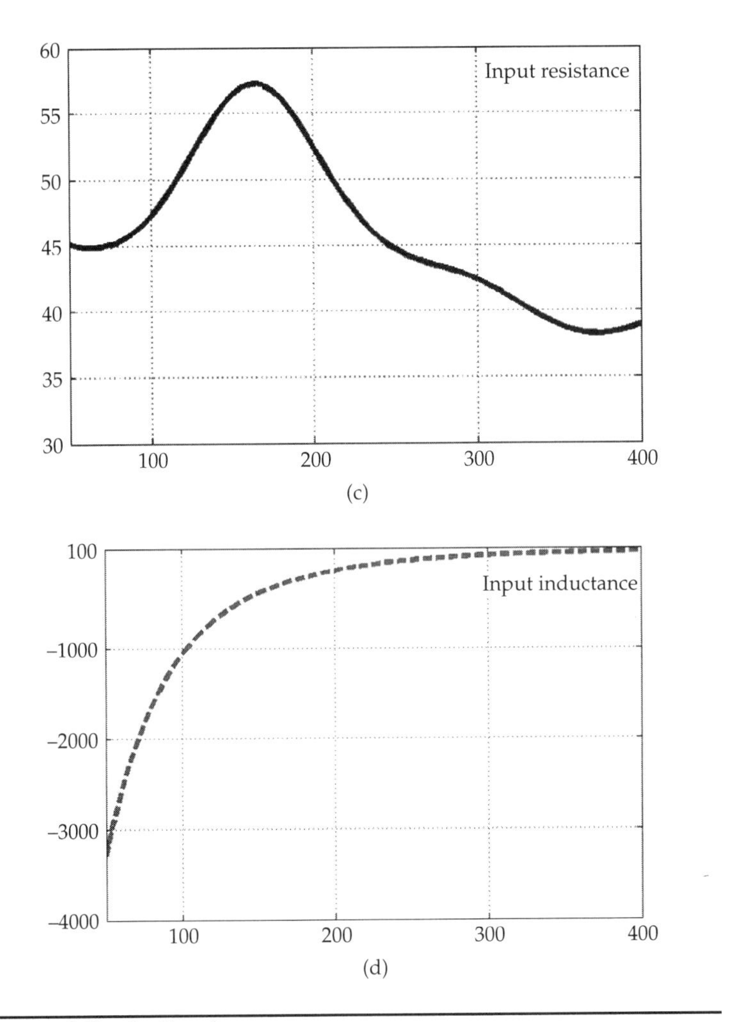

FIGURE 8.20 *Continued.*

8.5 Concluding Comments

It is possible to obtain size reduction by exploiting passive and active networks and even a combination of the two. Impedance matching does not alter the antenna structure; rather it can shift its resonance frequency to lower values. An optimization choice can be used to design broadband matching, and was shown that matching circuits can provide a 25% miniaturization.

Negative networks provide a powerful way to construct very small efficient antennas. The difficulty in non-Foster impedance matching

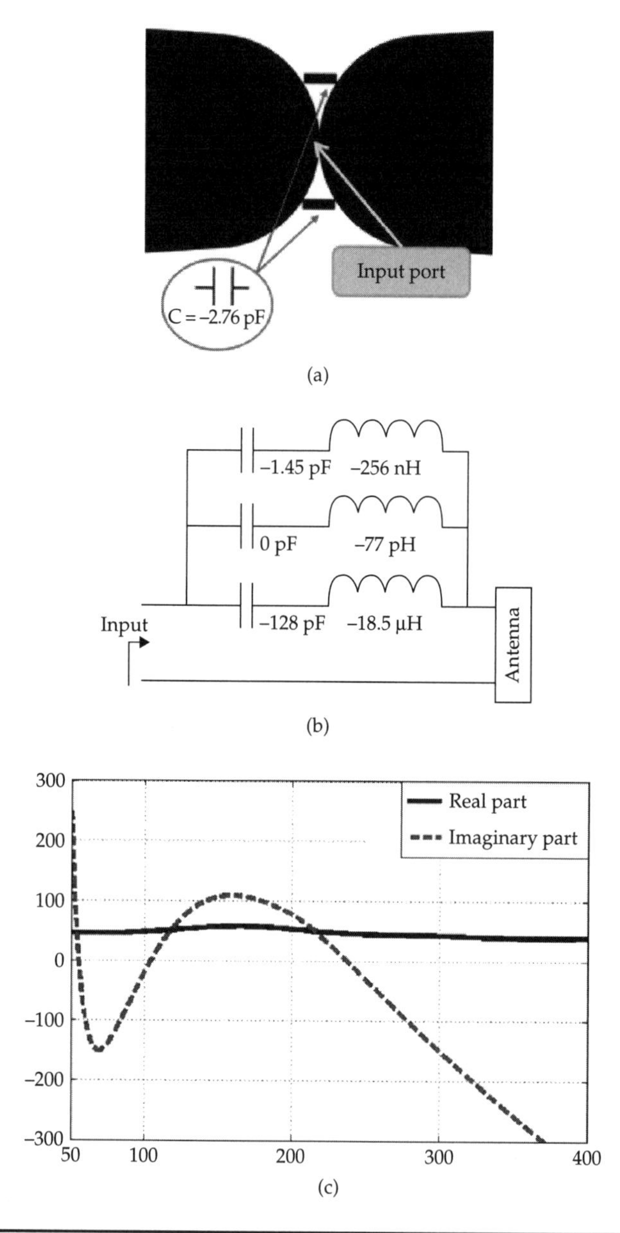

FIGURE 8.21 Placement of negative capacitors and non-Foster circuit at the flare dipole input port to improve impedance matching. (a) Negative capacitors location; (b) Impedance network inserted for matching at the antenna's input port; (c) Input resistance and reactance at low frequencies (note the achieved 50 Ω input resistance and low reactance); (d) Realized gain showing 3.5:1 bandwidth for the matched dipole having one-third of the original dipole size (the unmatched dipole had a 4.5:1 bandwidth, but was three times larger).

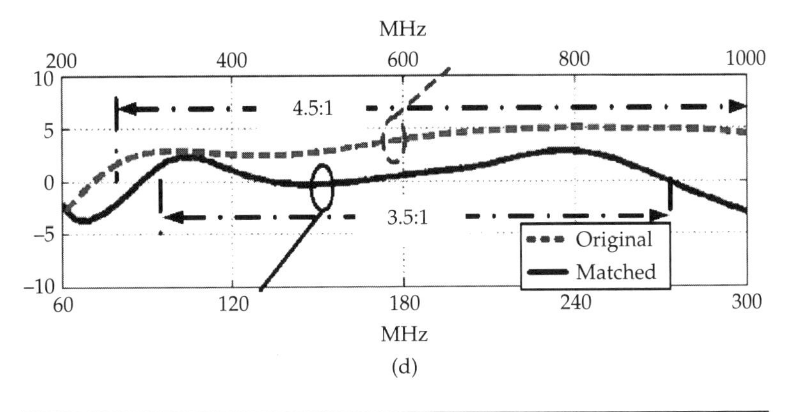

FIGURE 8.21 *Continued.*

lies in their realization as they are based on active elements. While many realistic topologies can be proposed, noise, narrow bandwidth, and instability have prevented their actual implementation. It is author's belief that these issues can be addressed in the future. When so, negative matching networks will become a powerful tool for antenna minimization.

References

1. S. Koulouridis, G. Kiziltas, Y. Zhou, D. J. Hansford, and J. L. Volakis, "Polymer-ceramic composites for microwave applications: fabrication and performance assessment," *IEEE Transactions on Microwave Theory and Techniques*, vol. 54, 2006, pp. 4202–4208.
2. G. Mumcu, K. Sertel, and J. L. Volakis, "Miniature antenna using printed coupled lines emulating degenerate band edge crystals," *IEEE Transactions on Antennas and Propagation*, vol. 57, 2009, pp. 1618–1624.
3. D. M. Pozar, *Microwave Engineering*, 3d ed., John Wiley & sons, Hoboken, N.J. 2005.
4. T. R. Cuthbert, Jr., *Circuit Design Using Personal Computers*, John Wiley, New York, 1983.
5. W. K. Chen and T. Chaisrakeo, "Explicit formulas for the synthesis of optimum bandpass butterworth and chebyshev impedance-matching networks," *IEEE Transactions on Circuits and Systems*, vol. 27, October 1980, pp. 928–942.
6. H. Dedieu, C. Dehollain, J. Neirynck, and G. Rhodes, "A new method for solving broadband matching problems," *IEEE Transactions on Circuits and Systems*, vol. 41, September 1994, pp. 561–571.
7. R. Rhea, "The Ying-Yang of matching (Part 1 and 2)," *High Frequency Electronics*, March 2006, pp. 16–40.
8. S. Koulouridis and J. L. Volakis, "Minimization of flare dipole via shape optimization and matching circuits," *in Proceedings of the 2007 IEEE Antenna and Propagation Society Symposium*, 2007, pp. 4785–4788.

9. M. A. Abramson, "Pattern search filter algorithms for mixed variable general constrained optimization problems," Ph.D. thesis, Rice University, Houston, Texas, 2002.
10. S. Koulouridis and J. L Volakis, "A novel planar conformal antenna designed with splines," *IEEE Antennas and Wireless Propagation Letters*, vol. 8, 2009, pp. 34–36.
11. J. Zhao, C. C. Chen, and J. L. Volakis, "Frequency-scaled UWB inverted-hat antenna," to appear in *IEEE Transactions on Antennas and Propagation*, 2010.
12. M. Lee, B. A. Kramer, C. C. Chen, and J. L. Volakis, "Distributed lumped loads and lossy transmission line model for wideband spiral antenna miniaturization and characterization," *IEEE Transactions on Antennas and Propagation*, vol. 55, October 2007, pp. 2671–2678.
13. J. G. Linvill, "RC active filters," *Proceedings of the IRE*, vol. 42, March 1954, pp. 555–564.
14. T. Yanagisawa, "RC active networks using current inversion type negative impedance converters," *IRE Transactions on Circuit Theory*, vol. 4, September 1957, pp. 140–144.
15. S. E. Sussman-Fort, "Matching network design using non-Foster impedances," *International Journal of RF and Microwave Computer-Aided Engineering*, no 16, 2006, pp. 135–142.
16. S. E. Sussman-Fort and R. M. Rudish, "Non-foster impedance matching for transmit applications," *Proceedings of 2006 IEEE International Workshop on Antenna Technology Small Antennas and Novel Metamaterials*, March 6–8, 2006, pp. 53–56.
17. J. T. Aberle and R. Loepsinger-Romak, "Antennas with Non-Foster Matching Networks," C. A. Balanis: Synthesis Lecture on Antennas Series, Morgan & Claypool, 2006.
18. S. E. Sussman-Fort and R. M. Rudish, "Non-foster impedance matching of electrically-small antennas," *IEEE Transactions on Antennas and Propagation*, vol. 57, August 2009, pp. 2230–2241.
19. A. J. Bahr, "On the use of active coupling networks with electrically small receiving antennas," *IEEE Transactions on Antennas and Propagation*, vol. 25, November 1977, pp. 841–845.
20. J. T. Aberle, "Two-port representation of an antenna with application to non-Foster matching networks," *IEEE Transactions on Antennas and Propagation*, vol. 56, May 2008, pp. 1218–1222.
21. A. Kaya and E. Y. Yuksel, "Investigation of a compensated rectangular microstrip antenna with negative capacitor and negative inductor for bandwidth enhancement," *IEEE Transactions on Antennas and Propagation*, vol. 55, May 2007, pp. 1275–1282.
22. G. Bit-Babik, C. Di Nallo, J. Svigelj, and A. Faraone, "Small wideband antenna with non-foster loading elements," *in Proceedings of the ICEAA 2007 International Conference on Electromagnetics in Advanced Applications*, Torino, Italy, September 17–21, 2007, pp. 105–107 (on ieeexplore.ieee.org).
23. J. L. Volakis, A. Chatterjee, and L. C. Kempel, *Finite Element Method for Electromagnetics*, IEEE Press, New York, 1998.

CHAPTER 9
Antennas for RFID Systems

Ugur Olgun, Kenneth E. Browne, William F. Moulder, and John L. Volakis

9.1 Historical Background

Radio frequency identification (RFID) is a wireless technology mainly used to communicate digital information between a stationary device and one or more movable objects. The stationary device typically employs a sophisticated reader with a large antenna emitting a signal periodically or on demand. These readers, also known as beacons or interrogators, are connected to a network or host computer. The RFIDs operate at a number of frequencies ranging from 100 KHz to 24 GHz, and include the industrial scientific and medical (ISM) bands at 125 KHz, 13.56 MHz, 433 MHz, 915 MHz, 2.45 GHz, 5.8 GHz, and 24.125 GHz. Table 9.1 provides some comparative information relating to various RFID system characteristics at different frequency bands.

The objects to be detected and identified have an attached tag (transponder) incorporating a CMOS chip for data storage, which is connected to an antenna or a surface acoustic wave (SAW) device. The tags can be powered via battery or rectification of the transmitted signal originating from the reader. Communication between the tag and reader allows for data to be transferred and stored in real-time with a high level of accuracy, enabling continuous identification and monitoring. Figure 9.1 provides basic information on how passive RFIDs (powered via rectification) work.

Over the past decades, the realm of RFIDs has grown exponentially and has become highly integrated within our daily lives. RFID

Frequency Band	LPF 125 KHz	HF 13.56 MHz	UHF 860–960 MHz	Microwave 2.45 GHz and Up
Read range (Passive tags)	<2 Feet	<3 Feet	<10–30 Feet	<10–20 Feet
Tag power source	Passive	Passive	Passive and Active	Passive and Active
Tag cost	Expensive	Expensive, but less than LF	Potential to be very cheap	Potential to be very cheap
Typical applications	Keyless entry, animal tracking, POS	Smart cards, item-level tracking, libraries	Pallet tracking, logistics, airline baggage handling	Electronic toll collection
Data rate	1 kbit/s	25 kbit/s	30 kbit/s	100 kbit/s
Performance near metal or liquid	Best	Better	Worse	Worst
Passive tag size	Largest	Larger	Smaller	Smallest
Power limitations	Inductively coupled	Inductively coupled	4W EIRP, USA 0.5 EIRP, Europe	4W EIRP, USA 0.5 EIRP, Europe

TABLE 9.1 RFID System Characteristics at Various Frequencies

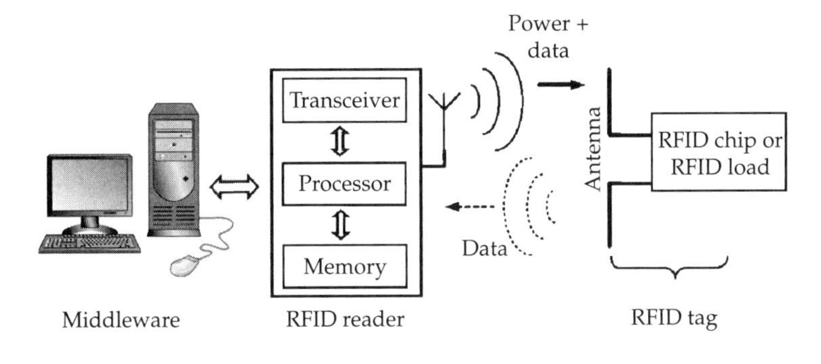

FIGURE 9.1 Passive RFID system layout.

technologies span many areas of research including antenna theory, radio propagation, microwave techniques, integrated circuit design, encryption, mechanical design, materials, network and systems engineering, software development, and circuit theory. Since RFID commercialization in the late 1980s, applications using RFIDs have grown dramatically to the point of ubiquity. These include RFIDs for theft prevention, automated toll collection, traffic management, border security, pet tracking, library book tracking, and temperature monitoring, among others [1–3].

A particular business area using RFID technology on a much larger scale, is that of supply chain management and logistics. The US Food and Drug Administration (FDA), US Department of Defense (DoD), and several large department stores have integrated RFID technology in product tracking to increase safety and reduce costs. Walmart was one of the first department stores to include RFID tags on its products in both retail stores and warehouses [4]. The DoD followed suit and issued an RFID mandate [5] requiring top suppliers to place RFID tags on all pallets and cases. Motivated by safety (rather than cost), the FDA is utilizing RFID technology to ensure prescription drug authenticity by implementing read only tags to identify their product's unique serial number. Suppliers track these serial numbers encoded into tags to verify product authenticity, shipment origin, and final destination.

Although RFID technology seems to be a relatively new concept, the idea dates back to the 1940s when radar was used to reflect coded signals to identify aircraft as friend or foe. Use of RFIDs for their present adaptation was to some degree envisioned by Harry Stockman in 1948 in his article, *Communication by Means of Reflected Power* [6]. However, 30 years passed before the concept was translated to practical form. RFID interest and research began when integrated transistor circuits were coupled with handheld wireless devices. Like cell phones [7], by the 1970s, interest in RFIDs grew dramatically. In fact, passive RFID tags were operational with a detection distance of more than 10 m [4]. By the late 1970s, interest in low-power and low-voltage CMOS circuits for RFIDs had also increased (see Fig. 9.2*a*). Tag memory was dramatically enhanced via switch or wire bond implementation along with the use of fusible link diode arrays.

More rapid RFID development came in the 1980s when the introduction of personal computers provided the means for organized data-collection management. Several microwave tags were developed and fabricated, integrating wireless components and customized CMOS circuits. The primary tag memory choice was the EEPROM, permitting manufacturing at a large scale with each tag having unique programming capability. Such mass scale production allowed for significant cost reduction, while concurrently increasing functionality and reducing size.

The rapid acceleration of RFID technology continued into the 1990s with significant advancements in Schottky diode fabrication on standard CMOS chips. This allowed for microwave RFID tags using a

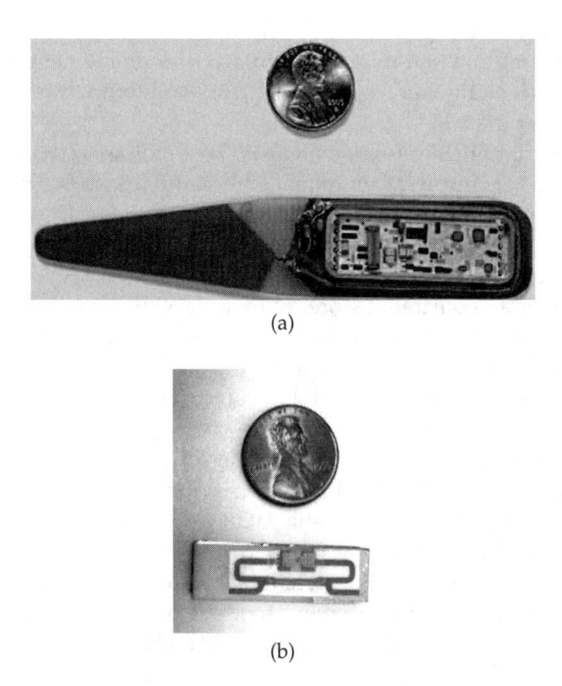

(a)

(b)

FIGURE 9.2 Evolution of RFID tags. (a) A 12-bit read-only tag built using CMOS logic chips and hybrid thick film for antenna printing (circa 1976); circuitry covers half of tag area. (*After [4] ©IEEE 2007.*); (b) A commercial 800-bit read/write tag from Omni-ID.

FIGURE 9.3 A 96-bit UHF-RFID tag from Alien Technologies fabricated in the form of sticky label.

single integrated circuit. Previously, such integration was attained via low-frequency inductively coupled tags (used primarily for keyless applications). In contrast, Schottky diodes attached to receiving antennas can allow detection and data transfers extending several meters.

Currently, RFID technology is being applied and used ubiquitously. The smallest tags are still realized by integrating two components: a CMOS chip (integrating memory, processor, and amplification) and an antenna (see Fig. 9.2*b*). RFID tags are of low cost and small enough to be placed on a variety of objects in the form of flat adhesive based labels (see Fig. 9.3). Currently, RFID deployment is planned for a variety of applications, ranging from product tracking (as a replacement of the familiar laser scanned bar codes) to safety applications. Thus, there is much interest to develop tags at lower cost, smaller size, and more functionality (i.e., storing not only product ID numbers but also product manufacturing location, date, delivery route, temperature, and vibration history). Several critical components to this effort include antenna miniaturization and capability for remote RFID powering (i.e., batteryless tags) [8,9].

9.2 Basic Operation of RFID Systems

9.2.1 Tag Categories

RFID tags are categorized based on the method used to power them. As such, they can be grouped in three categories.

- Active tags—These RFIDs have an integrated power source (viz., they are battery operated). Active tags typically communicate at 433 MHz, 2.45 GHz, and 5.8 GHz.

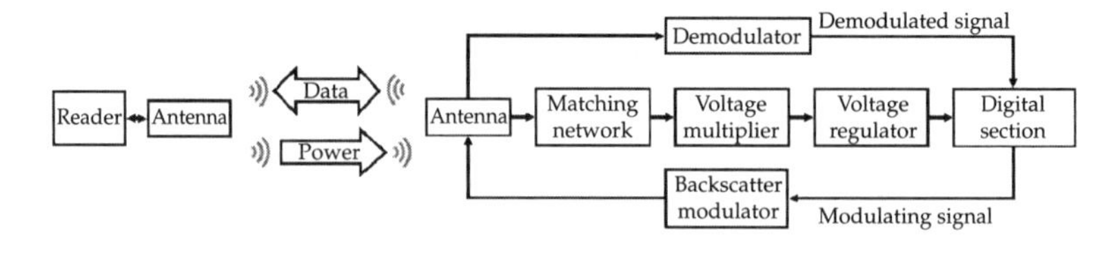

FIGURE 9.4 Passive RFID system overview.

- Passive tags—These RFIDs rely on power from the interrogating reader (see Fig. 9.4).
- Battery-assisted passive (BAP) tags—The BAP tags fill the gap between the shorter-range passive tags and high-cost active tags. Tag-reader communication for BAP tags is passive since it is done by reflecting the readers' RF energy. The tags in this category can often achieve remarkable read range due to the integrated battery that powers the onboard chip.

An active tag has the advantage of transmitting strong signals detectable from 20 m up to 100 m. However, active tags are generally much larger and more expensive when compared to passive tags. The latter rely on charging an internal capacitor from the reader's radiated RF energy. Not suprisingly, research focus has been directed toward extending read range from a few centimeters to tens of meters. A schematic of the integrated components and design layout of a typical RFID tag is given in Fig. 9.4. In the following section, we focus more on passive tags, including functionality, impedance matching, antennas, and power harvesting approaches.

9.2.2 Passive Radio Frequency Identifications

Passive RFID tags are comprised of three main components:

- An integrated circuit for modulating/demodulating the RF signal, storing information, and processing data.
- Power harvesting unit (rectifier or rectenna) and the impedance matching circuit.
- Transceiver antenna.

Given the topic of this book, the rest of the chapter will focus on the last two components consisting of the antenna and rectifier circuit. To obtain a better understanding we first describe the basic operation of typical passive tags.

One type of primitive passive RFID tag is based on using an inductive coil to code the RFID information. As depicted in Fig. 9.5,

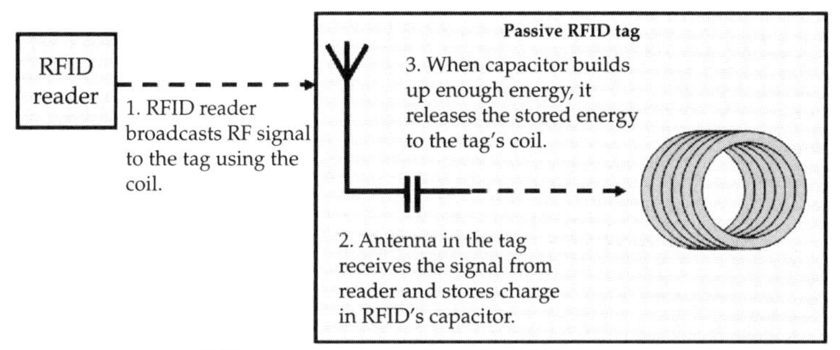

FIGURE 9.5 Simplified view of data transfer at low-frequencies using passive RFID tags (tag is enlarged for clarity).

the reader transmits an RF signal received by the tag's antenna. This received energy charges an onboard capacitor via a rectifier circuit (see Section 9.4). Once the capacitor is charged, it then powers the RFID tag circuitry to produce a modulated signal using the tag's coil. The latter is then retransmitted and demodulated by the reader.

At higher frequencies (>100 MHz), it is more practical to code the RFID signal using energy scattered directly back from the tag. A simple way to program a code or store some sort of information onto such an RFID tag, is to use chip circuitry to control the switches over a distinct time frame (see Fig. 9.6). The switches alter the impedance seen by the receiving tag antenna, which in turn alters the reflected signal back toward the reader's antenna. The modified backscatter signal can be individually coded by programming the chip's circuitry, leading to specific tag responses (see Fig. 9.7) that identify a product or some collected sensor information. It should be noted that the RFID's chip response is nonlinear, leading to input impedances that not only vary with frequency, but also with input power levels. Thus, for optimal performance (i.e., to maximize the tag's read range), it is important to match the antenna impedance to the chip at the lowest possible tag activating power level.

9.2.2.1 RFID Microchip

RFID microchip is the brain of the tag, as it houses the processor and the tag memory, both powered by the rectifier unit. RFID chips are small logic units that may consist of a few thousand logic gates (see Fig. 9.8). The simplest of these are on the order of 1500 gate equivalents [11].

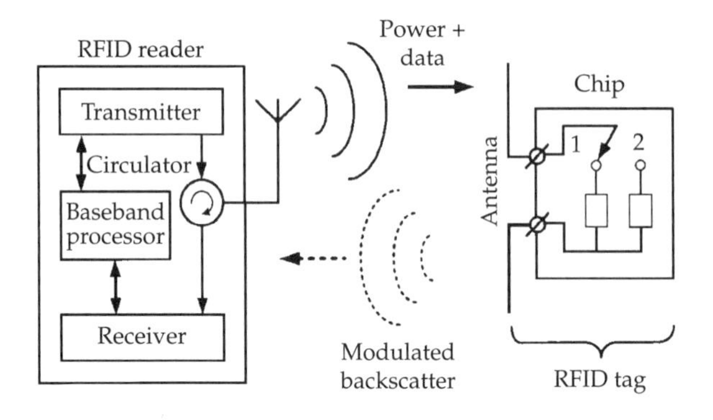

FIGURE 9.6 RFID system layout based on modulation of the backscatter tag signal. (*After [10] ©IEEE 2006.*)

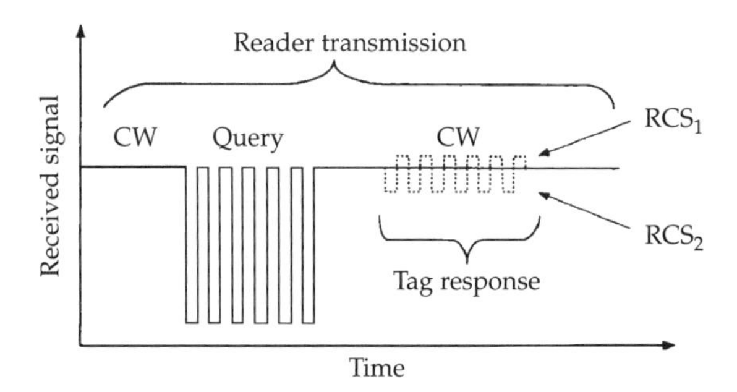

FIGURE 9.7 Example reader and tag coded response based on the modulation of the backscatter signal. (*After [10] ©IEEE 2006.*)

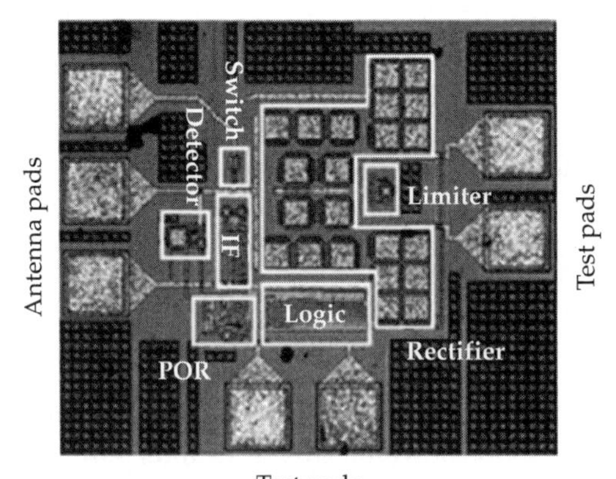

FIGURE 9.8 Die microphotograph of an UHF-RFID microchip. (*After [11] ©IEEE 2005.*)

The major blocks of the basic RFID microchip are outlined in Fig. 9.9. The logic/processor unit is responsible for implementing the communication protocol between the tag and the reader. The synchronization signal (i.e., clock signal) for the processor unit is typically extracted from the reader antenna signal. Thus, a clock is not included in the microchip, resulting in reduced size and cost.

The RFID chip memory is generally segmented (i.e., consists of several blocks or fields) and is addressable. Addressability implies that the memory can be read/erased and rewritten. A tag memory block

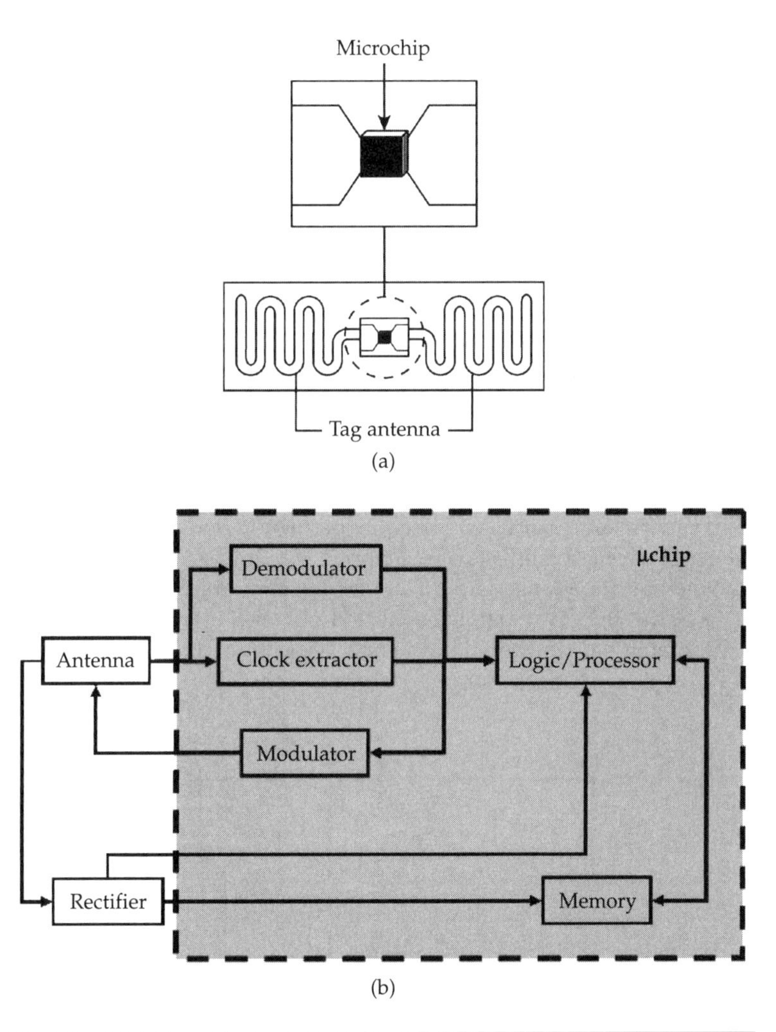

FIGURE 9.9 Basic components of an RFID microchip. (a) Components of a passive tag (microchip is enlarged for clarity); (b) Schematic of an RFID microchip.

can hold different data types, such as a portion of the tagged object identifier data, checksum [e.g., cyclic redundancy check (CRC)] bits for verifying transmitted data accuracy, and so on. Recent advances have shrunk the size of the microchip to less than that of a grain of sand. However, a tag's physical dimensions are not determined by the microchip size, rather by the antenna size.

In general, a commercial off-the-shelf (COTS) RFID microchip is integrated with more sophisticated components. A digital anticollision system is one such component. This component prevents tag data collisions when many tags are present in a small area. More complex RFID chips include security primitives, encryption, and tamperproofing hardware.

9.2.2.2 Impedance Matching

Proper impedance matching between the antenna and the chip is of paramount importance. Good impedance matching maximizes power transfer to the chip, directly impacting RFID performance characteristics such as maximum distance for read/write capability [10]. Tag orientation sensitivity, data link quality, RFID read reliability, and accuracy are also dependent upon impedance matching performance.

As already noted, RFIDs are nonlinear devices requiring a certain threshold voltage (i.e., power) to turn on. It also demonstrates a complex input impedance behavior dependent on frequency and input power. An equivalent circuit for the RFID tag is depicted in Fig. 9.10. The real part of the chip impedance varies with input power and depends on chip power consumption. However, the reactive part of the chip's impedance varies with frequency, mostly determined by parasitic and packaging effects of the chip.

Antenna impedance is typically matched to the high impedance state of the chip to maximize collected power [12]. High impedance state matching requires the antenna impedance to be matched to

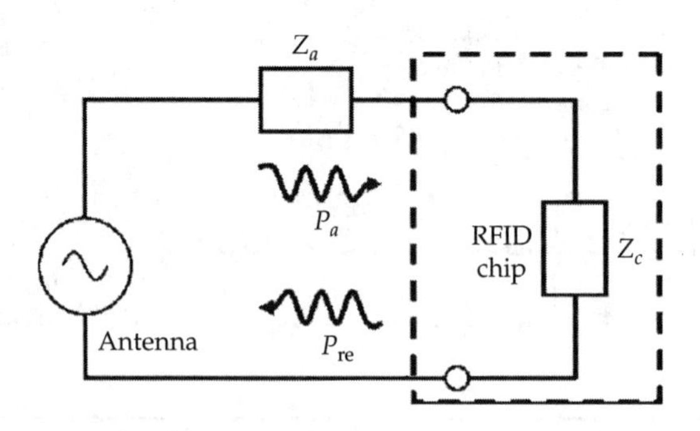

FIGURE 9.10 RFID tag equivalent circuit.

the chip impedance at the minimum power level required for the chip to turn on. By doing so, reliable and accurate tag operation is ensured at the maximum tag range. It is normally expected that the tag will operate well within the range. However, due to the microchip's nonlinear characteristics, the higher reader power may result in severe impedance mismatches, causing dead detection spots.

9.2.2.3 RFID Radar Cross Section

RFID radar cross section is important for reflecting passive tags [13, 14]. In this case, the RFID contains a resonating structure that reradiates when illuminated by the reader. The detectability of the tag depends on the RCS (or backscattered field) returned by the tag. The RFID reacts to the incident power density generated by the reader.

$$S = \frac{P_t G_t}{4\pi r^2} \tag{9.1}$$

Here S is the power density in Watts/m^2, P_t is the total transmitted power by the reader, G_t is the reader's antenna gain, and r is the distance between the reader and the RFID tag. If the tag's antenna has an effective aperture, A_e, its received power at the antenna terminals is then $P_a = SA_e$. This becomes the input power to the circuit powering the tag's circuit. Referring to the tag's equivalent circuit (see Fig. 9.10), we can then compute the reradiated power P_{re} due to P_a.

$$P_{\text{re}} = KP_a G_a = \frac{4R_a^2}{|Z_a + Z_c|^2} \cdot P_a G_a \tag{9.2}$$

Here, G_a is the tag's antenna gain and K is the ratio of reradiated to incident energy with $R_a = \Re\{Z_a\}$, where Z_a is the complex antenna input impedance. We observe that a short circuited tag ($Z_c = 0$) reradiates 4 times more power than a conjugate matched one ($Z_c = Z_a^*$). We also note that, in the case of conjugate matching, the tag absorbs and reradiates equal amounts of power. The RFID sends information back to the reader by taking advantage of the returned field variation as a function of Z_c. That is, by changing or modulating the RFID impedance, Z_c, in a known fashion, a specific code can be transmitted and identified by the reader.

In general, we use the tag radar cross section, σ, to measure the radiation effectiveness. Specifically, in evaluating the RFID's effectiveness, the ratio of the returned to transmitted power is of interest. This is also the formal definition of the RCS, σ. Thus, we have

$$\sigma = \frac{P_{\text{re}}}{S} = KA_e G_a = \frac{\lambda^2 G_a G_t R_a^2}{\pi |Z_a + Z_c|^2} |\hat{\mathbf{p}}_{\text{reader}} \cdot \hat{\mathbf{p}}_{\text{tag}}|^2 \tag{9.3}$$

Here $\hat{p}_{\text{reader}}$ and $\hat{p}_{\text{tag}}$ refer to the corresponding antenna polarizations. Clearly, the tag antenna gain plays a major factor. Therefore, a challenge is to design an efficient tag antenna.

9.3 Radio Frequency Identification Antennas

A plethora of RFID antenna configurations have been exploited to enhance RFID technology in terms of size, detection distance, and functional robustness. Given the enhancements in tag circuitry [15,16], the main bottleneck for RFID development continues to be the tag antenna. The main goal in RFID tag design is to have an antenna that is not only simple, physically small, and compatible with the tag's material, but also has a large effective aperture. Several forms and types of RFID antennas have been proposed. Some of these were discussed in Section 3.2.7 of Chap. 3. The proposed antennas include loop [17], slot, fractal [18], planar [19,20], inductively coupled [21], meander line [22, 23], and printed folded dipoles [24] (see also [25,26]). Of interest is the development of platform independent antennas to reduce cost due to material loading [3]. For example, Chaps. 5 through 7 discuss a variety of metamaterial antennas that are less sensitive (i.e., have less impedance variation) to platform characteristics [27–29]. The reader is also referred to a large body of references on RFID antennas [30–34]. A few of these not covered in Chap. 3 are discussed in the following sections.

9.3.1 Meander-Line Dipoles

A meander antenna is an extension of the basic folded dipole antenna as it is made of several folded elements arranged in a manner that leads to predefined resonance. Meander-line antennas (MLAs) are an attractive choice for UHF tags [22,23,35] due to the necessity to reduce size at these frequencies. As discussed in Chap. 7, folding the printed lines or meandering them produces a wire configuration having both capacitive and inductive reactance. Resonance then occurs when the capacitive and inductive reactance cancel each other, but at much lower frequency than that corresponding to the $\lambda/2$ physical length of the overall antenna. However, this occurs at the expense of lower gain, especially when the antenna surface is contained in a confined space [36].

MLAs exhibit a radiation pattern similar to a conventional half-wavelength dipole but across a broad bandwidth [37]. Indeed, many UHF-RFID applications require the tag antenna to have dipole-like radiation pattern and broad bandwidth to compensate for changes in the operating frequency due to different radio regulations in specific

FIGURE 9.11 Meander-line antenna for a UHF-RFID tag. (*Courtesy of Alien Technologies [38].*)

geographical areas. MLAs can satisfy these requirements and are, therefore, popular for commercial UHF-RFID tags. An example is depicted in Fig. 9.11 which refers to a commercial tag manufactured by Alien Technologies.

9.3.2 Patch Antennas

Patch antennas are highly popular in RFID applications due to their conformal, lightweight, and low-profile nature [39]. Historically, patches were primarily used for military applications. However, the recent reduced prices for the dielectric substrate has led to increased use of microstrip antennas for commercial RFID applications [40,41].

In addition to being simple to fabricate, patch antennas have an inherent ability for polarization diversity. A patch antenna can be easily designed to have any desired polarization that can be readily fabricated by etching the metal bonded to the dielectric substrate. Also, patch arrays can be used to deliver (at little additional cost) much higher gains, while preserving the lightweight and planar nature of the antenna. Example patch antenna arrays are depicted in Fig. 9.12. They work well for readers, bacause of the need to generate dual [42] or circular polarization [43].

9.3.3 Fractal Antennas

The term fractal, meaning broken or irregular fragments, was originally used to describe a family of complex shapes that possess an inherent self-similarity or self-affinity in their geometrical structure [44]. As noted in Chap. 3, fractals are useful in antenna engineering applications since their self-replicating characteristic allows for multiband behavior [45]. Fractal geometries allow more electrical length to be realized in a given area, leading to miniaturization [46]. Common replicating geometries for fractal antennas include dipoles, loops, and patches. Typical fractal curves for such antennas include the Koch, Minkowski, and Hilbert curves.

As an example, a tag antenna resonating at 868 MHz and 2.45 GHz based on a Minkowski fractal loop is depicted in Fig. 9.13*a*. Dual-band dipole antennas [48,49] and patches [47,50] for RFID have

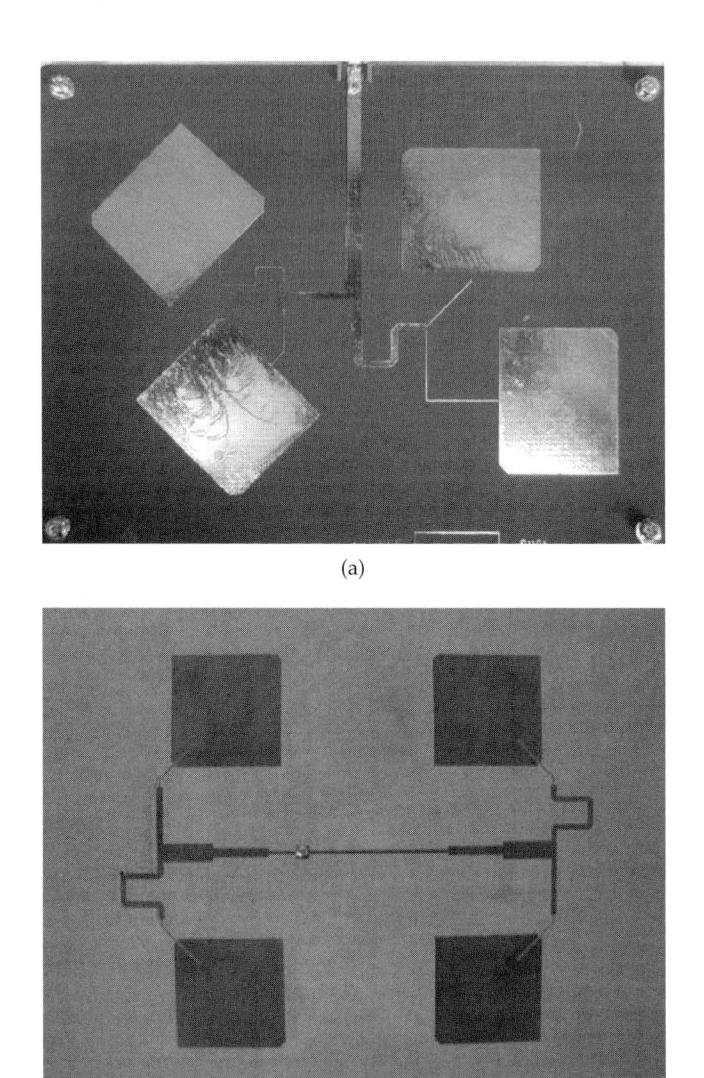

(a)

(b)

FIGURE 9.12 Patch antenna arrays for RFID readers. (a) Four-element patch array manufactured by Alien Technologies; (b) Four-element patch array manufactured at The Ohio State University ElectroScience Laboratory.

also been realized using fractal geometries. Figure 9.13*b* shows an example fractal RFID antenna operating at three distinct bands. Several other fractal antennas for RFIDs can be found in [47, 51–53], many of which exhibit miniaturization.

(a)

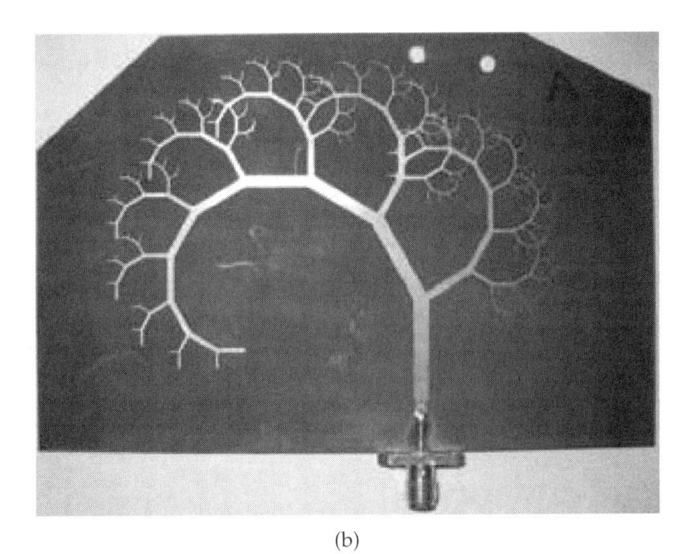

(b)

FIGURE 9.13 Fractal antennas for RFID applications. (a) Minkowski fractal loop tag operating at 868 MHz and 2.45 GHz [47]; (b) Fractal antenna operating at 3 RFID bands. (*After [45] ©IEEE 2008.*)

9.3.4 Planar Antennas

Planar antennas refer to those printed conformally on the RFID's structure or on adhesive strips and/or on flexible substrates [54, 55]. Cost is typically a driving factor in adapting these. Also, of interest is to print these antennas on a variety of surfaces, including paper, thin mylar, or polymer surfaces. Not suprisingly, they are common in commercial RFID systems. Some of the commercial tags which work with these systems are depicted in Fig. 9.14.

9.3.5 Slot Antennas

A slot antenna consists of a radiator formed by cutting narrow slots in a metal surface. An RFID tag with a slot antenna is depicted in Fig. 9.15. The shape and size of the slot(s) determine the operating frequency and antenna radiation pattern [56]. Slot antennas are often used instead of dipole and meander-line antennas when greater control of the radiation pattern is required, as is the case with some RFID applications. The slot antenna's main advantages are compact size,

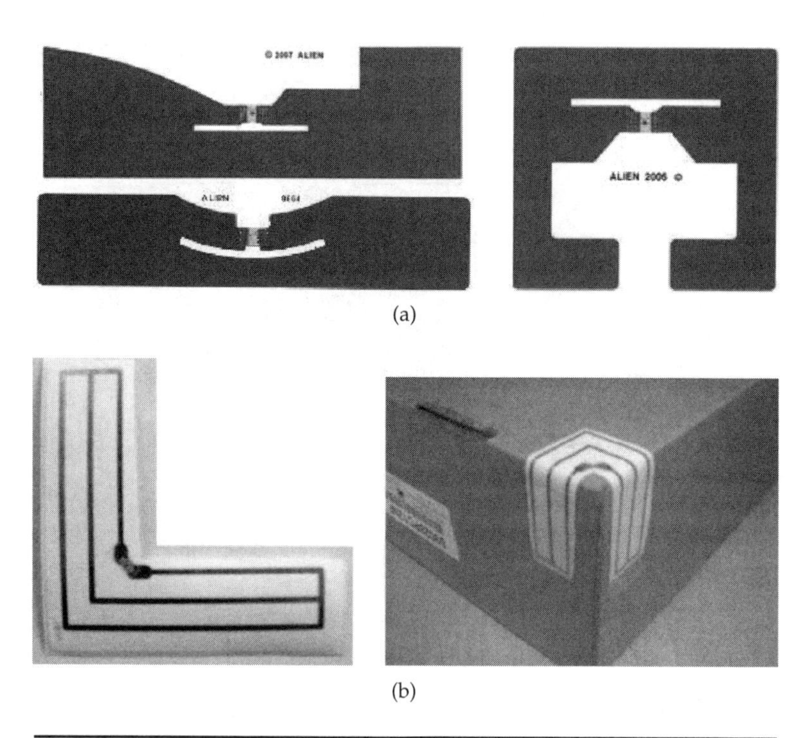

FIGURE 9.14 Planar antennas for RFID applications. (a) Planar RFID tags. (*Courtesy of Alien Technologies [57].*); (b) Planar RFID tag for attachment at cartboard surfaces or other packages. (*After [58] ©IEEE 2006.*)

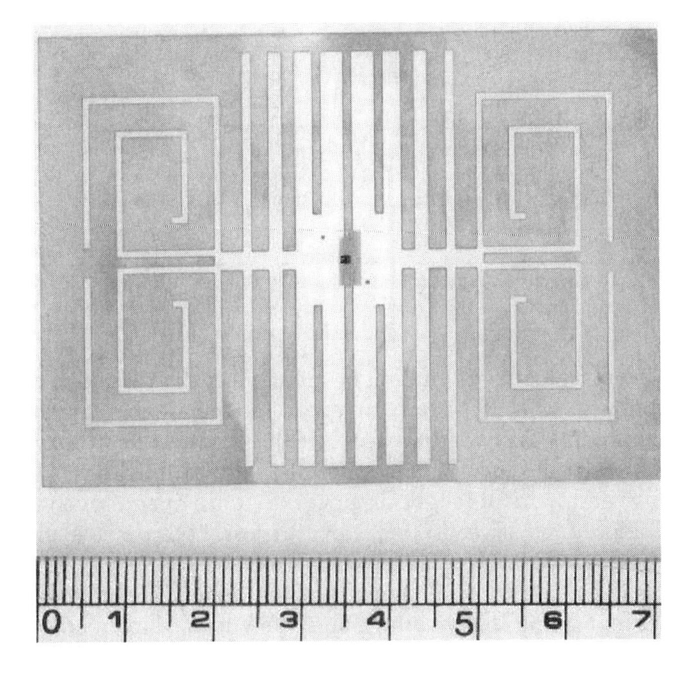

FIGURE 9.15 Miniaturized slot tag. (*After [59] ©IEEE 2005.*)

design simplicity, robustness, and convenient adaptation to mass production using PCB technology.

9.4 RFID Power Harvesting—Rectennas

Passive RFIDs employ a power harvesting unit (rectenna) to extract DC energy from the incoming RF wave, which is used to power the integrated circuit components. A rectenna is an antenna with a zero bias Schottky diode connected to its terminals [39] (see Fig. 9.16). As the RFID antenna receives a voltage, the Schottky diode allows for capacitor charging (but no discharging). The goal is to charge the capacitor to a level where a minimum required voltage can be delivered to the RFID circuitry over a prespecified minimum time period.

Off-the-shelf RFIDs use a single antenna for both communication and power transfer, a limiting factor in rectenna design. A number of investigators [60,61] have also proposed the addition of features like sensing, tracking, and on-tag data processing to conventional passive RFIDs. However, even with state-of-the-art low-power circuits, conventional RFIDs require a battery to achieve these ambitious goals. But, the battery itself is a major challenge. In many applications,

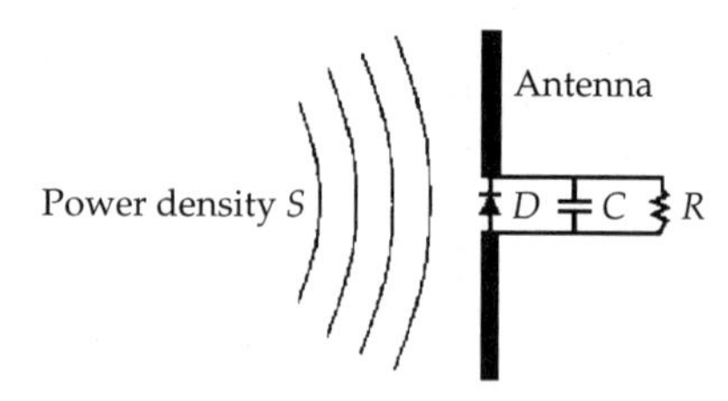

FIGURE 9.16 Schematic of the simplest power harvesting unit.

battery replacement is a limiting factor and a costly proposition. One idea (batteryless RFID) is to employ an independent rectenna optimized for wireless power transfer [62,63]. In contrast to ordinary passive RFIDs, batteryless RFIDs can handle complex functions (sensing, tracking) and work at longer distances (5–10 m) over a much longer time period (tens of years) [64]. Nevertheless, both passive and batteryless tags require a power harvesting unit in their package.

Typical tags require a voltage of 1 to 3 V (depending on the type of transistors used in the circuitry), and a few tens of microamps [61]. However, if the tag is a few meters away, the rectenna open-circuit voltage may only be 0.1 to 0.3 V. This is not large enough to operate the transistors in the circuitry. A common approach to boosting the voltages from a rectifier is to use a charge pump as part of the rectifier circuit (see Fig. 9.17). A charge pump incorporates a number of diodes

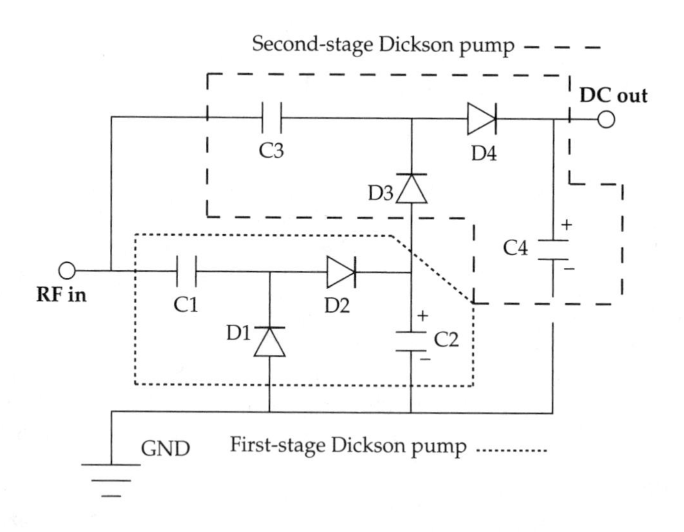

FIGURE 9.17 Schematic of a two-stage Dickson charge pump rectifier.

connected in series. This has the net effect of increasing the output voltage of the array.

A schematic of a two-stage Dickson charge pump [63] is shown in Fig. 9.17. As seen, two zero-bias Schottky diode pairs are employed and denoted as D1, D2, D3, and D4 in the schematic. The zero-bias diode pairs were chosen as they have excellent performance at UHF and do not require external biasing. This is crucial, as even a few microamperes of bias current is difficult to generate. Further, these diodes have relatively low-barrier height and high-saturation current when compared to externally biased detector diodes. This results in higher output voltage at low-power levels. However, a drawback is their higher series resistance, which inherently leads to higher losses.

Each stage of the Dickson charge pump is formed using two diodes and two capacitors. Focusing on the first-stage Dickson charge pump in Fig. 9.17, we note that C1 serves as a filter to prevent DC current. The capacitor being charged is C2 and the diodes D1 and D2 are intended to prevent discharging while current flow is oscillatory [63].

Specifically, when the RF voltage is negative, D1 is on and the current flows from the ground through D1. Alternatively, when the RF input is positive, D1 turns off and D2 turns on to continue charging C2. As such, we can expect the DC voltage on C2 (intermediate storage capacitor) will be twice the supplied peak RF voltage. The operation of the second-stage Dickson charge pump is similar, resulting in an output voltage across C4 (main storage capacitor), double that of the voltage across C2.

An impedance matching network is essential in providing maximum power transfer from the antenna to the rectifier circuit. One approach is to design the matching network as proposed in [65]. According to [65], the component models used in the simulations were not accurate enough to include all of the circuit parasitics. Therefore, it was concluded that the modeling of the rectifier circuit should be based on experimental characterization. This can be done by measuring the input impedance (S_{11}) of the rectifier circuit (consisting of the zero-bias diode pairs, capacitors, load resistance, and wires for output DC voltage) without a matching network. The results from the experimental characterization can then be used as a black box for the impedance-matching circuit design. Assuming that the rectenna is powering a load, R_{load}, the efficiency is given by

$$\eta = \frac{P_{\text{DC}}}{P_{\text{RF}}} = \frac{V_{\text{load}}^2}{R_{\text{load}} P_{\text{RF}}} \tag{9.4}$$

Referring to Fig. 9.1, V_{load} is the voltage across R_{load} and P_{RF} is the incident power received by the rectenna at the diode terminals. For the circuit shown in Fig. 9.18*a*, the conversion efficiency, η, for RF power

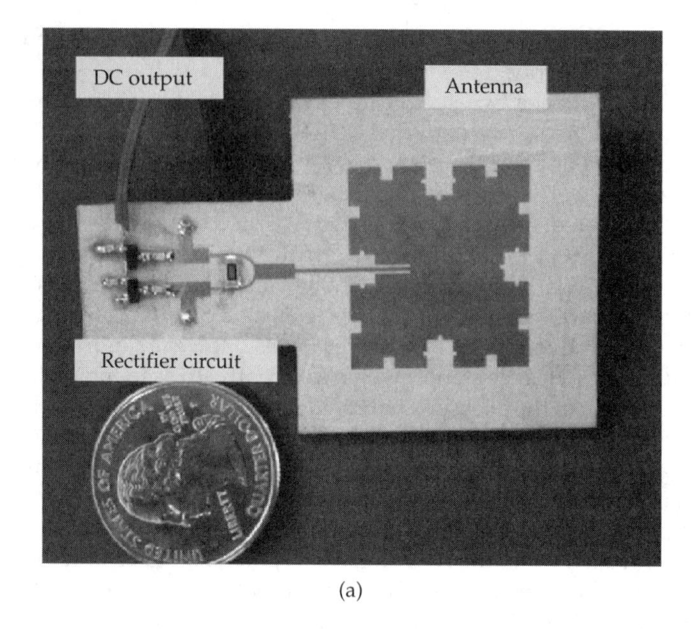

(a)

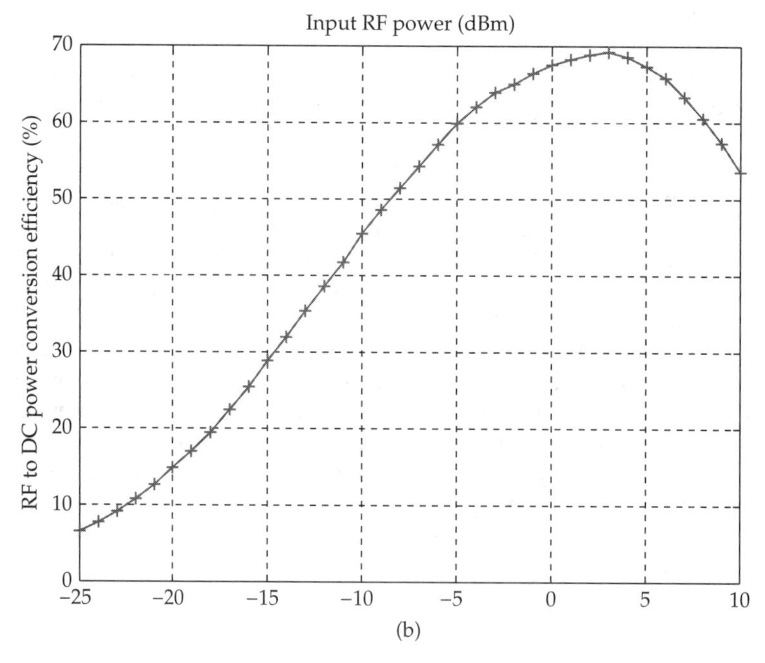

(b)

FIGURE 9.18 Typical rectenna for RFID sensing applications and its efficiency performance. (a) A rectenna composed of a Koch's fractal antenna and a two-stage Dickson charge pump; (b) Efficiency of a typical rectifier versus input RF power [65].

levels varying from −25 to 10 dBm is depicted in Fig. 9.18*b*. This plot was calculated using Eq. (9.4). It is apparent that a typical rectifier can realize an RF to DC conversion efficiency greater than 60% over a wide range of power levels, with a maximum efficiency close to 70%.

Comments

As interest in RFID and its adaptation continues to grow, cost and functionality will play a key role in influencing device design, fabrication, and marketing. Considering the number of RFID pilot trial announcements each month, the RFID industry is expected to experience a booming growth over the next decade in terms of sales and potential applications.

There are key issues still to be resolved with RFID deployment. Among them,

- Lack of globally accepted regulations/standards.
- Privacy issues: Many human habits and activities can be tracked with ubiquitous use of RFIDs.
- Lack of globally integrative software tools to track products from the manufacturing site to the consumer or sale point.
- Long-range detection, cost, wireless power harvesting, and multifunctionality will, of course, continue to motivate future design.

Regardless, RFIDs are here to stay just like cell phones, TV, and radios. RFID has already become a part of our daily lives through use in credit cards, passports, medical drugs, and mobile phones. Extensive research and investment in the technology will extend its capabilities to many more applications in the near future.

References

1. S. Basat, K. Lim, I. Kim, M. M. Tentzeris, and J. Laskar, "Design and development of a miniaturized embedded UHF-RFID tag for automotive tire applications," *Proceedings of the 55th Electronic Components and Technology Conference*, vol. 1, Lake Buena Vista, Florida, May–June 2005, pp. 867–870 (on ieeexplore.ieee.org).
2. W. Liu, H. Ning, and B. Wang, "RFID antenna design of highway ETC in ITS," *7th International Symposium on Antennas, Propagation & EM Theory*, Guilin, China, 2006, pp. 1–4 (on ieeexplore.ieee.org).
3. G. Marrocco, "RFID antennas for the UHF remote monitoring of human subjects," *IEEE Transactions on Antennas and Propagation*, vol. 55, June 2007, pp. 1862–1870.
4. J. Landt, "The history of RFID," *IEEE Potentials*, vol. 24, October–November 2005, pp. 8–11.
5. M. P. Peterson, "Defense federal acquisition regulation supplement; radio frequency identification," *Federal Register*, vol. 72, February 2007, pp. 6480–6484.

6. H. Stockman, "Communication by means of reflected power," in *Proceedings of the Institute of Radio Engineers (IRE)*. IEEE Press, October 1948, pp. 1196–1204.
7. M. Cooper, R. W. Dronsuth, and A. J. Mikulski, "Radio telephone system," U.S. Patent 403,725, 1973.
8. R. Weinstein, "RFID: a technical overview and its application to the enterprise," *IT Professional*, vol. 7, May–June 2005, pp. 27–33.
9. K. V. S. Rao, P. V. Nikitin, and S. F. Lam, "Antenna design for UHF-RFID tags: a review and a practical application," *IEEE Transactions on Antennas and Propagation*, vol. 53, no. 12, December 2005, pp. 3870–3876.
10. P. V. Nikitin and K. V. S. Rao, "Theory and measurement of backscattering from RFID tags," *IEEE Antennas and Propagation Magazine*, vol. 48, December 2006, pp. 212–218.
11. J.-P. Curty, N. Joehl, C. Dehollain, and M. J. Declercq, "Remotely powered addressable UHF-RFID integrated system," *IEEE Journal of Solid State Circuits*, vol. 40, November 2005, pp. 2193–2202.
12. K. V. S. Rao, P. V. Nikitin, and S. F. Lam, "Impedance matching concepts in RFID transponder design," *Fourth IEEE Workshop on Automatic Identification Advanced Technologies*, Buffalo, New York, October 2005, pp. 39–42.
13. S. S. Hossain and N. Karmakar, "An overview on RFID frequency regulations and antennas," *6th International Conference on Electrical and Computer Engineering*, Dhaka, Bangladesh, December 2006, pp. 424–427.
14. A. Harish, A. Gupta, H. Purohit, G. S. Lambda, and N. Grover, "Performance analysis of RFID tag antennas," *International Journal on Wireless & Opitical Communications*, vol. 1, 2007, pp. 1–8.
15. M. Usami, "An ultra-small RFID chip: Îij-chip," *Proceedings of 2004 IEEE Asia-Pacific Conference on Advanced System Integrated Circuits 2004*, August 2004, pp. 2–5.
16. R. Imura, "The world's smallest RFID μ-chip, bringing about new business and lifestyles," *2004 Symposium on VLSI Circuits, 2004. Digest of Technical Papers*, June 2004, pp. 120–123.
17. M. L. Ng, K. S. Leong, D. M. Hall, and P. H. Cole, "A small passive UHF-RFID tag for livestock identification," *IEEE International Symposium on Microwave, Antenna, Propagation and EMC Technologies for Wireless Communications*, vol. 1, August 2005, pp. 67–70.
18. S. K. Padhi, G. F. Swiegers, and M. E. Bialkowski, "A miniaturized slot ring antenna for RFID applications," *15th International Conference on Microwaves, Radar and Wireless Communications*, vol. 1, May 2004, pp. 318–321.
19. C. A. Diugwu, J. C. Batchelor, R. J. Langley, and M. Fogg, "Planar antenna for passive radio frequency identification (RFID) tags," *7th AFRICON Conference in Africa*, vol. 1, 2004, pp. 21–24.
20. V. Pillai, H. Heinrich, D. Dieska, P. V. Nikitin, R. Martinez, and K. V. S. Rao, "An ultra-low-power long-range battery / passive RFID tag for UHF and microwave bands with a current consumption of 700 na at 1.5 v," *IEEE Transactions on Circuits and Systems I: Fundamental Theory and Applications*, vol. 54, July 2007, pp. 1500–1512.
21. H.-W. Son and C.-S. Pyo, "Design of RFID tag antennas using an inductively coupled feed," *IEEE Electronics Letters*, vol. 41, September 2005, pp. 994–996.
22. G. Marrocco, "Gain-optimized self-resonant meander line antennas for RFID applications," *Antennas and Wireless Propagation Letters*, vol. 2, 2003, pp. 302–305.
23. C. T. Rodenbeck, "Planar miniature RFID antennas suitable for integration with batteries," *IEEE Transactions on Antennas and Propagation*, vol. 54, December 2006, pp. 3700–3706.
24. S. A. Delichatsios, D. W. Engels, L. Ukkonen, and L. Sydanheimo, "Albano multidimensional UHF passive RFID tag antenna designs," *International Journal of Radio Frequency Identification Technology and Applications*, vol. 1, no. 1, 2006, pp. 24–40.
25. Y. Yamada, W. H. Jung, and N. Michishita, "Extremely small antennas for RFID tags," *10th IEEE Singapore International Conference on Communication Systems*, October 2006, pp. 1–5.

26. A. Ibrahiem, T.-P. Vuong, A. Ghiotto, and S. Tedjini, "New design antenna for RFID-UHF tags," *IEEE Antennas and Propagation Society International Symposium*, July 2006, pp. 1355–1358.
27. L. S. Chan, K. W. Leung, R. Mittra, and K. V. S. Rao, "Platform-tolerant design for an RFID tag antenna," *IEEE International Workshop on Anti-counterfeiting, Security, Identification*, April 2007, pp. 57–60.
28. M. Stupf, R. Mittra, J. Yeo, and J. R. Mosig, "Some novel design for RFID antennas and their performance enhancement with metamaterials," *IEEE Antennas and Propagation Society International Symposium*, July 2006, pp. 1023–1026.
29. E. Irci, K. Sertel, and J. L. Volakis, "Antenna miniaturization for vehicular platforms using printed coupled lines emulating magnetic photonic crystals," to appear in *Elsevier Metamaterials Journal*, 2010.
30. R. Li, G. DeJean, M. M. Tentzeris, and J. Laskar, "Integrable miniaturized folded antennas for RFID applications," *IEEE Antennas and Propagation Society International Symposium*, vol. 2, June 2004, pp. 1431–1434.
31. L. Ukkonen, M. Schaffrath, D. Engels, L. Sydanheimo, and M. Kivikoski, "Operability of folded microstrip patch-type tag antenna in the UHF-RFID bands within 865–928 MHz," *IEEE Antennas and Wireless Propagation Letters*, vol. 5, no. 1, December 2006, pp. 414–417.
32. L. Ukkonen, L. Sydanheimo, and M. Kivikoski, "Patch antenna with ebg ground plane and two-layer substrate for passive RFID of metallic objects," *IEEE Antennas and Propagation Society International Symposium*, vol. 1, June 2004, pp. 93–96.
33. B. Gao, C. H. Cheng, M. M. F. Yuen, and R. D. Murch, "Low cost passive UHF-RFID packaging with electromagnetic band gap (ebg) substrate for metal objects," *Proceedings. 57th Electronic Components and Technology Conference*, May 2007, pp. 974–978.
34. P. Raumonen, M. Keskilammi, L. Sydanheimo, and M. Kivikoski, "A very low-profile cp ebg antenna for RFID reader," *IEEE Antennas and Propagation Society International Symposium*, vol. 4, June 2004, pp. 3808–3811.
35. K.-S. Min, T. V. Hong, and D.-W. Kim, "A design of a meander-line antenna using magneto-dielectric material for RFID system," *Asia-Pacific Conference Proceedings Microwave Conference Proceedings*, vol. 4, December 2005, pp. 1–4.
36. T. J. Warnagiris and T. J. Minardo, "Performance of a meandered line as an electrically small transmitting antenna," *Electronics Letters*, vol. 40, 2004, pp. 1516–1517.
37. H. Nakano, H. Tagami, A. Yoshizawa, and J. Yamauchi, "Shortening ratios of modified dipole antennas," *IEEE Transactions on Antennas and Propagation*, vol. 32, no. 4, April 1984, pp. 385–386.
38. A. Technology, "aln-9534 2x2 inlay product overview," *Product Datasheet*, 2008.
39. J. L. Volakis, *Antenna Engineering Handbook*, 4th ed. McGraw-Hill Professional, New York, 2007.
40. Z. N. Chen, X. Qing, and H. L. Chung, "A universal UHF-RFID reader antenna," *IEEE Transactions on Microwave Theory and Techniques*, vol. 57, 2009, pp. 1275–1282.
41. H. L. Chung, X. Qing, and Z. N. Chen, "Broadband circularly polarized stacked patch antenna for UHF-RFID applications," *IEEE Antennas and Propagation International Symposium*, 2007, pp. 1189–1192.
42. S. K. Padhi, N. C. Karmakar, C. L. Law, and S. Aditya, "A dual-polarized aperture coupled microstrip patch antenna with high isolation for RFID applications," *IEEE Antennas and Propagation Society International Symposium*, vol. 2, 2001, pp. 2–5.
43. R. Suwalak and C. Phongcharoenpanich, "Circularly polarized truncated planar antenna with single feed for UHF-RFID reader," *2007 Asia-Pacific Conference Communications*, 2007, pp. 103–106.
44. B. B. Mandelbrot, *The Fractal Geometry of Nature*, W. H. Freeman, New York, 1983.
45. S. Preradovic, I. Balbin, N. C. Karmakar, and G. Swiegers, "Chipless frequency signature based RFID transponders," *Proceedings of the 1st European Wireless Technology Conference*, 2008, pp. 302–305.

46. J. P. Gianvittorio and Y. Rahmat-Samii, "Fractal antennas: a novel miniaturization technique and applications," *IEEE Antennas and Propagation Magazine*, vol. 44, 2002, pp. 20–36.
47. M. Rusu, M. Hirvonen, H. Rahimi, P. Enoksson, C. Rusu, N. Pesonen, O. Vermesan, and H. Rustad, "Minkowski fractal microstrip antenna for RFID tags," *Proceedings of the 38th European Microwave Conference*, 2008, pp. 666–669.
48. Y. C. Lee and J. S. Sun, "Dual-band dipole antenna for RFID-tag applications," *Proceedings of the 38th European Microwave Conference*, 2008, pp. 995–997.
49. A. Ibrahiem, A. Ghiotto, T. P. Vuong, and S. Tedjini, "Bi-band fractal antenna design for RFID applications at UHF," *Proceedings of the 1st European Conference on Antennas and Propagation*, 2006, pp. 1–3.
50. H. Yang, S. Yan, L. Chen, and H. Shi, "Investagation and design of a modified aperture-couple fractal antenna for RFID applications," *ISECS International Colloquium on Computing, Communication, Control, and Management*, vol. 2, 2008, pp. 505–509.
51. J. Burnell, "Converter licenses innovative RFID antenna technology," *RFID Update*, June 15, 2007.
52. D. Psychoudakis, W. F. Moulder, C. C. Chen, H. Zhu, and J. L. Volakis, "A portable low-power harmonic radar system and conformal tag for insect tracking," *IEEE Antennas and Wireless Propagation Letters*, vol. 7, 2008, pp. 444–447.
53. N. A. Murad, M. Esa, M. F. M. Yusoff, and S. H. A. Ali, "Hilbert curve fractal antenna for RFID application," *2006 International RF and Microwave Conference*, 2006, pp. 182–186.
54. L. Yang, A. Rida, R. Vyas, and M. M. Tentzeris, "RFID tag and rf structures on a paper substrate using inkjet-printing technology," *IEEE Transactions on Microwave Theory and Techniques*, vol. 55, December 2007, pp. 2894–2902.
55. A. Rida, L. Yang, S. S. Basat, A. Ferrer-Vidal, S. Nikolaou, and M. M. Tentzeris, "Design, development, and integration of novel antennas for miniaturized UHF-RFID tags," *IEEE Transactions on Antennas and Propagation*, vol. 57, November 2009, pp. 3450–3458.
56. S.-Y. Chen and P. Hsu, "Cpw-fed folded-slot antenna for 5.8 GHz RFID tags," *IEEE Transactions on Microwave Theory and Techniques*, vol. 55, December 2007, pp. 2894–2902.
57. A Technology, "aln-9554 m inlay product overview," *Product Datasheet*, 2008.
58. A. Delichatsios and D. W. Engels, "The albano passive UHF-tag antenna: design and performance evaluation," *IEEE Antennas and Propagation Society International Symposium*, 2006, pp. 3209–3212.
59. Y. Tikhov, Y. Kim, and Y. H. Min, "A novel small antenna for passive RFID transponder," *35th European Microwave Conference*, vol. 1, 2005.
60. M. Philipose, K. P. Fishkin, M. Perkowitz, D. J. Patterson, D. Hahnel, D. Fox, and H. Kautz., "Inferring activities from interactions with objects," *IEEE Pervasive Computing: Mobile and Ubiquitous Systems*, vol. 3.4, 2004, pp. 50–57.
61. A. P. Sample, D. J. Yeager, P. S. Powledge, and J. R. Smith, "Design of a passively-powered, programmable sensing platform for UHF-RFID systems," *IEEE International Conference on RFID 2007*, 2007.
62. U. Olgun, C.-C. Chen, and J. Volakis, "Wireless power harvesting with planar rectennas," *2010 International Symposium on Electromagnetic Theory (EMT-S), 2010. URSI'10. IEEE*, August 2010, pp. 1–4.
63. D. M. Dobkin, *The RF in RFID: Passive UHF-RFID in Practice*. Elsevier, Burlington, MA & Oxford, UK, 2006.
64. J. Hagerty, F. Helmbrecht, W. McCalpin, R. Zane, and Z. Popovic, "Recycling ambient microwave energy with broadband rectenna arrays," *IEEE Transactions on Microwave Theory and Techniques*, vol. 52, no. 3, March 2004, pp. 1014–1024.
65. U. Olgun, C.-C. Chen, and J. Volakis, "Low-profile planar rectenna for batteryless RFID sensors," *Antennas and Propagation Society International Symposium, 2010. APSURSI'10. IEEE*, July 2010, pp. 1–4.

Index

A

H

I

K

S